AF595002

Bacterial Meningitis

Bacterial Meningitis: Epidemiology and Vaccination

Editor

James M. Stuart

MDPI • Basel • Beijing • Wuhan • Barcelona • Belgrade • Manchester • Tokyo • Cluj • Tianjin

Editor
James M. Stuart
University of Bristol, Bristol
UK

Editorial Office
MDPI
St. Alban-Anlage 66
4052 Basel, Switzerland

This is a reprint of articles from the Special Issue published online in the open access journal *Microorganisms* (ISSN 2076-2607) (available at: https://www.mdpi.com/journal/microorganisms/special_issues/meningitis).

For citation purposes, cite each article independently as indicated on the article page online and as indicated below:

LastName, A.A.; LastName, B.B.; LastName, C.C. Article Title. *Journal Name* **Year**, *Volume Number*, Page Range.

ISBN 978-3-0365-1807-7 (Hbk)
ISBN 978-3-0365-1808-4 (PDF)

Contents

About the Editor

James MacNaughton Stuart is an honorary professor in population health sciences at the University of Bristol and a WHO consultant. After qualifying in medicine, James worked for ten years as a clinical doctor in the UK and rural South Africa before specialising in public health and epidemiology of infectious diseases, particularly meningococcal meningitis. He has been involved in the investigation and control of outbreaks internationally and has published extensively on the epidemiology of meningococcal disease and carriage. In recent years, James has worked at the London School of Hygiene and Tropical Medicine on MenAfriCar, a major research project into meningococcal carriage during the introduction of a serogroup A conjugate vaccine across the meningitis belt of Africa. He co-ordinated revision of WHO outbreak response guidelines for the meningitis belt and helped to develop the WHO global strategy on "Defeating Meningitis by 2030".

Preface to "Bacterial Meningitis: Epidemiology and Vaccination"

The suddenness, severity and dire consequences of meningitis remain a challenge for all countries of the world. Although medical countermeasures to date, such as vaccines, diagnostics and therapies, are generalized to prevent, rapidly identify and treat acute bacterial meningitis, these still have major limitations. At the same time, support for people and their families living with long-term disability after meningitis is largely insufficient or non-existent. In addition, the occurrence of meningitis epidemics, which are difficult to predict, continues to pose a threat to communities in several countries. The unprecedented progress made in the fight against meningitis in recent decades has brought hope that the disease could be defeated. It is in this context that a call to action has gradually materialized and given rise to a collective and collaborative effort by many stakeholders to develop, under WHO leadership, the global roadmap to defeat meningitis by 2030. This roadmap, at the heart of the current strategy of the World Health Organization, is an essential element in achieving universal health coverage. It was approved by the 73rd World Health Assembly in November 2020, when Member States endorsed the first ever resolution on meningitis prevention and control. Collaboration among stakeholders from different fields and perspectives has enabled the development of an ambitious but achievable global roadmap, which is a powerful means, integrated with other initiatives, to advance primary health care, protection against health emergencies and enable more people to enjoy better health and well-being.

In the same spirit of a complete, global and multidisciplinary approach, the editor of this book, James Stuart, a long-standing expert in meningitis who has been closely involved in the development of the road map, has brought together a wide range of different experts in the field of bacterial meningitis. The articles included in this book cover various important and complementary aspects of the strategic goals of the global roadmap and, as such, they constitute valuable and timely documentation on the eve of the launch of this roadmap for 2030. I am very pleased to see publication of this book on bacterial meningitis and grateful to the editor and contributors to this broad range of substantive papers, that together recognize the importance of the roadmap and support our drive to defeat meningitis across the world and improve the care of those affected by this devastating infection.

Dr Marie-Pierre Preziosi
Meningitis lead, Division for Universal Health Coverage /
Life Course, WHO, Geneva, Switzerland

Editorial

Editorial for the Special Issue: Bacterial Meningitis—Epidemiology and Vaccination

James M. Stuart

Population Health Sciences, Bristol Medical School, University of Bristol, Bristol BS8 1QU, UK; james.stuart@bristol.ac.uk

Bacterial meningitis has serious health, economic, and social consequences with a high risk of death and lifelong disability. WHO has published the first global road map on meningitis "Defeating meningitis by 2030" to tackle the main causes of acute bacterial meningitis: *Neisseria meningitidis, Streptococcus pneumoniae, Haemophilus influenzae*, and *Streptococcus agalactiae* (Group B *Streptococcus* (GBS)) [1,2]. The road map was endorsed by the World Health Assembly in November 2020 [3].

The three main goals of the meningitis roadmap are to eliminate epidemics of bacterial meningitis, reduce cases and deaths from vaccine-preventable bacterial meningitis, and reduce disability and improve quality of life after meningitis of any cause. Proposed measures to achieve these goals include development of new vaccines, increased effectiveness of prevention and control strategies, efficient global surveillance with accurate estimates of disease burden including sequelae, and global availability of and access to rapid diagnosis and high-quality treatment of meningitis and its after-effects.

This Special Issue includes a wide range of original research articles and review articles on epidemiology and vaccination of bacterial meningitis that have direct relevance to advancing the goals of the road map.

Citation: Stuart, J.M. Editorial for the Special Issue: Bacterial Meningitis—Epidemiology and Vaccination. *Microorganisms* **2021**, *9*, 917. https://doi.org/10.3390/microorganisms9050917

Received: 22 April 2021
Accepted: 23 April 2021
Published: 24 April 2021

A fundamental step in establishing the importance of meningitis and in monitoring progress toward prevention and care is quantifying the burden from illness, death, and disability. Wright et al. [4] described wide variation in different modelled estimates of the global burden and advocated for alignment with improving surveillance data to improve the accuracy of model parameters. The consequences of meningitis are even harder to measure. Schiess et al. [5] underlined the social and economic costs of meningitis, the lack of recognition of more subtle sequelae, and the lack of knowledge on long-term effects, especially in low- and middle-income countries. Building care services for those affected by meningitis across the world will be a challenging objective for the meningitis strategy.

The principal means of achieving targets to reduce cases and deaths from meningitis will inevitably be through vaccination. Alderson et al. [6] gave a comprehensive overview of past and present developments in meningitis vaccines. They drew attention to the importance of low-cost vaccines for global introduction, the expanding range of conjugate vaccines and the more recently developed meningococcal protein vaccines, and the challenges in reaching prevention goals. As vaccines are developed and vaccination programmes expanded, Deghmane and Taha [7] made the case that preventing disease among those at higher risk will become increasingly important.

For meningococcal meningitis, the high-burden region of the meningitis belt in sub-Saharan Africa deserves particular attention. Karachaliou Prasinou et al. [8] showed how mathematical models can be used to optimise the effectiveness of vaccination programmes, with two key parameters being the duration of protection and age at vaccination. Such models are relevant both for the serogroup A vaccine currently being deployed in the meningitis belt and for the anticipated roll out of pentavalent (ACWXY) conjugate vaccines [6]. The need for broader-valency vaccines in the global control of invasive meningococcal disease was well demonstrated in the paper by Tzeng and Stephens [9], describing the changing epidemiology and emerging disease due to serogroups other than A, B, and C.

Slack [10] documented how meningitis due to *Haemophilus influenzae* fell rapidly with the introduction of conjugate Hib vaccines from the 1990s such that, by 2015, the burden of meningitis due to serotype b was limited to a few countries that had not introduced these vaccines into the national immunization programmes. However, as she pointed out, invasive disease due to non-typable organisms and other serotypes is of increasing concern. Serotype A has emerged as a significant cause of meningitis in indigenous populations of North America and has stimulated the development of a new conjugate vaccine.

A series of papers from the impressive PSERENADE project [11–13] demonstrated the substantial global impact of multivalent pneumococcal conjugate vaccines on invasive pneumococcal disease, including meningitis, after their introduction into infant immunisation programmes. Six or more years after introduction, they found a 95% reduction in serotype 1 disease in all age groups. Measuring the impact does depend on robust serotype surveillance systems, and they acknowledged the need for more data from the meningitis belt countries that are at high risk of pneumococcal meningitis and serotype 1 outbreaks.

Vaccines are in development but not yet available to protect against disease due to GBS [6]. Prevention of GBS in newborns currently relies on risk-based or microbiological screening for infection in pregnancy. However, a study of meningitis among infants under 90 days of age in a large paediatric hospital in the USA [14] showed that the majority of cases of bacterial meningitis were due to GBS, despite universal screening and intrapartum prophylaxis. This only emphasises the importance of a vaccine that could hopefully have more impact than prepartum screening with the additional protection of stillbirths due to GBS infection and late-onset GBS disease.

Tsang [15] focused on the molecular epidemiology of the four main bacterial causes of meningitis in the roadmap and the power of conjugate vaccines to both reduce the burden and drive the evolution of these bacteria, thus underlining the need for improved surveillance and expansion of whole-genome sequencing.

Zainel et al. [16] highlighted neurological complications from bacterial meningitis in children such as hearing loss, cognitive impairment, and epilepsy, as well as the importance of prompt effective treatment regimens in improving outcomes. A key component of prompt treatment is rapid accurate diagnosis of meningitis through bedside tests that can be applicable in low- and middle-income countries. Rondy et al. [17] reported on a field evaluation of a rapid test that should aid timely decisions on vaccine deployment in meningitis epidemics.

Meningitis can be caused by many infectious organisms: bacteria, viruses, fungi, and parasites. The focus in the "Defeating meningitis by 2030" strategy is on the main bacteria responsible for the overall global burden with potential for prevention by vaccination. Another major cause of bacterial meningitis, *Mycobacterium tuberculosis*, was given prominence in this issue by Basu-Roy et al. [18]. Their review highlighted how the "Defeating meningitis" roadmap can be applied to the prevention and control of tuberculosis in children, affirming the need for a collaborative endeavour and linking with activities of other initiatives such as the WHO TB roadmap [19]. The fact that many elements of the roadmap apply to TB and all other causes of meningitis must not be forgotten in the drive to defeat meningitis.

The global roadmap to defeat meningitis is an ambitious strategy, particularly in the context of the Covid pandemic. As shown by the contributions to this Special Issue, a concerted drive to reduce the burden of this illness is, without question, a worthy ambition. The theme of World Meningitis Day 2021 is "Take Action #DefeatMeningitis" [20,21]. Start now!

Funding: This work received no external funding.

Acknowledgments: I am very grateful for the contributions in time and effort from all the authors, and I am particularly thankful for the support and encouragement from Marie-Pierre Preziosi at WHO and Brian Greenwood at the London School of Hygiene and Tropical Medicine.

Conflicts of Interest: The author was employed as a consultant by WHO to help develop the meningitis roadmap.

References

1. A new roadmap for meningitis. *Lancet* **2020**, *395*, 1230. [CrossRef]
2. WHO. Defeating Meningitis by 2030: A Global Road Map. Available online: https://www.who.int/docs/default-source/immunization/meningitis/defeatingmeningitisroadmap.pdf (accessed on 20 April 2021).
3. WHO. World Health Assembly Endorses the 1st ever Resolution on Meningitis Prevention and Control. Available online: https://www.who.int/news/item/13-01-2021-world-health-assembly-endorses-the-1st-ever-resolution-on-meningitis-prevention-and-control (accessed on 20 April 2021).
4. Wright, C.; Blake, N.; Glennie, L.; Smith, V.; Bender, R.; Kyu, H.; Wunrow, H.Y.; Liu, L.; Yeung, D.; Knoll, M.D.; et al. The Global Burden of Meningitis in Children: Challenges with Interpreting Global Health Estimates. *Microorganisms* **2021**, *9*, 377. [CrossRef] [PubMed]
5. Schiess, N.; Groce, N.; Dua, T. The impact and burden of neurological sequelae following bacterial meningitis: A narrative review. *Microorganisms* **2021**, *9*, 900. [CrossRef]
6. Alderson, M.R.; Welsch, J.A.; Regan, K.; Newhouse, L.; Bhat, N.; Marfin, A.A. Vaccines to Prevent Meningitis: Historical Perspectives and Future Directions. *Microorganisms* **2021**, *9*, 771. [CrossRef]
7. Deghmane, A.E.; Taha, M.K. Invasive Bacterial Infections in Subjects with Genetic and Acquired Susceptibility and Impacts on Recommendations for Vaccination: A Narrative Review. *Microorganisms* **2021**, *9*, 467. [CrossRef] [PubMed]
8. Karachaliou Prasinou, A.; Conlan, A.J.K.; Trotter, C.L. Understanding the Role of Duration of Vaccine Protection with MenAfriVac: Simulating Alternative Vaccination Strategies. *Microorganisms* **2021**, *9*, 461. [CrossRef] [PubMed]
9. Tzeng, Y.L.; Stephens, D.S. A Narrative Review of the W, X, Y, E, and NG of Meningococcal Disease: Emerging Capsular Groups, Pathotypes, and Global Control. *Microorganisms* **2021**, *9*, 519. [CrossRef] [PubMed]
10. Slack, M.P.E. Long Term Impact of Conjugate Vaccines on Haemophilus influenzae Meningitis: Narrative Review. *Microorganisms* **2021**, *9*, 886. [CrossRef]
11. Knoll, M.D.; Bennett, J.; Garcia Quesada, M.; Kagucia, E.; Peterson, M.; Feikin, D.; Cohen, A.; Hetrich, M.; Yang, Y.; Sinkevitch, J.; et al. Global Landscape Review of Serotype-Specific Invasive Pneumococcal Disease Surveillance among Countries Using PCV10/13: The Pneumococcal Serotype Replacement and Distribution Estimation (PSERENADE) Project. *Microorganisms* **2021**, *9*, 742. [CrossRef]
12. Garcia Quesada, M.; Yang, Y.; Bennett, J.; Hayford, K.; Zeger, S.; Feikin, D.; Peterson, M.; Cohen, A.; Almeida, S.; Ampofo, K.; et al. Serotype Distribution of Remaining Pneumococcal Meningitis in the Mature PCV10/13 Period: Findings from the PSERENADE Project. *Microorganisms* **2021**, *9*, 738. [CrossRef]
13. Bennett, J.C.; Hetrich, M.K.; Quesada, M.G.; Sinkevitch, J.N.; Knoll, M.D.; Feikin, D.R.; Zeger, S.L.; Kagucia, E.W.; Cohen, A.L.; Ampofo, K.; et al. Changes in Invasive Pneumococcal Disease Caused by Streptococcus pneumoniae Serotype 1 Following Introduction of PCV10 and PCV13: Findings from the PSERENADE Project. *Microorganisms* **2021**, *9*, 696. [CrossRef] [PubMed]
14. Erickson, T.A.; Munoz, F.M.; Troisi, C.L.; Nolan, M.S.; Hasbun, R.; Brown, E.L.; Murray, K.O. The Epidemiology of Meningitis in Infants under 90 Days of Age in a Large Pediatric Hospital. *Microorganisms* **2021**, *9*, 526. [CrossRef] [PubMed]
15. Tsang, R.S.W. A Narrative Review of the Molecular Epidemiology and Laboratory Surveillance of Vaccine Preventable Bacterial Meningitis Agents: Streptococcus pneumoniae, Neisseria meningitidis, Haemophilus influenzae and Streptococcus agalactiae. *Microorganisms* **2021**, *9*, 449. [CrossRef] [PubMed]
16. Zainel, A.; Mitchell, H.; Sadarangani, M. Bacterial Meningitis in Children: Neurological Complications, Associated Risk Factors, and Prevention. *Microorganisms* **2021**, *9*, 535. [CrossRef] [PubMed]
17. Rondy, M.; Tamboura, M.; Sidikou, F.; Yameogo, I.; Dinanibe, K.; Sawadogo, G.; Kambire, C.; Mainassara, H.; Mahamane, A.E.; Bienvenu, B.; et al. Field Evaluation of the Performance of Two Rapid Diagnostic Tests for Meningitis in Niger and Burkina Faso. *Microorganisms* **2021**, *9*, 832. [CrossRef]
18. Basu Roy, R.; Bakeera-Kitaka, S.; Chabala, C.; Gibb, D.M.; Huynh, J.; Mujuru, H.; Sankhyan, N.; Seddon, J.A.; Sharma, S.; Singh, V.; et al. Defeating Paediatric Tuberculous Meningitis: Applying the WHO "Defeating Meningitis by 2030: Global Roadmap". *Microorganisms* **2021**, *9*, 857. [CrossRef]
19. WHO. Roadmap towards Ending TB in Children and Adolescents. Available online: https://www.who.int/tb/publications/2018/tb-childhoodroadmap/en/ (accessed on 20 April 2021).
20. Confederation of Meningitis Organisations. World Meningitis Day 2021. Available online: https://www.comomeningitis.org/world-meningitis-day-2021 (accessed on 20 April 2021).
21. WHO. World Meningitis Day. Available online: https://www.who.int/news-room/events/detail/2021/04/24/default-calendar/world-meningitis-day (accessed on 20 April 2021).

MDPI

Article

The Global Burden of Meningitis in Children: Challenges with Interpreting Global Health Estimates

Claire Wright [1,*], Natacha Blake [1], Linda Glennie [1], Vinny Smith [1], Rose Bender [2], Hmwe Kyu [2], Han Yong Wunrow [2], Li Liu [3], Diana Yeung [4], Maria Deloria Knoll [5], Brian Wahl [5], James M. Stuart [6,7] and Caroline Trotter [8]

1 Meningitis Research Foundation, Bristol BS1 5HX, UK; tachablake26@hotmail.co.uk (N.B.); Lindag@meningitis.org (L.G.); vinnys@meningitis.org (V.S.)
2 Institute for Health Metrics and Evaluation, University of Washington, Seattle, WA 98105, USA; rbender1@uw.edu (R.B.); hmwekyu@uw.edu (H.K.); hwunrow@uw.edu (H.Y.W.)
3 Department of Population, Family and Reproductive Health and Institute for International Programmes, Johns Hopkins Bloomberg School of Public Health, Baltimore, MD 21205, USA; lliu26@jhu.edu
4 Institute for International Programmes, Department of International Health, Johns Hopkins Bloomberg School of Public Health, Baltimore, MD 21205, USA; dyeung@jhu.edu
5 International Vaccine Access Center, Johns Hopkins Bloomberg School of Public Health, Baltimore, MD 21231, USA; mknoll2@jhu.edu (M.D.K.); bwahl@jhu.edu (B.W.)
6 Population Health Sciences, Bristol Medical School, University of Bristol, Bristol BS8 1QU, UK; James.Stuart@bristol.ac.uk
7 World Health Organization, 1211 Geneva, Switzerland
8 Disease Dynamics Unit, Department of Veterinary Medicine, University of Cambridge, Cambridge CB3 0ES, UK; clt56@cam.ac.uk
* Correspondence: clairew@meningitis.org

Citation: Wright, C.; Blake, N.; Glennie, L.; Smith, V.; Bender, R.; Kyu, H.; Wunrow, H.Y.; Liu, L.; Yeung, D.; Knoll, M.D.; et al. The Global Burden of Meningitis in Children: Challenges with Interpreting Global Health Estimates. *Microorganisms* **2021**, *9*, 377. https://doi.org/10.3390/microorganisms9020377

Academic Editor: Glenn S. Tillotson
Received: 11 January 2021
Accepted: 4 February 2021
Published: 13 February 2021

Abstract: The World Health Organization (WHO) has developed a global roadmap to defeat meningitis by 2030. To advocate for and track progress of the roadmap, the burden of meningitis as a syndrome and by pathogen must be accurately defined. Three major global health models estimating meningitis mortality as a syndrome and/or by causative pathogen were identified and compared for the baseline year 2015. Two models, (1) the WHO and the Johns Hopkins Bloomberg School of Public Health's Maternal and Child Epidemiology Estimation (MCEE) group's Child Mortality Estimation (WHO-MCEE) and (2) the Institute for Health Metrics and Evaluation (IHME) Global Burden of Disease Study (GBD 2017), identified meningitis, encephalitis and neonatal sepsis, collectively, to be the second and third largest infectious killers of children under five years, respectively. Global meningitis/encephalitis and neonatal sepsis mortality estimates differed more substantially between models than mortality estimates for selected infectious causes of death and all causes of death combined. Estimates at national level and by pathogen also differed markedly between models. Aligning modelled estimates with additional data sources, such as national or sentinel surveillance, could more accurately define the global burden of meningitis and help track progress against the WHO roadmap.

Keywords: meningitis; child mortality; neonatal sepsis; global health; global health estimates; modelling; *Streptococcus pneumoniae*; *Haemophilus influenzae*; *Neisseria meningitidis*

1. Introduction

The world saw great progress in reducing child mortality over the lifetime of the United Nations (UN) Millennium Development Goals (MDGs) with an estimated 54% decline in children under five years of age from 93 deaths per 1000 live births in 1990 to 43 per 1000 live births in 2015 [1]. The successor UN Sustainable Development Goals (SDGs) are more ambitious again, and urge that by 2030 we should "end preventable deaths of newborns and children under five years of age, with all countries aiming to reduce neonatal mortality to at least as low as 12 per 1000 live births and under-five mortality to at least as low as 25 per 1000 live births." However, with the majority of an estimated 38 deaths per 1000 live births in 2019 being caused by preventable and treatable diseases [1], we are a long way from achieving this target.

Among these preventable diseases, meningitis has one of the highest fatality rates and the potential to cause devastating epidemics. Since the turn of the century, we have seen advances as a result of widespread global introduction of *Haemophilus influenzae* type b (Hib) and pneumococcal vaccines as well as the roll out of the meningococcal A vaccine, MenAfriVac, across some of the highest incidence areas of sub-Saharan Africa. Despite this, recent estimates have identified that the global burden of meningitis in all age groups remains high and progress lags substantially behind that of other vaccine preventable diseases [2]. Whilst deaths from measles and tetanus in children under five years are estimated to have decreased by 86% and 92% respectively, between 1990 and 2017, over the same time period deaths from meningitis are estimated to have decreased by just 51% [3]. Despite its burden, meningitis is seldom, if at all, mentioned in key global and regional health documents [4–9].

In response to calls from governments, global health organisations, civil society, public health bodies, academia and the private sector, a World Health Organization (WHO)-led collaboration is developing a Defeating Meningitis by 2030 Global Roadmap [10]. The Roadmap focuses on the four leading global causes of bacterial meningitis; *Neisseria meningitidis* (meningococcus), *Streptococcus pneumoniae* (pneumococcus), *Haemophilus influenzae* (Hi), and *Streptococcus agalactiae* (group B streptococcus (GBS)).

To advocate for a global roadmap to defeat meningitis, the global burden of meningitis as a syndrome in relation to other infectious causes of death needs to be accurately described, and countries with the highest burden identified, so that efforts and resources can be targeted effectively. Estimates of pathogen-specific meningitis incidence and mortality at the global level can identify the need for new vaccines or support wider access to existing ones. Tracking trends in pathogen-specific meningitis and syndromic disease over time at the national and international level is vital to assess the impact of interventions such as vaccines implemented as part of the global roadmap to defeat meningitis.

Vital registration systems and disease surveillance platforms are limited across many geographies and regions, so there is a reliance on modelled estimates to get a complete global picture of disease across all settings but cause of death estimates have been found to differ across these different modelling efforts [11]. Modelled estimates also attempt to account for changes in causes of death over time, but to do so accurately they must be informed by reliable data to make accurate predictions where real data is lacking.

In this paper we aim to compare the available modelled estimates for cases and deaths from meningitis as a syndrome, by causative pathogen and the methods used, in order to assess whether these models can be used with confidence by decision makers to prioritise recommendations from a plan to defeat meningitis, and by those needing to track progress on the WHO Defeating Meningitis by 2030 Global Roadmap'.

2. Materials and Methods

2.1. Identification of Data Sources

Through attending key stakeholder meetings, we identified three modelling efforts that estimate the global burden of meningitis and neonatal sepsis: (1) WHO and the Johns Hopkins Bloomberg School of Public Health's Maternal and Child Epidemiology Estimation (MCEE) group's Child Mortality Estimation (WHO-MCEE), which estimates15 causes of death for children under five years of age [12]; (2) The Institute for Health Metrics and Evaluation (IHME) Global Burden of Disease Study (GBD 2017) which estimates age specific mortality for 282 causes of death in all ages [3]; and (3) The WHO's Global Health Estimates (WHO GHE) which estimates age specific mortality for 136 causes of death in all ages [13].

Two additional models were also identified that estimated disease burden caused by pathogens of particular relevance to the WHO Defeating Meningitis by 2030 Global Roadmap: (1) the WHO-MCEE group's estimates of the burden of pneumococcal and Hib disease in children [14], and (2) the London School of Hygiene and Tropical Medicine (LSHTM) Burden of Group B Streptococcus Worldwide for Pregnant Women, Stillbirths, and Children [15].

Not all of these efforts were directly comparable because they did not provide the same level of data or use the same indicators of burden (Table 1).

Table 1. Models estimating the global burden of meningitis and neonatal sepsis.

	GBD 2017	WHO GHE	WHO-MCEE Syndromic Model	WHO-MCEE Pathogen Model	LSHTM
Years	1990–2017	2000–2016	2000–2017	2000–2015	2015
Number of countries & territories	195	183	194	194	195
Global under five population estimate in 2015	678,053,340	673,253,870 **	671,355,776 **	657,127,399 ***	N/A
Age range	All ages (Including: Early neonatal: 0–6 days Late neonatal: 7–27 days Post neonatal: 28–364 days 1–4 years)	All ages (including: 0–28 days 1–59 months)	0–59 months (including: 0–28 days 1–59 months)	1–59 months	0–89 days
Relevant disease categories	Meningitis, neonatal sepsis and other neonatal infections	Meningitis *, neonatal sepsis and infections	Meningitis/encephalitis, sepsis and other infectious conditions of the newborn	Meningitis, Non-pneumonia/non-meningitis (which is primarily but not exclusively sepsis)	Meningitis, Sepsis
Outputs	Cases, Incidence rate, Prevalence, Deaths, Mortality rate, DALYs	Deaths, Mortality rate, DALYs	Deaths, Mortality rate	Cases, Incidence rate, Deaths, Mortality rate	Cases, Incidence rate, Deaths, Mortality rate
Published rate per population	Per 100,000 population	Per 100,000 population	Per 1000 livebirths	Per 100,000 population	Per 1000 livebirths
Aetiology	Nm, Spn, Hib, Other	No breakdown by aetiology	No breakdown by aetiology	Nonepidemic disease from: Spn, Hib, Nm	GBS

DALYs = Disability Adjusted Life Years; GBS = Group B streptococcus; Hib = *Haemophilus influenzae* type b; Nm = *Neisseria meningitidis* (meningococcus); Spn = *Streptococcus pneumoniae* (pneumococcus).* WHO GHE use a ratio of meningitis to encephalitis deaths obtained from IHME data to separate out MCEE under-five meningitis/encephalitis estimates. ** Estimates derived from UN World Population Prospects 2017. Differences between WHO GHE and WHO-MCEE population estimates likely due to draft estimates circulating prior to final publication. *** Derived from UN World Population Prospects 2015

As WHO GHE estimates were an amalgamation of historical models (WHO-MCEE's 2000–2016 and IHME's GBD 2016) we did not consider them further in our analysis. We did not include GBS estimates from LSHTM in our analysis because the age categorisation

(0–89 days) did not correspond with the disaggregated age categories of the other models and so did not allow for meaningful comparison.

2.2. Analysis of Data Sources

The scale of the global burden of meningitis deaths relative to all causes and leading infectious causes of death was assessed by comparing, death and mortality estimates from GBD 2017 and the WHO-MCEE's 2000–2017 model according to the following syndromic cause of death categories "All causes", "Infectious disease", "meningitis/encephalitis" and "neonatal sepsis".

We considered the burden of meningitis and neonatal sepsis together for the purposes of comparison with other leading infectious causes of death because distinguishing between these syndromes is almost impossible based on clinical signs alone in the neonate [16,17]. Lumbar puncture (LP) and analysis of the cerebrospinal fluid is the only reliable way of confirming a case of meningitis. However, in many countries there is a shortage of trained staff to perform LP [18], and in low-income settings as few as 2% of neonates with infection might have an LP or blood sample taken [19].

The WHO-MCEE have historically estimated sepsis and meningitis in the neonatal period within the same cause category because of difficulties in distinguishing between these clinical syndromes in this age group. These causes were estimated separately for the first time in their latest modelling round by using the ratio of neonatal meningitis and neonatal sepsis deaths derived from IHME estimates. Because WHO-MCEE estimate meningitis/encephalitis as one cause category, GBD 2017 meningitis and encephalitis deaths were amalgamated for the purpose of comparison.

Denominators used to report mortality rates were standardised across the models and, where necessary, recalculated to be expressed as deaths per 1000 live births in the neonatal period and deaths per 100,000 population in the post neonatal period. GBD 2017 mortality rates in the neonatal period were calculated from IHME live birth estimates for the year 2015. WHO-MCEE postneonatal mortality rates were calculated using UN population estimates for the year 2015 [20].

Priority geographical areas for targeting a plan to defeat meningitis were identified from country-specific GBD 2017 and WHO-MCEE meningitis/encephalitis mortality estimates for the year 2015 in children under five years.

Meningitis mortality and incidence estimates according to pathogen over time (2000–2015) were analysed using estimates produced by GBD 2017 and the WHO-MCEE pathogen model. Meningococcal meningitis is commonly associated with epidemics. As WHO-MCEE meningococcal meningitis estimates did not account for deaths and cases resulting from epidemics, estimates for 'Hib meningitis' and 'pneumococcal meningitis' mortality and incidence in the post neonatal period (28 days–<5 years) were the categories and age group used for comparison.

An analysis of the estimation methodology for each model was also undertaken in an attempt to explain any inconsistencies between models.

3. Results

3.1. Global Meningitis and Neonatal Sepsis Mortality Estimates in Children Aged Under Five Years

Overall, the WHO-MCEE estimated there to be approximately 100,000 fewer deaths in the under-five age group than the GBD 2017, with proportionally more under-five deaths occurring in the neonatal period (46% compared to GBD 2017's 43%).

The GBD 2017 estimated 34% more deaths from meningitis/encephalitis than the WHO-MCEE in the year 2015 (190,515 and 142,841, respectively) (Table 2). Meningitis made up the majority of the GBD 2017 combined meningitis/encephalitis category; 87% in under five-year-olds, 86% in 1–59 months and 93% in 0–28 days.

Table 2. Estimated deaths by cause and model for the year 2015 in children under five years of age.

		GBD 2017		WHO-MCEE Pathogen Model		Difference *
		n	Rate †	n	Rate †	%
All causes	Under 5	5,917,285 (5,723,776–6,120,099)	872.69 (844.15–902.60)	5,792,509 (5,573,633–6,123,477)	862.81 a	2%
	1–59 months	3,354,404 (3,231,491–3,483,015)	502.51 (484.10–521.78)	3,122,698 (2,700,899–3,581,030	473.02 a	7%
	0–28 days	2,562,881 (2,478,272–2,655,261)	18.40 (17.20–19.58)	2,669,811 (2,542,447–2,872,734)	19.01 (18.10–20.50)	−4%
Infectious diseases **	Under 5	2,519,567 (2,379,024–2,671,856)	371.59 (350.86–394.05)	2,426,882 (2,279,602–3,169,783)	361.49 a	4%
	1–59 months	1,967,826 (1,847,763–2,091,762)	294.79 (276.81–313.36)	1,810,771 (1,703,587–2,350,572)	274.29 a	8%
	0–28 days	551,740 (510,918–603,527	3.96 (3.60–4.38)	616,111 (605,290–877,610)	4.39 (4.31–6.25)	−11%
Meningitis & Encephalitis	Under 5	190,515 (163,374–217,259)	28.10 (24.09–32.04)	142,841 (87,427–178,552)	21.28 a	29%
	1–59 months	167,880 (143,529–192,447)	25.15 (21.50–28.83)	105,406 (87,188–145,213)	15.97 a	46%
	0–28 days	22,636 (18,532–25,642)	0.16 (0.13–0.19)	37,435 (157–51,299)	0.27 (0.001–0.37)	−49%
Neonatal sepsis	Under 5	211,273 (186,657–275,821)	31.16 (27.53–40.68)	364,188 (282,744–524,021)	54.25 a	−53%
	1–59 months	12,693 (10,626–16,586)	1.90 (1.59–2.48)	386 b (14–579)	0.06 a	188%
	0–28 days	198,580 (175,866–263,096)	1.43 (1.24–1.86)	363,802 (282,341–523,853)	2.59 (2.01–3.73)	−59%

* Percent difference (n) = (GBD 2017–WHO-MCEE)/((GBD 2017 + WHO-MCEE)/2) × 100. ** Sum of specific infectious diseases from WHO-MCEE cause list (HIV/AIDS; diarrhoeal diseases; tetanus; measles; meningitis/encephalitis; malaria; acute respiratory infections; sepsis and other infectious conditions of the newborn). † Rates per 100,000 population in 'Under 5' and '1–59 months', and per 1000 livebirths for '0–28 days'. a Uncertainty intervals not available–rate calculated using n and under-5 population statistic from UN WPP 2017 Revision–year 2015 (1–59 months calculated using 59/60 months population). b Figures only account for neonatal sepsis deaths in China.

However, the WHO-MCEE estimated >100,000 more deaths than the GBD 2017 when neonatal sepsis deaths were combined with meningitis/encephalitis, due to the WHO-MCEE's much higher estimate of neonatal sepsis deaths. Uncertainty intervals do not overlap between modelled estimates of deaths from neonatal sepsis in any of the age categories. This section may be divided by subheadings. It should provide a concise and precise description of the experimental results, their interpretation, as well as the experimental conclusions that can be drawn.

The WHO-MCEE model estimated meningitis/encephalitis and neonatal sepsis as the second largest infectious cause of death, co-ranked with diarrhoeal diseases, in children aged under five years in 2015, after acute respiratory infections (Figure 1). In contrast the GBD 2017 estimated this cause category to be the third largest infectious case of death after acute respiratory infections and diarrhoeal diseases.

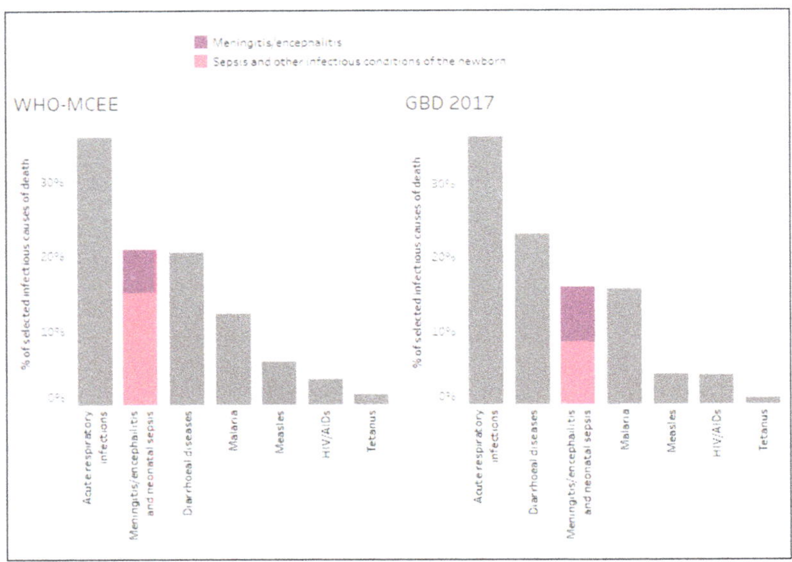

Figure 1. Meningitis/encephalitis and neonatal sepsis mortality burden estimates by model in relation to other selected infectious causes of death in children under five for the year 2015.

At the country level, there was considerable variability in estimates of burden per population. For meningitis/encephalitis mortality rates, the WHO-MCEE model ranked Somalia highest for the year 2015 (139.7 deaths per 100,000 population), whilst the GBD 2017 ranked Somalia 11th highest for the same year (68.6 deaths per 100,000).

For numbers of deaths, both models attribute approximately 70% of all meningitis/encephalitis deaths in children under five years to just 12 countries including India, Nigeria, Pakistan, Democratic Republic of Congo (DRC), Ethiopia, Niger, Afghanistan, Mali, Uganda and China. However, whilst Somalia and Chad feature in the top 12 (ranked 7th and 8th respectively) in the WHO-MCEE estimates, they did not feature in the GBD 2017 top 12, where Indonesia and Burkina Faso featured instead (ranked 8th and 9th highest respectively) (Figure 2).

3.2. *Meningitis Incidence and Mortality Estimates by Aetiology in Children Aged Under Five Years*

The GBD 2017 and WHO-MCEE's pathogen models both estimated pneumococcal and Hib meningitis mortality and incidence in children aged 1 to 59 months at the national and global levels.

A comparison of the global estimates for the year 2015 (Table 3) showed that both models agree there were more cases of pneumococcal meningitis than Hib meningitis in 2015. However, whilst the GBD 2017 estimated around twice as many deaths from Hib meningitis compared to pneumococcal meningitis, the WHO-MCEE estimated around five times more deaths from pneumococcal meningitis than from Hib meningitis in the same year.

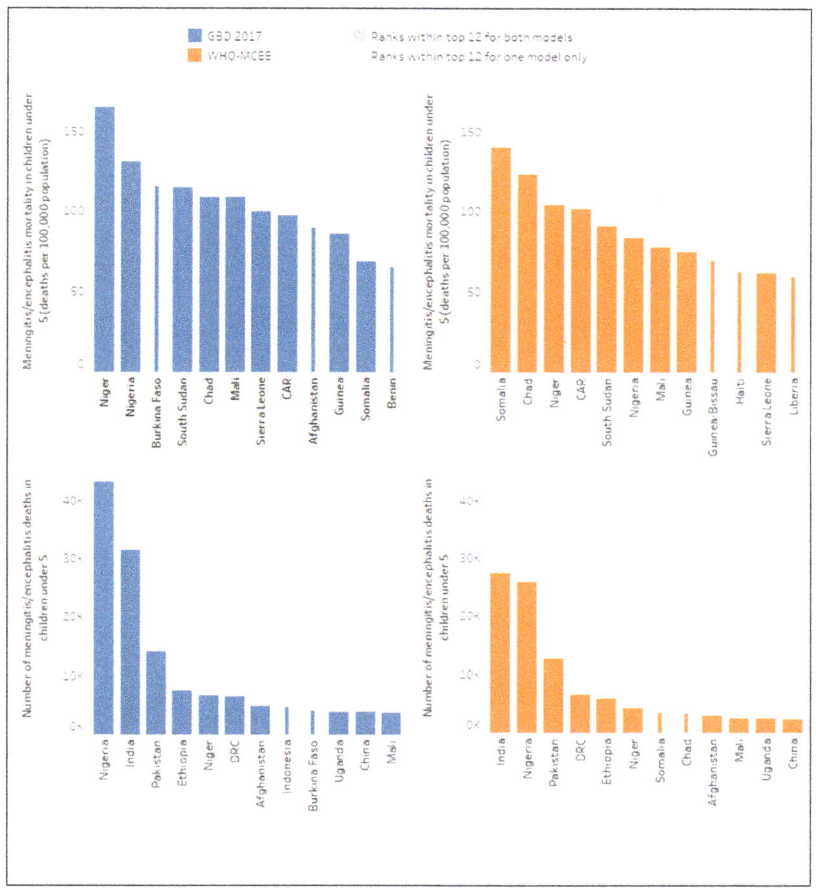

Figure 2. Top twelve ranking countries by meningitis/encephalitis mortality (number and rate) according to model for the year 2015.

Table 3. Global aetiology-specific meningitis deaths and cases, 2015, in children aged 1–59 months.

		GBD 2017		**WHO-MCEE Pathogen Model**		**Difference ***
		n	**Rate †**	**n**	**Rate †**	**%**
Pneumococcal meningitis	Cases	267,686 (179,314–374,902)	40.10 (26.86–56.16)	83,809 (36,160–168,500)	13 (5–26)	105%
	Deaths	20,156 (16,114–25,199)	3.02 (2.41–3.78)	37,964 (15,397–79,718)	5 (2–11)	−61%
Hib meningitis	Cases	208,658 (139,815–304,035)	31.26 (20.95–45.55)	31,243 (13,386–50,595)	5 (2–8)	148%
	Deaths	39,380 (31,782–48,754)	5.90 (4.76–7.30)	7156 (2707–11,320)	1 (0–2)	138%

* Percent difference = (GBD 2017–WHO-MCEE)/((GBD 2017 + WHO-MCEE)/2) × 100. † Rates per 100,000 population

Despite major differences in the relative proportions of meningitis deaths attributed to Hib and pneumococcal bacteria between models, both models agreed that Hib and pneumococcal meningitis combined were the underlying cause of approximately 40% of all meningitis/encephalitis deaths globally.

When comparisons were made between the modelled estimates for Hib and pneumococcal meningitis incidence and mortality over time (Figure 3), both models showed a steeper decline in Hib meningitis incidence and mortality compared to pneumococcal meningitis mortality, which is consistent with wider roll-out of Hib vaccination globally compared to pneumococcal vaccination. However, the GBD 2017 consistently reported much higher incidence of pneumococcal and Hib meningitis over time compared to the WHO-MCEE. The GBD 2017 estimated Hib and pneumococcal incidence to be 31 and 40 cases per 100,000, respectively, in 2015 compared to the WHO-MCEE estimates of around five and 13 cases per 100,000 for Hib and pneumococcal meningitis, respectively.

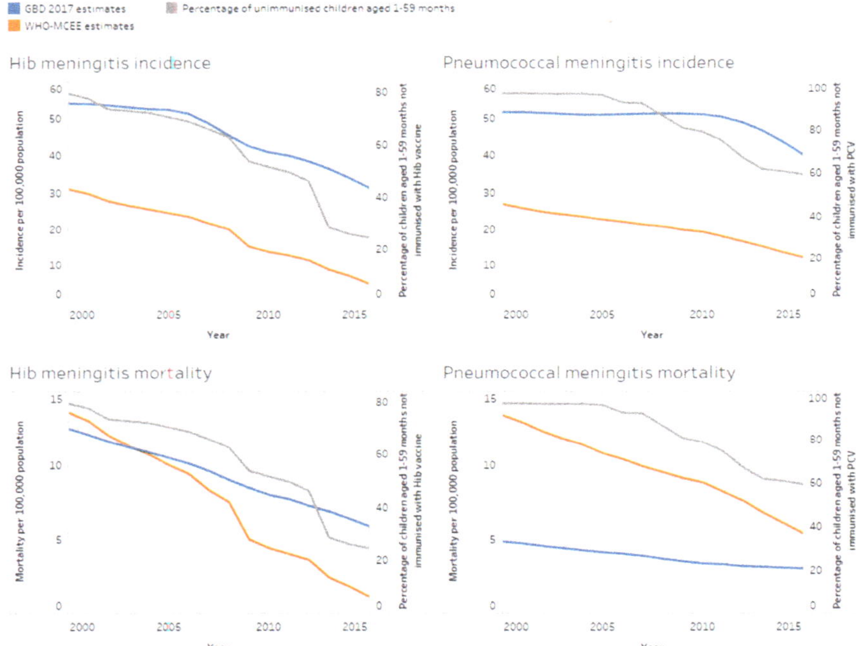

Figure 3. Estimated Hib/pneumococcal meningitis mortality and incidence amongst children aged 1–59 months according to model in relation to the proportion of children unimmunised with Hib vaccine and pneumococcal conjugate vaccine (PCV) over time.

Of note is that case fatality rates (CFRs) differed dramatically between the two sets of estimates. CFRs derived from WHO-MCEE global cases, and deaths estimates for Hib and pneumococcal meningitis in 2015, were 23% and 45%, respectively. However, CFRs calculated from GBD 2017 estimates were 8% for pneumococcal meningitis, 19% for Hib meningitis and 8% for meningococcal meningitis. Evidence from the literature closely agrees with the WHO-MCEE CFRs, consistently reporting higher CFRs from pneumococcal meningitis compared to Hib meningitis and meningococcal meningitis [21–26].

3.3. Modelling Methodology Which Could Account for Differences in Mortality and Pathogen Specific Estimates

Figure 4 depicts a simplified methodology for both modelling approaches. A more detailed explanation is provided in the appendix, and full methodological approaches are also outlined elsewhere [3,27]. When calculating the meningitis death envelope, both models used country-specific death data from vital registration and other sources and applied statistical modelling to fill gaps in the data using country-specific covariates and drawing on trends observed where data was more complete. Whilst the GBD included intervention covariates (such as vaccine coverage) within their cause of death ensemble modelling (CODEm) (Figure 4), the WHO-MCEE model used intervention covariates in both their modelling, and also in post hoc adjustments, to redistribute causes accounting for interventions. Details of the covariates used by the models are available in the Supplementary Materials. Both models ensured that the sum of deaths attributed to different causes fitted within a total all-cause mortality envelope calculated from surveys, censuses and vital registration data.

Whilst there was little difference between estimated mortality from all causes and infectious diseases in children under five years (2% and 4% difference in estimated deaths, respectively), between models there was a marked difference between meningitis/encephalitis and neonatal sepsis mortality estimates in this age group (29 and 53 percent difference, respectively) (Table 2).

Further investigation into the modelling methods and underlying data showed that countries with the highest meningitis burden have the lowest quality death registration data. Whilst this is also the case for all causes of death, a higher proportion of meningitis/encephalitis death estimates were based on extrapolating from low-quality underlying data compared to all-cause death estimates. For example, 77% of meningitis/encephalitis deaths came from countries with no or very low-quality death registration data (scaled 0 to 1) compared to 60% of deaths due to all causes in the GBD 2017 model. Likewise, in the WHO-MCEE model, 95% of meningitis/encephalitis deaths were estimated using modelling underpinned by verbal autopsy (VA) studies compared to 90% of all cause deaths due to these countries having poor quality death registration data (see Supplementary Materials). As would be expected, there were greater differences between estimates from countries with low-quality underlying data compared to those with higher quality data (Figure 5).

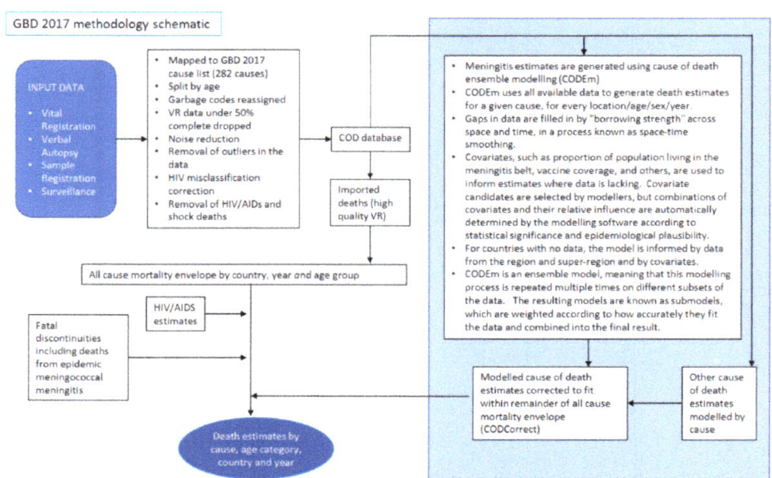

Figure 4. *Cont.*

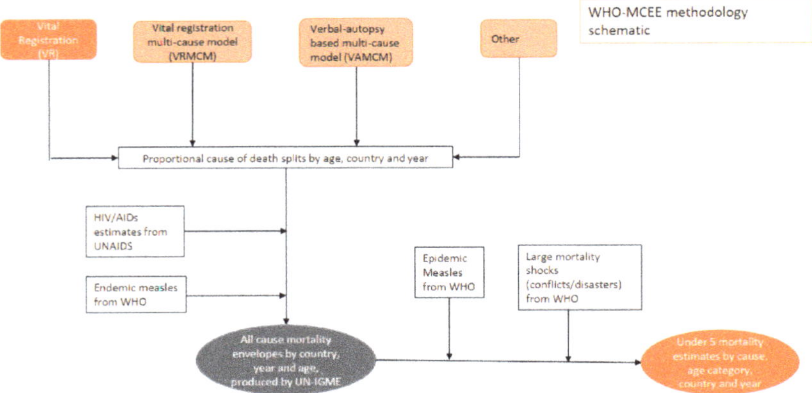

Figure 4. Simplified schematic of the different mortality modelling approaches. VR Data—Data from 76 countries with high quality VR data covering >80% of the population was mapped directly to cause of death categories (see appendix for ICD10 codes mapped to meningitis and sepsis and other severe infections in the neonatal period). VRMCM—Data from the countries with high quality VR data was used to fit a multinomial logistic regression model which was used to predict cause of death proportions in 38 low mortality countries (<35 deaths/1000 live births 2000–2010) with low quality VR data. Covariates used in the model are provided in appendix. VAMCM—In 78 high mortality countries (>35 deaths/1000 live births 2000–2010) verbal autopsy data from 119 research studies in 39 high mortality countries was used to fit a multinomial model to predict causes of death. Cause of death proportions for India were estimated using a combination of VAMCM for the neonatal period and data from the million deaths study and INDEPTH sites in India for the post neonatal period. See appendix for model covariates Other – Cause of death proportions for China were estimated using data from the China Maternal and Child Health Surveillance system. A complete explanation of methods used to produce WHO/MCEE estimates is outlined elsewhere [27].

To estimate meningitis mortality by aetiology, both models applied a proportional split by pathogen to the country-specific meningitis death envelope and adjusted for vaccine coverage. Whilst GBD 2017 pathogen specific mortality proportions were informed by vital registration (VR) data from data rich locations, the WHO-MCEE model based mortality proportions on studies reporting the distribution of pathogen-specific meningitis cases adjusted by pathogen-specific CFRs to derive proportions of deaths. This approach was used due to a lack of literature reporting meningitis mortality fractions by pathogen. To adjust pathogen-specific estimates according to vaccine coverage, IHME ran a metaregression model (DisMod-MR 2.1) with pneumococcal and Hib vaccine coverage as covariates driving down the proportions of disease attributed to those pathogens. The WHO-MCEE model used a deterministic approach to account for vaccine use by calculating the percentage reduction in disease as a result of vaccine efficacy, coverage and, in the case of PCV, the vaccine product and proportion of disease caused by vaccine-specific serotypes.

The models used very different approaches for estimating incidence by aetiology. The GBD 2017 calculated meningitis incidence independently from meningitis mortality using incidence data gathered from hospital records, claims data and a systematic review of the literature capturing incidence studies. The WHO-MCEE incidence estimates were derived by dividing pathogen-specific death estimates by literature-derived CFRs. The WHO-MCEE also published an update to a previous incidence-based model for Hib and pneumococcal meningitis [28], which predicted even lower incidence rates for pneumococcal meningitis and similar rates for Hib.

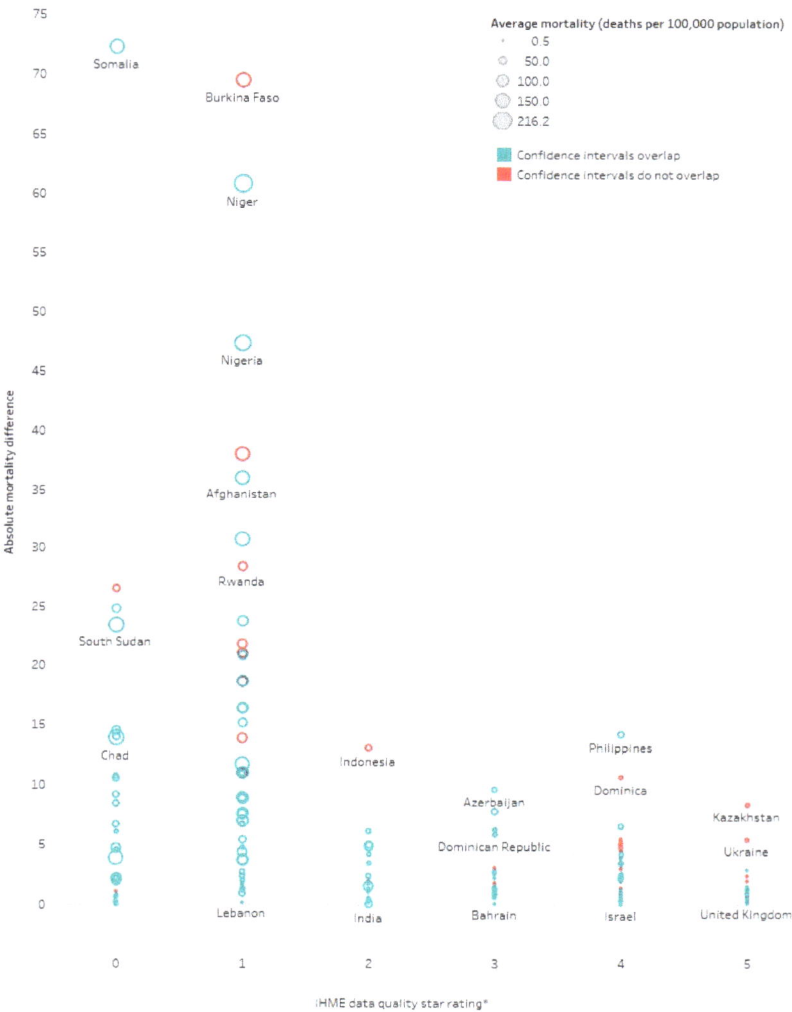

*IHME data quality star rating indicates the quality of data underpinning the estimates. A 5 star rating indicates the highest quality rating and 0 stars the lowest quality rating. 5 stars: 85%-100% well-certified, 4 stars: 65%-84% well-certified, 3 stars: 35%-64% well-certified, 2 stars: 10%-34% well-certified, 1 star: >0%-9% well-certified, 0 stars: No vital registration or verbal autopsy data available from 1980-2017. Per cent well-certified is a function of the completeness of the cause of deaths data multiplied by the quality of the data (the proportion of deaths registered to a well-defined cause.

Figure 5. Absolute difference between WHO-MCEE and GBD 2017 meningitis/encephalitis mortality estimates according to country.

4. Discussion

Despite major differences in the number of deaths attributed to meningitis, both models agree that there is a substantial burden of disease, with meningitis as either the 2nd or 3rd most important infectious syndrome. By far the biggest burden of meningitis is estimated to occur in countries with low quality or no death registration data where these models rely heavily on extrapolating from VA studies. Accurately attributing meningitis as a cause of death using VA is extremely challenging [29–31] and could lead to meningitis as a cause of death being underestimated. VA has a high specificity but low to moderate sensitivity for meningitis [32–34] and can easily attribute death from meningitis to a different cause, especially in malaria endemic regions where severe febrile illness is often assumed to be malaria [35–37].

If these syndromic models systematically underestimate deaths from meningitis, this would result in an underestimate of incidence by pathogen in the WHO-MCEE model because incidence is derived by dividing estimated deaths by CFR based on location and pathogen. The GBD 2017 estimated pathogen-specific incidence separately to pathogen-specific deaths and produced higher estimates than the WHO-MCEE model, but the incidence estimates were out of line with deaths when literature-derived CFRs were applied. Following a meeting where results from this analysis were presented to all modelling groups, the IHME amended their methodology for calculating pathogen-specific incidence. In the recently published GBD 2019 model [38], published studies and hospital data were used to estimate pathogen-specific CFRs as a function of healthcare access and quality. Pathogen specific mortality was then derived from estimates of pathogen-specific incidence and CFRs.

Using global health estimates to derive baseline numbers and targets against which progress can be measured is challenging. Estimates for the entire time series are updated with successive model iterations as new input data are considered and amendments are made to statistical modelling processes. This means that baseline estimates for a given year fluctuate with successive model iterations.

It is vital that the methods used to derive estimates are clearly communicated. Across models it was unclear from published methods exactly how neonatal meningitis as a cause was disaggregated from neonatal sepsis, and other infectious conditions of the newborn, when we know that the majority of the underlying input data does not distinguish between these two causes of death. Unless methods are made transparent, it is difficult for policy makers to understand, and therefore trust, model outputs [39].

Experts responsible for monitoring progress also need to know exactly how estimates were derived in order to assess whether they are capable of measuring progress against certain indicators. Whilst both models accounted for PCV and Hib vaccine impact, they did so using substantially different methods. The IHME's GBD 2017 study accounted for vaccine impact by finding existing relationships between vaccine coverage and the proportion of pathogen-specific meningitis targeted by the vaccine (from countries where data is available) and using these existing relationships to make predictions where data is unavailable. Whilst this approach has an advantage of using as much raw data as possible, it does not distinguish between differences in vaccine products and the varying efficacy associated with different dosing schedules between countries. Although incidence proportion models included data from some countries in sub-Saharan Africa and Asia, the use of VR data alone to determine proportional cause of death means that vaccine effects on pathogen-specific mortality in high mortality countries with no vital registration data are heavily reliant on effects demonstrated in data-rich low-mortality countries. The WHO-MCEE, on the other hand, make predictions where data is sparse/unavailable by simulating the effect of a given vaccine over time on a country specific basis. Assumptions about vaccine impact are transparent and take into account differences in vaccine formulations and dosing schedules, but they may be applied to a pathogen specific meningitis death estimate which is highly uncertain.

It is also important for decision makers to be aware that even in data-rich locations, global health estimates for the most recent year can be based on predictions rather than real underlying data. These estimates may, therefore, be unsuitable for tracking change as a result of a recent intervention, especially if the intervention has not been accounted for as a covariate in the model.

The IHME's GBD model is currently the only available complete source of information about the global and national burden of meningitis amongst all age groups and for most of the pathogens of interest to the global roadmap to defeat meningitis. The IHME have also improved some of their methods for the latest round of estimates by including more surveillance data from high mortality settings in the GBD 2019 than was included in the GBD 2017. Additionally, there are plans for future versions of the IHME's model to include estimates on the incidence and mortality from GBS meningitis, one of the major causes of meningitis in neonates worldwide. However, tracking outputs from multiple models in parallel has advantages in identifying areas of higher uncertainty, generating opportunities for modellers to improve methods and prioritising further primary data collection/strengthening surveillance. An interactive visualisation has been created to track progress using estimates from all of the major global health estimation models [40].

None of the models we assessed were able to accurately account for the fluctuating scales of periodic, large epidemics of meningitis, which are irregular and unpredictable in nature. Whilst GBD 2017 attempted to account for epidemic meningococcal meningitis deaths by adding these to the meningitis death envelope, they did not use equivalent methodology to account for epidemic meningococcal meningitis cases. The WHO-MCEE syndromic model attempted to account for epidemic disease by estimating the average increase in deaths in epidemic years relative to nonepidemic years and adding these to estimates in years with epidemics identified by WHO surveillance reports and published literature. This increases estimates during an epidemic year, but the underlying data from the country are not always reliable, and it does not accurately reflect the variation in the size of the epidemic for a given year. The WHO-MCEE pathogenic model only estimated pathogen-specific deaths for endemic disease, removing the simulated effects of epidemics from the syndromic model before applying proportional splits to the remaining meningitis envelope. Therefore, neither model estimating pathogen specific causes of meningitis was able to account for epidemic pneumococcal meningitis, yet this is an important consideration because it has been demonstrated as having a significant mortality burden [41].

Considering the current limitations of modelled meningitis estimates, it is desirable to track progress alongside additional data where possible. Countries across the African meningitis belt experience the highest burden of meningitis globally because they are susceptible to large and devastating outbreaks of meningococcal disease linked to climatic factors such as dry winds, low humidity and high levels of dust in the air [42]. Whilst many of these countries have poor death registration systems, they have relatively rich and complementary meningitis surveillance systems. Since 2003 an enhanced meningitis surveillance network has been established across the meningitis belt to strengthen outbreak detection and enable a rapid response to outbreaks of meningococcal disease across the region [43]. The network now covers 24 countries, reporting suspected cases and deaths from meningitis to the WHO intercountry support team (WHO/IST) each week during the meningitis season and every month for the rest of the year [44]. Case-based surveillance systems have been established in five countries within the region allowing for comprehensive information on CFRs by age [45].

Triangulating modelled estimates against surveillance data provides the opportunity to reality-check modelled outputs. Utilising surveillance data in combination with evidence of age and regionally specific CFRs has already successfully been used by experts wishing to monitor global progress towards the 2005 measles mortality reduction goal because measles mortality estimates calculated from vital registration data were considered an unreliable way to track progress [46]. Surveillance data for meningitis is not currently available for every country worldwide. However, comprehensive roll out of Hib and pneumococcal vaccines is driving down incidence and mortality from meningitis caused by these pathogens across the globe. Improved pathogen-specific surveillance informed by accurate and timely laboratory diagnosis is required to adequately assess the impact of these important life-saving interventions. This is particularly important for countries transitioning out of Gavi support which need to justify national investments in these vaccines. Additionally, all member states of the UN have committed to achieving universal health coverage by signing up to the SDGs, so there is reason to believe that the availability of good quality surveillance data will improve over time as health systems are strengthened.

More work is required to provide credible meningitis burden estimates for measuring progress. Currently meningitis mortality estimates are highly uncertain because the models rely heavily on death registration data, which is largely missing or incomplete in countries with the highest meningitis burden. Additionally, since postmortem examination is rarely performed in countries without vital registration systems, and the symptoms of meningitis can easily be mistaken for other diseases, there is a risk that the mortality burden of meningitis could be underestimated. Encouragingly, better data on cause of death are becoming available in regions where child mortality rates are the highest through the use of minimally invasive tissue sampling [47,48] and inclusion of these data in future models could considerably improve the reliability of their outputs.

5. Conclusions

Global meningitis estimates should be interpreted with caution. Tracking progress towards controlling this disease should also include analysis of real surveillance data where available. The WHO Defeating Meningitis by 2030 Global Roadmap will improve awareness, diagnosis and surveillance of meningitis. As the roadmap drives more comprehensive data on meningitis, a convergence in modelled estimates and a more reliable picture of reductions in the burden of meningitis are anticipated.

Supplementary Materials: The following are available online at https://www.mdpi.com/2076-2607/9/2/377/s1 [49–51].

Author Contributions: Conceptualization, C.W., N.B., L.G., V.S., J.M.S. and C.T.; methodology, C.W., N.B., L.G., J.M.S., C.T.; validation, R.B., H.K., H.Y.W., L.L., D.Y., M.D.K., B.W.; formal analysis, C.W. and N.B.; writing—original draft preparation, C.W. and N.B.; writing—review and editing, C.W., L.G., V.S., R.B., H.K., H.Y.W., L.L., D.Y., M.D.K., B.W., J.M.S., C.T.; visualization, C.W.; supervision, L.G., J.M.S. and C.T. All authors have read and agreed to the published version of the manuscript.

Funding: This research received no specific funding, but Meningitis Research Foundation has received unrestricted educational grants from GSK, Pfizer and Sanofi which allowed this study to take place.

Institutional Review Board Statement: Not applicable.

Informed Consent Statement: Not applicable.

Data Availability Statement: Publicly available datasets were analyzed in this study. GBD 2017 estimates are available from http://ghdx.healthdata.org/gbd-2017; WHO-MCEE syndromic meningitis estimates are available from http://158.232.12.119/healthinfo/global_burden_disease/estimates/en/index2.html and WHO-MCEE syndromic estimates are available from their publication [14].

Acknowledgments: This work arose from a meeting to evaluate the global burden of meningitis estimates convened by the Gates Foundation in November 2018.

Conflicts of Interest: C.T. received a consulting payment from GSK in 2018 outside the submitted work. C.W., N.B., L.G., V.S. took part in this research as employees of Meningitis Research Foundation, which has received unrestricted educational grants from GSK, Pfizer and Sanofi. All other authors declare no conflict of interest. GSK, Pfizer and Sanofi had no role in the design of this study; in the collection, analyses, or interpretation of data; in the writing of the manuscript, or in the decision to publish the results.

References

1. United Nations Inter-agency Group for Child Mortality Estimation (UN IGME). Levels & Trends in Child Mortality: Report 2020, Estimates Developed by the UN Inter-Agency Group for Child Mortality Estimation. 2020. Available online: https://www.un.org/development/desa/pd/news/levels-and-trends-child-mortality-2020-report (accessed on 1 December 2020).
2. Zunt, J.R.; Kassebaum, N.J.; Blake, N.; Glennie, L.; Wright, C.; Nichols, E.; Abd-Allah, F.; Abdela, J.; Abdelalim, A.; A Adamu, A.; et al. Global, regional, and national burden of meningitis, 1990–2016: A systematic analysis for the Global Burden of Disease Study 2016. *Lancet Neurol.* **2018**, *17*, 1061–1082. [CrossRef]

3. GBD Causes of Death Collaborators. Global, regional, and national age-sex-specific mortality for 282 causes of death in 195 countries and territories, 1980–2017: A systematic analysis for the Global Burden of Disease Study 2017. *Lancet* **2018**, *392*, 1736–1788. [CrossRef]
4. World Health Organization. Global Action Plan on Antimicrobial Resistance. Available online: https://www.who.int/antimicrobial-resistance/publications/global-action-plan/en/ (accessed on 3 February 2020).
5. Global Strategy for Women's, Children's, and Adolescents' Health 2016–2030. Available online: https://www.who.int/life-course/partners/global-strategy/globalstrategyreport2016-2030-lowres.pdf (accessed on 3 February 2020).
6. World Health Organization/The United Nations Children's Fund (UNICEF). Global Action Plan for Prevention and Control of Pneumonia (GAPP). Available online: https://www.unicef.org/media/files/GAPP3_web.pdf (accessed on 3 February 2020).
7. World Health Organization/The United Nations Children's Fund (UNICEF). Ending Preventable Child Deaths from Pneumonia and Diarrhoea by 2025. The Integrated Global Action Plan for Pneumonia and Diarrhoea (GAPPD). Available online: https://www.unicef.org/media/files/Final_GAPPD_main_Report-_EN-8_April_2013.pdf (accessed on 3 February 2020).
8. Transforming Our World: The 2030 Agenda for Sustainable Development. Goal 3. Ensure Healthy Lives and Promote Well-Being for All at All Ages. Available online: https://sustainabledevelopment.un.org/post2015/transformingourworld (accessed on 3 February 2020).
9. Working Group of the Specialised Technical Committee on Health, Population and Drug Control, Africa Health Strategy 2016–2030. Available online: https://au.int/sites/default/files/newsevents/workingdocuments/27513-wd-sa16951_e_africa_health_strategy-1.pdf (accessed on 3 February 2020).
10. World Health Organization. Defeating meningitis by 2030. Available online: https://www.who.int/initiatives/defeating-meningitis-by-2030 (accessed on 8 February 2021).
11. Rudan, I.; Campbell, H.; Marusic, A.; Sridhar, D.; Nair, H.; Adeloye, D.; Theodoratou, E.; Chan, K.Y. Assembling GHERG: Could "academic crowd–sourcing" address gaps in global health estimates? *J. Glob. Heal.* **2015**, *5*, 010101. [CrossRef]
12. World Health Organization. Disease Burden and Mortality Estimates. Child Causes of Death, 2000–2017. Available online: https://www.who.int/healthinfo/global_burden_disease/estimates/en/index2.html (accessed on 14 October 2019).
13. World Health Organization. Global Health Estimates 2016: Deaths by Cause, Age, Sex, by Country and by Region, 2000–2016. Available online: http://www.who.int/healthinfo/global_burden_disease/estimates/en/ (accessed on 10 June 2019).
14. Wahl, B.; O'Brien, K.L.; Greenbaum, A.; Majumder, A.; Liu, L.; Chu, Y.; Lukšić, I.; Nair, H.; A McAllister, D.; Campbell, H.; et al. Burden of Streptococcus pneumoniae and Haemophilus influenzae type b disease in children in the era of conjugate vaccines: Global, regional, and national estimates for 2000–15. *Lancet Glob. Heal.* **2018**, *6*, e744–e757. [CrossRef]
15. Seale, A.C.; Bianchi-Jassir, F.; Russell, N.J.; Kohli-Lynch, M.; Tann, C.J.; Hall, J.; Madrid, L.; Blencowe, H.; Cousens, S.; Baker, C.J.; et al. Estimates of the Burden of Group B Streptococcal Disease Worldwide for Pregnant Women, Stillbirths, and Children. *Clin. Infect. Dis.* **2017**, *65*, S200–S219. [CrossRef]
16. Okike, I.O.; Johnson, A.P.; Henderson, K.L.; Blackburn, R.M.; Muller-Pebody, B.; Ladhani, S.N.; Anthony, M.; Ninis, N.; Heath, P.T. Incidence, Aetiology and Outcome of Bacterial Meningitis in Infants Aged <90 days in the UK and Republic of Ireland: Prospective, enhanced, national population-based surveillance. *Clin. Infect. Dis* **2014**. [CrossRef]
17. Ku, L.C.; Boggess, K.A.; Cohen-Wolkowiez, M. Bacterial Meningitis in Infants. *Clin. Perinatol.* **2015**, *42*, 29–45. [CrossRef]
18. English, M.; Esamai, F.; Wasunna, A.; Were, F.; Ogutu, B.; Wamae, A.; Snow, R.W.; Peshu, N. Assessment of inpatient paediatric care in first referral level hospitals in 13 districts in Kenya. *Lancet* **2004**, *363*, 1948–1953. [CrossRef]
19. Okomo, U.; Dibbasey, T.; Kassama, K.; E Lawn, J.; Zaman, S.M.A.; Kampmann, B.; Howie, S.R.C.; Bojang, K. Neonatal admissions, quality of care and outcome: 4 years of inpatient audit data from The Gambia's teaching hospital. *Paediatr. Int. Child Heal.* **2015**, *35*, 252–264. [CrossRef]
20. United Nations Department of Economic and Social Affairs Population Division. Population Databases. Available online: https://www.un.org/en/development/desa/population/publications/database/index.asp (accessed on 10 June 2019).
21. Thigpen, M.C.; Whitney, C.G.; Messonnier, N.E.; Zell, E.R.; Lynfield, R.; Hadler, J.L.; Harrison, L.H.; Farley, M.M.; Reingold, A.; Bennett, N.M.; et al. Bacterial meningitis in the United States, 1998–2007. *N. Engl. J. Med.* **2011**, *364*, 2016–2025. [CrossRef] [PubMed]
22. Swartz, M.N. Bacterial meningitis—A view of the past 90 years. *N. Engl. J. Med.* **2004**, *351*, 1826–1828. [CrossRef]
23. van de Beek, D.; de Gans, J.; Spanjaard, L.; Weisfelt, M.; Reitsma, J.B.; Vermeulen, M. Clinical features and prognostic factors in adults with bacterial meningitis. *N. Engl. J. Med.* **2004**, *351*, 1849–1859. [CrossRef] [PubMed]
24. Boisier, P.; Maïnassara, H.B.; Sidikou, F.; Djibo, S.; Kairo, K.K.; Chanteau, S. Case-fatality ratio of bacterial meningitis in the African meningitis belt: We can do better. *Vaccine* **2007**, *25*, A24–A29. [CrossRef] [PubMed]
25. Dickinson, F.; E Pérez, A. Bacterial Meningitis in children and adolescents: An observational study based on the national surveillance system. *BMC Infect. Dis.* **2005**, *5*, 103. [CrossRef] [PubMed]
26. Gurley, E.S.; Montgomery, S.P.; Mayer, L.W.; Uddin, A.K.M.R.; Petersen, L.R.; Hossain, M.J.; Luby, S.P.; Rahman, M.E.; Breiman, R.F.; Whitney, A.; et al. Etiologies of bacterial meningitis in Bangladesh: Results from a hospital-based study. *Am. J. Trop. Med. Hyg.* **2009**, *81*, 475–483. [CrossRef] [PubMed]
27. Department of Evidence, Information and Research (WHO, Geneva) and Maternal Child Epidemiology Estimation (MCEE). MCEE-WHO Methods and Data Sources for Child Causes of Death 2000–2017. Available online: https://www.who.int/healthinfo/global_burden_disease/childcod_methods_2000_2017.pdf?ua=1 (accessed on 4 February 2019).

28. O'Brien, K.L.; Wolfson, L.J.; Watt, J.P.; Henkle, E.; Deloria-Knoll, M.; McCall, N.; Lee, E.; Mulholland, K.; Levine, O.S.; Cherian, T.; et al. Burden of disease caused by Streptococcus pneumoniae in children younger than 5 years: Global estimates. *Lancet* **2009**, *374*, 893–902. [CrossRef]
29. Benara, S.K.; Singh, P. Validity of causes of infant death by verbal autopsy. *Indian J. Pediatr.* **1999**, *66*, 647–650. [CrossRef] [PubMed]
30. World Health Organization. A Standard Verbal Autopsy Method for Investigating Causes of Death in Infants and Children. Available online: https://www.who.int/csr/resources/publications/surveillance/whocdscsrisr994.pdf?ua=1 (accessed on 9 September 2019).
31. Butler, D. Verbal autopsy methods questioned. *Nature* **2010**, *467*, 1015. [CrossRef]
32. Misganaw, A.; Mariam, D.H.; Araya, T.; Anteneh, A. Validity of verbal autopsy method to determine causes of death among adults in the urban setting of Ethiopia. *BMC Med. Res. Methodol.* **2012**, *12*, 130. [CrossRef]
33. Mpimbaza, A.; Filler, S.; Katureebe, A.; Kinara, S.O.; Nzabandora, E.; Quick, L.; Ratcliffe, A.; Wabwire-Mangen, F.; Chandramohan, D.; Staedke, S.G. Validity of Verbal Autopsy Procedures for Determining Malaria Deaths in Different Epidemiological Settings in Uganda. *PLoS ONE* **2011**, *6*, e26892. [CrossRef]
34. Snow, R.W.; Armstrong, J.R.; Forster, D.; Winstanley, M.T.; Marsh, V.M.; Newton, C.R.; Waruiru, C.; Mwangi, I.; Winstanley, P.A.; Marsh, K. Childhood deaths in Africa: Uses and limitations of verbal autopsies. *Lancet* **1992**, *340*, 351–355. [CrossRef]
35. Anglewicz, P.; Kohler, H.P. Overestimating HIV infection: The construction and accuracy of subjective probabilities of HIV infection in rural Malawi. *Demogr. Res.* **2009**, *20*, 65–96. [CrossRef] [PubMed]
36. Chandler, C.I.; Jones, C.; Boniface, G.; Juma, K.; Reyburn, H.; Whitty, C.J. Guidelines and mindlines: Why do clinical staff over-diagnose malaria in Tanzania? A qualitative study. *Malar. J.* **2008**, *7*, 53. [CrossRef]
37. Gwer, S.; Newton, C.R.; Berkley, J.A. Over-diagnosis and co-morbidity of severe malaria in African children: A guide for clinicians. *Am. J. Trop. Med. Hyg.* **2007**, *77*, 6–13. [CrossRef] [PubMed]
38. GBD 2019 Diseases Injuries Collaborators. Global burden of 369 diseases and injuries in 204 countries and territories, 1990–2019: A systematic analysis for the Global Burden of Disease Study 2019. *Lancet* **2020**, *396*, 1204–1222. [CrossRef]
39. Mathers, C.D. History of global burden of disease assessment at the World Health Organization. *Arch. Public Health* **2020**, *78*, 1–13. [CrossRef]
40. Meningitis Research Foundation. Meningitis Progress Tracker. Available online: https://public.tableau.com/profile/meningitis.research.foundation#!/vizhome/MeningitisandNeonatalSepsisTracker_15543731247200/CasesandDeaths (accessed on 14 September 2020).
41. Traore, Y.; Tameklo, T.A.; Njanpop-Lafourcade, B.-M.; Lourd, M.; Yaro, S.; Niamba, D.; Drabo, A.; E Mueller, J.; Koeck, J.-L.; Gessner, B.D. Incidence, Seasonality, Age Distribution, and Mortality of Pneumococcal Meningitis in Burkina Faso and Togo. *Clin. Infect. Dis.* **2009**, *48*, S181–S189. [CrossRef] [PubMed]
42. Palmgren, H. Meningococcal disease and climate. *Glob. Health Action* **2009**, *2*. [CrossRef] [PubMed]
43. Lingani, C.; Bergeron-Caron, C.; Stuart, J.M.; Fernandez, K.; Djingarey, M.H.; Ronveaux, O.; Schnitzler, J.C.; Perea, W.A. Meningococcal Meningitis Surveillance in the African Meningitis Belt, 2004–2013. *Clin. Infect. Dis.* **2015**, *61*, S410–S415. [CrossRef]
44. World Health Organization. Meningitis Weekly Bulletin. Available online: https://www.who.int/emergencies/diseases/meningitis/epidemiological/en/ (accessed on 16 September 2020).
45. Patel, J.C.; Soeters, H.M.; Diallo, A.O.; Bicaba, B.W.; Kadadé, G.; Dembélé, A.Y.; Acyl, M.A.; Nikiema, C.; Lingani, C.; Hatcher, C.; et al. MenAfriNet: A Network Supporting Case-Based Meningitis Surveillance and Vaccine Evaluation in the Meningitis Belt of Africa. *J. Infect. Dis.* **2019**, *220*, S148–S154. [CrossRef]
46. Wolfson, L.J.; Strebel, P.M.; Gacic-Dobo, M.; Hoekstra, E.J.; McFarland, J.W.; Hersh, B.S.; Measles, I. Has the 2005 measles mortality reduction goal been achieved? A natural history modelling study. *Lancet* **2007**, *369*, 191–200. [CrossRef]
47. Castillo, P.; Ussene, E.; Ismail, M.R.; Jordao, D.; Lovane, L.; Carrilho, C.; Lorenzoni, C.; Lacerda, M.V.; Palhares, A.; Rodríguez-Carunchio, L.; et al. Pathological Methods Applied to the Investigation of Causes of Death in Developing Countries: Minimally Invasive Autopsy Approach. *PLoS ONE* **2015**, *10*, e0132057. [CrossRef] [PubMed]
48. Bassat, Q.; Ordi, J.; Vila, J.; Ismail, M.R.; Carrilho, C.; Lacerda, M.; Munguambe, K.; Odhiambo, F.; Lell, B.; Sow, S.; et al. Development of a post-mortem procedure to reduce the uncertainty regarding causes of death in developing countries. *Lancet Glob. Health* **2013**, *1*, e125–e126. [CrossRef]
49. World Health Organization. World Health Statistics 2017. Monitoring Health for the SDGs. Available online: https://apps.who.int/iris/bitstream/handle/10665/255336/9789241565486-eng.pdf;jsessionid=C00C94A637FA197A27B773D735A711A4?sequence=1 (accessed on 19 September 2019).
50. Thatte, N.; Kalter, H.D.; Baqui, A.H.; Williams, E.M.; Darmstadt, G.L. Ascertaining causes of neonatal deaths using verbal autopsy: Current methods and challenges. *J. Perinatol.* **2009**, *29*, 187–194. [CrossRef] [PubMed]
51. GBD 2017 Diseases Injuries Collaborators. Global, regional, and national incidence, prevalence, and years lived with disability for 354 diseases and injuries for 195 countries and territories, 1990–2017: A systematic analysis for the Global Burden of Disease Study 2017. *Lancet* **2018**, *392*, 1789–1858. [CrossRef]

Review

The Impact and Burden of Neurological Sequelae Following Bacterial Meningitis: A Narrative Review

Nicoline Schiess [1,*], Nora E. Groce [2] and Tarun Dua [1]

1 Brain Health Unit, Department of Mental Health and Substance Use, World Health Organization (WHO), 1202 Geneva, Switzerland; duat@who.int
2 UCL International Disability Research Centre, Department of Epidemiology and Health Care, University College London, London WC1E 7HB, UK; nora.groce@ucl.ac.uk
* Correspondence: schiessn@who.int

Abstract: The burden, impact, and social and economic costs of neurological sequelae following meningitis can be devastating to patients, families and communities. An acute inflammation of the brain and spinal cord, meningitis results in high mortality rates, with over 2.5 million new cases of bacterial meningitis and over 236,000 deaths worldwide in 2019 alone. Up to 30% of survivors have some type of neurological or neuro-behavioural sequelae. These include seizures, hearing and vision loss, cognitive impairment, neuromotor disability and memory or behaviour changes. Few studies have documented the long-term (greater than five years) consequences or have parsed out whether the age at time of meningitis contributes to poor outcome. Knowledge of the socioeconomic impact and demand for medical follow-up services among these patients and their caregivers is also lacking, especially in low- and middle-income countries (LMICs). Within resource-limited settings, the costs incurred by patients and their families can be very high. This review summarises the available evidence to better understand the impact and burden of the neurological sequelae and disabling consequences of bacterial meningitis, with particular focus on identifying existing gaps in LMICs.

Keywords: meningitis; burden; social and economic costs; neurological sequelae; WHO meningitis roadmap; tuberculous meningitis; disability

Citation: Schiess, N.; Groce, N.E.; Dua, T. The Impact and Burden of Neurological Sequelae Following Bacterial Meningitis: A Narrative Review. *Microorganisms* **2021**, *9*, 900. https://doi.org/10.3390/microorganisms9050900

Academic Editor: James Stuart

Received: 16 March 2021
Accepted: 19 April 2021
Published: 22 April 2021

1. Introduction

Many different bacteria can cause meningitis; however, *Streptococcus pneumoniae* (Sp or pneumococcus), *Haemophilus influenzae* type b (Hib) and *Neisseria meningitidis* (Nm or meningococcus) are the most common pathogens other than those in infants, who are most commonly affected by *Streptococcus agalactiae* (group B streptococcus or GBS) [1]. Prior to the advent of widespread vaccination campaigns, bacterial meningitis outbreaks imparted a significant toll, with some pathogens, such as group A *Neisseria meningitidis*, having meningitis rates as high as 1% of the population during major African epidemics in the last century [2]. Tuberculosis, which affects millions of people each year worldwide, predominantly in low- and middle-income countries (LMICs) [3] affects the central nervous system in approximately 1% of cases [4] yet can also result in profound mortality and morbidity [5]. Multiple factors contribute to the impact or severity of different pathogens causing meningitis. Meningococcus and pneumococcus can cause severe central nervous system damage and have the propensity to cause sepsis, a significant cause of mortality. However, other comorbid conditions can also impact the severity and sequelae of meningitis-causing pathogens. These include malnutrition, immunocompromising conditions, and delays in diagnosis and treatment.

Globally, the epidemiology of bacterial meningitis has changed dramatically with the introduction of conjugate vaccines [6,7]. The Hib conjugate vaccine has essentially eradicated Hib meningitis [6,8–10], and with widespread use of the meningococcal serogroup

A conjugate vaccine (MACV), the overall burden of suspected meningococcal meningitis cases has been reduced by almost 60% in high-risk countries across northern Africa ("Meningitis belt") with near-complete elimination of confirmed serogroup A disease [11]. Pneumococcal conjugate vaccines (PCVs) have also resulted in a slight decrease in pneumococcal disease [12], and in many countries, this pathogen has overtaken *H. influenzae* as the most common cause of meningitis [6,13,14]. Despite these advances, there were still over 2.5 million new cases of bacterial meningitis and over 236,000 deaths worldwide in 2019 alone [15].

Meningitis survivors can be left with disabling neurological sequelae such as seizures, hearing and vision loss, neuromotor disability and hydrocephalus. Cognitive and behavioural sequelae following bacterial meningitis have also been reported [16,17]; however, it is likely that these more subtle sequelae may sometimes go undiagnosed and can have profoundly detrimental effects on school and work performance. The burden of disabling sequelae is highest in low- and middle-income countries (LMICs) as these countries have high rates of meningitis [16].

Over the past several years, the expansion of meningitis-related vaccination programs, increasing research and intervention efforts, and growing advocacy on behalf of meningitis survivors and their families have presented significant possibilities for both meningitis prevention and life improvement for survivors. However, coordination of these advances has been lacking. In response to this, a new international response to meningitis is now underway; WHO's Defeating Meningitis by 2030 Global Roadmap [18] intends to address the global issues around bacterial meningitis (meningococcus, pneumococcus, *Haemophilus influenzae* and group B streptococcus), with one of the main goals focusing on the long-term sequelae of meningitis and quality of life. A key activity proposed in the meningitis roadmap is to conduct research on the socioeconomic impact of sequelae on children, adults and their families/carers and on the availability and effectiveness of aftercare/support interventions.

In this review, we summarise the evidence to better understand the impact and burden of the neurological sequelae and disability of bacterial meningitis, with a focus on LMICs and with particular attention to the long-term impact of meningitis on those who survive, thus advising the third goal of the Defeating Meningitis Roadmap.

2. Global Burden of Meningitis

In 2019, worldwide mortality from all causes of meningitis (excluding tuberculous and cryptococcal meningitis) was over 236,000 deaths, with approximately 2.5 million new cases [15]. Additionally, in 2019, meningitis ranked sixth in the top causes of disability adjusted life years (DALYs) in children under 10 years of age [19]. While global deaths due to meningitis decreased between 1990 and 2016, the 21% decrease pales in comparison to the dramatic reductions in mortality from other diseases such as measles (93%) and tetanus (91%) [20].

Meningitis Belt

In 2019, there were over 22,000 suspected cases of meningitis, with 1261 deaths reported to the WHO in African countries sharing data [21]. A disproportionally high rate of bacterial meningitis occurs in Africa due to elevated endemic disease, a younger population and regularly occurring epidemics across the "meningitis belt"—a span of countries between Ethiopia and Senegal that includes Nigeria, Burkina Faso and Sudan (See Figure 1). Outbreaks in these countries are characterised by sporadic seasonal infections, with periodically superimposed larger epidemics. Although the burden of meningitis in this region has declined following the introduction of a MACV in 2010, other meningococcal serogroups and bacterial pathogens continue to cause endemic and epidemic disease [22,23]. As these epidemics have a profound effect on the population, meningitis is considered a priority disease in the WHO integrated disease surveillance and response platform [24].

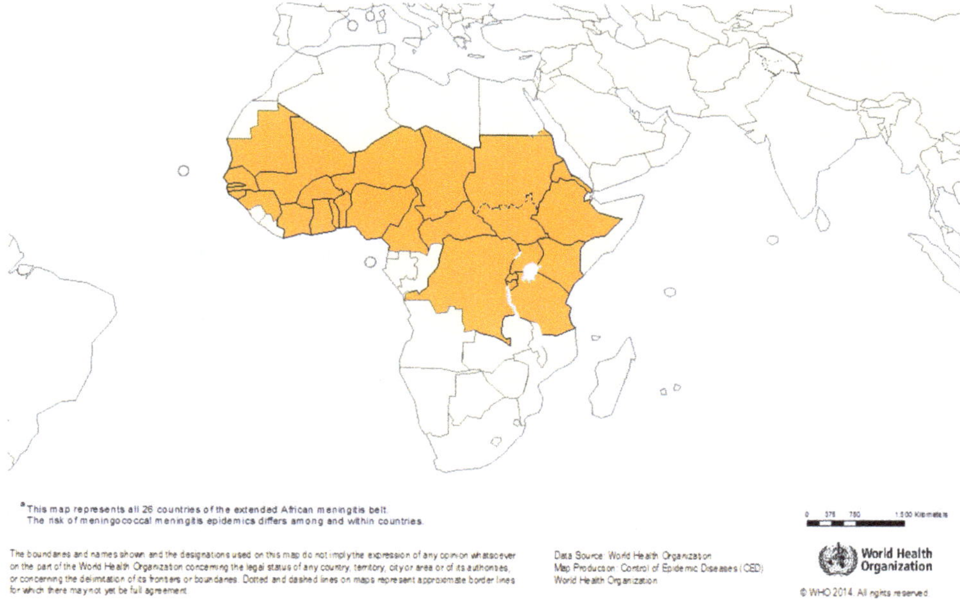

Figure 1. African meningitis belt. Source: "Meningitis outbreak response in sub-Saharan Africa: WHO Guideline. Geneva: World Health Organization; 2014".

The costs of meningitis outbreaks for governments and ministries of health are significant as well. A Colombini et al. review of the Burkina Faso public health response [25] estimated the total cost for the 2006–2007 epidemic season to be 9.4 million USD—three quarters of which was covered by the government and the Ministry of Health's financial and technical partners. The remaining cost was absorbed by the families of meningitis victims. The review noted challenges that included medicine shortages, a paucity of healthcare workers and a lack of government funding for medication. [25]. The highest cost were the vaccine and injection supplies themselves. Vaccine transportation and personnel costs were the next highest cost although they were a fraction of that of the vaccines and injection supplies. The cost of potential long-term neurological sequelae and the associated expenses of rehabilitation were not evaluated.

3. Neurological Sequelae

3.1. Frequency and Types of Neurological Sequelae Following Meningitis

Acute bacterial meningitis can have severe complications with long-term neurological sequelae resulting in disability even in high-income countries (HICs) with appropriate antibiotic therapy and vaccine availability [26]. For example, a recent study in the United States on paediatric bacterial meningitis demonstrated a 45.9% complication rate at 30 days for community-acquired bacterial meningitis, with hydrocephalus (20.8%), intracranial abscess (8.8%) and cerebral oedema (8.1%) being the most common short-term neurological sequelae [27].

A large systematic review and meta-analysis by Edmond et al. estimated the risks of neurological sequelae globally by region and socioeconomic status from 1980 to 2008 and determined that the risk of suffering from some type of sequelae after bacterial meningitis was 20%. The risk was almost threefold higher in Africa and Asia compared to Europe [16]. Treatment delay [28,29], length of travel to receive care, lower immune defences as a result of chronic malnutrition and cost of hospital care [30] have all been reported to contribute to this elevated risk in low-income countries.

While hearing loss and seizures were the most common sequelae among the 132 studies in the Edmond et al. meta-analysis, cognitive impairment clearly affects a large proportion of survivors and, in LMICs, is no doubt underestimated considering that only two studies from Africa and Asia specifically evaluated cognitive domains. In many studies from LMICs, standardised assessment tools and thorough neurological examinations are not utilised and therefore do not capture possible subtle manifestations such as neurocognitive impairment or behavioural changes [16].

In addition, most studies do not compare the rates of sequelae among children with or without a history of meningitis. This method might provide a more accurate picture of the risk of sequelae after bacterial meningitis by controlling for baseline rates of neurological disorders within a population. This method was utilised in a prospective cohort study in Senegal that used standardised assessment tools on both the control and affected groups, making comparison and categorisation more reliable. The affected children in Senegal were found to have 3 times higher odds of major disability (such as cognitive/motor deficits, hearing loss or seizures) after suffering from bacterial meningitis when compared with a community control group. Multiple domains were often involved, the most frequent being cognitive and motor deficits with seizures [31]. Almost 40% of affected children did not attend pre-school or school compared with 16.7% of the control group. The importance of including a control group is underscored by the results in this study that showed that, while 51.8% of children with prior meningitis had hearing loss, a substantial number (30.3%) of children in the control group also had hearing loss, possibly due to untreated otitis media within the population.

3.2. *Persistence of Sequelae over Time*

The study follow-up time after acute infection is also an important component as subtle deficits, including poor school performance, behavioural issues and undiagnosed attention deficit disorder, may not be appreciated initially and can affect survivors for many years [32,33]. A survey of parents and teachers in the United Kingdom on 739 infantile meningitis cases and 606 matched controls was conducted years later when the subjects were teenagers. The results of the study showed that 46% of parents of affected children reported behavioural problems compared to 21% in the control group. The percentages of behavioural problems reported by their teachers were 37% and 23%, respectively [34].

A 2011 systematic literature review by Chandran et al. focused specifically on neurological sequelae five years or more after the acute attack. Searching all globally published articles of the consequences or sequelae of bacterial meningitis in children (one month to 18 years), they identified that almost one-half of survivors five years out or longer suffered from some type of sequelae, with over three fourths having intellectual or behavioural problems [32]. This study is particularly important as it defined "long-term" as five years or longer in contrast to other observational studies that either specified "long-term" as any time post-discharge or had no defined follow up [33,35–37].

Control-based studies examining the sequelae of meningitis ten years or longer after infection also have the potential to parse out the risks of sequelae according to age of infection. In other words, does the age of meningitis onset contribute to the severity of long-term sequelae or predict outcome? This question was examined by Anderson et al. in a longitudinal, prospective study that focused specifically on the age of illness and long-term sequelae in meningitis survivors 12 years later [38]. Reassuringly, those who had had meningitis did not show progressive deterioration when compared to healthy controls, indicating the ability to developmentally compensate in executive functioning. However, a clear difference showed that those who had had meningitis prior to one year of age had poorer performances in certain domains such as language and executive functioning compared to those who had meningitis after 12 months.

3.3. Neurological Sequelae in LMICs

Few studies in LMICs have examined the long-term neurological sequelae following bacterial meningitis. The large systematic review and meta-analysis of all sequelae post discharge by Edmond et al. [16] revealed that the number of studies published on disabling sequelae was much higher in regions such as Europe (40%) and the Americas (24%) versus Asia (6%) and Africa (10%). A different systematic review in 2009 by Ramakrishnan et al. included 6029 African children under age 15 years with confirmed meningitis in 21 African countries and revealed that nearly 20% of bacterial meningitis survivors experienced neurological sequelae while in the acute hospitalised setting [35]. Notably, only seven of these countries had post-discharge follow-up studies with the follow-up time ranging from 3 to 90 months. The total number of patients included in these studies was much lower (Table 1. Significantly, the analysis found that 10% of children died after discharge and that 25% (range 3–47%) had neurological sequelae 3–60 months after diagnosis based on clinical exam alone [35].

Table 1. The post-discharge sequelae in children with all causes of bacterial meningitis for studies with >25 subjects.

Country	Year Published, Reference	Total No. Assessed for Sequelae	Ave Follow Up Time (Months)	Post-Discharge Neurological Sequelae	Bacterial Pathogens
Cameroon	1995, [39]	67	14	25%	Spn, Hib, Nm, others
Egypt	1989, [40]	367	3	3%	Spn, Hib, Nm
	1991, [41]	78	2–24	24%	Tuberculosis
	1998, [42]	289	12	32%	Tuberculosis
Ethiopia	2003, [43]	53	Not specified	34%	Spn, Hib, Nm, others
The Gambia	1990, [44]	48	8	13%	Hib
	2000, [45]	73	11 to 90	47%	Spn, Hib
Nigeria	1999, [46]	47	Not specified	23%	Spn, Hib, Nm, *Klebsiella* and others
Sudan	1990, [47]	27	3–48	33%	Spn, Hib, Nm and others
Tunisia	1992, [48]	82	60	13%	Spn, Hib, Nm

Neurological sequelae defined as behavioural problem, cognitive delay, speech or language disorder, seizures or vision loss. Hib = *Haemophilus influenzae* type b; Nm = *Neisseria meningitidis*; Spn = *Streptococcus pneumoniae*; adapted from Ramakrishnan et al., 2009 [35].

Similarly, in Bangladesh, a study on children with pneumococcal meningitis showed that many survivors had hearing (33%), vision (8%), mental (41%) and psychomotor deficits (49%) within 40 days post-discharge. A second group of pneumococcal meningitis survivors in the study were followed up at 12–24 months and showed deficits in hearing (18%), vision (4%), and mental (41%) and psychomotor development (35%) [49].

4. Social and Economic Burden of Neurological Sequelae

Globally, but particularly in Africa, there are limited data on the long-term social and economic burden of neurological sequelae among meningitis survivors and their families. Social and economic factors can dramatically affect survivors' ability across the life course to perform in school or to obtain gainful employment, particularly as the risk of sequelae in children under five years has been found to be double that for children older than five [16]. While some children have very severe sequelae, there are many other children who are less severely affected. Neurodevelopmental delays can often be subtle and may not be adequately diagnosed in routine clinical exams. Cognitive and behavioural difficulties may only be noticed once a child has started school [17], or they may remain undiagnosed. Whether undiagnosed or simply unable to access adequate resources for help and support, these children may struggle to keep up, be labelled as delayed, and drop or flunk out, setting themselves up for a lifetime of limited opportunities.

For children in particular, downstream consequences of neurological sequelae can be dire for the whole family, with studies showing that caregivers are often forced to choose care for their disabled child versus working to generate an income or provide for other siblings [49,50]. A study of 107 South African children with TB meningitis who lived in low socioeconomic environments showed that 19% of all mothers reported experiencing financial difficulty after their child fell ill [50]. A reported case in Bangladesh painfully illustrates what a profound impact a disabled child can have on the whole family. A young boy, initially misdiagnosed and thus treated late for pneumococcal meningitis, lost key developmental milestones. The family's socioeconomic status underwent a dramatic change as a result of his disabilities. To pay for medical bills, the father was forced to sell his small piece of land and to work several jobs, barely earning enough to feed the family. The mother attends to all his needs, neglecting care for the rest of the family. An elder sibling's education was disrupted since they could not afford school supplies [49]. As noted by the authors:

"In countries like Bangladesh, (the impact of impairments) is quite different from that in developed parts of the world, because of very limited facilities for the education of these children and almost no priority for the facilitation of a normal life. As a result, most of these children cannot have an independent life, are unable to participate in any social activities, and remain confined at home. All these factors have psychological, social, and financial impacts on the entire family and on society." [49].

Even if patients have access to public health services, along with care in a timely manner and a reasonable distance, costs associated with medical care can be financially devastating to families and communities as personal household earnings or savings cover many expenses of medical care in countries such as Kenya [51] and Burkina Faso [52]. A study in Burkina Faso estimated the total average cost for each family to treat a child with a meningitis episode to be approximately 34% of the GDP per capita. For children with additional neurological sequelae, the total cost over the course of the two-year 2006–2007 epidemic was near the GDP per capita level. With little or no disposable income, most households were forced to sacrifice one or more basic necessities to pay for care [52].

5. Neurological Disability, Quality of Life and Access to Care

The challenges of those living with disability—as a permanent sequela from meningitis or from any other cause—are coming to the forefront of discussion within the scientific and public health community [53,54]. Stigma, restriction to education or employment opportunities, and a lack of specialised follow-up healthcare further result in a negative feedback loop that creates an economic gap between households with a disabled member and those without [55]. To illustrate this, a prospective cohort study of disabling meningitis sequelae in Senegal revealed that 40% of children who had had meningitis did not attend school compared to 17% of children with no history of meningitis [31]. Another study of 112 confirmed meningitis patients admitted to a children's hospital between 1992 and 2007 in the United Kingdom revealed that, 8 years after acute meningitis, both parents (32%) and teachers (19%) reported behavioural problems and lowered health-related quality of life (HRQoL) on Pediatric Quality of Life inventory (PedsQL) measurements. The authors of this review highly recommended that meningitis survivors be specifically screened for psychiatric and neurobehavioral difficulties at certain stages of development [56].

As limited as the data are on children in LMICs regarding the long-term impact of meningitis, even more striking is the lack of data related to how adolescents and adults fare in the aftermath of meningitis. HRQoL studies assessing the emotional, psychological, social and behavioural effects of meningitis are lacking in both HICs and LMICs. A 2018 systematic review of the quality-of-life impact on both patients and carers following invasive meningococcal disease in HICs found no studies describing HRQoL for patients who had meningitis-induced sequelae [57]. However, in survivors, particularly adolescents and young adults, self-esteem, friendships, well-being and school performance are important aspects of a good quality of life and problems in these areas also affect caregivers and the community. The implications for someone disabled as a child are profoundly different than when disabled as an adult.

Recognition of those suffering from meningitis-induced disability and their access to (or lack of) resources is an important first step in order to provide equal opportunities for care, rehabilitation, specialised education and employment. For example, a study looking at 107 South African children with TB meningitis showed that, overall, less than half of children with documented neurological sequelae attended specialty clinics for follow-up care and that those in rural settings did not have access to these services [50]

The ramifications of meningitis in adults is no less significant. A range of short- and long-term sequelae including vision loss, neurological (cranial nerve palsies, aphasia, paresis and seizures) or neurobehavioral sequelae and cognitive impairment are found in adults [58,59], even among those considered to have made a "good" recovery from bacterial meningitis [60]. One of the few large studies looking at cognitive sequelae in adults was conducted by van de Beek et al. in Denmark in 2002. Fifty-one adult survivors "with good recovery" after bacterial meningitis were evaluated 6–24 months following meningitis. Cognitive disorders and lower scores in general health and quality of life were found in 27% of cases [60]. The social and economic impacts on individuals thus affected by the disease are profound even following a reported recovery. A study in the UK focusing on tuberculous meningitis in adults found that over one-third of survivors had residual neurological sequelae one year later [5].

A significant number of children and adults permanently affected by meningitis will live with one or more permanent disabilities. Increasingly, it is recognised that, in addition to medical and (neuro-focused) rehabilitative supports, where available, the lives of these individuals and their families can be dramatically improved by ensuring that they are also linked to a rapidly evolving global disability rights effort to improve the lives of persons with disabilities. Improving access to care by strengthening referral systems and health systems can subsequently also improve care for people who have disability from other types of meningitis or even other nervous system diseases.

In addition to services and support that may be available to children and adults disabled by meningitis, it is important to emphasise that additional resources for people with disability are often available and overlooked by individuals and clinical services that are wholly focused on meningitis. This includes Disabled People's Organizations (DPOs), organisations run by and for persons with disabilities, and disability-focused government services and charities that are available to all disabled members of the community. Such organisations can be found at both the local and national levels in both HICs and LMICs. Such support services often can help with education, employment and advice on social services and economic support programmes available through government agencies and local charities. Importantly, such organisations can advise people disabled by meningitis on their rights and entitlements designated under local and national disability law. For example, currently 164 countries are signatories to the United Nations Convention on the Rights of Persons with Disabilities, which means that their national laws should be in alignment with this international human rights declaration [61]. These advances are not limited to only improved access to health care and social services but have broader educational and socioeconomic implications. For example, the identification and inclusion of disabled people and their households into development efforts has been a significant part of this new global disability effort, with initiatives underway towards improving disabled children's right to education and efforts for adults to improve their socioeconomic status and involvement in the workforce, their right to self-determination, and their right to equal involvement in their communities and their societies. The resources available to DPOs and disability-focused services vary from one country to the next and, in LMICs, are often limited, but these organisations are an important and growing resource for people with disabilities around the world. Those disabled as the result of meningitis and those involved in providing care and support to those disabled by meningitis should be aware of the potential benefits that links with the DPOs, government services, charities and the broader Disability Rights Movement can provide.

6. Conclusions

The burden, impact, and social and economic costs of neurological sequelae following meningitis can be devastating to patients, families and communities. Severe sequelae can present as seizures, hearing and vision loss, and neuromotor disability; however, it is likely that more subtle effects such as cognitive impairment, memory and behaviour changes are often overlooked and can have detrimental effects on school and work performance. Importantly, the majority of studies have not followed patients after five years. The long-term consequences, socioeconomic impact and demand for medical follow-up services for these patients and their caregivers is essentially unknown in many LMICs such as those located in the meningitis belt of Africa. More research on the care and support needs of patients and families would be valuable, and early recognition, improved management, support services, and access to care should be priority areas for research and funding programs. Building links to local, regional and global organisations that advocate on behalf of broader disability issues also provides additional support for improving the lives of children and adults with long-term sequelae of meningitis and their families.

Author Contributions: Conceptualization, N.S., N.E.G., T.D. writing—original draft preparation, N.S., N.E.G.; writing—review and editing, T.D.; supervision, T.D. All authors have read and agreed to the published version of the manuscript.

Funding: This research received no external funding.

Institutional Review Board Statement: Not applicable.

Informed Consent Statement: Not applicable.

Data Availability Statement: Not applicable.

Acknowledgments: The authors gratefully acknowledge the edits and insightful comments to the manuscript provided by Lucy McNamara and Christine Dubray.

Conflicts of Interest: The authors declare no conflict of interest.

Disclaimer: The authors are responsible for the views expressed in this article. Those views do not necessarily represent the views, decisions, or policies of the institutions with which they are affiliated.

References

1. World Health Organization (WHO). Bacterial Meningitis. Available online: https://www.who.int/immunization/monitoring_surveillance/burden/vpd/surveillance_type/sentinel/meningitis_surveillance/en/ (accessed on 11 January 2020).
2. Greenwood, B.M. Meningococcal meningitis in Africa. *Trans. R. Soc. Trop. Med. Hyg.* **1999**, *93*, 341–353. [CrossRef]
3. World Health Organization (WHO). Tuberculosis. 2020. Available online: https://www.who.int/health-topics/tuberculosis#tab=tab_1 (accessed on 12 March 2021).
4. Phypers, M.; Harris, T.; Power, C. CNS tuberculosis: A longitudinal analysis of epidemiological and clinical features. *Int. J. Tuberc. Lung Dis.* **2006**, *10*, 99–103. [PubMed]
5. Logan, C.; Mullender, C.; Mirfenderesky, M.; Feasey, N.; Cosgrove, C.; Riley, P.; Houston, A.; Harrison, T.; Bicanic, T.; Rich, P.; et al. Presentations and outcomes of central nervous system TB in a UK cohort: The high burden of neurological morbidity. *J. Infect.* **2021**, *82*, 90–97. [CrossRef]
6. Koelman, D.L.H.; van Kassel, M.N.; Bijlsma, M.W.; Brouwer, M.C.; van de Beek, D.; van der Ende, A. Changing Epidemiology of Bacterial Meningitis Since Introduction of Conjugate Vaccines: Three Decades of National Meningitis Surveillance in The Netherlands. *Clin. Infect. Dis.* **2020**. [CrossRef] [PubMed]
7. Brouwer, M.C.; Tunkel, A.R.; Van De Beek, D. Epidemiology, Diagnosis, and Antimicrobial Treatment of Acute Bacterial Meningitis. *Clin. Microbiol. Rev.* **2010**, *23*, 467–492. [CrossRef] [PubMed]
8. Schuchat, A.; Robinson, K.; Wenger, J.D.; Harrison, L.H.; Farley, M.; Reingold, A.L.; Lefkowitz, L.; Perkins, B.A. Bacterial Meningitis in the United States in 1995. *N. Engl. J. Med.* **1997**, *337*, 970–976. [CrossRef]
9. Cowgill, K.D.; Ndiritu, M.; Nyiro, J.; Slack, M.P.E.; Chiphatsi, S.; Ismail, A.; Kamau, T.; Mwangi, I.; English, M.; Newton, C.R.J.C.; et al. Effectiveness of Haemophilus influenzae Type b Conjugate Vaccine Introduction Into Routine Childhood Immunization in Kenya. *JAMA* **2006**, *296*, 671–678. [CrossRef]

10. Adegbola, R.A.; Secka, O.; Lahai, G.; Lloyd-Evans, N.; Njie, A.; Usen, S.; Oluwalana, C.; Obaro, S.; Weber, M.; Corrah, T.; et al. Elimination of Haemophilus influenzae type b (Hib) disease from The Gambia after the introduction of routine immunisation with a Hib conjugate vaccine: A prospective study. *Lancet* **2005**, *366*, 144–150. [CrossRef]
11. Trotter, C.L.; Lingani, C.; Fernandez, K.; Cooper, L.V.; Bita, A.; Tevi-Benissan, C.; Ronveaux, O.; Préziosi, M.-P.; Stuart, J.M. Impact of MenAfriVac in nine countries of the African meningitis belt, 2010–15: An analysis of surveillance data. *Lancet Infect. Dis.* **2017**, *17*, 867–872. [CrossRef]
12. Iwata, S.; Takata, M.; Morozumi, M.; Miyairi, I.; Matsubara, K.; Ubukata, K. Drastic reduction in pneumococcal meningitis in children owing to the introduction of pneumococcal conjugate vaccines: Longitudinal analysis from 2002 to 2016 in Japan. *J. Infect. Chemother.* **2021**, *27*, 604–612. [CrossRef]
13. Klugman, K.P.; Madhi, S.A.; Huebner, R.E.; Kohberger, R.; Mbelle, N.; Pierce, N. A Trial of a 9-Valent Pneumococcal Conjugate Vaccine in Children with and Those without HIV Infection. *N. Engl. J. Med.* **2003**, *349*, 1341–1348. [CrossRef]
14. El Kareh, A.; El Hage, S.; Safi, S.; Assouad, E.; Mokled, E.; Salameh, P. Epidemiology of bacterial meningitis in Lebanon from 2011 to 2019. *J. Clin. Neurosci.* **2020**, *81*, 32–36. [CrossRef]
15. Global Health Data Exchanage. IHME Data. 2020. Available online: http://ghdx.healthdata.org/gbd-results-tool (accessed on 20 January 2021).
16. Edmond, K.; Clark, A.; Korczak, V.S.; Sanderson, C.; Griffiths, U.K.; Rudan, I. Global and regional risk of disabling sequelae from bacterial meningitis: A systematic review and meta-analysis. *Lancet. Infect. Dis.* **2010**, *10*, 317–328. [CrossRef]
17. Koomen, I.; van Furth, A.M.; Kraak, M.A.; Grobbee, D.E.; Roord, J.J.; Jennekens-Schinkel, A. Neuropsychology of academic and behavioural limitations in school-age survivors of bacterial meningitis. *Dev. Med. Child. Neurol.* **2004**, *46*, 724–732. [CrossRef]
18. WHO. Defeating meningitis by 2030. Available online: https://www.who.int/initiatives/defeating-meningitis-by-2030 (accessed on 21 April 2021).
19. GBD 2019 Diseases and Injuries Collaborators. Global burden of 369 diseases and injuries in 204 countries and territories, 1990–2019: A systematic analysis for the Global Burden of Disease Study 2019. *Lancet* **2020**, *396*, 1204–1222. [CrossRef]
20. GBD 2016 Disease and Injury Incidence and Prevalence Collaborators. Global, regional, and national incidence, prevalence, and years lived with disability for 328 diseases and injuries for 195 countries, 1990–2016: A systematic analysis for the Global Burden of Disease Study 2016. *Lancet* **2017**, *390*, 1211–1259. [CrossRef]
21. World Health Organization (WHO). *Meningitis Weekly Bulletin 2019 (2–29 December 2019)*; WHO: Geneva, Switzerland, 2019.
22. Fernandez, K.; Lingani, C.; Aderinola, O.M.; Goumbi, K.; Bicaba, B.; Edea, Z.A.; Glèlè, C.; Sarkodie, B.; Tamekloe, A.; Ngomba, A.; et al. Meningococcal Meningitis Outbreaks in the African Meningitis Belt After Meningococcal Serogroup A Conjugate Vaccine Introduction, 2011–2017. *J. Infect. Dis.* **2019**, *220*, S225–S232. [CrossRef]
23. Soeters, H.M.; Diallo, A.O.; Bicaba, B.W.; Kadadé, G.; Dembélé, A.Y.; Acyl, M.A.; Sadji, A.Y.; Poy, A.N.; Lingani, C.; Tall, H.; et al. Bacterial Meningitis Epidemiology in Five Countries in the Meningitis Belt of Sub-Saharan Africa, 2015–2017. *J. Infect. Dis.* **2019**, *220* (Suppl. S4), S165–S174. [CrossRef]
24. World Health Organization (WHO). Control of Epidemic Meningococcal Disease. WHO Practical Guidelines, 2nd Edition. Available online: https://www.who.int/csr/resources/publications/meningitis/WHO_EMC_BAC_98_3_EN/en/ (accessed on 18 November 2020).
25. Colombini, A.; Badolo, O.; Gessner, B.D.; Jaillard, P.; Seini, E.; Da Silva, A. Costs and impact of meningitis epidemics for the public health system in Burkina Faso. *Vaccine* **2011**, *29*, 5474–5480. [CrossRef]
26. Jit, M. The risk of sequelae due to pneumococcal meningitis in high-income countries: A systematic review and meta-analysis. *J. Infect.* **2010**, *61*, 114–124. [CrossRef]
27. Adil, S.M.; Hodges, S.E.; Charalambous, L.T.; Kiyani, M.; Liu, B.; Lee, H.-J.; Parente, B.A.; Perfect, J.R.; Lad, S.P. Paediatric bacterial meningitis in the USA: Outcomes and healthcare resource utilization of nosocomial versus community-acquired infection. *J. Med. Microbiol.* **2021**, *70*, 001276. [CrossRef]
28. Kilpi, T.; Anttila, M.; Kallio, M.J.T.; Peltola, H. Length of prediagnostic history related to the course and sequelae of childhood bacterial meningitis. *Pediatr. Infect. Dis. J.* **1993**, *12*, 184–188. [CrossRef] [PubMed]
29. Kilpi, T.; Anttila, M.; Kallio, M.; Peltola, H. Severity of childhood bacterial meningitis and duration of illness before diagnosis. *Lancet* **1991**, *338*, 406–409. [CrossRef]
30. Victora, C.G.; Wagstaff, A.; Schellenberg, J.A.; Gwatkin, D.; Claeson, M.; Habicht, J.-P. Applying an equity lens to child health and mortality: More of the same is not enough. *Lancet* **2003**, *362*, 233–241. [CrossRef]
31. Edmond, K.; Dieye, Y.; Griffiths, U.K.; Fleming, J.; Ba, O.; Diallo, N.; Mulholland, K. Prospective Cohort Study of Disabling Sequelae and Quality of Life in Children With Bacterial Meningitis in Urban Senegal. *Pediatr. Infect. Dis. J.* **2010**, *29*, 1023–1029. [CrossRef] [PubMed]
32. Chandran, A.; Herbert, H.; Misurski, D.; Santosham, M. Long-term sequelae of childhood bacterial meningitis: An underappreciated problem. *Pediatr. Infect. Dis. J.* **2011**, *30*, 3–6. [CrossRef] [PubMed]
33. Berg, S.; Trollfors, B.; Hugosson, S.; Fernell, E.; Svensson, E. Long-term follow-up of children with bacterial meningitis with emphasis on behavioural characteristics. *Eur. J. Nucl. Med. Mol. Imaging* **2002**, *161*, 330–336. [CrossRef]
34. Halket, S.; de Louvois, J.; Holt, D.E.; Harvey, D. Long term follow up after meningitis in infancy: Behaviour of teenagers. *Arch. Dis. Child.* **2003**, *88*, 395–398. [CrossRef]

35. Ramakrishnan, M.; Ulland, A.J.; Steinhardt, L.C.; Moïsi, J.C.; Were, F.; Levine, O.S. Sequelae due to bacterial meningitis among African children: A systematic literature review. *BMC Med.* **2009**, *7*, 47. [CrossRef]
36. Baraff, L.J.; Lee, S.I.; Schriger, D.L. Outcomes of bacterial meningitis in children: A meta-analysis. *Pediatr. Infect. Dis. J.* **1993**, *12*, 389–394. [CrossRef]
37. Hodgson, A.; Smith, T.; Gagneux, S.; Akumah, I.; Adjuik, M.; Pluschke, G.; Binka, F.; Genton, B. Survival and sequelae of meningococcal meningitis in Ghana. *Int. J. Epidemiol.* **2001**, *30*, 1440–1446. [CrossRef]
38. Anderson, V.; Anderson, P.; Grimwood, K.; Nolan, T. Cognitive and Executive Function 12 Years after Childhood Bacterial Meningitis: Effect of Acute Neurologic Complications and Age of Onset. *J. Pediatr. Psychol.* **2004**, *29*, 67–81. [CrossRef]
39. Mbonda, E.; Masso-Misse, P.; Ntaplia, A.; Mefo Sile, H. Séquelles neurologiques des méningites bactériennes chez le nourrisson et l'enfant à Yaoundé. *Med. Afr. Noire* **1995**, *42*, 39–45.
40. Girgis, N.I.; Farid, Z.; Mikhail, I.A.; Farrag, I.; Sultan, Y.; Kilpatrick, M.E. Dexamethasone treatment for bacterial meningitis in children and adults. *Pediatr. Infect. Dis. J.* **1989**, *8*, 848–851. [CrossRef]
41. Girgis, N.I.; Farid, Z.; Kilpatrick, M.E.; Sultan, Y.; Mikhail, I.A. Dexamethasone adjunctive treatment for tuberculous meningitis. *Pediatr. Infect. Dis. J.* **1991**, *10*, 179–182. [CrossRef]
42. Girgis, N.; Erian, M.W.; Farid, Z.; Mansour, M.M.; Sultan, Y.; Mateczun, A.J.; Hanna, L.S. Tuberculosis meningitis, Abbassia Fever Hospital-Naval Medical Research Unit No. 3-Cairo, Egypt, from 1976 to 1996. *Am. J. Trop. Med. Hyg.* **1998**, *58*, 28–34. [CrossRef]
43. Melaku, A. Sensorineural hearing loss in children with epidemic meningococcal meningitis at Tikur Anbessa Hospital. *Ethiop. Med. J.* **2003**, *41*, 113–121.
44. Bijlmer, H.A.; Van Alphen, L.; Greenwood, B.M.; Brown, J.; Schneider, G.; Hughes, A.; Menon, A.; Zanen, H.C.; Valkenburg, H.A. The Epidemiology of Haemophilus influenzae Meningitis in Children under Five Years of Age in The Gambia, West Africa. *J. Infect. Dis.* **1990**, *161*, 1210–1215. [CrossRef]
45. Goetghebuer, T.; West, T.E.; Wermenbol, V.; Cadbury, A.L.; Milligan, P.; Lloyd-Evans, N.; Adegbola, R.A.; Mulholland, E.K.; Greenwood, B.M.; Weber, M.W. Outcome of meningitis caused by Streptococcus pneumoniae and Haemophilus influenzae type b in children in The Gambia. *Trop. Med. Int. Health* **2000**, *5*, 207–213. [CrossRef]
46. Akpede, G.O.; Akuhwa, R.T.; Ogiji, E.O.; Ambe, J.P. Risk factors for an adverse outcome in bacterial meningitis in the tropics: A reappraisal with focus on the significance and risk of seizures. *Ann. Trop. Paediatr.* **1999**, *19*, 151–159. [CrossRef]
47. Salih, M.A.M.; El Hag, A.I.; Ahmed, H.S.; Bushara, M.; Yasin, I.; Omer, M.I.A.; Hofvander, Y.; Olcen, P. Endemic bacterial meningitis in Sudanese children: Aetiology, clinical findings, treatment and short-term outcome. *Ann. Trop. Paediatr.* **1990**, *10*, 203–210. [CrossRef] [PubMed]
48. Mhirsi, A.; Boussoffara, R.; Ayadi, A.; Soua, H.; Driss, N.; Belkhir, Y.; Dalel, H.; Braham, H.; Pousse, H.; Essghairi, K. Prognosis of purulent meningitis in the infant and young child. *Tunis. Med.* **1992**, *70*, 79–85.
49. Saha, S.K.; Khan, N.Z.; Ahmed, A.S.M.N.U.; Amin, M.R.; Hanif, M.; Mahbub, M.; Anwar, K.S.; Qazi, S.A.; Kilgore, P.; Baqui, A.H.; et al. Neurodevelopmental sequelae in pneumococcal meningitis cases in Bangladesh: A comprehensive follow-up study. *Clin. Infect. Dis.* **2009**, *48* (Suppl. S2), S90–S96. [CrossRef]
50. Krauss-Mars, A.H.; Lachman, P.I. Social factors associated with tuberculous meningitis. A study of children and their families in the western Cape. *South Afr. Med. J.* **1992**, *81*, 16–19.
51. Ayieko, P.; Akumu, A.O.; Griffiths, U.K.; English, M. The economic burden of inpatient paediatric care in Kenya: Household and provider costs for treatment of pneumonia, malaria and meningitis. *Cost Eff. Resour. Alloc.* **2009**, *7*. [CrossRef] [PubMed]
52. Colombini, A.; Bationo, F.; Zongo, S.; Ouattara, F.; Badolo, O.; Jaillard, P.; Seini, E.; Gessner, B.D.; Da Silva, A. Costs for Households and Community Perception of Meningitis Epidemics in Burkina Faso. *Clin. Infect. Dis.* **2009**, *49*, 1520–1525. [CrossRef] [PubMed]
53. Groce, N.E. Global disability: An emerging issue. *Lancet Glob. Health* **2018**, *6*, e724–e725. [CrossRef]
54. Scior, K.; Hamid, A.; Hastings, R.; Werner, S.; Belton, C.; Laniyan, A.; Patel, M.; Groce, N.; Kett, M. Consigned to the margins: A call for global action to challenge intellectual disability stigma. *Lancet Glob. Health* **2016**, *4*, e294–e295. [CrossRef]
55. Mitra, S.; Palmer, M.; Kim, H.; Mont, D.; Groce, N. Extra costs of living with a disability: A review and agenda for research. *Disabil. Health J.* **2017**, *10*, 475–484. [CrossRef]
56. Sumpter, R.; Brunklaus, A.; McWilliam, R.; Dorris, L. Health-related quality-of-life and behavioural outcome in survivors of childhood meningitis. *Brain Inj.* **2011**, *25*, 1288–1295. [CrossRef]
57. Olbrich, K.J.; Müller, D.; Schumacher, S.; Beck, E.; Meszaros, K.; Koerber, F. Systematic Review of Invasive Meningococcal Disease: Sequelae and Quality of Life Impact on Patients and Their Caregivers. *Infect. Dis. Ther.* **2018**, *7*, 421–438. [CrossRef]
58. Zelano, J.; Westman, G. Epilepsy after brain infection in adults: A register-based population-wide study. *Neurology* **2020**, *95*, e3213–e3220. [CrossRef]
59. Van de Beek, D.; de Gans, J.; Spanjaard, L.; Weisfelt, M.; Reitsma, J.B.; Vermeulen, M. Clinical features and prognostic factors in adults with bacterial meningitis. *N. Engl. J. Med.* **2004**, *351*, 1849–1859. [CrossRef] [PubMed]
60. Van de Beek, D.; Schmand, B.; de Gans, J.; Weisfelt, M.; Vaessen, H.; Dankert, J.; Vermeulen, M. Cognitive impairment in adults with good recovery after bacterial meningitis. *J. Infect. Dis.* **2002**, *186*, 1047–1052.
61. United Nations Department of Economic and Social Affairs Disability. Convention on the Rights of Persons with Disabilities (CRPD). Available online: https://www.un.org/development/desa/disabilities/convention-on-the-rights-of-persons-with-disabilities.html (accessed on 30 January 2021).

MDPI

Review

Vaccines to Prevent Meningitis: Historical Perspectives and Future Directions

Mark R. Alderson *, Jo Anne Welsch, Katie Regan, Lauren Newhouse, Niranjan Bhat and Anthony A. Marfin

Center for Vaccine Innovation and Access, PATH, Seattle, WA 98121, USA; jwelsch@path.org (J.A.W.); kregan@path.org (K.R.); lnewhouse@path.org (L.N.); nbhat@path.org (N.B.); aamarfin@path.org (A.A.M.)
* Correspondence: malderson@path.org; Tel.: +1-206-302-4855

Abstract: Despite advances in the development and introduction of vaccines against the major bacterial causes of meningitis, the disease and its long-term after-effects remain a problem globally. The Global Roadmap to Defeat Meningitis by 2030 aims to accelerate progress through visionary and strategic goals that place a major emphasis on preventing meningitis via vaccination. Global vaccination against *Haemophilus influenzae* type B (Hib) is the most advanced, such that successful and low-cost combination vaccines incorporating Hib are broadly available. More affordable pneumococcal conjugate vaccines are becoming increasingly available, although countries ineligible for donor support still face access challenges and global serotype coverage is incomplete with existing licensed vaccines. Meningococcal disease control in Africa has progressed with the successful deployment of a low-cost serogroup A conjugate vaccine, but other serogroups still cause outbreaks in regions of the world where broadly protective and affordable vaccines have not been introduced into routine immunization programs. Progress has lagged for prevention of neonatal meningitis and although maternal vaccination against the leading cause, group B streptococcus (GBS), has progressed into clinical trials, no GBS vaccine has thus far reached Phase 3 evaluation. This article examines current and future efforts to control meningitis through vaccination.

Keywords: meningitis; meningococcus; pneumococcus; *Haemophilus influenzae*; Hib; group B streptococcus; conjugate vaccine

Citation: Alderson, M.R.; Welsch, J.A.; Regan, K.; Newhouse, L.; Bhat, N.; Marfin, A.A. Vaccines to Prevent Meningitis: Historical Perspectives and Future Directions. *Microorganisms* **2021**, *9*, 771. https://doi.org/10.3390/microorganisms9040771

Academic Editor: James Stuart

Received: 13 March 2021
Accepted: 2 April 2021
Published: 7 April 2021

1. Introduction

Despite advances against individual pathogens, bacterial meningitis and sepsis remain public health challenges globally. Meningitis, characterized by inflammation of the meninges, is swift and severe and is associated with significant morbidity and mortality. Low- and middle-income countries (LMICs) suffer the greatest burden, with the African Meningitis Belt, a string of 26 countries from Senegal and The Gambia in the west to Ethiopia in the east, experiencing a disproportionate share of disease [1]. Bacterial meningitis epidemics are common in this region and many have been large-scale, threatening economic stability alongside human life. However, outbreaks and epidemics can occur globally [2,3]. There are an estimated 5 million cases of meningitis each year, with up to 300,000 deaths—nearly half of which are in children younger than five years of age (u5) [4]. Survivors are not always spared; a high proportion suffer long-term after affects including hearing loss, visual, physical, and cognitive impairment, and limb loss. Despite this sobering reality, progress against meningitis lags that of other vaccine preventable diseases [4].

The Global Roadmap to Defeat Meningitis by 2030, an initiative to raise awareness of bacterial meningitis as a public health problem and create a framework for addressing it, aims to reverse this trend. Critical goals include eliminating bacterial meningitis epidemics and reducing cases and deaths from the most significant causes of bacterial meningitis: *Haemophilus influenzae* type B (Hib), *Neisseria meningitidis* (meningococcus), *Streptococcus pneumoniae* (pneumococcus), and *Streptococcus agalactiae* (group B streptococcus (GBS)) [5].

Vaccines will play an essential role in preventing these diseases and fulfilling the roadmap vision. Effective vaccines exist and have been in use for years against some of these pathogens, and while there have been significant successes, there also remain significant challenges. As recognized in the World Health Organization (WHO) Immunization Agenda 2030, too many children have insufficient access to vaccines, driven in part by high prices for some of the most effective conjugate vaccines—resulting in limited availability in LMICs [6]. Moreover, existing vaccine formulations do not necessarily reflect the disease serogroups and serotypes most prevalent in the highest burden countries. And even in countries where vaccines are accessible, there is no standard approach to vaccination.

To defeat meningitis, it is critical that we advance new and better vaccines that will be affordable and accessible globally. This will not be easy, but past vaccine development efforts offer direction for future ones. From development of the first conjugate vaccine for humans to the groundbreaking Meningitis Vaccine Project (MVP) that effectively eliminated serogroup A meningococcal meningitis in Africa, the vaccine development landscape is rife with important lessons for developing and delivering vaccines to prevent meningitis [7].

Progress against the four key pathogens identified in the roadmap spans the vaccine development and delivery lifecycle. Hib conjugate vaccines (HibCVs), pneumococcal conjugate vaccines (PCVs), and meningococcal conjugate vaccines (NmCVs) have been in use for decades; vaccines against GBS are on the horizon. This article will explore the challenges, successes, and lessons learned through the development and introduction of meningitis vaccines—lessons critical for successful implementation of the roadmap and the strategy to defeat meningitis by 2030.

2. History and Status of Meningitis Vaccines

Hib, pneumococcus, meningococcus, and GBS are encapsulated bacteria that cause sepsis, meningitis, and other invasive and mucosal diseases [8]. Capsular polysaccharides are important virulence factors and have become the major vaccine target for all four pathogens. HibCVs, PCVs, and NmCVs are highly successful at preventing meningitis and other disease manifestations caused by these organisms. Conjugate vaccines against these bacteria not only protect against disease in multiple age groups, but also confer herd protection via reductions in pharyngeal carriage [9]. GBS is amenable to conjugate vaccine development but development thus far has targeted maternal immunization, given that the greatest disease burden occurs in the first three months of life [10].

HibCV was the prototype for targeting capsular polysaccharides. Polysaccharide-alone vaccines (purified Hib polysaccharide) were certified for use in the US in 1985 but suffered from an inability to elicit immunological memory, poor persistence of immunity, and poor immunogenicity in children under 2 years of age. Covalently coupling the polysaccharide to a protein carrier transformed the vaccine into T-dependent antigens and as such elicited strong immune responses, immunological memory, and immune responses in infants. The first approved HibCV in the US was manufactured using polyribosylribitol phosphate (PRP) conjugated to diphtheria toxoid (DT), though it was eventually replaced by more effective vaccines using meningococcal outer membrane protein (OMP), cross-reactive material 197 (CRM_{197}), or tetanus toxoid (TT) carriers. HibCVs are usually used in combination with other pediatric vaccines including tetanus, diphtheria, hepatitis B, and pertussis (Table 1). Importantly, many developing country vaccine manufacturers (DCVMs) have licensed and secured WHO prequalification for low-cost Hib-containing vaccines, resulting in wide-spread introduction globally. The development, licensure, and introduction of HibCVs paved the way for other conjugate vaccines.

The prototypic pneumococcal vaccine was also polysaccharide-based, covering 23 serotypes, and was licensed in 1983 primarily for use in high-risk adults [11]. The first PCV (Prevnar®, PCV7) was licensed in the US in 2000 and was designed to protect against the seven most prevalent invasive disease serotypes in the US and Europe. PCV7 did not, however, protect against the serotypes responsible for considerable disease in LMICs, such as serotypes 1 and 5. Additionally, the introduction of PCV7 led to the emergence of non-vaccine serotypes, a phenomenon referred to as serotype replacement. This experience prompted the development of 10- and 13-valent PCVs that offer broader coverage (Table 1). Next generation PCVs that extend coverage up to 24 serotypes are currently in mid- to late-stage development.

The first licensed meningococcal vaccines were also polysaccharide based. More recently, NmCVs containing various combinations of serotypes A, C, W, and Y have been licensed and introduced (Table 1). There are two licensed vaccines for serotype B, both protein-based, though neither is WHO prequalified. Vaccines are in development that combine either serotypes A, C, W, X, and Y or serotypes A, B, C, W, and Y [12,13].

There are currently no licensed GBS vaccines, but several candidates are in early- to mid-stage clinical assessment, including a hexavalent version formulated with serotypes Ia, Ib, II, III, IV, and V undergoing Phase 1/2 clinical study [14]. A protein based GBS vaccine has advanced into multiple clinical studies and has demonstrated encouraging safety and immunogenicity data [15,16].

Conjugate vaccines are not without limitations, though; limited serotype coverage and serotype replacement have resulted in the need to make higher valency vaccines for pneumococcus and meningococcus. This, in turn, contributes to manufacturing complexity and difficulty in ensuring affordability for LMICs. Protein-based vaccines are a possible alternative for meningococcus and GBS, but the vaccine against serogroup B is the only protein-based vaccine (OMP/outer membrane vesicle [OMV]) currently licensed [17–22]. This approach is not needed for Hib as the current conjugate vaccines are effective, and it has been difficult to develop a protein-based vaccine for pneumococcus.

Table 1. World Health Organization (WHO) prequalified Hib, pneumococcal, and meningococcal conjugate vaccines [23].

Disease	Vaccine	Manufacturer
Hib	Monovalent (Hib)	Centro de Ingenieria Genetica y Biotecnologia, Sanofi Pasteur, Serum Institute of India, Pvt. Ltd. (SIIPL)
	Quadrivalent (DTP, Hib)	SIIPL
	Pentavalent (DTP, Hep B, Hib)	SIIPL, BioFarma, Biological E, LG Chem, Panacea, Sanofi India (Shantha)
	Pentavalent (DTP, polio, Hib)	Sanofi Pasteur
	Hexavalent (DTP, Hep B, polio, Hib)	Sanofi Pasteur
Pneumococcal	13-valent	Pfizer
	10-valent	GlaxoSmithKline (GSK), SIIPL
Meningococcal	Men A monovalent	SIIPL
	Men A, C, W, Y quadrivalent	GSK, Pfizer, Sanofi Pasteur

3. Early-Stage Meningitis Vaccine Development

The development of monovalent meningitis vaccines paved the way for newer licensed multivalent vaccines that have broader coverage, including the multivalent GBS vaccines currently under development. Multivalent conjugate vaccines (PCVs, NmCVs, and GBS conjugate vaccines) must be fit for purpose and are highly complex products from a manufacturing perspective—and, as such, are challenging from development and cost-effectiveness perspectives.

3.1. Considerations

The Target Product Profile or Preferred Product Characteristics are critical to guide early strategic decisions for all stages of meningitis vaccine development, from drug substance formulation to presentation, preclinical through clinical studies, product licensure, and introduction (Table 2). Additionally, WHO publishes technical report series (TRS) documents that provide guidance for assuring the quality, safety, and efficacy of vaccines—including for HibCVs, PCVs, and serogroup A and C NmCVs. The TRS includes recommendations on vaccine manufacturing, nonclinical evaluation, clinical evaluation, and for national regulatory authorities.

Table 2. Components of a meningitis vaccine Target Product Profile [24–27].

Attribute	General Considerations
Indication	Prevention of invasive disease (including meningitis) by HibCV, PCV, NmCV, and GBS. PCVs also indicated for pneumonia and otitis media.
Target population/age groups	HibCV: Infants and children u5. PCV: Generally infants and children u5. NmCV: Infants, children, and adolescents. GBS: Pregnant women to protect infants $\leq$ 3 months of age.
Serotypes	Invasive disease serotypes based on epidemiology of countries/populations targeted.
Immunogenicity	Assays used by licensed vaccines to measure IgG and functional responses (SBA [a] and OPK [b]). Clinical trials should include a persistence timepoint and ensure the data package is rigorous per TRS recommendations.
Safety, reactogenicity, and contraindications	Similar to other licensed conjugate vaccines; incorporating TRS recommendations.
Schedule	As recommended by WHO and countries' national immunization program schedules.
Interference and co-administration with other vaccines	Phase 2/3 studies should assess safety and immune responses to vaccines co-administered in target population per WHO and country EPI requirements.
Route of administration	Typically intramuscular. Other routes (e.g., intradermal and mucosal) can be considered.
Product presentation	Useable in target countries (prefilled syringes/vials/liquid/lyophilized). Multidose could reduce cost per WHO recommendations.
Product formulation	Attributes include stability, consistency in quality, manufacturability. Preservative may be necessary for multidose formulations targeted toward LMICs. Aluminum based adjuvants used for some conjugate vaccines.
Storage and cold chain requirements	Address storage and cold chain options in target countries; shelf life, temperature, ability to stockpile.
Packaging and labeling	Translate to local language.
Product registration and WHO prequalification	Advanced planning essential to ensure data package is appropriate for the regulatory agency and WHO (if prequalification is the goal).
Post marketing surveillance	Monitor safety, protection of target population, potential serotype replacement, and herd immunity.
Value proposition	Marketing attributes that contribute to the product's business case (commercial interest, advantage over licensed products, cost to produce, etc.).

[a] serum bactericidal activity, [b] opsonophagocytic killing.

3.2. Manufacturing

Chemistry, Manufacturing and Controls (CMC) are well described for conjugate vaccines. Many vaccine attributes must be considered during development, including selection of the serotypes and carrier protein conjugation technology, formulation, presentation (Table 3), LMIC needs, and the cost of goods sold (COGS) [28]. The considerations to plan for include saccharide antigen, carrier protein, preservation of immunogenic epitopes, conjugation chemistry, stability of both drug substance and drug product, formulation, consistency of quality, analytics, preclinical models, and commercially viable manufacturing process. Due to the safety considerations for vaccines used in healthy humans, carrier protein choice is currently limited to CRM_{197}, TT, OMP, DT, and *H. influenzae* protein D, each of which has nuances with antibody avidity and quantity of antibodies elicited.

Table 3. CMC general considerations [28–31].

Manufacturing Recommendations	Comments
Polysaccharide (s): • Strains • Seed lots • Culture medium • Purification • Release testing	Source and identity. High yield strains. Master and working cell banks. Animal product free medium highly desirable.
Carrier protein (s): • Source • Purity • Release testing	Commonly tetanus toxoid, diphtheria toxoid, CRM_{197}, and OMP/OMV. CRM_{197} can either be native (expressed in *Corynebacterium diphtheriae*) or recombinant.
Monovalent bulk (s): • Conjugation chemistry • Release testing	Efficient conjugation contributes to lower COGS.
Final bulk: • Adjuvant formulation (as needed)	
Filling and containers	Multidose liquid presentations require a preservative.
Control tests on final product	Stability indicating assays are key—typically free polysaccharide and size distribution.

Other considerations include the optimal methodologies for the fermentation and purification of polysaccharides and whether they should be native or size-reduced for conjugation. Several technologies have been used to conjugate the polysaccharides to the proteins in currently licensed meningitis vaccines, the most common of which involve reductive amination or cyanylation chemistry [28]. Newer technologies under development are designed to increase conjugation efficiency, simplify the manufacturing processes, and better preserve immunological epitopes on both the saccharide and protein components [32–34]. The use of an adjuvant is an important consideration and is often driven by either licensed vaccines or clinical assessment, as preclinical models are not good indicators of adjuvant benefits on immunogenicity.

3.3. Nonclinical Assessment

Evaluating meningitis conjugate vaccines in animal models provides an initial assessment prior to clinical evaluation. For licensed vaccines such as Hib, NmCVs, and PCVs, however, demonstrating protection against disease in a preclinical model is not required and assessment focuses on immunogenicity. Preclinical animal models usually differentiate between the antibody responses of the formulations being tested [28]. Cost, availability, study duration, cross-reactivity, and applicability to humans contribute to animal model selection, though ultimately the choice relies on published work on similar vaccines and compares the responses of the candidate vaccine to a licensed vaccine for the serotypes they have in common. In vivo experiments may not predict the human response but are the best way to distinguish between vaccine formulations. In vitro assays to measure antibody responses in the animal models are, ideally, identical to the assays used in human antibody evaluation. Non-human primates are sometimes used to evaluate immunogenicity for advanced candidates; however, they may not predict immune responses in humans [35,36]. Preclinical animal immunogenicity assessments and toxicology study data (to indicate safety if the product does not elicit a toxic response) are required by regulatory authorities prior to the first in-human clinical study.

If the vaccine is intended for maternal immunization, as is the case for GBS vaccines currently in development, or may be used in a campaign setting that includes pregnant women, as is the case for meningococcal vaccines, a developmental and reproductive toxicology study is required to understand the impact of the vaccine or vaccine candidate on fertility and developmental toxicity. Pre- and post-natal development studies are also necessary to understand the full spectrum of potential reproductive impacts.

3.4. Importance of Functional Assays

Preclinical and clinical measurements of immune responses to Hib, pneumococcus, meningococcus, and GBS conjugate vaccines have focused on binding and/or functional assays. Binding assays (typically enzyme linked immunosorbent assays) are simple, can be multiplexed, and are highly quantitative in nature. Additionally, it is critical to measure functional antibody responses, whether serum bactericidal activity (SBA) titers for NmCVs, or opsonophagocytic killing assay (OPK/OPA) titers for PCVs and GBS conjugate vaccines. Both SBA and OPK assays demonstrate the ability of vaccine-elicited antibodies to kill live bacteria and are considered to correlate better with clinical efficacy than IgG binding assays.

The use of standardized assays and reagents for both pre-clinical and clinical trial assessment is essential for comparing data between trials, establishing a correlate of protection, and understanding results in the absence of a comparator vaccine. Standardized assays and reagents exist for HibCVs, PCVs, and NmCVs and are in development for GBS conjugate vaccines [37–39].

3.5. Phase 1 Clinical Trials

Phase 1 trials obtain initial safety, reactogenicity, and immunogenicity data in healthy adults. When licensed vaccines exist, such as for HibCVs, PCVs, and NmCVs, the candidate vaccine is measured against a licensed one. Phase 1 studies may provide initial assessment of different dose levels and formulations both with and without adjuvant, though for HibCVs, PCVs and NmCVs these parameters are becoming well defined with multiple licensed products (Table 1). Notably, for conjugate vaccines, aluminum adjuvants are sometimes incorporated for vaccine stabilization rather than to enhance immune responses. Phase 1 trials are usually small (<100 subjects) so the dose range and adjuvant must be definitively assessed in a Phase 2 trial.

3.6. Phase 2 Clinical Trials

Phase 2 trials assess the dose selection, adjuvant need, safety, and antibody response to a licensed vaccine (when available) in a larger number of subjects in the target age group. This ensures sufficient statistical power to determine whether the vaccine is promising enough to advance to the next phase of clinical study. For GBS vaccines in development, immunogenicity will be assessed in pregnant women, in cord blood, and in the newborns to determine whether there is adequate transplacental transfer of antibodies and how well they persist.

4. Late-Stage Clinical Development

HibCVs, PCVs and NmCVs have followed distinct scientific and regulatory pathways in the late stages of their clinical development. However, their licensure strategies have certain aspects in common, based on similarities shared across the three targets, including the type of pathogen, the vaccine platform, and the clinical outcomes targeted. For instance, experience with conjugate vaccine technology allows developers to make initial assumptions regarding dose range and schedules for early clinical development and likely methods for immunological assessment. Similarly, all of these pathogens exhibit a wide spectrum of clinical disease, ranging from asymptomatic carriage to invasive disease, including sepsis and meningitis. Protection against these more severe conditions formed the basis for initial licensure of the early vaccine candidates—but the rare occurrence of these conditions in the population has had similar implications for subsequent vaccine development.

This last consideration has been one of the more consequential factors in shaping late-stage development of recent Hib, meningococcal, and pneumococcal vaccines. Licensure of the earliest conjugate vaccines was based on clinical efficacy trials against invasive bacterial disease outcomes, including meningitis, whose relatively low incidence required tens of thousands of participants. For instance, the efficacy of HibCVs was initially established through several randomized placebo-controlled clinical trials conducted in the late 1980s and early 1990s with invasive disease as the primary endpoint [40,41]. Conducted in both high-resource (California, UK, Finland) and lower-resource (Chile, The Gambia, US Alaskan Natives and Navajo) settings, these trials established the clinical efficacy of PRP conjugate vaccines based on four different protein carriers [41]. Having established the presence of safe and efficacious vaccines to protect against invasive Hib disease, it was considered unethical to conduct subsequent placebo-controlled efficacy trials that would leave a subset of participating infants unprotected. However, conducting a comparative efficacy trial between a new and an established vaccine would have been prohibitively large, given the low incidence of vaccine failures likely to occur in either arm. Therefore, later trials of HibCVs, either as new products, newer formulations (such as in combination vaccines), or in alternate schedules have relied on immunologic outcomes (anti-PRP serum IgG levels) for licensure.

In the case of meningococcal vaccines, the low incidence and sporadic epidemiology of disease in industrialized countries pushed this concept even further. The clinical efficacy of meningococcal vaccination was initially established with polysaccharide A and A/C vaccines more than 40 years ago. Effectiveness was demonstrated in closed populations of high-risk adults, demonstrating the vaccines' utility in controlling outbreaks [42–44]. Later, when the UK became the first country to introduce NmCV (against serogroup C) in 1999, licensure was not granted on the basis of clinical efficacy, but rather on the demonstration of adequate immunogenicity [45]. The licensure of all subsequent NmCVs has been granted based on immunogenicity relative to an accepted surrogate of protection, with later demonstration of protection against clinical disease achieved following broader use [45]. Notably, this approach was used for vaccines containing additional meningococcal serogroups, including W and Y, despite having no studies linking specific antibody levels to clinical protection. Licensure was nevertheless granted based on the assumption that these conjugate vaccines would behave similarly, given the infeasibility of conducting efficacy trials for these serogroups. In contrast, serogroup B meningococcal vaccines were relatively delayed, as similarities between group B capsular polysaccharides and host epitopes prevented use of the polysaccharide conjugate platform. Instead, vaccines based on protein subunits were developed. Nevertheless, licensure was still granted based on the induction of serum bactericidal antibody, an immunological outcome, with a post-marketing commitment to demonstrate clinical benefit [46].

MenAfriVac®, a monovalent group A meningococcal conjugate vaccine (NmCV-A) developed through MVP (a partnership between WHO, PATH, and SIIPL), has been deployed through two strategies, first a series of national mass vaccination campaigns throughout the African meningitis belt covering a broad age group (1 to 29 years of age), followed by incorporation of the vaccine into the routine infant immunization (EPI) schedules of the affected countries. To accomplish this, the vaccine's licensure strategy involved two stages. Initial licensure and WHO prequalification was based on a series of clinical trials in individuals 1 to 34 years of age demonstrating the safety and immunologic superiority of a full dose (10 μg PsA-TT) to a group A-containing polysaccharide vaccine [47–49], thus allowing the start of mass campaigns. Subsequently, an indication for a 5 μg single-dose regimen in children 3 to 24 months of age was achieved based on demonstration of immunologic non-inferiority to the 10 μg dose in two trials in infants [50].

More recently, the licensure strategy for a new pentavalent NmCV containing serogroups A, C, W, X and Y has followed a parallel path relying on demonstration of immunologic non-inferiority to established quadrivalent conjugate vaccines. Two ongoing Phase 3 trials, one in 2- to 29-year-old individuals in Mali and The Gambia [51–54], and another in adult and elderly individuals in India, both using Menactra as the comparator, are intended to gain licensure for use in mass campaigns and travelers. Another Phase 3 trial is planned for younger infants and toddlers in Mali to allow use in routine infant immunization. This trial will use Nimenrix as the comparator because, unlike Menactra, Nimenrix is licensed for use as a single dose down to 6 months of age.

Finally, for PCVs, the clinical efficacy of initial 7- and 9-valent vaccines against invasive pneumococcal disease (IPD) was established in four large-scale trials conducted in the late 1990s and early 2000s in both high- and low-income settings [52–55]. The observed efficacy in these studies ranged between 76.8 and 97.4 percent for IPD caused by serotypes contained in the vaccine, with higher efficacy seen in more industrialized settings. A later 10-valent vaccine was initially licensed using immune correlates of protection, with effectiveness subsequently established through two randomized double-blind controlled trials in the late 2000s in Finland (in a cluster-randomized design) and Latin America [56,57]. Vaccine development expanding the initial 7-valent vaccine to a 13-valent formulation and comparisons for different immunization schedules for the PCV13 and PCV10 vaccines subsequently relied on immunologic endpoints [58], as did the development and licensure of a newer 10-valent PCV in India [59].

In the evaluations of efficacy noted above, the clinical endpoints were chosen by balancing a need for the specificity and clinical relevance of laboratory-confirmed severe disease with the practicality of measuring relatively uncommon outcomes in a population. By necessity, meningococcal vaccine trials were limited to evaluation of protection against meningitis in the case of polysaccharide vaccines, and immunologic outcomes for conjugate vaccines. For Hib and pneumococcal vaccines, initial clinical trials assessed efficacy against all invasive disease, including bacteremia, bacteremic pneumonia, and meningitis, typically in such low numbers that these presentations were not differentiated in their reporting. The effectiveness of these vaccines in the prevention of meningitis specifically has been demonstrated in multiple later studies following implementation in various countries.

An important consideration for the overall clinical development plan as specified in WHO TRSs is the incorporation of antibody persistence studies to inform vaccine implementation strategies and schedules that may potentially require booster doses. For example, in the case of NmCV-A, antibody persistence analysis was used to estimate that protective immune responses would persist for at least 10 years following immunization [60]. As mentioned earlier, a critical feature of conjugate vaccines is their ability to invoke herd protection. The ability to prevent acquisition of carriage, an indicator for herd immunity, can be assessed in Phase 3 trials or in post-licensure studies.

Immunological Correlates of Protection

Despite the similarities among these vaccines, there are also aspects that were unique or assumed special prominence for each pathogen. Ideally, the reliance on immunologic endpoints for regulatory or policy decision-making should be based on a true immune correlate of protection. However, such a correlate is not always available. In the case of Hib vaccines, two immunologic correlates were established. Based on initial experimental data, an anti-PRP IgG level of 0.15 μg/mL indicated ongoing protection from invasive Hib disease, while field studies indicated that a peak post-vaccination response level of 1.0 μg/mL was needed for long-term protection (Table 4). As a result, both thresholds were ultimately considered for regulatory approval and post-licensure evaluation of new vaccines and schedules [41]. The presence of immune correlates proved to be particularly useful for assessing the adequacy of different infant schedules, especially those that were accelerated (2, 3, and 4 months) or early (6 weeks) [61]. Immune correlates were also instrumental in evaluating potential immunological interference between Hib and other childhood vaccines. For instance, a resurgence of Hib cases in the UK in the early 2000s was attributed to interference between Hib vaccine and the recently adopted acellular pertussis vaccines. Evaluation of antibody levels in cohorts receiving both vaccines revealed lower anti-PRP IgG levels in later toddler years compared to prior cohorts, prompting the addition of a Hib booster dose at school entry [41]. Benchmarking antibody levels to short- and long-term thresholds became prominent again in subsequent years, as more complex combination infant vaccines were developed. Immunologic evaluation of these formulations revealed not only interactions between Hib, other antigens, and their carrier proteins, but also incompatibilities among adjuvants [62]; nevertheless, multiple Hib-containing pentavalent and hexavalent vaccines have ultimately come to market.

Table 4. Immunological correlates of protection for Hib, meningococcal, pneumococcal, and group B streptococcus (GBS) vaccines [37–39,63,64].

Vaccine	Correlate of Protection	Notes
PCV	IgG concentration of ≥0.35 µg/mL.	Weighted data across serotypes from 3 efficacy studies. Individual serotypes vary (Goldblatt).
HibCV	IgG concentration of ≥0.15 and ≥1 µg/mL.	Immediate and long-term protection.
NmC	hSBA [a] of ≥4 or Rsba [b] ≥8.	
Other Nm serogroups	Correlates not defined but thresholds the same as NmC often used.	
GBS	Proposed to be between 1 and 10 µg/mL in pregnant women.	Assay standardization in progress.

[a] Human complement serum bactericidal activity, [b] rabbit complement serum bactericidal activity.

For meningococcal vaccines, maintaining adequate levels of circulating serum antibody is considered most important, as the onset of severe clinical disease upon exposure is too rapid to allow time for generation of an immune memory recall response [45,65]. Therefore, assuring serum antibody persistence has been an important feature of meningococcal vaccine evaluation. The immunological evaluation of NmCVs has focused on functional immune responses, namely SBAs. In comparison with HibCVs or PCVs, NmCVs require only one or two doses for durable protection, which may be partly due to the older ages at which they are generally given [42].

In the case of pneumococcal vaccines, a meta-analysis of humoral responses using pooled results from three of the original efficacy trials was conducted, allowing the scientific community to establish a non-inferiority threshold of 0.35 µg/mL capsular polysaccharide antibody against each serotype for the evaluation of newer PCVs. While this threshold is not serotype-specific, and true correlates of protection for specific serotypes may ultimately vary [66], this benchmark has allowed the development of later PCV formulations with higher valency based on immunologic outcomes [67].

Among the major causes of bacterial meningitis, GBS has remained a challenge for vaccine developers. Notably, the early age at which this pathogen acts indicates the best approach to vaccination would be administration during pregnancy to transfer protection to the infant through maternal antibody. While regulatory guidance has been proposed for this novel indication, no "maternal" vaccine has yet been licensed for this purpose, and several uncertainties remain, particularly regarding late-stage development [68].

Several GBS vaccine candidates are currently in Phase 2 development, and progression to licensure will follow one of two main pathways: efficacy trials demonstrating protection against specific clinical outcomes, or immunogenicity trials that target immunologic correlates of protection. Each developmental program has its own strengths and challenges.

Demonstration of clinical efficacy through randomized controlled trials would be the most direct route to licensure. As with the other pathogens discussed in this review, GBS is associated with a wide spectrum of disease, with laboratory-confirmed invasive disease (early- and late-onset meningitis being particularly prominent) the most likely clinical endpoint, given its specificity and relevance to clinical care and public health [69]. Similarly, this outcome is relatively uncommon, particularly if focused on neonatal disease alone, and thus would require relatively large clinical trials to establish efficacy. For this reason, composite endpoints that incorporate additional important laboratory-confirmed fetal and obstetric outcomes, such as stillbirth and maternal sepsis, have been proposed to reduce study size [69]. Neonatal invasive GBS disease occurs at a rate of 1 to 3 per 1000 live births in many geographies, and in those areas, best practices associated with prenatal and perinatal care and intrapartum antibiotic prophylaxis can reduce this rate to 0.5–1.0 per 1000 live births. Given these incidence rates, an efficacy trial could require between 30,000 and 1.8 million mother-infant pairs [39]. While some infant vaccine trials have included up to 70,000 participants, evaluating maternal immunization would also be more resource-intensive on a per-subject basis by comparison. Other clinical endpoints could be considered, including maternal urinary tract infection and colonization, but are unlikely to be included, as they do not directly correlate with invasive disease and otherwise do not pose a significant clinical or public health burden.

Given the impracticality of conducting clinical trials of this size, developers must consider pathways that utilize an immunologic endpoint. However, without prior vaccine efficacy trials, a correlate of protection must be established through sero-epidemiological studies that examine naturally occurring disease. Since the 1970s, serotype-specific maternal capsular antibodies were known to correlate with a reduced risk of invasive GBS disease. However, differences in methodology prevented the establishment of protective thresholds. More recently, larger-scale studies have been initiated in South Africa and the UK using a standardized approach to more definitively establish these associations. These efforts, along with data from animal models, will hopefully produce suitable criteria for pivotal Phase 3 vaccine trials based on immunologic endpoints [39].

Several aspects of the immune response to vaccination are particularly relevant to the maternal immunization model. Since fetal and infant protection is primarily generated through passive transfer of IgG antibody through the placenta during gestation, achieving a high peak maternal serum IgG antibody response to maximize infant levels by the time of birth is a key objective. Therefore, longevity of the immune response, generation of durable immune memory, and even protection of the mother, are secondary—although important—goals. In addition, since this model involves adult vaccine recipients who likely have been previously exposed to GBS, a single vaccine dose to boost pre-existing memory responses is likely to be sufficient. Finally, either before or after licensure, vaccine manufacturers will need to demonstrate a lack of immune interference between their GBS vaccine and other vaccines currently given to pregnant women, including tetanus, pertussis, and influenza, or under development, such as respiratory syncytial virus. Moreover, compatibility studies among these vaccines could allow their incorporation into a combination maternal vaccine, which could greatly improve affordability and access.

5. Accelerating Vaccine Introduction to Prevent Meningitis

Introducing HibCVs, NmCVs, and PCVs and optimizing their coverage in affected populations has been critical for reducing meningitis morbidity and mortality in the last 20 years. However, the availability of effective and safe vaccines alone is insufficient to increase LMIC uptake. Despite the success of these vaccines in high-income countries, overcoming barriers to introduction and sustaining vaccine delivery in LMICs—where the greatest meningitis burden persists—remains a major challenge to global meningitis control [70].

5.1. HibCV: Developing New Approaches to Increase Meningitis Vaccine Uptake

In 2000, 13 years after HibCV was licensed, Hib still caused 8 million meningitis cases and about 400,000 deaths in u5 children in LMICs [71,72]. No Asian countries and only one sub-Saharan African country had introduced HibCV. By 2008, 70 percent of WHO members had introduced HibCV; Hib deaths in u5 children were cut in half [73]. Despite this remarkable impact, HibCV uptake remained low in LMICs. New LMIC introduction approaches were needed.

In the late 1990s, public-private partnerships started to develop new policies, strategies, and priorities for vaccine introduction and to financially support HibCV procurement—dramatically increasing uptake in LMICs [74]. In 2006, combining HibCV into WHO-prequalified quadri-, penta-, and hexavalent vaccines accelerated uptake and contributed to sustain HibCV use in Gavi-eligible countries [75]. These strategies and approaches would be replicated to increase uptake of NmCV-A and PCVs (Table 5). Incorporating similar approaches will lead to successful introduction of GBS vaccines and boost uptake of multivalent NmCVs and higher valency PCVs. In addition, as countries become ineligible for Gavi support, three approaches (vaccine procurement groups; lower-price, high-quality, WHO-prequalified vaccines from DCVMs; and combination vaccines) will allow middle-income countries (MICs) to continue to introduce new meningitis vaccines.

Table 5. Introducing polysaccharide conjugate vaccines to prevent meningitis due to Hib, meningococcus, and pneumococcus in LMICs [1,29,58,76].

Vaccine	Meningitis Epidemiology	WHO Introduction Recommendation	Specific Vaccination Strategy			Activities to Accelerate Uptake and Sustained Use
			Countries	Within countries	EPI schedule	
HibCV	Peak incidence: <2 years of age. Endemic transmission.	All children, all countries. Initially, RI [1] for children <4 mo. of age.	All	National	Multiple, multi-dose RI schedules; first dose critical by 6 weeks to 2 mo. of age.	(1) Highly proscriptive WHO recommendation. (2) Vaccine procurement through The Vaccine Fund and Gavi. (3) Use of vaccine probe studies. (4) Develop HibCV containing penta- and hexavalent combination vaccines.
NmCV-A	Peak incidence: 9 to 14 years of age. Low-level endemicity with periodic outbreaks.	Meningitis Belt residents. Initially, SIA [2] for persons 1 to 29 years of age, then RI in children 9 to 18 mo. of age.	Epidemic prone, African countries.	Mix, national and subnational.	1-dose primary in children 9 to 18 mo. of age. Need for booster dose not yet determined. Strategy to use NmCV-5 is being developed.	(1) Enhanced surveillance linking emergency vaccine requests to define meningitis burden. (2) Develop low-price, high-quality, 1-dose vaccine through DCVM.
PCV	Peak incidence: ~85 percent of pneumococcal meningitis cases occur in children <2 years of age. Endemic transmission.	Initially, RI for <6 mo. of age.	All	National	Currently, for children <6 mo. of age: (1) 3-dose primary/no booster. (2) 2-dose primary/booster at 9 to 18 mo.	Unique financing options (e.g., Advance Market Commitment, PAHO Revolving fund).

[1] RI–routine administration within EPI schedule; [2] SIA–supplementary immunization activity (mass campaign).

5.2. Defining Meningitis Burden to Justify Vaccine Introduction

Poor understanding of the meningitis burden is a major hurdle that requires considerable time, effort, and resources to overcome. Laboratory-based meningitis surveillance to identify at-risk populations, detect outbreaks, and define the potential impact of meningitis control shows the public health value of these vaccines. Meningitis surveillance is challenging and requires significant technical capacity to culture blood/cerebrospinal fluid and identify serogroups/serotypes of meningitis pathogens. However, without evidence that a specific pathogen is a public health problem, countries will be slow to commit to vaccine introduction.

The highest meningococcal disease burden is in the 26 countries of the African meningitis belt. From 1970 through 2010, recurring explosive serogroup A meningococcal (Nm-A) meningitis epidemics in sub-Saharan Africa increased in frequency and magnitude [77]. In 1992, WHO country offices, UNICEF, and non-governmental organizations (NGOs) began submitting outbreak data to WHO to justify release of stockpiled meningococcal vaccines. Because these periodic Nm-A epidemics largely defined the meningitis burden, these surveillance data-containing requests yielded data to support NmCV-A introduction. In 2014, MenAfriNet, a case-based meningitis surveillance system, began monitoring meningitis outbreaks, which will be important in future decisions to introduce a multivalent NmCV.

Hib meningitis results from endemic transmission. Because u5 children accounted for 90 percent of Hib meningitis cases and parents often seek hospital care for ill children, hospital-based surveillance of 0- to 59-month-old children was used to define disease burden [78]. Because only one serotype caused disease, the laboratory demands were much less than those for meningococcus and pneumococcus. This surveillance resulted in high quality burden data in Africa, where Hib was well recognized as a meningitis pathogen. However, most Asian countries did not show sufficient burden to justify HibCV introduction; that changed when a landmark vaccine probe study in Indonesia showed that Hib accounted for a large portion of meningitis and pneumonia not found in routine surveillance [79,80]. Subsequent vaccine probe studies showed significant reduction of meningitis was possible through vaccination and greatly accelerated HibCV uptake [81].

In LMICs that successfully introduce HibCV and NmCV-A, pneumococcus becomes the most common cause of meningitis in all age groups—yet, defining pneumococcal meningitis burden can be difficult [82]. Because of the disease's endemic transmission, broad age distribution, and multiple serotypes, defining the best surveillance is a challenge. As a result, pneumococcal meningitis burden data are often underestimated and insufficient alone to justify PCV introduction. It is better justified by the much higher burden of community-acquired pneumococcal pneumonia and PCVs' cost-effectiveness in preventing pneumonia. Compared to the pneumonia burden, except for periodic serotype 1 pneumococcus meningitis epidemics in Africa, pneumococcal meningitis surveillance has played a small role in accelerating PCV uptake.

5.3. Highly Directive Policies from Global Public Health Authorities

Global public health authorities have highly influential voices that can be used to advance vaccine introduction. Because of the challenges in diagnosing Hib meningitis, its high treatment costs, high mortality, and the severe neurologic impacts in survivors, HibCV was clearly cost-effective in most LMICs [83]. Yet, decisions to introduce HibCV lagged for many reasons, including inadequate in-country technical capacity to assess the value and potential impact of vaccines [84]. In 2006, WHO overcame this barrier when it universally recommended the implementation of Hib vaccination in all infant immunization programs worldwide without accumulating more surveillance data [76]. Such a statement was possible because the global risk of Hib meningitis was roughly the same for all children, the potential impact of vaccination was similar globally, and HibCV had an excellent safety and efficacy profile [81]. This statement was critical in the LMIC decisions to introduce HibCV [85].

5.4. Structuring Vaccination Strategies for Success

Successful vaccine introduction strategies can have a high impact in a short time and can motivate decision-makers in other countries to introduce new vaccines. Successful introduction strategies begin by clearly defining target populations. Several factors come into play when defining this target, such as peak-incidence age, opportunities to vaccinate, persistence of immunity, and need for a booster vaccine. HibCV and NmCV-A introduction showed that well-targeted strategies can quickly achieve near elimination of disease and that successful introduction in early-adopting countries led to decisions to introduce in other countries.

The vaccine introduction strategy for HibCV was relatively straight-forward. Because WHO recommended vaccination for all children, there was no need to develop surveillance systems to identify at-risk countries, districts, or populations or to develop subnational introduction plans. Because peak-incidence age was in the first two years of life, vaccine had to be delivered to infants, and national immunization programs had well-developed opportunities to vaccinate 6-, 10-, and 14-week-old infants.

In contrast, NmCV-A does not universally benefit all children because Nm-A meningitis is not equally distributed globally [86]. Although epidemics were reported globally until the 1940s, Nm-A meningitis outbreaks had become restricted to African meningitis belt countries. Moreover, Nm-A meningitis was not equally distributed within countries. Consequently, highly granular disease surveillance data was needed to allow subnational NmCV-A introduction. Whereas the goal of HibCV was to prevent endemic disease, the goal of NmCV-A was to prevent periodic epidemics driven by meningococcal nasal carriage in children 10 to 14 years of age. The decision to conduct introduction campaigns in 1- to 29-year-olds was a strategy that stopped outbreaks and prevented meningitis in young adolescents. However, this strategy had to be balanced by the fact that routine vaccine delivery to school-aged children is not well-developed. Currently, NmCV-A vaccination occurs in children younger than 2 years of age. Whether bactericidal antibodies persist beyond 10–12 years at a level that will later suppress nasal carriage and prevent Nm-A epidemics is unknown.

5.5. Public-Private Initiatives to Provide Vaccine and Improve Vaccination Practices

Public-private partnerships have been critical to HibCV, NmCV-A, and PCV introduction by providing support for vaccine delivery, procurement, and technical assistance to low-income countries (LICs). These partnerships will remain critical for new vaccine introduction going forward.

In 1998, the William H. Gates Foundation donated $100 million to establish the Children's Vaccine Initiative (CVI) to improve vaccine delivery to LICs [87]. Prior, many LICs used funds intended to support delivery to buy HibCV. To reverse this, CVI proposed funding to make vaccines more available and to improve the quality of vaccine delivery, rather than to procure vaccines. Through partnerships with WHO, UNICEF, PATH, and other international NGOs, CVI funded guideline development, vaccination worker training, model immunization programs, cost-effectiveness studies, and advocacy and communication programs to increase HibCV acceptance.

In 2001, the Bill & Melinda Gates Foundation funded MVP, which successfully developed, tested, licensed, WHO-prequalified, and introduced MenAfriVac®, an affordable NmCV-A. The keys to MVP's success included developing strong public private partnerships [7,88,89]; engaging SIIPL, a DCVM, to develop a low-cost, high-quality NmCV-A; providing technical assistance to SIIPL to acquire WHO prequalification; conducting clinical trials in Africa alongside African researchers; and supporting operational costs of introduction. Since 2010, more than 340 million Africans have been vaccinated with NmCV-A and Nm-A meningitis has been eliminated from this region.

In 2002, Gavi and key partners, including Johns Hopkins University, established the Pneumococcal Vaccines Accelerated Development and Introduction Plan (PneumoADIP) to increase uptake of PCVs in Gavi-eligible countries [90]. The keys to PneumoADIP's success included supporting PCV procurement and the operational cost of vaccine introduction, standardizing pneumococcal disease surveillance, developing advocacy and education activities to inform country decision-makers within national immunization programs regarding PCV and HibCV introduction, and providing technical assistance for vaccine introduction campaigns and the transition to routine immunization. As a result of PneumoADIP, between 2000 and 2018, 59 of 73 Gavi-eligible countries introduced PCV.

In 2005, Gavi's Hib Initiative (GHI), a consortium of WHO, Johns Hopkins University, London School of Hygiene and Tropical Medicine, and the Centers for Disease Control and Prevention, was funded to help Gavi-eligible countries make evidence-based decisions regarding HibCV introduction [85]. Through these CVI and GHI activities, the number of LICs introducing HibCV increased from 13 in 2004 to 66 in 2008 [91]. Currently, all Gavi-eligible countries use HibCV-containing vaccines.

5.6. Developing Innovative Vaccine Financing Options

Defining the cost-effectiveness of HibCV, NmCV-A, and PCVs has been important for new vaccine decision-makers and has accelerated the uptake of these vaccines in LMICs. Studies have shown that HibCV is cost saving or highly cost-effective in essentially all settings. Cost-effectiveness has further increased due to the recent decline in HibCV prices, integration of HibCV into quadri-, penta- and hexavalent combination vaccines, and data showing the loss of productivity in meningitis survivors [92]. Similarly, compared with a reactive vaccination strategy, prevention strategies using NmCV-A were shown to be significantly cost saving in Burkina Faso [93]. Such analyses will be important for decision-makers considering whether the higher price of the next generation of meningococcal or pneumococcal vaccines or the price of new vaccines to prevent GBS meningitis are justified by their benefits [94].

Prior to 2000, vaccine cost was often the greatest barrier to meningitis vaccine introduction. Since then, LICs have greatly benefited from Gavi's vaccine investment strategy and procurement of meningitis vaccines through The Vaccine Fund [95,96]. Unfortunately, many MICs that procure their own vaccines face financial challenges to introduction. In addition, LIC decision-makers are more widely considering the long-term costs of vaccination, not just the initial introduction costs.

To address this, in 2009, the Advance Market Commitment (AMC) for pneumococcal vaccines was launched. In the AMC, donors commit funds to guarantee the price of vaccines once they have been developed. In exchange, manufacturers make a legally binding commitment to provide the vaccines at a price affordable to LICs [97]. Although the AMC has been recognized as a valuable way to make effective and affordable pneumococcal vaccines available, it has also been criticized for not encouraging innovation, discouraging competition from new market entrants, and raising vaccine costs [98,99].

Another financing option is multi-country procurement groups, such as the Pan American Health Organization (PAHO) Revolving Fund. Since 1977, this fund has pooled the resources of 41 mostly middle-income Latin American countries to procure vaccines at a lower cost through consolidated ordering [100]. Currently, the fund is used to procure HibCV-containing vaccines and PCV, which has resulted in sustained use of these vaccines throughout Central and South America.

Finally, for countries that purchase their own vaccines, the availability of lower-cost, high-quality, WHO-prequalified vaccines produced by DCVMs has been an important alternative to vaccines produced by multi-national vaccine manufacturers.

5.7. Implications for New Meningitis Vaccines

The lessons learned from HibCV, NmCV-A, and PCV introduction will likely be applied to the introduction of new meningitis vaccines. For example, there has been development and successful use of several other meningococcal vaccines, including monovalent meningococcal vaccines against serogroups C and B and multivalent NmCVs against serogroups A, C, W, X and Y. WHO has stated the decision to use other meningococcal vaccines or to replace NmCV-A with a multivalent NmCV will depend on the locally prevalent meningococcal serogroup(s), identification of the best target group for vaccination, and opportunities to vaccinate within national immunization programs [42]. This underscores the importance of meningitis surveillance. Discussion is ongoing regarding the use of new multivalent NmCVs being developed by DCVMs.

New vaccines are being developed against GBS to prevent meningitis in neonates and young infants [27]. Some of the approaches described above will likely be used to increase uptake (e.g., combination vaccines, support for vaccine procurement) [27]. However, because the goal of a GBS vaccine is to prevent invasive disease in neonates and infants, the target group for vaccination is pregnant women. Given the challenges of accessing obstetric care in LICs and the lower emphasis on vaccination in antenatal care clinics compared to EPI clinics, new approaches will be needed with special attention to advocacy and communication and antenatal healthcare worker training to introduce a GBS vaccine.

6. Conclusions and Future Directions

The development and global introduction of low-cost vaccines to prevent Hib and pneumococcus has had a significant impact on meningitis and other disease manifestations caused by these pathogens. DCVMs have become the major suppliers of affordable Hib combination vaccines and the recent licensure and WHO prequalification of a 10-valent PCV by SIIPL, in partnership with PATH, is poised to increase availability of low-cost PCVs for LMICs, notably in those countries that have not introduced PCVs into their routine immunization programs. Like with Hib vaccines, it is anticipated other DCVMs will license PCVs and increase the global supply of affordable vaccines. Despite the considerable success in reducing the burden of pneumococcal disease globally, serotype replacement and emergence has resulted in significant residual disease burden. Higher valency (15–24 serotypes) PCVs are in development, though there are considerable manufacturing and licensing challenges for such vaccines and LMIC affordability is uncertain.

Meningococcal vaccines present a dichotomy: Quadrivalent NmCV-ACWY and meningococcal serogroup B protein vaccines manufactured by multinational vaccine manufacturers are cost prohibitive for widespread use in LMICs, while a low-cost NmCV-A that has had incredible impact in the African meningitis belt has limited utility in other parts of the world. The development and licensure of low-cost NmCV-ACWY(X) and meningococcal B vaccines has the potential for broad appeal and to greatly reduce the burden of meningococcal meningitis globally.

In addition to reducing the per dose cost of meningitis vaccines, strategies to increase cost-effectiveness by minimizing the number of doses administered are in development. For example, WHO currently recommends a single dose of NmCV-A at 9 to 18 months of age for routine immunization and studies to assess whether a 2-dose schedule (1 + 1) instead of 3-dose schedule for PCVs may be sufficient to maintain adequate herd immunity are underway [101,102].

What about other meningitis pathogens that are potentially vaccine preventable? *Haemophilius influenzae* type A (Hia) causes meningitis in certain regions and populations globally, including indigenous populations in North America and Australia. Development of a Hia vaccine should be technically feasible but a limited market would likely require donor support to incentivize a manufacturer. *Klebsiella pneumoniae* is becoming increasingly recognized as an important cause of sepsis and meningitis in neonates in LMICs and as such could be targeted for maternal vaccine together with GBS. The relatively high number of *K. pneumoniae* capsular serotypes makes this a challenging approach, although targeting a more limited number of O antigens or protein antigens is also being considered [103].

Defeating meningitis is an ambitious undertaking that will require significant time, effort, and resources—particularly when it comes to developing new or improved meningitis vaccines. There are hurdles along the vaccine development and delivery spectrum but well-established vaccines like HibCV, PCV, and NmCV offer lessons for what does and does not work, how to successfully advance products toward market, and how to ensure they reach the populations in need—and where gaps remain that need to be filled. Despite their challenges, vaccines are a public health best-buy and have been critical to the progress we have made against meningitis thus far. Vaccines have saved millions of lives around the world and new entrants are poised to take that success further to make the vision of defeating meningitis by 2030 a reality.

Author Contributions: Conceived and plotted the manuscript, M.R.A., J.A.W., K.R., L.N., N.B. and A.A.M.; drafted the original manuscript, M.R.A., J.A.W., K.R., N.B. and A.A.M.; reviewed, edited, and formatted the manuscript, K.R. and L.N. All authors have read and agreed to the published version of the manuscript.

Funding: This work was supported, in part, by the Bill & Melinda Gates Foundation, Seattle, WA, USA (OPP1116594, OPP1150281, OPP1054296) and the Foreign, Commonwealth, & Development Office, London, UK (PATH MVPP 300341-108). Under the grant conditions of the Foundation, a Creative Commons Attribution 4.0 Generic License has already been assigned to the Author Accepted Manuscript version that might arise from this submission. The findings and conclusions contained within are those of the authors and do not necessarily reflect positions or policies of the Foundation or the Foreign, Commonwealth, & Development Office.

Institutional Review Board Statement: Not applicable.

Informed Consent Statement: Not applicable.

Acknowledgments: The authors thank all their global partners in PATH's work on pneumococcal, meningococcal, and group B streptococcal vaccines.

Conflicts of Interest: The authors declare no conflict of interest.

References

1. World Health Organization. Meningococcal Meningitis. Available online: https://www.who.int/en/news-room/fact-sheets/detail/meningococcal-meningitis (accessed on 4 March 2021).
2. van Kessel, F.; van den Ende, C.; Oordt-Speets, A.M.; Kyaw, M.H. Outbreaks of meningococcal meningitis in non-African countries over the last 50 years: A systematic review. *J. Glob. Health* **2019**, *9*, 10411. [CrossRef]
3. World Health Organization. International Coordination Group on Vaccine Provision for Epidemic Meningitis. In Proceedings of the Report of the Annual Meeting, Geneva, Switzerland, 18 September 2018; Available online: https://apps.who.int/iris/bitstream/handle/10665/279828/WHO-WHE-IHM-2019.1-eng.pdf?sequence=1&isAllowed=y (accessed on 4 March 2021).
4. GBD 2016 Meningitis Collaborators. Global, regional, and national burden of meningitis, 1990–2016: A systematic analysis for the Global Burden of Disease Study 2016. *Lancet Neuro* **2018**, *17*, P1061–P1082. [CrossRef]
5. Centers for Disease Control and Prevention. Bacterial Meningitis. Available online: https://www.cdc.gov/meningitis/bacterial.html (accessed on 4 March 2021).
6. World Health Organization. Immunization Agenda 2030: A Global Strategy to Leave No One Behind. Available online: https://www.who.int/publications/m/item/immunization-agenda-2030-a-global-strategy-to-leave-no-one-behind (accessed on 4 March 2021).
7. LaForce, F.M.; Djingarey, M.; Viviani, S.; Preziosi, M. Lessons from the Meningitis Vaccine Project. *Viral Immunol* **2018**, *31*, 109–113. [CrossRef]
8. Sadarangani, M. Protection Against Invasive Infections in Children Caused by Encapsulated Bacteria. *Front Immunol* **2018**, *9*, 2674. [CrossRef]
9. Stephens, D.S. Protecting the herd: The remarkable effectiveness of the bacterial meningitis polysaccharide-protein conjugate vaccines in altering transmission dynamics. *Trans. Am. Clin. Climatol. Assoc.* **2011**, *122*, 115–123. [PubMed]
10. Kobayashi, M.; Schrag, S.J.; Alderson, M.R.; Madhi, S.A.; Baker, C.J.; Sobanjo-Ter Meulen, A.; Kaslow, D.C.; Smith, P.G.; Moorthy, V.S.; Vekemans, J. WHO consultation on group B Streptococcus vaccine development: Report from a meeting held on 27–28 April 2016. *Vaccine* **2019**, *37*, 7307–7314. [CrossRef] [PubMed]
11. World Health Organization. 23-valent pneumococcal polysaccharide vaccine. WHO position paper. *Wkly. Epidemiol. Rec.* **2008**, *83*, 373–384.
12. Pizza, M.; Bekkat-Berkani, R.; Rappuoli, R. Vaccines against Meningococcal Diseases. *Microorganisms* **2020**, *8*, 1521. [CrossRef]
13. Zahlanie, Y.C.; Hammadi, M.M.; Ghanem, S.T.; Dbaibo, G.S. Review of meningococcal vaccines with updates on immunization in adults. *Hum. Vaccin. Immunother.* **2014**, *10*, 995–1007. [CrossRef]
14. Absalon, J.; Segall, N.; Block, S.L.; Center, K.J.; Scully, I.L.; Giardina, P.C.; Peterson, J.; Watson, W.J.; Gruber, W.C.; Jansen, K.U.; et al. Safety and immunogenicity of a novel hexavalent group B streptococcus conjugate vaccine in healthy, non-pregnant adults: A phase 1/2, randomised, placebo-controlled, observer-blinded, dose-escalation trial. *Lancet Infect. Dis.* **2021**, *21*, 263–274. [CrossRef]
15. ClinicalTrials.gov. Safety and Immunogenicity of a Group B Streptococcus Vaccine in Non-Pregnant Women 18–40 Years of Age. (MVX13211). Available online: https://clinicaltrials.gov/ct2/show/NCT02459262?term=NCT02459262&draw=2&rank=1 (accessed on 4 March 2021).

16. Rose, F.; Roovers, S.; Fano, M.; Harloff-Helleberg, S.; Kirkensgaard, J.J.K.; Hejnaes, K.; Fischer, P.; Foged, C. Temperature-Induced Self-Assembly of the Group B Streptococcus (GBS) Fusion Antigen GBS-NN. *Mol. Pharm.* **2018**, *15*, 2584–2593. [CrossRef] [PubMed]
17. Burman, C.; Alderfer, J.; Snow, V.T. A review of the immunogenicity, safety and current recommendations for the meningococcal serogroup B vaccine, MenB-FHbp. *J. Clin. Pharm. Ther.* **2020**, *45*, 270–281. [CrossRef] [PubMed]
18. Rappuoli, R.; Pizza, M.; Masignani, V.; Vadivelu, K. Meningococcal B vaccine (4CMenB): The journey from research to real world experience. *Expert Rev. Vaccines* **2018**, *17*, 1111–1121. [CrossRef]
19. Paoletti, L.C.; Kasper, D.L. Surface Structures of Group B Streptococcus Important in Human Immunity. *Microbiol. Spectr.* **2019**, *7*, 201–227. [CrossRef] [PubMed]
20. Maeland, J.A.; Afset, J.E.; Lyng, R.V.; Radtke, A. Survey of immunological features of the alpha-like proteins of Streptococcus agalactiae. *Clin. Vaccine Immunol.* **2015**, *22*, 153–159. [CrossRef] [PubMed]
21. Lindahl, G. Subdominance in Antibody Responses: Implications for Vaccine Development. *Microbiol. Mol. Biol. Rev.* **2020**, *85*. [CrossRef]
22. Stålhammar-Carlemalm, M.; Waldemarsson, J.; Johnsson, E.; Areschoug, T.; Lindahl, G. Nonimmunodominant regions are effective as building blocks in a streptococcal fusion protein vaccine. *Cell Host Microbe* **2007**, *2*, 427–434. [CrossRef]
23. World Health Organization. List of Prequalified Vaccines. Available online: https://extranet.who.int/pqweb/vaccines/list-prequalified-vaccines (accessed on 4 March 2021).
24. World Health Organization. Haemophilus Influenzae Type b (Hib). Available online: https://www.who.int/teams/health-product-and-policy-standards/standards-and-specifications/vaccine-standardization/hib (accessed on 4 March 2021).
25. World Health Organization. Meningococcal Meningitis. Available online: https://www.who.int/biologicals/vaccines/Meningococcal_Meningitis/en/ (accessed on 4 March 2021).
26. World Health Organization, *Target Product Profile (TPP) for the Advance Market Commitment (AMC) for Pneumococcal Conjugate Vaccines: Parts I and II*; Geneva, Switzerland, 2008. Available online: https://www.who.int/immunization/sage/target_product_profile.pdf (accessed on 4 March 2021).
27. World Health Organization, *WHO Preferred Product Characteristics for Group B Streptococcus Vaccines*; Geneva, Switzerland, 2017. Available online: https://apps.who.int/iris/bitstream/handle/10665/258703/WHO-IVB-17.09-eng.pdf;jsessionid=BBF3B32A411AE301336F68C794BF0310?sequence=1 (accessed on 4 March 2021).
28. Prasad, A.K. *Carbohydrate-Based Vaccines: From Concept to Clinic*; American Chemical Society: Washington, DC, USA, 2018; ISBN 9780841233379.
29. Granoff, D.M.; Rathore, M.H.; Holmes, S.J.; Granoff, P.D.; Lucas, A.H. Effect of immunity to the carrier protein on antibody responses to Haemophilus influenzae type b conjugate vaccines. *Vaccine* **1993**, *11*, S46–S51. [CrossRef]
30. Rappuoli, R. Glycoconjugate vaccines: Principles and mechanisms. *Sci. Transl. Med.* **2018**, *10*, eaat4615. [CrossRef]
31. World Health Organization. *Recommendations for the Production and Control of Meningococcal Group C Conjugate Vaccines*; WHO Technical Report No. 924; Geneva, Switzerland, 2004. Available online: https://www.who.int/biologicals/publications/trs/areas/vaccines/meningococcal/WHO_TRS_924_MeningA2_102-128.pdf?ua=1 (accessed on 4 March 2021).
32. Gao, N.J.; Uchiyama, S.; Pill, L.; Dahesh, S.; Olson, J.; Bautista, L.; Maroju, S.; Berges, A.; Liu, J.Z.; Zurich, R.H.; et al. Site-Specific Conjugation of Cell Wall Polyrhamnose to Protein SpyAD Envisioning a Safe Universal Group A Streptococcal Vaccine. *Infect. Microbes Dis.* **2020**. Publish Ahead of Print. [CrossRef]
33. Wacker, M.; Wang, L.; Kowarik, M.; Dowd, M.; Lipowsky, G.; Faridmoayer, A.; Shields, K.; Park, S.; Alaimo, C.; Kelley, K.A.; et al. Prevention of Staphylococcus aureus infections by glycoprotein vaccines synthesized in Escherichia coli. *J. Infect. Dis.* **2014**, *209*, 1551–1561. [CrossRef] [PubMed]
34. Zhang, F.; Lu, Y.-J.; Malley, R. Multiple antigen-presenting system (MAPS) to induce comprehensive B- and T-cell immunity. *Proc. Natl. Acad. Sci. USA* **2013**, *110*, 13564–13569. [CrossRef] [PubMed]
35. Buurman, E.T.; Timofeyeva, Y.; Gu, J.; Kim, J.-H.; Kodali, S.; Liu, Y.; Mininni, T.; Moghazeh, S.; Pavliakova, D.; Singer, C.; et al. A Novel Hexavalent Capsular Polysaccharide Conjugate Vaccine (GBS6) for the Prevention of Neonatal Group B Streptococcal Infections by Maternal Immunization. *J. Infect. Dis.* **2019**, *220*, 105–115. [CrossRef] [PubMed]
36. Xie, J.; Zhang, Y.; Caro-Aguilar, I.; Indrawati, L.; Smith, W.J.; Giovarelli, C.; Winters, M.A.; MacNair, J.; He, J.; Abeygunawardana, C.; et al. Immunogenicity Comparison of a Next Generation Pneumococcal Conjugate Vaccine in Animal Models and Human Infants. *Ped. Infect. Dis. J.* **2020**, *39*, 70–77. [CrossRef]
37. Borrow, R.; Balmer, P.; Miller, E. Meningococcal surrogates of protection—Serum bactericidal antibody activity. *Vaccine* **2005**, *23*, 2222–2227. [CrossRef] [PubMed]
38. Balmer, P.; Borrow, R. Serologic correlates of protection for evaluating the response to meningococcal vaccines. *Expert Rev. Vaccines* **2004**, *3*, 77–87. [CrossRef] [PubMed]

39. Vekemans, J.; Crofts, J.; Baker, C.J.; Goldblatt, D.; Heath, P.T.; Madhi, S.A.; Le Doare, K.; Andrews, N.; Pollard, A.J.; Saha, S.K.; et al. The role of immune correlates of protection on the pathway to licensure, policy decision and use of group B Streptococcus vaccines for maternal immunization: Considerations from World Health Organization consultations. *Vaccine* **2019**, *37*, 3190–3198. [CrossRef] [PubMed]
40. Obonyo, C.O.; Lau, J. Efficacy of Haemophilus influenzae type b vaccination of children: A meta-analysis. *Eur. J. Clin. Microbiol. Infect. Dis.* **2006**, *25*, 90–97. [CrossRef]
41. World Health Organization. *The Immunological Basis for Immunization Series: Module 9: Haemophilus Influenzae Type B*; Geneva, Switzerland, 2007. Available online: https://apps.who.int/iris/bitstream/handle/10665/43799/9789241596138_eng.pdf;jsessionid=B894BD4A69F655646CC7163BADCA3D91?sequence=1 (accessed on 4 March 2021).
42. World Health Organization. Meningococcal vaccines: WHO position paper, November 2011. *Wkly. Epidemiol. Rec.* **2011**, *86*, 521–539.
43. Biselli, R.; Fattorossi, A.; Matricardi, P.M.; Nisini, R.; Stroffolini, T.; D'Amelio, R. Dramatic reduction of meningococcal meningitis among military recruits in Italy after introduction of specific vaccination. *Vaccine* **1993**, *11*, 578–581. [CrossRef]
44. Greenwood, B.M.; Hassan-King, M.; Whittle, H.C. Prevention of secondary cases of meningococcal disease in household contacts by vaccination. *Br. Med. J.* **1978**, *1*, 1317–1319. [CrossRef]
45. World Health Organization. *The Immunological Basis for Immunization Series: Module 15: Meningitis*; Geneva, Switzerland, 2010; Available online: https://www.who.int/immunization/research/development/Meningococcal_disease_module15.pdf (accessed on 4 March 2021).
46. Mbaeyi, S.A.; Bozio, C.H.; Duffy, J.; Rubin, L.G.; Hariri, S.; Stephens, D.S.; MacNeil, J.R. Meningococcal Vaccination: Recommendations of the Advisory Committee on Immunization Practices, United States, 2020. *MMWR Recomm. Rep.* **2020**, *69*, 1–41. [CrossRef]
47. Hirve, S.; Bavdekar, A.; Pandit, A.; Juvekar, S.; Patil, M.; Preziosi, M.-P.; Tang, Y.; Marchetti, E.; Martellet, L.; Findlow, H.; et al. Immunogenicity and safety of a new meningococcal A conjugate vaccine in Indian children aged 2–10 years: A phase II/III double-blind randomized controlled trial. *Vaccine* **2012**, *30*, 6456–6460. [CrossRef] [PubMed]
48. Sow, S.O.; Okoko, B.J.; Diallo, A.; Viviani, S.; Borrow, R.; Carlone, G.; Tapia, M.; Akinsola, A.K.; Arduin, P.; Findlow, H.; et al. Immunogenicity and safety of a meningococcal A conjugate vaccine in Africans. *N. Engl. J. Med.* **2011**, *364*, 2293–2304. [CrossRef] [PubMed]
49. Kshirsagar, N.; Mur, N.; Thatte, U.; Gogtay, N.; Viviani, S.; Préziosi, M.-P.; Elie, C.; Findlow, H.; Carlone, G.; Borrow, R.; et al. Safety, immunogenicity, and antibody persistence of a new meningococcal group A conjugate vaccine in healthy Indian adults. *Vaccine* **2007**, *25*, A101–A107. [CrossRef] [PubMed]
50. World Health Organization. *Results from the MenA Conjugate Vaccine (PsA-TT) Randomized Controlled Trials in Infants and Young Children*; Geneva, Switzerland, October 2014. Available online: https://www.who.int/immunization/sage/meetings/2014/october/3_MenA_vaccine_trials_SAGE_01Oct2014.pdf?ua=1 (accessed on 4 March 2021).
51. ClinicalTrials.gov. Study to Assess Immunogenicity & Safety of Pentavalent Meningococcal Vaccine (NmCV-5). Available online: https://clinicaltrials.gov/ct2/show/NCT03964012 (accessed on 4 March 2021).
52. O'Brien, K.L.; Moulton, L.H.; Reid, R.; Weatherholtz, R.; Oski, J.; Brown, L.; Kumar, G.; Parkinson, A.; Hu, D.; Hackell, J.; et al. Efficacy and safety of seven-valent conjugate pneumococcal vaccine in American Indian children: Group randomised trial. *Lancet* **2003**, *362*, 355–361. [CrossRef]
53. Klugman, K.P.; Madhi, S.A.; Huebner, R.E.; Kohberger, R.; Mbelle, N.; Pierce, N. A trial of a 9-valent pneumococcal conjugate vaccine in children with and those without HIV infection. *N. Engl. J. Med.* **2003**, *349*, 1341–1348. [CrossRef]
54. Cutts, F.T.; Zaman, S.M.A.; Enwere, G.; Jaffar, S.; Levine, O.S.; Okoko, J.B.; Oluwalana, C.; Vaughan, A.; Obaro, S.K.; Leach, A.; et al. Efficacy of nine-valent pneumococcal conjugate vaccine against pneumonia and invasive pneumococcal disease in The Gambia: Randomised, double-blind, placebo-controlled trial. *Lancet* **2005**, *365*, 1139–1146. [CrossRef]
55. Black, S.; Shinefield, H.; Fireman, B.; Lewis, E.; Ray, P.; Hansen, J.R.; Elvin, L.; Ensor, K.M.; Hackell, J.; Siber, G.; et al. Efficacy, safety and immunogenicity of heptavalent pneumococcal conjugate vaccine in children. Northern California Kaiser Permanente Vaccine Study Center Group. *Pediatr. Infect. Dis. J.* **2000**, *19*, 187–195. [CrossRef]
56. Tregnaghi, M.W.; Sáez-Llorens, X.; López, P.; Abate, H.; Smith, E.; Pósleman, A.; Calvo, A.; Wong, D.; Cortes-Barbosa, C.; Ceballos, A.; et al. Efficacy of pneumococcal nontypable Haemophilus influenzae protein D conjugate vaccine (PHiD-CV) in young Latin American children: A double-blind randomized controlled trial. *PLoS Med.* **2014**, *11*, e1001657. [CrossRef]
57. Palmu, A.A.; Jokinen, J.; Borys, D.; Nieminen, H.; Ruokokoski, E.; Siira, L.; Puumalainen, T.; Lommel, P.; Hezareh, M.; Moreira, M.; et al. Effectiveness of the ten-valent pneumococcal Haemophilus influenzae protein D conjugate vaccine (PHiD-CV10) against invasive pneumococcal disease: A cluster randomised trial. *Lancet* **2013**, *381*, 214–222. [CrossRef]
58. World Health Organization. Pneumococcal conjugate vaccines in infants and children under 5 years of age: WHO position paper—February 2019. *Wkly. Epidemiol. Rec.* **2019**, *94*, 85–104.
59. Clarke, E.; Bashorun, A.O.; Okoye, M.; Umesi, A.; Badjie Hydara, M.; Adigweme, I.; Dhere, R.; Sethna, V.; Kampmann, B.; Goldblatt, D.; et al. Safety and immunogenicity of a novel 10-valent pneumococcal conjugate vaccine candidate in adults, toddlers, and infants in The Gambia-Results of a phase 1/2 randomized, double-blinded, controlled trial. *Vaccine* **2020**, *38*, 399–410. [CrossRef]

60. White, M.; Idoko, O.; Sow, S.; Diallo, A.; Kampmann, B.; Borrow, R.; Trotter, C. Antibody kinetics following vaccination with MenAfriVac: An analysis of serological data from randomised trials. *Lancet Infect. Dis.* **2019**, *19*, 327–336. [CrossRef]
61. World Health Organization. Haemophilus influenzae type b (Hib) Vaccination Position Paper—July 2013. *Wkly. Epidemiol. Rec.* **2013**, *88*, 413–426.
62. Dagan, R.; Poolman, J.; Siegrist, C.-A. Glycoconjugate vaccines and immune interference: A review. *Vaccine* **2010**, *28*, 5513–5523. [CrossRef] [PubMed]
63. World Health Organization. *Correlates of Vaccine-Induced Protection: Methods and Implications*; WHO/IVB/13.01; Geneva, Switzerland, May 2013. Available online: https://apps.who.int/iris/bitstream/handle/10665/84288/WHO_IVB_13.01_eng.pdf (accessed on 4 March 2021).
64. Le Doare, K.; Kampmann, B.; Vekemans, J.; Heath, P.T.; Goldblatt, D.; Nahm, M.H.; Baker, C.; Edwards, M.S.; Kwatra, G.; Andrews, N.; et al. Serocorrelates of protection against infant group B streptococcus disease. *Lancet Infect. Dis.* **2019**, *19*, e162–e171. [CrossRef]
65. Auckland, C.; Gray, S.; Borrow, R.; Andrews, N.; Goldblatt, D.; Ramsay, M.; Miller, E. Clinical and immunologic risk factors for meningococcal C conjugate vaccine failure in the United Kingdom. *J. Infect. Dis.* **2006**, *194*, 1745–1752. [CrossRef]
66. Andrews, N.J.; Waight, P.A.; Burbidge, P.; Pearce, E.; Roalfe, L.; Zancolli, M.; Slack, M.; Ladhani, S.N.; Miller, E.; Goldblatt, D. Serotype-specific effectiveness and correlates of protection for the 13-valent pneumococcal conjugate vaccine: A postlicensure indirect cohort study. *Lancet Infect. Dis.* **2014**, *14*, 839–846. [CrossRef]
67. World Health Organization. *Annex 3: Recommendations to Assure the Quality, Safety and Efficacy of Pneumococcal Conjugate Vaccines: Replacement of WHO Technical Report Series*; No. 927, Annex 2; WHO Technical Report Series; Geneva, Switzerland. Available online: https://www.who.int/biologicals/vaccines/TRS_977_Annex_3.pdf (accessed on 11 March 2011).
68. Roberts, J.N.; Gruber, M.F. Regulatory considerations in the clinical development of vaccines indicated for use during pregnancy. *Vaccine* **2015**, *33*, 966–972. [CrossRef]
69. Seale, A.C.; Baker, C.J.; Berkley, J.A.; Madhi, S.A.; Ordi, J.; Saha, S.K.; Schrag, S.J.; Sobanjo-Ter Meulen, A.; Vekemans, J. Vaccines for maternal immunization against Group B Streptococcus disease: WHO perspectives on case ascertainment and case definitions. *Vaccine* **2019**, *37*, 4877–4885. [CrossRef] [PubMed]
70. Segal, S.; Pollard, A.J. Vaccines against bacterial meningitis. *Br. Med. Bull.* **2004**, *72*, 65–81. [CrossRef] [PubMed]
71. Hajjeh, R.; Mulholland, K.; Schuchat, A.; Santosham, M. Progress towards demonstrating the impact of Haemophilus influenzae type b conjugate vaccines globally. *J. Pediatr.* **2013**, *163*, S1–S3. [CrossRef]
72. Watt, J.P.; Wolfson, L.J.; O'Brien, K.L.; Henkle, E.; Deloria-Knoll, M.; McCall, N.; Lee, E.; Levine, O.S.; Hajjeh, R.; Mulholland, K.; et al. Burden of disease caused by Haemophilus influenzae type b in children younger than 5 years: Global estimates. *Lancet* **2009**, *374*, 903–911. [CrossRef]
73. World Health Organization. Estimated Hib and Pneumococcal Deaths for Children under 5 Years of Age. 2008. Available online: https://www.who.int/immunization/monitoring_surveillance/burden/estimates/Pneumo_hib/en/ (accessed on 1 January 2021).
74. Watt, J.P.; Chen, S.; Santosham, M. Haemophilus Influenzae Type B Conjugate Vaccine: Review of Observational Data on Long Term Vaccine Impact to Inform Recommendations for Vaccine Schedules. Available online: https://www.who.int/immunization/sage/meetings/2012/november/5_Review_observational_data_long_term_vaccine_impact_recommendations_vaccine_schedules_Watt_J_et_al_2012.pdf (accessed on 4 March 2021).
75. Gavi the Vaccine Alliance. Pentavalent Vaccine Support: Pentavalent Vaccine Protects Against Five Major Diseases. Available online: https://www.gavi.org/types-support/vaccine-support/pentavalent#:~{}:text=Gavi%20began%20offering%20pentavalent%20vaccine,part%20of%20the%20pentavalent%20vaccine (accessed on 4 January 2021).
76. World Health Organization. WHO position paper on Haemophilus influenzae type b conjugate vaccines. *Wkly. Epidemiol. Rec.* **2006**, *81*, 445–452.
77. World Health Organization. *WHO Report on Global Surveillance of Epidemic-Prone Infectious Diseases*; WHO/CDS/CSR/ISR/2000.1; Geneva, Switzerland, 2000. Available online: https://www.who.int/csr/resources/publications/surveillance/WHO_Report_Infectious_Diseases.pdf?ua=1 (accessed on 4 March 2021).
78. World Health Organization. *WHO-Recommended Standards for Surveillance of Selected Vaccine-Preventable Diseases*; WHO/EPI/GEN/98.1 REV.2; Geneva, Switzerland, 1999. Available online: https://apps.who.int/iris/bitstream/handle/10665/64165/WHO_EPI_GEN_98.01_Rev.2.pdf?sequence=1&isAllowed=y (accessed on 4 March 2021).
79. Gessner, B.D.; Sutanto, A.; Linehan, M.; Djelantik, I.G.G.; Fletcher, T.; Gerudug, I.K.; Ingerani; Mercer, D.; Moniaga, V.; Moulton, L.H.; et al. Incidences of vaccine-preventable Haemophilus influenzae type b pneumonia and meningitis in Indonesian children: Hamlet-randomised vaccine-probe trial. *Lancet* **2005**, *365*, 43–52. [CrossRef]
80. Feikin, D.R.; Scott, J.A.G.; Gessner, B.D. Use of vaccines as probes to define disease burden. *Lancet* **2014**, *383*, 1762–1770. [CrossRef]
81. Gessner, B.D.; Adegbola, R.A. The impact of vaccines on pneumonia: Key lessons from Haemophilus influenzae type b conjugate vaccines. *Vaccine* **2008**, *26* (Suppl. 2), B3–B8. [CrossRef]
82. Mulholland, K. Strategies for the control of pneumococcal diseases. *Vaccine* **1999**, *17*, S79–S84. [CrossRef]
83. Bärnighausen, T.; Bloom, D.E.; Canning, D.; Friedman, A.; Levine, O.S.; O'Brien, J.; Privor-Dumm, L.; Walker, D. Rethinking the benefits and costs of childhood vaccination: The example of the Haemophilus influenzae type b vaccine. *Vaccine* **2011**, *29*, 2371–2380. [CrossRef]

84. Hajjeh, R. Accelerating introduction of new vaccines: Barriers to introduction and lessons learned from the recent Haemophilus influenzae type B vaccine experience. *Philos. Trans. R. Soc. Lond. B Biol. Sci.* **2011**, *366*, 2827–2832. [CrossRef]
85. Hajjeh, R.A.; Privor-Dumm, L.; Edmond, K.; O'Loughlin, R.; Shetty, S.; Griffiths, U.K.; Bear, A.P.; Cohen, A.L.; Chandran, A.; Schuchat, A.; et al. Supporting new vaccine introduction decisions: Lessons learned from the Hib Initiative experience. *Vaccine* **2010**, *28*, 7123–7129. [CrossRef] [PubMed]
86. Greenwood, B. Meningococcal meningitis in Africa. *Trans. R. Soc. Trop. Med. Hyg.* **1999**, *93*, 341–353. [CrossRef]
87. Altman, L.K. Gates Gives $100 Million to Children's Vaccine Initiative. *New York Times*. 2 December 1998. Available online: https://archive.nytimes.com/www.nytimes.com/library/tech/98/12/biztech/articles/02gates-donation.html (accessed on 4 March 2021).
88. Aguado, M.T.; Jodar, L.; Granoff, D.; Rabinovich, R.; Ceccarini, C.; Perkin, G.W. From Epidemic Meningitis Vaccines for Africa to the Meningitis Vaccine Project. *Clin. Infect. Dis.* **2015**, *61* (Suppl. 5), S391–S395. [CrossRef] [PubMed]
89. Djingarey, M.H.; Diomandé, F.V.K.; Barry, R.; Kandolo, D.; Shirehwa, F.; Lingani, C.; Novak, R.T.; Tevi-Benissan, C.; Perea, W.; Preziosi, M.-P.; et al. Introduction and Rollout of a New Group A Meningococcal Conjugate Vaccine (PsA-TT) in African Meningitis Belt Countries, 2010–2014. *Clin. Infect. Dis.* **2015**, *61* (Suppl. 5), S434–S441. [CrossRef] [PubMed]
90. International Vaccine Access Center/Johns Hopkins Bloomberg School of Public Health. Our Story. Available online: https://www.jhsph.edu/ivac/about/our-story/ (accessed on 4 December 2020).
91. O'Loughlin, R.; Hajjeh, R. Worldwide introduction and coverage of Haemophilus influenzae type b conjugate vaccine. *Lancet Infect. Dis.* **2008**, *8*, 736. [CrossRef]
92. Griffiths, U.K.; Clark, A.; Hajjeh, R. Cost-effectiveness of Haemophilus influenzae type b conjugate vaccine in low- and middle-income countries: Regional analysis and assessment of major determinants. *J. Pediatr.* **2013**, *163*, S50–S59.e9. [CrossRef]
93. Colombini, A.; Trotter, C.; Madrid, Y.; Karachaliou, A.; Preziosi, M.-P. Costs of Neisseria meningitidis Group A Disease and Economic Impact of Vaccination in Burkina Faso. *Clin. Infect. Dis.* **2015**, *61* (Suppl. 5), S473–S482. [CrossRef]
94. Arifin, S.M.N.; Zimmer, C.; Trotter, C.; Colombini, A.; Sidikou, F.; LaForce, F.M.; Cohen, T.; Yaesoubi, R. Cost-Effectiveness of Alternative Uses of Polyvalent Meningococcal Vaccines in Niger: An Agent-Based Transmission Modeling Study. *Med. Decis. Mak.* **2019**, *39*, 553–567. [CrossRef]
95. Martin, J. New tendencies and strategies in international immunisation: GAVI and The Vaccine Fund. *Vaccine* **2003**, *21*, 587–592. [CrossRef]
96. Gavi the Vaccine Alliance. Vaccine Investment Strategy. Available online: https://www.gavi.org/our-alliance/strategy/vaccine-investment-strategy (accessed on 4 December 2020).
97. Cernuschi, T.; Furrer, E.; Schwalbe, N.; Jones, A.; Berndt, E.R.; McAdams, S. Advance market commitment for pneumococcal vaccines: Putting theory into practice. *Bull. World Health Organ.* **2011**, *89*, 913–918. [CrossRef] [PubMed]
98. Medecins San Frontieres. Gavi Must Stop Giving Millions in Subsidies to Pfizer and GSK for Pneumonia Vaccine. Available online: https://www.doctorswithoutborders.org/what-we-do/news-stories/news/gavi-must-stop-giving-millions-subsidies-pfizer-and-gsk-pneumonia (accessed on 1 April 2021).
99. Usher, A.D. Dispute over pneumococcal vaccine initiative. *Lancet* **2009**, *374*, 1879–1880. [CrossRef]
100. Pan American Health Organization. PAHO Revolving Fund. Available online: https://www.paho.org/hq/index.php?option=com_content&view=article&id=1864:paho-revolving-fund&Itemid=4135&lang=en (accessed on 4 March 2021).
101. Madhi, S.A.; Mutsaerts, E.A.; Izu, A.; Boyce, W.; Bhikha, S.; Ikulinda, B.T.; Jose, L.; Koen, A.; Nana, A.J.; Moultrie, A.; et al. Immunogenicity of a single-dose compared with a two-dose primary series followed by a booster dose of ten-valent or 13-valent pneumococcal conjugate vaccine in South African children: An open-label, randomised, non-inferiority trial. *Lancet Infect. Dis.* **2020**, *20*, 1426–1436. [CrossRef]
102. Goldblatt, D.; Southern, J.; Andrews, N.J.; Burbidge, P.; Partington, J.; Roalfe, L.; Valente Pinto, M.; Thalasselis, V.; Plested, E.; Richardson, H.; et al. Pneumococcal conjugate vaccine 13 delivered as one primary and one booster dose (1 + 1) compared with two primary doses and a booster (2 + 1) in UK infants: A multicentre, parallel group randomised controlled trial. *Lancet Infect. Dis.* **2018**, *18*, 171–179. [CrossRef]
103. Opoku-Temeng, C.; Kobayashi, S.D.; DeLeo, F.R. Klebsiella pneumoniae capsule polysaccharide as a target for therapeutics and vaccines. *Comput. Struct. Biotechnol. J.* **2019**, *17*, 1360–1366. [CrossRef] [PubMed]

microorganisms

MDPI

Review

Invasive Bacterial Infections in Subjects with Genetic and Acquired Susceptibility and Impacts on Recommendations for Vaccination: A Narrative Review

Ala-Eddine Deghmane * and Muhamed-Kheir Taha

Invasive Bacterial Infections Unit, Institut Pasteur, 28 Rue du Dr. Roux, CEDEX 15, 75724 Paris, France; muhamed-kheir.taha@pasteur.fr

* Correspondence: ala-eddine.deghmane@pasteur.fr; Tel.: +33-1-44-38-95-90

Abstract: The WHO recently endorsed an ambitious plan, "Defeating Meningitis by 2030", that aims to control/eradicate invasive bacterial infection epidemics by 2030. Vaccination is one of the pillars of this road map, with the goal to reduce the number of cases and deaths due to *Neisseria meningitidis*, *Streptococcus pneumoniae*, *Haemophilus influenzae* and *Streptococcus agalactiae*. The risk of developing invasive bacterial infections (IBI) due to these bacterial species includes genetic and acquired factors that favor repeated and/or severe invasive infections. We searched the PubMed database to identify host risk factors that increase the susceptibility to these bacterial species. Here, we describe a number of inherited and acquired risk factors associated with increased susceptibility to invasive bacterial infections. The burden of these factors is expected to increase due to the anticipated decrease in cases in the general population upon the implementation of vaccination strategies. Therefore, detection and exploration of these patients are important as vaccination may differ among subjects with these risk factors and specific strategies for vaccination are required. The aim of this narrative review is to provide information about these factors as well as their impact on vaccination against the four bacterial species. Awareness of risk factors for IBI may facilitate early recognition and treatment of the disease. Preventive measures including vaccination, when available, in individuals with increased risk for IBI may prevent and reduce the number of cases.

Keywords: susceptibility; invasive bacterial infections; complement; genetic factors; *Neisseria meningitidis*; *Streptococcus pneumoniae*; *Haemophilus influenzae*; *Streptococcus agalactiae*; group B streptococci

check for updates

Citation: Deghmane, A.-E.; Taha, M.-K. Invasive Bacterial Infections in Subjects with Genetic and Acquired Susceptibility and Impacts on Recommendations for Vaccination: A Narrative Review. *Microorganisms* **2021**, *9*, 467. https://doi.org/10.3390/microorganisms9030467

Academic Editor: James Stuart

Received: 15 January 2021
Accepted: 20 February 2021
Published: 24 February 2021

1. Introduction

Invasive bacterial infections (IBI) usually refer to those infections provoked by *Neisseria meningitidis* ((Nm), meningococcus), *Streptococcus pneumoniae* ((Spn), pneumococcus), *Haemophilus influenzae* (Hi) and *Streptococcus agalactiae* (group B *Streptococcus* (GBS)). The major form of these invasive infections is acute bacterial meningitis. However, other clinical forms are also encountered. The term "bacterial meningitis" is frequently used to refer to all invasive infections due to these agents. In 2020, a road map, "Defeating Meningitis by 2030" was endorsed by WHO. This road map includes an ambitious and broad multidisciplinary plan that includes five pillars to control and eradicate invasive bacterial infection epidemics by 2030: (i) diagnosis and treatment; (ii) prevention and epidemic control; (iii) disease surveillance; (iv) support and aftercare for people affected; and (v) advocacy and information. Actions to achieve the specific goal of prevention and epidemic control include the introduction of vaccines against the four causative agents, achieving equal access to these vaccines and maintaining high coverage of targeted population [1].

Risk factors for developing IBI are linked to bacterial factors (virulence factors). Certain genotypes of these bacterial agents have been reported to be more significantly associated to IBI. The virulence traits are frequently associated with growth in the host, evasion of host immunity, persistence in the host and transmission between hosts [2–5]. Next, there are factors linked to the host that increase its susceptibility to IBI by enhancing acquisition and/or reducing the clearance of bacterial agents. IBI are often due to underlying anatomical or immune disorders, either of which may be inherited or acquired. Improving surveillance and implementation of vaccines will continue to reduce the incidence of IBI in the general population. However, the burden of these infections among subjects with enhanced susceptibility to IBI will increase proportionally. Another factor that also requires analysis is the severity of invasive bacterial infections. Better knowledge of these two facets (susceptibility and severity) of IBI is therefore warranted. Several aspects of these infections require exploring, for instance, little is known about the genotypes of the involved bacterial isolates and whether they differ from bacterial isolates encountered in the general population. Moreover, response to vaccination and vaccine failure in these subjects are less explored than in the general population. The need for special vaccination schedules also requires analysis. In this narrative review, we aim to summarize the genetic and acquired risk factors that increase the susceptibility to and severity of invasive infections related to the four above-mentioned pathogens and to discuss preventive measures under these conditions.

2. Method

We performed a search of PubMed with the objective of summarizing the inherited and acquired host factors associated with susceptibility of patients to invasive meningococcal, pneumococcal, *Haemophilus influenzae* and group B streptococci disease. The following Mesh terms were used: ((*Neisseria meningitidis*) OR ((*Streptococcus pneumoniae*) OR (*Haemophilus influenzae*) OR (*Streptococcus agalactiae*) OR (group B streptococc*)) AND (((invasive) AND ((disease*) OR (infection*))) OR (bacterial meningitis) OR (meningitis) AND ((genetic) OR (acquired) OR (immunocompromised)* or (deficien*) OR (immunodeficient*) OR (susceptibility) OR (predispose*) OR (recurrent infection*)). A built-in PubMed filter was used to limit the search to papers published in English or French up until 31 October 2020. Both authors independently screened titles and abstracts. Studies lacking outcomes of interest were considered not relevant to the aim of our review and were excluded. Relevant publications matching the criteria applied to the search results were identified, and the full text of each was reviewed by both authors separately.

3. Susceptibility to Invasive Meningococcal Infections

Nm is a human-restricted, Gram-negative encapsulated bacterium that is usually encountered as a member of the nasopharyngeal microbiota, which acts as a carriage. However, a few genotypes (hyper-invasive clonal complexes) are associated with invasiveness of the bloodstream and are responsible for most of the cases of invasive meningococcal disease (IMD). Carriage and hyper-invasive isolates differ genetically and phenotypically. Unlike invasive isolates, carriage isolates are more frequently non-capsulated and do not belong to hyperinvasive genotypes [6]. The incidence of IMD varies according to age, with three peaks: in infants < 1 year of age, in adolescents and young adults and in the elderly. This incidence also varies geographically and the epidemiology of IMD is continuously changing [7,8].

The meningococcal capsule is a polysaccharide, and when present, it determines the serogroup. Twelve serogroups have been described with serogroups A, B, C, W, Y and X being responsible for virtually all cases of IMD [8]. Capsular polysaccharide-based vaccines are available against Nm of serogroups A, C, W and Y, while subcapsular protein-based vaccines are available against Nm of serogroup B. Recommendations exist to use these vaccines in subjects with increased susceptibility to IMD. However, rational support for these recommendations may require clarification.

3.1. Genetic and Acquired Susceptibilitiesy to IMD

The ability of Nm to invade, to survive and to spread in the bloodstream is linked to its pathogenesis, which is correlated to the complement-dependent clearance of meningococci.

Factors that lead to the absence of bactericidal activity in complement-dependent serum increase the susceptibility to IMD [9,10]. These factors can be inherited and/or acquired.

3.1.1. Inherited Factors of Susceptibility to IMD

The three pathways of the complement system (the classical, the lectin and the alternative pathways) are major actors in the innate immune response. Activation of complement is tightly controlled with several regulators. Complement is activated through the early complement components of these three pathways to first form C3 convertases, then, they converge to form the C5 convertase, and subsequently, the membrane attack complex (MAC) through the activation of the late complement components (LCC) (C5 to C9). The MAC ultimately leads to the lysis of the targeted cell. Moreover, complement activation leads to the opsonization of the bacterial surface [11]. These two events (lysis and opsonophagocytosis) are directly responsible for efficient bacterial clearance [12]. For Nm, bactericidal activity (in the absence of blood inflammatory cells) is able to lyse bacteria through the insertion of the MAC at the bacterial surface [9,13]. Deficiencies in these late components of the complement system lead, therefore, to enhanced susceptibility to IMD, which can result in repeated IMD [13–15]. This is particularly the case in subjects with late components of complement deficiencies (LCCD), deficits of properdin deficiency or deficits of factor D deficiency [15,16]. Polymorphism of Factor H (a negative regulator of the complement) is also associated with an increased risk of IMD while deficiencies in the early components (such as C1) were not reported to be specifically associated with increased susceptibility to IMD [17,18]. The incidence of IMD among LCCD patients, in regard to number and proportion, will increase due to the decreasing incidence of IMD in immune-competent subjects upon implementation of vaccination strategies. The incidence of IMD is 1000 to 10,000 times higher among LCCD patients than among the general population [15]. The frequency of hereditary complement deficiencies varies according to their type, age, sex and geographical/ethnic distribution [15]. Terminal complement pathway, properdin and factor D deficiencies seem to lead specifically to an increased susceptibility to IMD. LCCD are the most frequent but seem to be associated with a low fatality rate (1%), and are usually detected in adolescents and young adults [15,19]. About 45% of these patients developed more than one IMD episode with a median interval of 6 years between episodes of IMD [19]. Meningococcal isolates from IMD in patients with LCCD are often of serogroup Y, non-groupeable isolates or serogroups/genotypes that are rare in typical cases of IMD. Moreover, IMD disease among LCCD patients seems to be less severe with lower mortality than IMD in the general population [15,19,20]. The median age for the detection of LCCD is 17 years and it is frequently suspected due to repeated IMD episodes, while the detection of properdin deficiencies occurs earlier [15]. Moreover, fulminant and fatal IMD in patients with properdin deficiencies has been frequently reported [21–24]. However, properdin deficiencies are not all complete and there are three types: total deficiency (type I), partial deficiency (type II), and deficiency due to a dysfunctional molecule (type III).

3.1.2. Acquired Factors of Susceptibility to IMD

The complement system has two facets and it plays the role of the two characters in the Dr Jekyll and Mister Hyde story. Indeed, complement is a major and beneficial actor in immune response and host defense, however, its over-activation may lead to systemic effects such as systemic lupus erythematosus (SLE, a systemic autoimmune disorder in which multiple autoantibodies against cell nuclear constituents form immune complexes that effectively activate the classical complement pathway and cause tissue damage) [25], paroxysmal nocturnal hemoglobinuria (PNH, an X-linked hematological disorder that results from somatic loss-of-function mutations impairing membrane expression of two complement inhibitors, CD55 and CD59, on red blood cells, resulting in erythrocytes-complement mediated lysis) [26], age-related macular degeneration (AMD, characterized by the progressive destruction of neurosensory retina in the macular area, and which contributes to vision loss) [27] and atypical hemolytic uremic syndrome (aHUS, a disorder related to mutations in complement regulators (such as the factor H), and that result in a renal disease that encompasses the triad of microangiopathic hemolytic anemia, thrombocytopenia, and acute renal failure) [28]. Several of these systemic diseases may benefit from anti-complement drugs, and in particular, monoclonal antibodies (Mabs) that inhibit the late complement components. This inhibition of the complement can therefore increase susceptibility to IMD. Mabs that inhibit the C5 (Eculizumab and Ravulizumab) have reached the market and are used to treat aHUS and PNH. Other drugs are under development, targeting other components such as C3, factor B and factor D [20,29]. Treating COVID-19 with compstatin-based complement C3 inhibitor (AMY-101) has also been reported [30]. The use of anti-complement drugs in the management of various pathologies is growing [31], including the treatment of COVID-19 to control the inflammatory response [32]. IMD frequency in these patients should therefore be kept under tight surveillance.

Other acquired susceptibilities to IMD are encountered in cases of anatomic or functional asplenia. The spleen plays a central role in mounting innate and adaptive immune responses against encapsulated pathogens such as Nm. Asplenia/hyposplenia (including sickle-cell disease) were reported as a recognized risk factor of IMD in a large case-control study (odds ratio, 6.7; 95% confidence interval (CI), 3.0–14.7). Patients with hematopoietic stem cell transplantation (hSCT) are also at high risk for IMD as well as HIV patients [33,34]. hSCT is a procedure in which the immune system is transferred from the donor to the recipient. This transfer is at best incomplete and vaccine protection from the donor is usually lost. This loss is observed in particular, when the patient suffers from a graft-versus-host disease (GVHD) that requires the administration of immunosuppressive treatments [33].

hSCT transplant recipients are at risk of IMD due to total body irradiation, which induces a hyposplenism, and especially the progressive loss of specific antibodies, which has been documented in the literature for meningococci [35]. Solid organ transplant recipients may also be at risk for IMD due to immunosuppressive treatment [36].

3.2. Host Factors of Severity of IMD

The severity of IMD is frequently linked to hyperinvasive clonal complexes, and particularly, the clonal complex 11 [37]. However, several host factors are reported to be associated with severity and/or bad evolution of the disease. The deficiency of either protein C or its cofactor, protein S (anticoagulant proteins) has been reported as being associated with an increased risk of severe meningococcal sepsis [38]. Moreover, high levels of the plasminogen activator inhibitor-1 (PAI-1) have been associated with poor outcome of IMD with high sequelae and mortality rates [39]. The exacerbated inflammatory response may lead to complications such as pachymeningitis, which can be linked to promoter variants in genes involved in the inflammatory response (IL6, PAI-1 and macrophage migration inhibitory factor, MIF) [40].

3.3. Impact on Anti-Meningococcal Vaccination Strategies

Exploring the complement is highly recommended in patients who develop recurrent/chronic forms and/or mild infections provoked by unusual serogroups/genotypes of Nm. This exploration should include assays for C3, C4, CH50 and AP50 in order to detect deficiencies in early and late components and alternative pathways. When detected in a patient, the investigation should be extended to the siblings. LCCD are inherited in an autosomal recessive manner while properdin deficiencies are usually inherited as an X-linked disorder.

These patients (with acquired or hereditary complement deficiencies) are increasing due to increasing detection and new indications for anti-complement drugs such as Mabs. These drugs are being investigated in the treatment of COVID-19 [41]. Moreover, the number of patients with spleen disorders is substantial, for example, 6000 to 9000 patients are splenectomized each year in France [42].

These patients with increased susceptibility to IMD require particular management strategies including:

- Large-spectrum vaccination against meningococci using conjugate vaccines against serogroups ACWY (with a booster dose every 5 years) and protein-based vaccines targeting serogroup B isolates.
- Exploration of the siblings in case of genetic deficiency (the same management should be proposed for each case detected).
- Reinforcing protection around the patient by vaccination of household contacts (cocooning or barrier) strategy.
- Prophylactic antibiotic treatment is also required using oral penicillin V. For example, penicillin V is recommended in several countries in addition to vaccination for patients receiving anti-C5 treatment.
- Teaching patients to seek immediate medical help if they feel unwell (fever).

The immunogenicity of meningococcal vaccines in these patients requires more exploration in order to adapt vaccination schemes. For example, in a study on adult asplenic patients, they were able to achieve protective bactericidal titers after vaccination against serogroup C meningococci. However, they showed a significantly lower geometric mean titer (GMT) (157.8; 95% CI, 94.5 to 263.3) of bactericidal antibody in serum (SBA) than an age-matched control group (1448.2; 95% CI, 751.1 to 2792.0). The primary vaccination schemes may require several doses in these patients in addition to repeated boosters [43]. Immunogenicity after one dose of tetravalent conjugated ACWY vaccine was also poor in recipients of allogeneic hematopoietic stem cell transplantation [44]. The administration of two primary doses of polysaccharide conjugated anti-meningococcal vaccines is therefore recommended in several countries for patients with asplenia, HIV, or complement disorders [31,45]. No immunogenicity data on vaccines against meningococcal B are available among these subjects.

4. Susceptibility to Invasive *H. influenzae* Infections

Like Nm, *H. influenzae* is also a Gram-negative human-restricted encapsulated bacterium that is a member of the nasopharyngeal microbiota. Hi is highly polymorphic with six different capsular types (serotypes a to f) as well as non-capsulated isolates (nontypeable isolates, HiNT). The incidence of Hib infection has been drastically reduced since the introduction of a vaccination against this serotype. Invasive disease due to other serotypes as well as non-typeable isolates persists and no vaccine is available against these non-Hib isolates.

4.1. Genetic and Acquired Susceptibilities to Invasive Haemophilus influenzae Disease

As for Nm, disorders that affect the immune defense mechanisms and mainly the complement system are expected to increase susceptibility to invasive *H. influenzae*. The frequency of Hi infection in patients with early component deficiencies (C1, C2, C4) seems to be similar to that of meningococcal infections. However, this frequency is lower in infections in patients with C3 deficiencies and LCCD, suggesting that functions other than the lytic functions of the MAC are involved in the defense against invasive Hi infections. However, Hi invasive infections are still higher among patients with complement deficiencies (including factors P or D) than in the general population [15].

Disorders that influence the efficiency of IgG2 binding, the main isotype produced in response to encapsulated bacteria may also increase susceptibility to Hi infections. For example, the His131Arg allele encoding Fcgamma RIIa receptor (rs1801274) binds IgG2 poorly, and therefore, increases the risk of Hi infections [46]. Patients with a single nucleotide polymorphism (SNP) in the *TIRAP* gene (Toll-interleukin 1 receptor domain containing adaptor protein, an adapter molecule associated with Toll-like receptor) (rs1893352) was reported to be strongly associated with non-meningitis cases of Hib in vaccinated children. Another SNP (rs1554286, a promoter SNP in the interleukin-10 encoding gene) was associated with epiglottitis [47]. Patients with asplenia, hSCT, HIV are also at high risk for invasive Hi disease [48].

4.2. Impact on Anti-Hi Vaccination Strategies

There is an unmet medical need in the field of vaccination against *H. influenzae* among patients at high risk due to the absence of vaccines against non-Hib isolates, and particularly, non-typeable Hi (NTHi) isolates. Unlike Nm, only vaccines against serotype B are available. New vaccines, immunogenicity knowledge and vaccination strategies are therefore needed. Non-Hib invasive infections can be more prevalent in patients at risk for Hi invasive infections, underlying the need for vaccines against other serotypes and non-typeable isolates of Hi. Moreover, studies on the immunogenicity of Hib vaccine in these patients are lacking; however, the implications of genetic traits on vaccine efficacy have been suggested [49].

5. Susceptibility to Invasive Pneumococcal Infections

The Gram-positive bacterium *Streptococcus pneumoniae* is an endemic global pathogen that causes a wide range of non-invasive and potentially life-threatening invasive diseases in children and adults. Invasive pneumococcal disease (IPD) implies invasion of pneumococcus into a normally sterile site, leading to several forms of IPD such as bacteremia, empyema, meningitis, endocarditis, and osteomyelitis [50,51]. The incidence of IPD, which ranges from 11 to 27 per 100,000 in Europe, is highest in younger children and the elderly [52–54]. Mortality rates for IPD vary from 12% to 22% in adults in developed countries and are substantially higher in low-income countries. Neurological sequelae, including hearing loss, focal neurological deficits, and cognitive impairment occur in 30–52% of surviving patients [55–58]. Susceptibility to IPD relates to both the virulence of the pathogen and to host factors. The most relevant host factors responsible for the increased risk of IPD are related to defects involving the immune system [59].

5.1. Genetic and Acquired Susceptibilities to IPD

Several inherited and acquired host factors have been shown to confer predisposition to IPD. In particular, primary immunodeficiency states, dysfunction or absence of the spleen and human immunodeficiency virus (HIV) infection, confer a high degree of susceptibility to IPD [60]. Recently, increasing evidence supports a central role of the NF-κB pathway in susceptibility to severe IPD [61].

5.1.1. Inherited Factors of Susceptibility to IPD

Congenital Deficiencies in Immunoglobulins

In contrast to *N. meningitidis* and *H. influenzae* (Gram negative bacteria), the thick cell wall of *S. pneumoniae* (Gram positive) renders it resistant to lysis by insertion of the complement MAC. Furthermore, the presence of a polysaccharide capsule (that can have a thickness of 175 nm in some serotypes) makes them even harder targets for complement-mediated lysis. Antibody-initiated complement-dependent opsonization (opsonophagocytosis), which activates the classic complement pathway, is thought to be the major immune mechanism of pneumococcal killing. Opsonization, refers to the coating of bacteria with antibodies and complement ligands, mainly C3b and iC3b, to facilitate their elimination through phagocytosis by cells bearing complement receptors. Therefore, the production of specific polysaccharide antibodies (IgA, IgM and IgG) and complement activation are the cornerstones to trigger complement-mediated opsonophagocytosis of pneumococci and proper T-B lymphocyte cooperation for an efficient antibody response. Specific antibody deficiencies to *S. pneumoniae* contribute to the increased rates of invasive infection [62]. Although specific rates are not available, patients with agammaglobulinemia (absence of B cell immunoglobulins due to a defect in maturation of B cells) or hypogammaglobulinemia (characterized by reduced serum levels of immunoglobulins and a diminished vaccinal response) are susceptible to invasive *S. pneumoniae* infection [63–65]. Specifically, as IgG antibody responses to bacterial capsular polysaccharide antigens are mostly restricted to IgG2, patients with IgG2 deficiency are more susceptible to infections with *S. pneumoniae*, presumably because of the proposed unique ability of IgG2 to support neutrophil phagocytosis of pneumococci in the absence of complement [66,67]. Moreover, hyper-IgM syndromes (HIGM) are a group of hereditary immune system pathologies, characterized by ineffective immunoglobulin class switching, resulting from interrupted B cell co-stimulation. Patients with hyper-IgM have ineffective production of specific IgG and are susceptible to IPD and sepsis [68].

Congenital Deficiencies in Complement

Only a few clinically defined groups of patients experiencing pneumococcal disease have been systematically examined for the frequency of complement deficiencies [69]. In particular, it has been shown that certain complement deficiencies predispose patients to pneumococcal infections with, in decreasing order of frequency, the C3, the C2 and the C4 defects [63]. Sporadic pneumococcal infections have been diagnosed in patients with C1 and alternative pathway defects (properdin, factor D or factor I deficiencies) [70]. Findings on the role and the link between Mannose-binding lectin (MBL) deficiency and increased susceptibility to pneumococcal infections are conflicting [71–73]. Nevertheless, Eisen et al. analyzed the association between MBL deficiency and the outcome of IPD using data pooled from five studies with adults and one study with children and concluded that the risk of death was increased among MBL-deficient patients with *S. pneumoniae* infection (odds ratio, 5.62; 95% confidence interval, 1.27–24.92) after adjustment for bacteremia, comorbidities and age [74]. MBL deficiency may therefore be considered as a factor of severity instead of a risk factor for developing IPD.

Toll-Like Receptor Signaling Deficiencies

TLR signaling is critically important in the first unspecific meeting between host and microbe. Specific defects of molecules in the TLR signaling pathway including interleukin-1-receptor associated kinase-4 deficiency (IRAK-4), myeloid differentiation factor 88 (MYD88) and nuclear factor-κB essential modulator deficiency (NEMO) [63,75–78] have recently been defined. IRAK-4, a serine threonine kinase, is essential for signal transduction downstream in TLR canonical pathways. IRAK-4 deficiencies are inherited in an autosomal recessive manner [79,80]. Selective susceptibility to *S. pneumoniae* infections is high and many experience recurrent IPD in early childhood. High mortality (40%) is reported before the age of 8 years; however, among survivors, clinical phenotype of patients with IRAK-4 and MyD88 deficiencies tend to improve with age [79].

NF-κB essential modulator (NEMO), encoded by the X-linked *IKBKG* gene, is a regulatory protein essential for activation of the ubiquitous transcription factor NF-κB [81,82]. Children with NEMO-related defects present variable levels of impaired host defenses, with severe susceptibility to IPD [83–86]. Patients with these disorders mount a weak inflammatory response with delayed fever or minimal change in inflammatory markers (e.g., leukocytosis and C reactive protein levels in serum), which may explain the mild inflammatory response elicited in vivo in these patients [87]. It is worth noting that patients with NEMO defects have persistent absence of anti-pneumococcal polysaccharides antibodies after naturally occurring pneumococcal infections and after challenge with polyvalent pneumococcal polysaccharide vaccine, whereas some IRAK-4-deficient patients do [82,87,88].

5.1.2. Acquired Factors of Susceptibility to IPD

S. pneumoniae is overwhelmingly the most common infecting organism in functional or anatomic asplenic patients, accounting for 50–90% of isolates from blood cultures in many cohorts of patients, particularly in younger patients with sickle cell anemia [89]. Mortality from IPD in asplenic patients is more than 50% [90]. As the major site for T-cell independent antibody responses to bacteria and splenic mononuclear phagocytes, the spleen plays a critical role in controlling pneumococcal infection. Patients with asplenia have reduced levels of IgM memory B cells and IgM anti-pneumococcal antibodies, causing reduced ability to produce protective antibodies against polysaccharide antigens, and hence, possible vaccine failure [91,92].

Several studies have shown that HIV-infected individuals and adults have a significantly higher risk of acquiring *S. pneumoniae* and developing recurrent IPD [93,94]. Although active antiretroviral therapy significantly reduces the overall burden of IPD in HIV-positive populations, the risk of IPD remains 35 times higher in HIV-infected individuals than in non-HIV-infected adults [95]. Several studies have underlined the increased susceptibility to IPD in respiratory viruses infected patients, including influenza and respiratory syncytial viruses, especially in children [96–98]. Moreover, patients being treated for underlying solid or hematologic malignancies have high rates of invasive pneumococcal disease, although, interestingly, less than one-fifth of these infections occur during periods of neutropenia [99,100].

5.2. Impact on Anti-Pneumococcal Vaccination Strategies

Systematic immunological exploration in patients hospitalized for recurrent IPD is advocated. Levels of plasma Ig and IgG subclasses should be determined, especially in children who have a history of recurrent infections. In addition, screening of component complement deficiencies can be accomplished by an assessment of total complement function (CH50). Splenic function should be evaluated. In case of inherited immune deficiencies, siblings should also be examined. When detected, prophylactic measures are required to prevent infection. Based on the type of abnormality detected, these prophylactic measures fall into the following major axes:

- Vaccination. Vaccination against pneumococcal disease is safe and strongly recommended. Patients should receive sequential pneumococcal vaccination. Two types of vaccine against invasive pneumococcal disease are available, the pneumo-13V-conjugate vaccine (PCV-13) and the pneumo-polysaccharide-23V (PPV-23). Because these distinct types of vaccine stimulate immune responses somewhat differently, the criteria for protection from invasive pneumococcal disease are not the same for both. It is now recommended that initial vaccination with PCV-13 in children at high risk for severe pneumococcal infection should be followed by PPV-23 immunization starting at 24 months of age. This immunization should be given at least 8 weeks after the last PCV. A second dose of PPV-23 is recommended 5 years after. In patients older than 65 years, one dose of PCV-13 should be followed by PPV-23 at 6 to 12 months later. If PPV-23 was given first, PCV-13 is recommended to be given at least 12 months later. These approaches take advantage of the priming effect of PCV-13 and avoid the hypo-responsiveness to vaccination that might be caused by the PPV-23 [101]. However, hypo-responsiveness has been suggested to occur when plain polysaccharide vaccine is used regardless of the order of administration [102]. Household and other close contacts of persons with altered immunocompetence should also receive age-appropriate *S. pneumoniae* vaccines to minimize the risk of transmission to the immunocompromised contact [103,104]. *S. pneumoniae* has more than 90 serotypes. Although immunization may induce cross-protection against serotypes responsible for the majority of invasive infections, the vaccination fails to protect against other serotypes.
- Prophylactic antibiotics. Penicillin V is the most frequently used antibiotic [105]. Nevertheless, there is no international consensus on when to discontinue prophylaxis [106]. Furthermore, poor adherence to taking daily medications, the global spread and the potential for selection of penicillin-resistant organisms remain unresolved problems [105,107].
- Immunoglobulin replacement therapy. In most forms of antibody deficiency, the mainstay of therapy can be categorized by immunoglobulin (Ig) replacement to provide a protective serum IgG level [108]. Therapeutic IgG, which is usually needed for the

duration of the patient's life, are administered by intravenous (400 to 600 mg/kg every 3 to 4 weeks) or subcutaneous (100 to 150 mg/kg per week) routes to regularly ensure IgG trough levels in the normal range [109].
- Patient education. It is of utmost importance that individuals with altered immune competence be informed and educated about their increased risk for serious, life-threatening infections and understand the importance of seeking prompt medical attention should situations of risk arise (e.g., high fever). When traveling, especially to high-risk geographic areas, a prior consultation is necessary to receive recommendations and update vaccinations.

6. Susceptibility to Invasive GBS Infections

Group B *streptococcus* (GBS) is a leading cause of neonatal and infant sepsis and meningitis globally [110,111]. GBS can also cause stillbirths, prematurity and disease in pregnant women, immunocompromised adults and the elderly, but the highest incidence of disease is in neonates and young infants [112].

6.1. Genetic and Acquired Susceptibilities to Invasive GBS Disease

The susceptibility of neonates to GBS is correlated with a deficiency of maternal (transplacental)-specific antibody and the intrinsically immature immune system of neonates [113]. Moreover, GBS infections in nonpregnant adults typically present when the host is in an immunocompromised or relatively compromised state, such as diabetes, cancer, HIV, with diabetes being the predominating underlying condition [114–116]. The search for monogenetic immunodeficiency disorders underlying susceptibility to invasive GBS infections has only been partially successful so far. One patient with very late-onset GBS sepsis suffering from IRAK-4 deficiency has been reported, supporting that cellular innate immunity and the TLR system are important for resistance against GBS [117].

The severity of disease can be attributed, at least in part, to the virulence of the strain and its ability to avoid immunological clearance and adapt to changing environments throughout disease progression. Indeed, the ST-17 lineage responsible for severe neonatal disease, has a number of ST-17-specific genes that may contribute to its ability to cause meningitis [118].

6.2. Impact on Preventive Strategies

Intrapartum antibiotic prophylaxis (IAP) is the only preventive strategy currently available for the prevention of perinatal GBS early-onset disease (occurring from day 0 to day 6 of life) [117,119,120].

However, IAP coverage has no impact on late onset disease (LOD, which occurs from day 7 to 90 of life), stillbirths and prematurity due to GBS, as well as a limited impact on disease in pregnant women and it might be an issue for antimicrobial resistance [121,122]. Implementing a suitable vaccine for pregnant women could provide effective protection to those forms of invasive disease that cannot be prevented with IAP or where IAP is not feasible. This preventive strategy has been identified as a priority by WHO. Based on specific capsular polysaccharide antigens, 10 serotypes of GBS have been described. A hexavalent GBS glycoconjugate vaccine that covers the major six serotypes responsible for 99% of GBS infections is the most advanced vaccine candidate. Preclinical and human phase I and II studies have been completed, revealing the safety and immunogenicity of these vaccines [123–125]. However, a large number of participants would be required to undertake Phase III clinical efficacy trials. Protein vaccines that might confer protection irrespective of serotype, are in earlier stages of development. Future use of these vaccines raises the question of the adherence of pregnant women to routine vaccination.

7. Conclusions

Several inherited or acquired risk factors are responsible for increased susceptibility to invasive bacterial diseases (Table 1). The investigation of patients with repeated invasive bacterial diseases and patients who developed these infections with unusual isolates is recommended. The genetic dissection of inherited factors will shed light on the molecular and cellular mechanisms underlying protective immunity to bacterial pathogens, and will improve our knowledge on the interaction of the pathogen with the human immune system to pave the way for the development of new, more appropriate treatments. Furthermore, early diagnosis and proper management of immune deficiencies are essential to avoid permanent damage and serious infectious complications. In addition to vaccination, antibiotic chemoprophylaxis (including intrapartum antibiotic prophylaxis for GBS infections) should be strongly considered. However, prolonged chemoprophylaxis using broad-spectrum antibiotics may select resistant bacterial isolates, increasing the risk of selective colonization with resistant isolates. Avoiding, when possible, the use of large-spectrum antibiotics and using vaccines, when available, can contribute to reducing antimicrobial resistance by reducing the selective pressure and preventing transmission of resistant isolates. Safe vaccination, when available, should be encouraged among high-risk patients and their close contacts to prevent these infectious diseases.

Table 1. Congenital and acquired deficiencies and anatomic conditions that may predispose to meningococcal, pneumococcal, *H. influenzae* or GBS invasive infections requiring prevention strategies against invasive bacterial diseases (adapted from references cited in the text).

Involved Deficiency	Transmission (if Known/Applicable)	Estimated Frequency	Clinical Aspects (if Known)	Risk for Invasive Bacterial Infections
Early complement components (C1–C4)	Mendelian recessive but dominant for C1q inhibitor deficiency	C1q and C2 (1:20,000 to 100,000). Rare for the other components	Hereditary angioedema (C1q), systemic lupus erythematosus, glomerulonephritis	Yes, IMD, IPD, IHiD, GBS
Late complement components (C5–C9)	Mendelian recessive. Acquired with anti C-5 treatment	Variable ethnically C6 in Africans and Afro-Americans (1:20 000). C9 in Japanese (1:1000)		Yes, in particular, repeated IMD
Mannose-binding lectin	Non- Mendelian	5% in Caucasian subjects		Debated
Complement regulators (Properdin, Factors B, D, I and H)	Mendelian recessive (X-linked for properdin)	Rare	Atypical hemolytic uremic syndrome, paroxysmal nocturnal hemoglobinuria (PNH), age-related macular degeneration (AMD)	Yes
Antibody (B cell) immunodeficiencies	Heterogeneous		Primary and secondary impairment of antibody production	Yes, in particular IPD, IHiD
Asplenia			Functional or anatomical	Yes. Repeated infections with capsulated bacteria
Toll-like receptor signaling (IRAK-4, MyD88, NEMO)			Innate immunity signaling and Immunodeficiency	Yes, in particular IPD, GBS
Other polymorphisms (IL-10 promoter, Fc-gamma RIIa receptor, TIRAP)			Innate immunity signaling and Immunodeficiency	Yes, in particular IHiD

IPD Invasive Pneumococcal Disease. IHiD Invasive *Haemophilus influenzae* Disease. IMD Invasive Meningococcal Disease. GBS Group B Streptococci. IRAK-4 Interleukin-1 receptor-associated kinase 4. MyD88 Myeloid differentiation primary response 88. NEMO Nuclear factor-kappa B Essential Modulator. TIRAP Toll-interleukin 1 receptor (TIR) domain containing adaptor protein.

Author Contributions: A.-E.D. and M.-K.T. contributed to the conceptualization, research strategy, writing, reviewing and editing. All authors have read and agreed to the published version of the manuscript.

Funding: This research received no external funding.

Institutional Review Board Statement: Not applicable.

Informed Consent Statement: Not applicable.

Data Availability Statement: Not applicable.

Acknowledgments: We would like to thank all members of our laboratory for helpful discussions and support.

Conflicts of Interest: M-K.T reports grants from GSK, grants from Pfizer, grants from Sanofi, outside the submitted work, M.-K.T. and A.-E.D. have a patent 630133 issued.

References

1. World Health Organization (WHO). Defeating meningitis by 2030. In Proceedings of the First Meeting of the Technical Taskforce, World Health Organization, Geneva, Switzerland, 18–19 July 2018. WHO: 2019.
2. Pinto, M.; Gonzalez-Diaz, A.; Machado, M.P.; Duarte, S.; Vieira, L.; Carrico, J.A.; Marti, S.; Bajanca-Lavado, M.P.; Gomes, J.P. Insights into the population structure and pan-genome of *Haemophilus influenzae*. *Infect. Genet. Evol. J. Mol. Epidemiol. Evol. Genet. Infect. Dis.* **2019**, *67*, 126–135. [CrossRef]
3. Dore, N.; Bennett, D.; Kaliszer, M.; Cafferkey, M.; Smyth, C.J. Molecular epidemiology of group B streptococci in Ireland: Associations between serotype, invasive status and presence of genes encoding putative virulence factors. *Epidemiol. Infect.* **2003**, *131*, 823–833. [CrossRef]
4. Sjöstrom, K.; Spindler, C.; Ortqvist, A.; Kalin, M.; Sandgren, A.; Kuhlmann-Berenzon, S.; Henriques-Normark, B. Clonal and capsular types decide whether pneumococci will act as a primary or opportunistic pathogen. *Clin. Infect. Dis.* **2006**, *42*, 451–459. [CrossRef]
5. Zarantonelli, M.L.; Lancellotti, M.; Deghmane, A.E.; Giorgini, D.; Hong, E.; Ruckly, C.; Alonso, J.M.; Taha, M.K. Hyperinvasive genotypes of *Neisseria meningitidis* in France. *Clin. Microbiol. Infect.* **2008**, *14*, 467–472. [CrossRef]
6. Yazdankhah, S.P.; Kriz, P.; Tzanakaki, G.; Kremastinou, J.; Kalmusova, J.; Musilek, M.; Alvestad, T.; Jolley, K.A.; Wilson, D.J.; McCarthy, N.D.; et al. Distribution of Serogroups and Genotypes among Disease-Associated and Carried Isolates of *Neisseria meningitidis* from the Czech Republic, Greece, and Norway. *J. Clin. Microbiol.* **2004**, *42*, 5146–5153. [CrossRef] [PubMed]
7. Harrison, L.H.; Pelton, S.I.; Wilder-Smith, A.; Holst, J.; Safadi, M.A.; Vazquez, J.A.; Taha, M.K.; LaForce, F.M.; von Gottberg, A.; Borrow, R.; et al. The Global Meningococcal Initiative: Recommendations for reducing the global burden of meningococcal disease. *Vaccine* **2011**, *29*, 3363–3371. [CrossRef] [PubMed]
8. Acevedo, R.; Bai, X.; Borrow, R.; Caugant, D.A.; Carlos, J.; Ceyhan, M.; Christensen, H.; Climent, Y.; De Wals, P.; Dinleyici, E.C.; et al. The Global Meningococcal Initiative meeting on prevention of meningococcal disease worldwide: Epidemiology, surveillance, hypervirulent strains, antibiotic resistance and high-risk populations. *Expert Rev. Vaccines* **2019**, *18*, 15–30. [CrossRef] [PubMed]
9. Goldschneider, I.; Gotschlich, E.C.; Artenstein, M.S. Human immunity to the meningococcus. I. The role of humoral antibodies. *J. Exp. Med.* **1969**, *129*, 1307–1326. [CrossRef]
10. Heist, G.D.; Solis-Cohen, S.; Solis-Cohen, M. A study of the virulence of meningococci for man and of human susceptibility to meningococcic infection. *J. Immunol.* **1922**, *7*, 1–33.
11. Ricklin, D.; Lambris, J.D. Complement in immune and inflammatory disorders: Therapeutic interventions. *J. Immunol.* **2013**, *190*, 3839–3847. [CrossRef] [PubMed]
12. Zipfel, P.F.; Skerka, C. Complement regulators and inhibitory proteins. *Nat. Rev. Immunol.* **2009**, *9*, 729–740. [CrossRef]
13. Lewis, L.A.; Ram, S. Meningococcal disease and the complement system. *Virulence* **2014**, *5*, 98–126. [CrossRef] [PubMed]
14. Botto, M.; Kirschfink, M.; Macor, P.; Pickering, M.C.; Wurzner, R.; Tedesco, F. Complement in human diseases: Lessons from complement deficiencies. *Mol. Immunol.* **2009**, *46*, 2774–2783. [CrossRef]
15. Figueroa, J.E.; Densen, P. Infectious diseases associated with complement deficiencies. *Clin. Microbiol. Rev.* **1991**, *4*, 359–395. [CrossRef] [PubMed]
16. Sprong, T.; Roos, D.; Weemaes, C.; Neeleman, C.; Geesing, C.L.; Mollnes, T.E.; van Deuren, M. Deficient alternative complement pathway activation due to factor D deficiency by 2 novel mutations in the complement factor D gene in a family with meningococcal infections. *Blood* **2006**, *107*, 4865–4870. [CrossRef]
17. El Sissy, C.; Rosain, J.; Vieira-Martins, P.; Bordereau, P.; Gruber, A.; Devriese, M.; de Pontual, L.; Taha, M.K.; Fieschi, C.; Picard, C.; et al. Clinical and Genetic Spectrum of a Large Cohort With Total and Sub-total Complement Deficiencies. *Front. Immunol.* **2019**, *10*, 1936. [CrossRef]

18. Davila, S.; Wright, V.J.; Khor, C.C.; Sim, K.S.; Binder, A.; Breunis, W.B.; Inwald, D.; Nadel, S.; Betts, H.; Carrol, E.D.; et al. Genome-wide association study identifies variants in the CFH region associated with host susceptibility to meningococcal disease. *Nat. Genet.* **2010**, *42*, 772–776. [PubMed]
19. Rosain, J.; Hong, E.; Fieschi, C.; Martins, P.V.; El Sissy, C.; Deghmane, A.E.; Ouachee, M.; Thomas, C.; Launay, D.; de Pontual, L.; et al. Strains Responsible for Invasive Meningococcal Disease in Patients With Terminal Complement Pathway Deficiencies. *J. Infect. Dis.* **2017**, *215*, 1331–1338. [CrossRef]
20. Ladhani, S.N.; Campbell, H.; Lucidarme, J.; Gray, S.; Parikh, S.; Willerton, L.; Clark, S.A.; Lekshmi, A.; Walker, A.; Patel, S.; et al. Invasive meningococcal disease in patients with complement deficiencies: A case series (2008–2017). *BMC Infect. Dis.* **2019**, *19*, 522. [CrossRef]
21. Braconier, J.H.; Sjoholm, A.G.; Soderstrom, C. Fulminant meningococcal infections in a family with inherited deficiency of properdin. *Scand. J. Infect. Dis.* **1983**, *15*, 339–345. [CrossRef]
22. Sjoholm, A.G.; Braconier, J.H.; Soderstrom, C. Properdin deficiency in a family with fulminant meningococcal infections. *Clin. Exp. Immunol.* **1982**, *50*, 291–297.
23. Genel, F.; Atlihan, F.; Gulez, N.; Sjoholm, A.G.; Skattum, L.; Truedsson, L. Properdin deficiency in a boy with fulminant meningococcal septic shock. *Acta Paediatr.* **2006**, *95*, 1498–1500. [CrossRef] [PubMed]
24. Soderstrom, C.; Sjoholm, A.G.; Svensson, R.; Ostenson, S. Another Swedish family with complete properdin deficiency: Association with fulminant meningococcal disease in one male family member. *Scand. J. Infect. Dis.* **1989**, *21*, 259–265. [CrossRef] [PubMed]
25. Berden, J.H.; Licht, R.; van Bruggen, M.C.; Tax, W.J. Role of nucleosomes for induction and glomerular binding of autoantibodies in lupus nephritis. *Curr. Opin. Nephrol. Hypertens.* **1999**, *8*, 299–306. [CrossRef]
26. Risitano, A.M.; Marotta, S.; Ricci, P.; Marano, L.; Frieri, C.; Cacace, F.; Sica, M.; Kulasekararaj, A.; Calado, R.T.; Scheinberg, P.; et al. Anti-complement Treatment for Paroxysmal Nocturnal Hemoglobinuria: Time for Proximal Complement Inhibition? A Position Paper From the SAAWP of the EBMT. *Front. Immunol.* **2019**, *10*, 1157. [CrossRef] [PubMed]
27. Park, D.H.; Connor, K.M.; Lambris, J.D. The Challenges and Promise of Complement Therapeutics for Ocular Diseases. *Front. Immunol.* **2019**, *10*, 1007. [CrossRef] [PubMed]
28. Kavanagh, D.; Goodship, T.H.; Richards, A. Atypical hemolytic uremic syndrome. *Semin. Nephrol.* **2013**, *33*, 508–530. [CrossRef] [PubMed]
29. Harris, C.L.; Pouw, R.B.; Kavanagh, D.; Sun, R.; Ricklin, D. Developments in anti-complement therapy; from disease to clinical trial. *Mol. Immunol.* **2018**, *102*, 89–119. [CrossRef]
30. Mastaglio, S.; Ruggeri, A.; Risitano, A.M.; Angelillo, P.; Yancopoulou, D.; Mastellos, D.C.; Huber-Lang, M.; Piemontese, S.; Assanelli, A.; Garlanda, C.; et al. The first case of COVID-19 treated with the complement C3 inhibitor AMY-101. *Clin. Immunol.* **2020**, *215*, 108450. [CrossRef]
31. Zuber, J.; Frimat, M.; Caillard, S.; Kamar, N.; Gatault, P.; Petitprez, F.; Couzi, L.; Jourde-Chiche, N.; Chatelet, V.; Gaisne, R.; et al. Use of Highly Individualized Complement Blockade Has Revolutionized Clinical Outcomes after Kidney Transplantation and Renal Epidemiology of Atypical Hemolytic Uremic Syndrome. *J. Am. Soc. Nephrol.* **2019**, *30*, 2449–2463. [CrossRef]
32. Cugno, M.; Meroni, P.L.; Gualtierotti, R.; Griffini, S.; Grovetti, E.; Torri, A.; Panigada, M.; Aliberti, S.; Blasi, F.; Tedesco, F.; et al. Complement Activation in Patients with Covid-19: A Novel Therapeutic Target. *J. Allergy Clin. Immunol.* **2020**, *146*, 215–217. [CrossRef] [PubMed]
33. Ljungman, P.; Lewensohn-Fuchs, I.; Hammarstrom, V.; Aschan, J.; Brandt, L.; Bolme, P.; Lonnqvist, B.; Johansson, N.; Ringden, O.; Gahrton, G. Long-term immunity to measles, mumps, and rubella after allogeneic bone marrow transplantation. *Blood* **1994**, *84*, 657–663. [CrossRef] [PubMed]
34. Taha, M.K.; Weil-Olivier, C.; Bouéec, S.; Emeryc, C.; Nachbaurd, G.; Pribild, C.; Loncle-Provotd, V. Risk factors for invasive meningococcal disease: A retrospective analysis of the French national public health insurance database. *Hum. Vacc. Immunother.* **2021**, *17*, 1–9.
35. van Veen, K.E.; Brouwer, M.C.; van der Ende, A.; van de Beek, D. Bacterial meningitis in hematopoietic stem cell transplant recipients: A population-based prospective study. *Bone Marrow Transplant.* **2016**, *51*, 1490–1495. [CrossRef]
36. van Veen, K.E.; Brouwer, M.C.; van der Ende, A.; van de Beek, D. Bacterial meningitis in solid organ transplant recipients: A population-based prospective study. *Transpl. Infect. Dis.* **2016**, *18*, 674–680. [CrossRef]
37. Levy, C.; Taha, M.K.; Weil Olivier, C.; Quinet, B.; Lecuyer, A.; Alonso, J.M.; Aujard, Y.; Bingen, E.; Cohen, R. Groupe des pediatres et microbiologistes de l'Observatoire National des M: Association of meningococcal phenotypes and genotypes with clinical characteristics and mortality of meningitis in children. *Pediatric Infect. Dis. J.* **2010**, *29*, 618–623. [CrossRef]
38. Faust, S.N.; Levin, M.; Harrison, O.B.; Goldin, R.D.; Lockhart, M.S.; Kondaveeti, S.; Laszik, Z.; Esmon, C.T.; Heyderman, R.S. Dysfunction of endothelial protein C activation in severe meningococcal sepsis. *N. Engl. J. Med.* **2001**, *345*, 408–416. [CrossRef] [PubMed]
39. Hermans, P.W.; Hibberd, M.L.; Booy, R.; Daramola, O.; Hazelzet, J.A.; de Groot, R.; Levin, M. 4G/5G promoter polymorphism in the plasminogen-activator-inhibitor-1 gene and outcome of meningococcal disease. Meningococcal Research Group. *Lancet* **1999**, *354*, 556–560. [CrossRef]
40. Toubiana, J.; Heilbronner, C.; Gitiaux, C.; Oualha, M.; Taha, M.K.; Rousseau, C.; Picard, C.; Mira, J.P.; Gendrel, D. Pachymeningitis after meningococcal infection. *Lancet* **2013**, *381*, 1596. [CrossRef]

41. Risitano, A.M.; Mastellos, D.C.; Huber-Lang, M.; Yancopoulou, D.; Garlanda, C.; Ciceri, F.; Lambris, J.D. Complement as a target in COVID-19? *Nat. Rev. Immunol.* **2020**, *20*, 343–344. [CrossRef]
42. Dahyot-Fizelier, C.; Debaene, B.; Mimoz, O. Management of infection risk in asplenic patients. *Ann. Fr. Anesth. Reanim.* **2013**, *32*, 251–256. [CrossRef] [PubMed]
43. Balmer, P.; Falconer, M.; McDonald, P.; Andrews, N.; Fuller, E.; Riley, C.; Kaczmarski, E.; Borrow, R. Immune response to meningococcal serogroup C conjugate vaccine in asplenic individuals. *Infect. Immun.* **2004**, *72*, 332–337. [CrossRef] [PubMed]
44. Mahler, M.B.; Taur, Y.; Jean, R.; Kernan, N.A.; Prockop, S.E.; Small, T.N. Safety and Immunogenicity of the Tetravalent Protein-Conjugated Meningococcal Vaccine (MCV4) in Recipients of Related and Unrelated Allogeneic Hematopoietic Stem Cell Transplantation. *Biol. Blood Marrow Transplant.* **2012**, *18*, 145–149. [CrossRef]
45. Centers for Disease Control and Prevention (CDC). Updated recommendations for use of meningococcal conjugate vaccines—Advisory Committee on Immunization Practices (ACIP), 2010. *MMWR Morb. Mortal. Wkly. Rep.* **2011**, *60*, 72–76.
46. Flesch, B.K.; Nikolaus, S.; El Mokhtari, N.E.; Schreiber, S.; Nebel, A. The FCGR2A–Arg131 variant is no major mortality factor in the elderly–evidence from a German centenarian study. *Int. J. Immunogenet.* **2006**, *33*, 277–279. [CrossRef]
47. Ladhani, S.N.; Davila, S.; Hibberd, M.L.; Heath, P.T.; Ramsay, M.E.; Slack, M.P.; Pollard, A.J.; Booy, R. Association between single-nucleotide polymorphisms in Mal/TIRAP and interleukin-10 genes and susceptibility to invasive *Haemophilus influenzae* serotype b infection in immunized children. *Clin. Infect. Dis.* **2010**, *51*, 761–767. [CrossRef] [PubMed]
48. Steinhart, R.; Reingold, A.L.; Taylor, F.; Anderson, G.; Wenger, J.D. Invasive *Haemophilus influenzae* infections in men with HIV infection. *JAMA* **1992**, *268*, 3350–3352. [CrossRef]
49. Murthy, B.N. Implications of genetic traits on vaccine efficacy. *Stat. Med.* **2003**, *22*, 1989–1998. [CrossRef]
50. Brown, A.O.; Mann, B.; Gao, G.; Hankins, J.S.; Humann, J.; Giardina, J.; Faverio, P.; Restrepo, M.I.; Halade, G.V.; Mortensen, E.M.; et al. *Streptococcus pneumoniae* translocates into the myocardium and forms unique microlesions that disrupt cardiac function. *PLoS Pathog.* **2014**, *10*, e1004383. [CrossRef] [PubMed]
51. Feldman, C.; Normark, S.; Henriques-Normark, B.; Anderson, R. Pathogenesis and prevention of risk of cardiovascular events in patients with pneumococcal community-acquired pneumonia. *J. Intern. Med.* **2019**, *285*, 635–652. [CrossRef] [PubMed]
52. Marrie, T.J.; Tyrrell, G.J.; Majumdar, S.R.; Eurich, D.T. Effect of Age on the Manifestations and Outcomes of Invasive Pneumococcal Disease in Adults. *Am. J. Med.* **2018**, *131*, 100.e1–100.e7. [CrossRef]
53. van de Beek, D.; Brouwer, M.; Hasbun, R.; Koedel, U.; Whitney, C.G.; Wijdicks, E. Community-acquired bacterial meningitis. *Nat. Rev. Dis Primers* **2016**, *2*, 16074. [CrossRef] [PubMed]
54. Lynch, J.P.; Zhanel, G.G., 3rd. *Streptococcus pneumoniae*: Epidemiology, risk factors, and strategies for prevention. *Semin. Respir. Crit. Care Med.* **2009**, *30*, 189–209. [CrossRef] [PubMed]
55. Lynch, J.P.; Zhanel, G.G., 3rd. *Streptococcus pneumoniae*: Does antimicrobial resistance matter? *Semin. Respir. Crit. Care Med.* **2009**, *30*, 210–238. [CrossRef]
56. Bijlsma, M.W.; Brouwer, M.C.; Kasanmoentalib, E.S.; Kloek, A.T.; Lucas, M.J.; Tanck, M.W.; van der Ende, A.; van de Beek, D. Community-acquired bacterial meningitis in adults in the Netherlands, 2006-14: A prospective cohort study. *Lancet Infect. Dis.* **2016**, *16*, 339–347. [CrossRef]
57. LeBlanc, J.J.; ElSherif, M.; Ye, L.; MacKinnon-Cameron, D.; Li, L.; Ambrose, A.; Hatchette, T.F.; Lang, A.L.; Gillis, H.; Martin, I.; et al. Burden of vaccine-preventable pneumococcal disease in hospitalized adults: A Canadian Immunization Research Network (CIRN) Serious Outcomes Surveillance (SOS) network study. *Vaccine* **2017**, *35*, 3647–3654. [CrossRef]
58. Thomas, K.; Mukkai Kesavan, L.; Veeraraghavan, B.; Jasmine, S.; Jude, J.; Shubankar, M.; Kulkarni, P.; Steinhoff, M. Network ISGI: Invasive pneumococcal disease associated with high case fatality in India. *J. Clin. Epidemiol.* **2013**, *66*, 36–43. [CrossRef]
59. Brouwer, M.C.; de Gans, J.; Heckenberg, S.G.; Zwinderman, A.H.; van der Poll, T.; van de Beek, D. Host genetic susceptibility to pneumococcal and meningococcal disease: A systematic review and meta-analysis. *Lancet Infect. Dis.* **2009**, *9*, 31–44. [CrossRef]
60. Adriani, K.S.; Brouwer, M.C.; van de Beek, D. Risk factors for community-acquired bacterial meningitis in adults. *Neth. J. Med.* **2015**, *73*, 53–60.
61. Waterer, G.W.; Wunderink, R.G. Genetic susceptibility to pneumonia. *Clin. Chest Med.* **2005**, *26*, 29–38. [CrossRef]
62. Butters, C.; Phuong, L.K.; Cole, T.; Gwee, A. Prevalence of Immunodeficiency in Children With Invasive Pneumococcal Disease in the Pneumococcal Vaccine Era: A Systematic Review. *JAMA Pediatr.* **2019**, *173*, 1084–1094. [CrossRef]
63. Picard, C.; Puel, A.; Bustamante, J.; Ku, C.L.; Casanova, J.L. Primary immunodeficiencies associated with pneumococcal disease. *Curr. Opin. Allergy Clin. Immunol.* **2003**, *3*, 451–459. [CrossRef] [PubMed]
64. Oksenhendler, E.; Gerard, L.; Fieschi, C.; Malphettes, M.; Mouillot, G.; Jaussaud, R.; Viallard, J.F.; Gardembas, M.; Galicier, L.; Schleinitz, N.; et al. Infections in 252 patients with common variable immunodeficiency. *Clin. Infect. Dis.* **2008**, *46*, 1547–1554. [CrossRef] [PubMed]
65. Winkelstein, J.A.; Marino, M.C.; Lederman, H.M.; Jones, S.M.; Sullivan, K.; Burks, A.W.; Conley, M.E.; Cunningham-Rundles, C.; Ochs, H.D. X-linked agammaglobulinemia: Report on a United States registry of 201 patients. *Medicine* **2006**, *85*, 193–202. [CrossRef] [PubMed]
66. Umetsu, D.T.; Ambrosino, D.M.; Quinti, I.; Siber, G.R.; Geha, R.S. Recurrent sinopulmonary infection and impaired antibody response to bacterial capsular polysaccharide antigen in children with selective IgG-subclass deficiency. *N. Engl. J. Med.* **1985**, *313*, 1247–1251. [CrossRef]

67. Lortan, J.E.; Kaniuk, A.S.; Monteil, M.A. Relationship of in vitro phagocytosis of serotype 14 *Streptococcus pneumoniae* to specific class and IgG subclass antibody levels in healthy adults. *Clin. Exp. Immunol.* **1993**, *91*, 54–57. [CrossRef]
68. Winkelstein, J.A.; Marino, M.C.; Ochs, H.; Fuleihan, R.; Scholl, P.R.; Geha, R.; Stiehm, E.R.; Conley, M.E. The X-linked hyper-IgM syndrome: Clinical and immunologic features of 79 patients. *Medicine* **2003**, *82*, 373–384. [CrossRef] [PubMed]
69. Ekdahl, K.; Truedsson, L.; Sjoholm, A.G.; Braconier, J.H. Complement analysis in adult patients with a history of bacteremic pneumococcal infections or recurrent pneumonia. *Scand. J. Infect. Dis* **1995**, *27*, 111–117. [CrossRef] [PubMed]
70. Hosea, S.W.; Brown, E.J.; Frank, M.M. The critical role of complement in experimental pneumococcal sepsis. *J. Infect. Dis.* **1980**, *142*, 903–909. [CrossRef]
71. Roy, S.; Knox, K.; Segal, S.; Griffiths, D.; Moore, C.E.; Welsh, K.I.; Smarason, A.; Day, N.P.; McPheat, W.L.; Crook, D.W.; et al. MBL genotype and risk of invasive pneumococcal disease: A case-control study. *Lancet* **2002**, *359*, 1569–1573. [CrossRef]
72. Moens, L.; Van Hoeyveld, E.; Peetermans, W.E.; De Boeck, C.; Verhaegen, J.; Bossuyt, X. Mannose-binding lectin genotype and invasive pneumococcal infection. *Hum. Immunol.* **2006**, *67*, 605–611. [CrossRef] [PubMed]
73. Kronborg, G.; Weis, N.; Madsen, H.O.; Pedersen, S.S.; Wejse, C.; Nielsen, H.; Skinhoj, P.; Garred, P. Variant mannose-binding lectin alleles are not associated with susceptibility to or outcome of invasive pneumococcal infection in randomly included patients. *J. Infect. Dis.* **2002**, *185*, 1517–1520. [CrossRef] [PubMed]
74. Eisen, D.P.; Dean, M.M.; Boermeester, M.A.; Fidler, K.J.; Gordon, A.C.; Kronborg, G.; Kun, J.F.; Lau, Y.L.; Payeras, A.; Valdimarsson, H.; et al. Low serum mannose-binding lectin level increases the risk of death due to pneumococcal infection. *Clin. Infect. Dis.* **2008**, *47*, 510–516. [CrossRef]
75. von Bernuth, H.; Picard, C.; Jin, Z.; Pankla, R.; Xiao, H.; Ku, C.L.; Chrabieh, M.; Mustapha, I.B.; Ghandil, P.; Camcioglu, Y.; et al. Pyogenic bacterial infections in humans with MyD88 deficiency. *Science* **2008**, *321*, 691–696. [CrossRef] [PubMed]
76. Enders, A.; Pannicke, U.; Berner, R.; Henneke, P.; Radlinger, K.; Schwarz, K.; Ehl, S. Two siblings with lethal pneumococcal meningitis in a family with a mutation in Interleukin-1 receptor-associated kinase 4. *J. Pediatr.* **2004**, *145*, 698–700. [CrossRef] [PubMed]
77. Ku, C.L.; Yang, K.; Bustamante, J.; Puel, A.; von Bernuth, H.; Santos, O.F.; Lawrence, T.; Chang, H.H.; Al-Mousa, H.; Picard, C.; et al. Inherited disorders of human Toll-like receptor signaling: Immunological implications. *Immunol. Rev.* **2005**, *203*, 10–20. [CrossRef]
78. Picard, C.; Puel, A.; Bonnet, M.; Ku, C.L.; Bustamante, J.; Yang, K.; Soudais, C.; Dupuis, S.; Feinberg, J.; Fieschi, C.; et al. Pyogenic bacterial infections in humans with IRAK-4 deficiency. *Science* **2003**, *299*, 2076–2079. [CrossRef] [PubMed]
79. Ku, C.L.; von Bernuth, H.; Picard, C.; Zhang, S.Y.; Chang, H.H.; Yang, K.; Chrabieh, M.; Issekutz, A.C.; Cunningham, C.K.; Gallin, J.; et al. Selective predisposition to bacterial infections in IRAK-4-deficient children: IRAK-4-dependent TLRs are otherwise redundant in protective immunity. *J. Exp. Med.* **2007**, *204*, 2407–2422. [CrossRef]
80. Ku, C.L.; Picard, C.; Erdos, M.; Jeurissen, A.; Bustamante, J.; Puel, A.; von Bernuth, H.; Filipe-Santos, O.; Chang, H.H.; Lawrence, T.; et al. IRAK4 and NEMO mutations in otherwise healthy children with recurrent invasive pneumococcal disease. *J. Med. Genet.* **2007**, *44*, 16–23. [CrossRef]
81. Orange, J.S.; Jain, A.; Ballas, Z.K.; Schneider, L.C.; Geha, R.S.; Bonilla, F.A. The presentation and natural history of immunodeficiency caused by nuclear factor kappaB essential modulator mutation. *J. Allergy Clin. Immunol.* **2004**, *113*, 725–733. [CrossRef]
82. Uzel, G. The range of defects associated with nuclear factor kappaB essential modulator. *Curr. Opin. Allergy Clin. Immunol.* **2005**, *5*, 513–518. [CrossRef] [PubMed]
83. Doffinger, R.; Smahi, A.; Bessia, C.; Geissmann, F.; Feinberg, J.; Durandy, A.; Bodemer, C.; Kenwrick, S.; Dupuis-Girod, S.; Blanche, S.; et al. X-linked anhidrotic ectodermal dysplasia with immunodeficiency is caused by impaired NF-kappaB signaling. *Nat. Genet.* **2001**, *27*, 277–285. [CrossRef] [PubMed]
84. Jain, A.; Ma, C.A.; Liu, S.; Brown, M.; Cohen, J.; Strober, W. Specific missense mutations in NEMO result in hyper-IgM syndrome with hypohydrotic ectodermal dysplasia. *Nat. Immunol.* **2001**, *2*, 223–228. [CrossRef]
85. Zonana, J.; Elder, M.E.; Schneider, L.C.; Orlow, S.J.; Moss, C.; Golabi, M.; Shapira, S.K.; Farndon, P.A.; Wara, D.W.; Emmal, S.A.; et al. A novel X-linked disorder of immune deficiency and hypohidrotic ectodermal dysplasia is allelic to incontinentia pigmenti and due to mutations in IKK-gamma (NEMO). *Am. J. Hum. Genet.* **2000**, *67*, 1555–1562. [CrossRef]
86. Dupuis-Girod, S.; Corradini, N.; Hadj-Rabia, S.; Fournet, J.C.; Faivre, L.; Le Deist, F.; Durand, P.; Doffinger, R.; Smahi, A.; Israel, A.; et al. Osteopetrosis, lymphedema, anhidrotic ectodermal dysplasia, and immunodeficiency in a boy and incontinentia pigmenti in his mother. *Pediatrics* **2002**, *109*, e97. [CrossRef] [PubMed]
87. von Bernuth, H.; Puel, A.; Ku, C.L.; Yang, K.; Bustamante, J.; Chang, H.H.; Picard, C.; Casanova, J.L. Septicemia without sepsis: Inherited disorders of nuclear factor-kappa B-mediated inflammation. *Clin. Infect. Dis.* **2005**, *41* (Suppl. 7), S436–S439. [CrossRef] [PubMed]
88. Medvedev, A.E.; Lentschat, A.; Kuhns, D.B.; Blanco, J.C.; Salkowski, C.; Zhang, S.; Arditi, M.; Gallin, J.I.; Vogel, S.N. Distinct mutations in IRAK-4 confer hyporesponsiveness to lipopolysaccharide and interleukin-1 in a patient with recurrent bacterial infections. *J. Exp. Med.* **2003**, *198*, 521–531. [CrossRef]
89. Holdsworth, R.J.; Irving, A.D.; Cuschieri, A. Postsplenectomy sepsis and its mortality rate: Actual versus perceived risks. *Br. J. Surg.* **1991**, *78*, 1031–1038. [CrossRef]

90. Halasa, N.B.; Shankar, S.M.; Talbot, T.R.; Arbogast, P.G.; Mitchel, E.F.; Wang, W.C.; Schaffner, W.; Craig, A.S.; Griffin, M.R. Incidence of invasive pneumococcal disease among individuals with sickle cell disease before and after the introduction of the pneumococcal conjugate vaccine. *Clin. Infect. Dis.* **2007**, *44*, 1428–1433. [CrossRef]
91. Kruetzmann, S.; Rosado, M.M.; Weber, H.; Germing, U.; Tournilhac, O.; Peter, H.H.; Berner, R.; Peters, A.; Boehm, T.; Plebani, A.; et al. Human immunoglobulin M memory B cells controlling *Streptococcus pneumoniae* infections are generated in the spleen. *J. Exp. Med.* **2003**, *197*, 939–945. [CrossRef]
92. Weller, S.; Braun, M.C.; Tan, B.K.; Rosenwald, A.; Cordier, C.; Conley, M.E.; Plebani, A.; Kumararatne, D.S.; Bonnet, D.; Tournilhac, O.; et al. Human blood IgM "memory" B cells are circulating splenic marginal zone B cells harboring a prediversified immunoglobulin repertoire. *Blood* **2004**, *104*, 3647–3654. [CrossRef] [PubMed]
93. Kourtis, A.P.; Bansil, P.; Posner, S.F.; Johnson, C.; Jamieson, D.J. Trends in hospitalizations of HIV-infected children and adolescents in the United States: Analysis of data from the 1994-2003 Nationwide Inpatient Sample. *Pediatrics* **2007**, *120*, e236–e243. [CrossRef] [PubMed]
94. Molyneux, E.M.; Tembo, M.; Kayira, K.; Bwanaisa, L.; Mweneychanya, J.; Njobvu, A.; Forsyth, H.; Rogerson, S.; Walsh, A.L.; Molyneux, M.E. The effect of HIV infection on paediatric bacterial meningitis in Blantyre, Malawi. *Arch. Dis. Child.* **2003**, *88*, 1112–1118. [CrossRef]
95. Heilmann, C. Human B and T lymphocyte responses to vaccination with pneumococcal polysaccharides. *APMIS Suppl.* **1990**, *15*, 1–23.
96. Peltola, V.; Waris, M.; Hyypia, T.; Ruuskanen, O. Respiratory viruses in children with invasive pneumococcal disease. *Clin. Infect. Dis.* **2006**, *43*, 266–268. [CrossRef]
97. O'Brien, K.L.; Walters, M.I.; Sellman, J.; Quinlisk, P.; Regnery, H.; Schwartz, B.; Dowell, S.F. Severe pneumococcal pneumonia in previously healthy children: The role of preceding influenza infection. *Clin. Infect. Dis.* **2000**, *30*, 784–789. [CrossRef] [PubMed]
98. Stensballe, L.G.; Hjuler, T.; Andersen, A.; Kaltoft, M.; Ravn, H.; Aaby, P.; Simoes, E.A. Hospitalization for respiratory syncytial virus infection and invasive pneumococcal disease in Danish children aged < 2 years: A population-based cohort study. *Clin. Infect. Dis.* **2008**, *46*, 1165–1171.
99. Kyaw, M.H.; Rose, C.E.; Fry, A.M., Jr.; Singleton, J.A.; Moore, Z.; Zell, E.R.; Whitney, C.G. Active Bacterial Core Surveillance Program of the Emerging Infections Program N: The influence of chronic illnesses on the incidence of invasive pneumococcal disease in adults. *J. Infect. Dis.* **2005**, *192*, 377–386. [CrossRef]
100. Kumashi, P.; Girgawy, E.; Tarrand, J.J.; Rolston, K.V.; Raad, I.I.; Safdar, A. *Streptococcus pneumoniae* bacteremia in patients with cancer: Disease characteristics and outcomes in the era of escalating drug resistance (1998–2002). *Medicine* **2005**, *84*, 303–312. [CrossRef]
101. Whitney, C.G.; Farley, M.M.; Hadler, J.; Harrison, L.H.; Bennett, N.M.; Lynfield, R.; Reingold, A.; Cieslak, P.R.; Pilishvili, T.; Jackson, D. Decline in invasive pneumococcal disease after the introduction of protein-polysaccharide conjugate vaccine. *N. Engl. J. Med.* **2003**, *348*, 1737–1746. [CrossRef]
102. Poolman, J.; Borrow, R. Hyporesponsiveness and its clinical implications after vaccination with polysaccharide or glycoconjugate vaccines. *Expert Rev. Vaccines* **2011**, *10*, 307–322. [CrossRef] [PubMed]
103. Medical Advisory Committee of the Immune Deficiency Foundation; Shearer, W.T.; Fleisher, T.A.; Buckley, R.H.; Ballas, Z.; Ballow, M.; Blaese, R.M.; Bonilla, F.A.; Conley, M.E.; Cunningham-Rundles, C.; et al. Recommendations for live viral and bacterial vaccines in immunodeficient patients and their close contacts. *J. Allergy Clin. Immunol.* **2014**, *133*, 961–966.
104. (HCSP) HCdlSP. Vaccination des personnes immunodéprimées ou aspléniques. Recommandations actualisées. In *Collections et Avis*; Haut Conseil de la Santé Publique (HCSP): Paris, France, 2014.
105. Steele, R.W.; Warrier, R.; Unkel, P.J.; Foch, B.J.; Howes, R.F.; Shah, S.; Williams, K.; Moore, S.; Jue, S.J. Colonization with antibiotic-resistant *Streptococcus pneumoniae* in children with sickle cell disease. *J. Pediatr.* **1996**, *128*, 531–535. [CrossRef]
106. Styrt, B. Infection associated with asplenia: Risks, mechanisms, and prevention. *Am. J. Med.* **1990**, *88*, 33N–42N. [PubMed]
107. Waghorn, D.J. Overwhelming infection in asplenic patients: Current best practice preventive measures are not being followed. *J. Clin. Pathol.* **2001**, *54*, 214–218. [CrossRef] [PubMed]
108. Lucas, M.; Lee, M.; Lortan, J.; Lopez-Granados, E.; Misbah, S.; Chapel, H. Infection outcomes in patients with common variable immunodeficiency disorders: Relationship to immunoglobulin therapy over 22 years. *J. Allergy Clin. Immunol.* **2010**, *125*, 1354–1360. [CrossRef] [PubMed]
109. Ballow, M. Immunoglobulin therapy: Methods of delivery. *J. Allergy Clin. Immunol.* **2008**, *122*, 1038–1039. [CrossRef] [PubMed]
110. O'Sullivan, C.P.; Lamagni, T.; Patel, D.; Efstratiou, A.; Cunney, R.; Meehan, M.; Ladhani, S.; Reynolds, A.J.; Campbell, R.; Doherty, L.; et al. Group B streptococcal disease in UK and Irish infants younger than 90 days, 2014-15: A prospective surveillance study. *Lancet Infect. Dis.* **2019**, *19*, 83–90. [CrossRef]
111. Seale, A.C.; Bianchi-Jassir, F.; Russell, N.J.; Kohli-Lynch, M.; Tann, C.J.; Hall, J.; Madrid, L.; Blencowe, H.; Cousens, S.; Baker, C.J.; et al. Estimates of the Burden of Group B Streptococcal Disease Worldwide for Pregnant Women, Stillbirths, and Children. *Clin. Infect. Dis.* **2017**, *65* (Suppl. 2), S200–S219. [CrossRef] [PubMed]
112. Raabe, V.N.; Shane, A.L. Group B Streptococcus (Streptococcus agalactiae). *Microbiol. Spectr.* **2019**, *7*, 228–238. [CrossRef]
113. Baker, C.J.; Kasper, D.L. Correlation of maternal antibody deficiency with susceptibility to neonatal group B streptococcal infection. *N. Engl. J. Med.* **1976**, *294*, 753–756. [CrossRef] [PubMed]

114. Sendi, P.; Johansson, L.; Norrby-Teglund, A. Invasive group B Streptococcal disease in non-pregnant adults: A review with emphasis on skin and soft-tissue infections. *Infection* **2008**, *36*, 100–111. [CrossRef]
115. Skoff, T.H.; Farley, M.M.; Petit, S.; Craig, A.S.; Schaffner, W.; Gershman, K.; Harrison, L.H.; Lynfield, R.; Mohle-Boetani, J.; Zansky, S.; et al. Increasing burden of invasive group B streptococcal disease in nonpregnant adults, 1990–2007. *Clin. Infect. Dis.* **2009**, *49*, 85–92. [CrossRef]
116. Smith, E.M.; Khan, M.A.; Reingold, A.; Watt, J.P. Group B streptococcus infections of soft tissue and bone in California adults, 1995–2012. *Epidemiol. Infect.* **2015**, *143*, 3343–3350. [CrossRef] [PubMed]
117. Picard, C.; von Bernuth, H.; Ghandil, P.; Chrabieh, M.; Levy, O.; Arkwright, P.D.; McDonald, D.; Geha, R.S.; Takada, H.; Krause, J.C.; et al. Clinical features and outcome of patients with IRAK-4 and MyD88 deficiency. *Medicine* **2010**, *89*, 403–425. [CrossRef]
118. Tazi, A.; Bellais, S.; Tardieux, I.; Dramsi, S.; Trieu-Cuot, P.; Poyart, C. Group B Streptococcus surface proteins as major determinants for meningeal tropism. *Curr. Opin. Microbiol.* **2012**, *15*, 44–49. [CrossRef]
119. Krause, J.C.; Ghandil, P.; Chrabieh, M.; Casanova, J.L.; Picard, C.; Puel, A.; Creech, C.B. Very late-onset group B Streptococcus meningitis, sepsis, and systemic shigellosis due to interleukin-1 receptor-associated kinase-4 deficiency. *Clin. Infect. Dis.* **2009**, *49*, 1393–1396. [CrossRef] [PubMed]
120. Picard, C.; Casanova, J.L.; Puel, A. Infectious diseases in patients with IRAK-4, MyD88, NEMO, or IkappaBalpha deficiency. *Clin. Microbiol. Rev.* **2011**, *24*, 490–497. [CrossRef]
121. Schrag, S.J.; Verani, J.R. Intrapartum antibiotic prophylaxis for the prevention of perinatal group B streptococcal disease: Experience in the United States and implications for a potential group B streptococcal vaccine. *Vaccine* **2013**, *31* (Suppl. 4), D20–D26. [CrossRef]
122. Le Doare, K.; O'Driscoll, M.; Turner, K.; Seedat, F.; Russell, N.J.; Seale, A.C.; Heath, P.T.; Lawn, J.E.; Baker, C.J.; Bartlett, L.; et al. Intrapartum Antibiotic Chemoprophylaxis Policies for the Prevention of Group B Streptococcal Disease Worldwide: Systematic Review. *Clin. Infect. Dis.* **2017**, *65* (Suppl. 2), S143–S151. [CrossRef] [PubMed]
123. Madhi, S.A.; Cutland, C.L.; Jose, L.; Koen, A.; Govender, N.; Wittke, F.; Olugbosi, M.; Meulen, A.S.; Baker, S.; Dull, P.M.; et al. Safety and immunogenicity of an investigational maternal trivalent group B streptococcus vaccine in healthy women and their infants: A randomised phase 1b/2 trial. *Lancet Infect. Dis.* **2016**, *16*, 923–934. [CrossRef]
124. Dzanibe, S.; Madhi, S.A. Systematic review of the clinical development of group B streptococcus serotype-specific capsular polysaccharide-based vaccines. *Expert Rev. Vaccines* **2018**, *17*, 635–651. [CrossRef] [PubMed]
125. Hillier, S.L.; Ferrieri, P.; Edwards, M.S.; Ewell, M.; Ferris, D.; Fine, P.; Carey, V.; Meyn, L.; Hoagland, D.; Kasper, D.L.; et al. A Phase 2, Randomized, Control Trial of Group B Streptococcus (GBS) Type III Capsular Polysaccharide-tetanus Toxoid (GBS III-TT) Vaccine to Prevent Vaginal Colonization With GBS III. *Clin. Infect. Dis.* **2019**, *68*, 2079–2086. [CrossRef] [PubMed]

 microorganisms

MDPI

Article

Understanding the Role of Duration of Vaccine Protection with MenAfriVac: Simulating Alternative Vaccination Strategies

Andromachi Karachaliou Prasinou *, Andrew J. K. Conlan and Caroline L. Trotter

Disease Dynamics Unit, Department of Veterinary Medicine, University of Cambridge, Cambridge CB3 0ES, UK; ajkc2@cam.ac.uk (A.J.K.C.); clt56@cam.ac.uk (C.L.T.)
* Correspondence: ak889@cam.ac.uk

Abstract: We previously developed a transmission dynamic model of *Neisseria meningitidis* serogroup A (NmA) with the aim of forecasting the relative benefits of different immunisation strategies with MenAfriVac. Our findings suggested that the most effective strategy in maintaining disease control was the introduction of MenAfriVac into the Expanded Programme on Immunisation (EPI). This strategy is currently being followed by the countries of the meningitis belt. Since then, the persistence of vaccine-induced antibodies has been further studied and new data suggest that immune response is influenced by the age at vaccination. Here, we aim to investigate the influence of both the duration and age-specificity of vaccine-induced protection on our model predictions and explore how the optimal vaccination strategy may change in the long-term. We adapted our previous model and considered plausible alternative immunization strategies, including the addition of a booster dose to the current schedule, as well as the routine vaccination of school-aged children for a range of different assumptions regarding the duration of protection. To allow for a comparison between the different strategies, we use several metrics, including the median age of infection, the number of people needed to vaccinate (NNV) to prevent one case, the age distribution of cases for each strategy, as well as the time it takes for the number of cases to start increasing after the honeymoon period (resurgence). None of the strategies explored in this work is superior in all respects. This is especially true when vaccine-induced protection is the same regardless of the age at vaccination. Uncertainty in the duration of protection is important. For duration of protection lasting for an average of 18 years or longer, the model predicts elimination of NmA cases. Assuming that vaccine protection is more durable for individuals vaccinated after the age of 5 years, routine immunization of older children would be more efficient in reducing disease incidence and would also result in a fewer number of doses necessary to prevent one case. Assuming that elimination does not occur, adding a booster dose is likely to prevent most cases but the caveat will be a more costly intervention. These results can be used to understand important sources of uncertainty around MenAfriVac and support decisions by policymakers.

Keywords: meningitis; vaccine; Africa; mathematical modelling

Citation: Karachaliou Prasinou, A.; Conlan, A.J.K.; Trotter, C.L. Understanding the Role of Duration of Vaccine Protection with MenAfriVac: Simulating Alternative Vaccination Strategies. *Microorganisms* **2021**, *9*, 461. https://doi.org/10.3390/microorganisms9020461

Academic Editor: James Stuart

Received: 14 January 2021
Accepted: 18 February 2021
Published: 23 February 2021

1. Introduction

Countries in the meningitis belt of sub-Saharan Africa have been repeatedly devastated by meningitis epidemics since the early 1900s. Primarily, these epidemics are caused by the bacterium *Neisseria meningitidis* and a number of circulating meningococcal serogroups are responsible for causing disease in the meningitis belt [1]. Until 2010, the predominant serogroup responsible for frequent epidemic cycles was *N. meningitidis* serogroup A (NmA) [2]. Since the introduction of a tailor made vaccine, MenAfriVac in 2010, over 300 million 1–29 year olds have been vaccinated against NmA, resulting in a more than 99% decline in the number of confirmed group A cases in fully vaccinated populations [3].

We previously developed a transmission dynamic model of NmA with the aim of forecasting the relative benefits of different immunisation strategies [4]. The model highlighted the importance of a long-term vaccination strategy following the introductory mass campaigns of 1–29 year olds. Of the long-term strategies we investigated, a combination strategy of routine immunisation within the Expanded Programme on Immunisation (EPI) together with a mini catch-up, targeting children born after the introductory campaign, was the most effective. After reviewing the model findings and additional comprehensive information from clinical trials, the World Health Organisation's recommendation for the countries of the African meningitis belt is to introduce MenAfriVac into routine immunisation programmes within 5 years after completion of the mass campaigns. The vaccine regimen is a 1-dose schedule given at 9–18 months of age. At the time of introduction into EPI, it is recommended that countries should also include a one-time catch-up campaign to immunise those born since the introductory campaigns [5].

One of the key assumptions in our previous work was that the duration of vaccine induced protection is the same for all ages. Due to limited data at the time, we assumed that MenAfriVac offered protection for an average of 10 years. Since then, several studies have investigated the persistence of vaccine-induced antibodies and the influence of age at vaccination. These studies provide empirical evidence on the duration of the immune response to MenAfriVac, which may be used as a proxy to the duration of protection. Correlates of protection for meningococcal disease are based upon serum bactericidal activity (SBA) [6]. The studies by White et al. [7] and Yaro et al. [8] suggest that vaccine protection is age-dependent and lasts longer for individuals targeted after the age of 2 years or 5 years, respectively. These new studies were consistent in suggesting that the duration may be age-specific, but inconsistent in their estimates of the duration.

The aim of this paper is to investigate the influence of both the duration and age-specificity of vaccine-induced protection on our model predictions and explore how the optimal vaccination strategy may change in the long-term. More specifically, we consider four scenarios that could be plausible alternative strategies to the current.

2. Materials and Methods

2.1. Model Structure

Details of the model structure have been previously published [4]. In brief, it is a compartmental model that divides the population into: susceptible state, carrier of NmA, disease due to NmA, and recovered and immune, with each of these states replicated for vaccinated and unvaccinated.

In a modification for this paper, instead of having broad age groups, we now divided the population into annual age cohorts. A model modification was also necessary in order to simulate an age-dependent duration of protection. We added four new compartments to the previous model. These four states represent the susceptible, carriers, diseased, and recovered/immune who receive vaccination before the age of five years. A table with all of the compartments and their descriptions can be found in Appendix A, together with a flow diagram of the model.

2.2. Model Parameters

Demographic data for Chad were used to estimate parameters for the model. Epidemics of NmA in Chad in the pre-vaccine era occurred every 8–12 years, which is representative of the epidemiology of NmA in the African meningitis belt [9]. The introduction of MenAfriVac in that country was completed in two phases during 2011 and 2012 [10]. In the model, because there is no geographic sub-division, we assume that 50% of the target population were vaccinated in 2011 while the remaining 50% received the vaccine in 2012.

We assumed that the coverage in Chad for routine immunisation of infants starts at 75% in 2017 and continues with annual increments of 1% until it reaches 90%. It is then assumed to stay constant for the remaining years. There is no vaccine currently being administered at 5 or 10 years of age. Hence, we explored the impact of vaccinations, assuming a coverage of 80% at these ages. Other parameters were based on the available literature wherever possible, as previously described [4].

To account for the uncertainty around the duration of protection, we ran each scenario outlined below under the following four different assumptions: (1) an average of 5 years duration of protection for all ages; (2) 10 years duration of protection for all ages; (3) 20 years duration of protection for all ages; and (4) 5 years duration of protection for <5 year olds and 10 years duration of protection for children at 5 years of age or older.

2.3. Vaccination Strategies

We considered a range of vaccination strategies (Table 1) that were elucidated through informal discussions with colleagues at WHO, PATH, and CDC to be of interest.

Table 1. Vaccination strategies considered in the model.

Strategy	Introduction	Catch-Up Campaign	Routine Immunisation	Assumed Coverage for EPI
EPI@12m (Current strategy)	2011–2012: 1–29 years old	2017: 1–6 years old	2017–2060: at 12 months	75% in 2017 and annual increments of 1% until it reaches 90%
EPI@5y	2011–2012: 1–29 years old	None	2017–2060: at 5 years	80%
EPI@10y	2011–2012: 1–29 years old	None	2022–2060: at 10 years	80%
Booster	2011–2012: 1–29 years old	None	2017–2060: at 12 months and 5 years	• 75% in 2017 and annual increments of 1% until it reaches 90% for 12 month olds • 80% for 5 year olds
Switch	2011–2012: 1–29 years old	2017: 1–6 years old	2017–2021: at 12 months 2022–2026: at 12 months and 5 years 2027–2060: at 5 years	• 75% in 2017 and annual increments of 1% until it reaches 90% for 12 month olds • 80% for 5 year olds

2.4. Model Implementation

The model is run for the time period 2010–2060 using a daily time step. For each model run, the number of cases by age is calculated per year. The average number of cases and the percentage of cases prevented by each of the strategies is calculated over 200 simulation runs per strategy. To account for the uncertainty due to the stochastic nature of the model, 95% confidence intervals were calculated using a student-t distribution as implemented by the *t*-test function in R version 3.4.2 [11].

To allow for a comparison between the different strategies, we report the time to resurgence, median age of infection, and the age distribution of cases for each strategy. As time to resurgence, we define the year in which the number of cases exceeds the threshold of 1 case per 100,000 population following the preventive campaigns. The comparison of each metric is based on non-overlapping confidence intervals. Due to the large range from 10 years to 20 years duration of protection for all ages, we also investigate the effect of all the intermediate years on the time to resurgence in a sensitivity analysis. As an additional measure of efficiency, we calculated the number of people needed to vaccinate (NNV) to prevent one case [12]. We define NNV as the total number of doses administered divided by the total number of cases prevented under each vaccination strategy over the time period under consideration. The total number of doses given for each scenario was calculated by multiplying the total number of people targeted with the assumed age-specific vaccine uptake.

3. Results

3.1. Baseline Scenario (10 Years Duration of Protection)

The model results suggest that if the assumed duration of protection is 10 years for all ages, routine immunization aimed at schoolchildren is not better than routine immunization of 1-year-old children. However, switching the age at vaccination from 12 months to 5 years is the single dose strategy with the lowest average number of cases predicted, albeit there is a 5-year period with two doses. The numerical results for the different strategies are given in Appendix B. The strategy that leads to the largest number of cases averted is the Booster strategy with a 79.3% (CI: 78.7–79.8%) predicted overall reduction, relative to a 66.8% (66.2–67.4%) predicted reduction if the current strategy remains unchanged until 2060, but the NNV is much higher (Table 2).

3.2. Time to Resurgence

The model predicts that when the assumed duration of protection is 12 years or shorter for all ages, a resurgence always follows the initial mass campaigns (Figure 1). The size of the peak as well as the year of resurgence both depend on the schedule and the duration of protection. The longer the duration of protection, the longer the honeymoon period is. No resurgence was seen in model runs (i.e., 100% of the 200 simulations result in elimination) when vaccine-induced protection was assumed to last for an average of 18 years or longer. For an assumed duration of protection of 16 years for all ages, 69.5% of the simulation runs resulted in elimination after the mass campaigns. Summary statistics showing the year the disease incidence exceeds the threshold of one case per 100,000 population for duration of protection between 10 and 20 years can be seen in Table 3 in Appendix B.

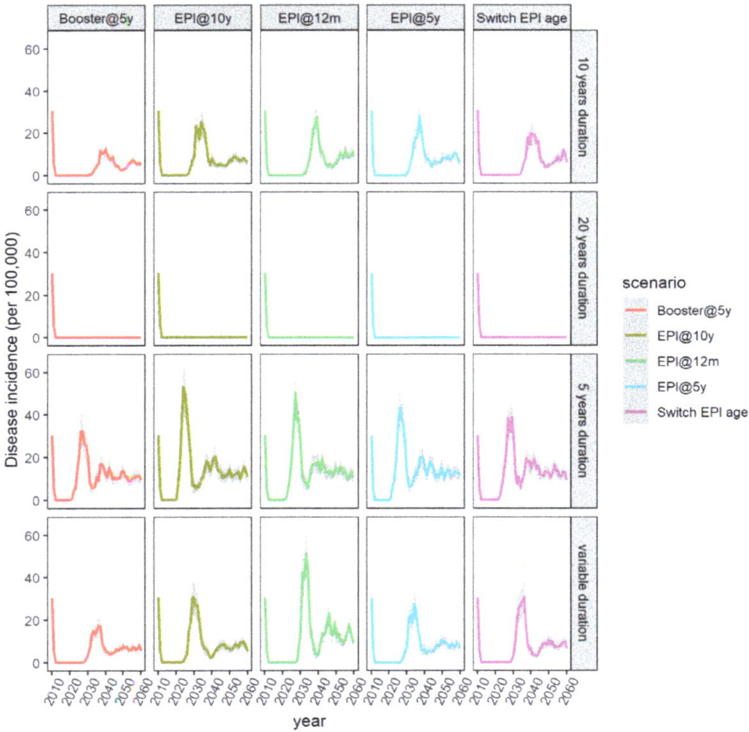

Figure 1. Average disease incidence across the different vaccination scenarios and across the different assumptions regarding the duration of vaccine-induced protection. Shaded areas represent the 95% confidence intervals.

If we assume that duration of protection is 5 years regardless of the age at vaccination, the model predicts that the number of cases will start increasing only 10 years after the introduction of MenAfriVac, compared to ~17 years of honeymoon period when duration of protection is 10 years. Earlier resurgence does not necessarily translate to a larger number of total cases (Figure 2).

3.3. *Burden of Disease*

Of the strategies considered, the Booster strategy resulted in the fewest cases across all different assumptions regarding the duration of protection (Figure 3, Table 2). Taking into consideration only the single-dose schedules, the model results suggest that if the duration of protection is assumed to be the same for everyone regardless of at what age they are targeted, routine immunization at 12 months of age (EPI@12m) is similar to routine immunization at older ages (EPI@5y and EPI@10y). There is considerable overlap in the results but strategy Switch is the strategy with the lowest average number of total cases predicted (Figure 3). However, assuming that vaccination of 1-year-old leads to a shorter duration of antibody persistence compared to vaccination at older ages, strategies EPI@5y and EPI@10y result in a lower number of predicted cases compared to the EPI@12m strategy.

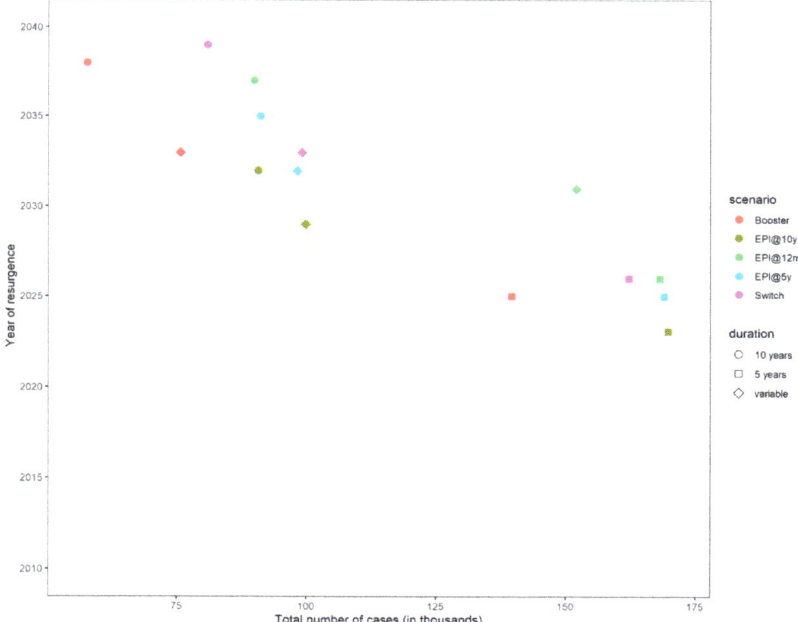

Figure 2. Total number of cases plotted against the year of resurgence across all scenarios and all assumptions regarding duration of protection and coverage. Each strategy is represented with a different colour and each assumption about the duration of protection is represented with a different symbol shape. Note that 20 years duration of protection is not shown. Error bars show the 95% confidence interval.

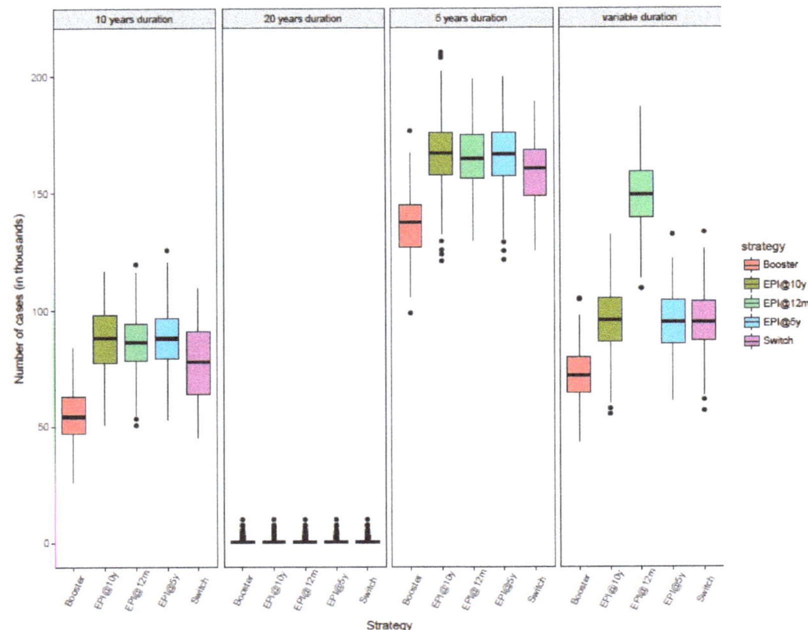

Figure 3. Box plot showing the median, interquartile range, and full range of the predicted total number of cases for different immunisation strategies in the time period 2010–2060 from 200 simulation runs.

Table 2. Number of doses given, number of cases predicted and averted, and number of doses needed to prevent one case for each immunization strategy for the time period 2011–2060. All of the numbers, apart from the number of people needed to vaccinate (NNV), are in the thousands. Averages across 200 simulation runs.

Strategy	Duration of Protection	# of Doses	Cases	Cases Prevented	NNV
EPI@12m	10 years	42,034	86.36	174.39	241
EPI@5y	10 years	33,878	87.62	173.13	196
EPI@10y	10 years	30,024	87.17	173.58	173
Switch	10 years	39,961	77.34	183.41	218
Booster	10 years	63,882	54.09	206.66	309
EPI@12m	20 years	42,034	0.65	260	162
EPI@5y	20 years	33,878	0.65	260	130
EPI@10y	20 years	30,024	0.71	260	115
Switch	20 years	39,961	0.65	260	154
Booster	20 years	63,882	0.65	260	246
EPI@12m	5 years	42,034	164.53	96.22	437
EPI@5y	5 years	33,878	165.41	95.34	355
EPI@10y	5 years	30,024	166.19	94.56	317
Switch	5 years	39,961	158.59	102.16	391
Booster	5 years	63,882	135.92	124.83	512
EPI@12m	Age-specific	42,034	148.6	112.3	374
EPI@5y	Age-specific	33,878	94.66	166.1	204
EPI@10y	Age-specific	30,024	96.28	164.47	183
Switch	Age-specific	39,961	95.56	165.19	242
Booster	Age-specific	63,882	72.09	188.66	339

Vaccination programmes raise the average age of infection since vaccinated children are protected against disease. Routine immunization at 10 years (EPI@10y) is associated with the lowest median age of infection as it results in a large number of unprotected children at a very young age leading to a large number of cases in the under 10-year-olds (Figure 4).

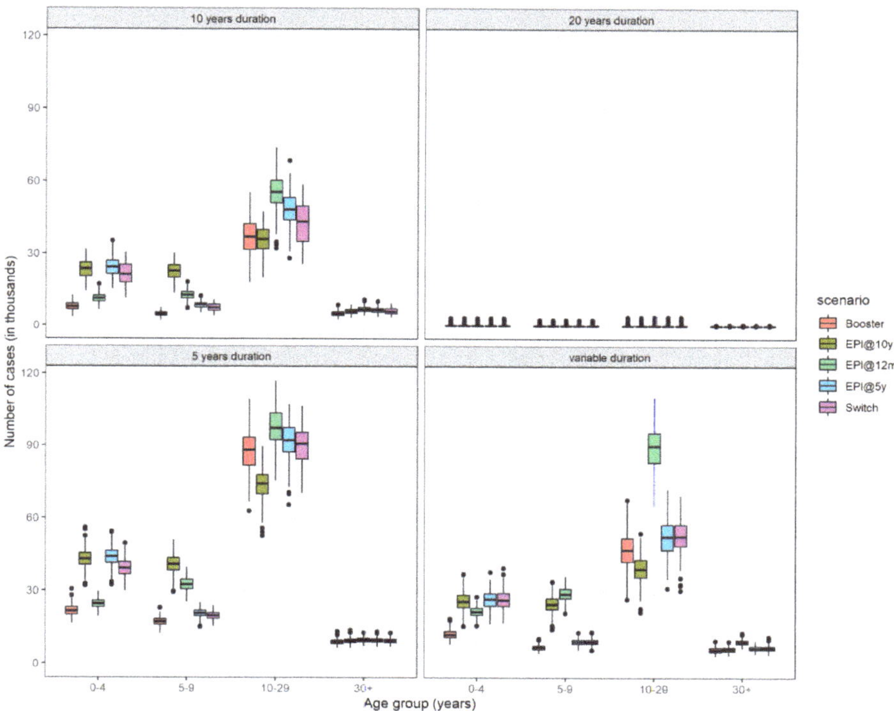

Figure 4. Box plot showing the median, interquartile range, and full range of the total number of cases by age group from 200 simulation runs aggregated over the time period 2010–2060.

The strategy with the highest median age of infection is the routine immunization targeting children on their first birthday, with or without a booster dose when they turn 5 years of age. Anyone can develop invasive meningococcal disease, but rates of disease are higher in children under the age of 5 years [13]. However, carriage prevalence is higher in individuals aged 5–19 years [14]. Routine immunization of 1-year-old children leads to waning of vaccine protection by the teenage years, when there is still heightened risk of meningitis. A booster dose extends the protection until the individuals age into a lower risk age group, which in turn results in a decreased transmission.

4. Discussion

At the time of developing our previous model, data on the duration of vaccine-induced protection was limited. We based our assumption of an average of 10 years duration of protection on findings from unpublished trials and expert opinion. The initial mass campaigns in the countries of the meningitis belt started taking place in 2010, but vaccination in children under the age of 12 months did not start before 2016. Here, we update our previous model to take into account findings from two recent studies, suggesting that protection lasts longer in individuals receiving MenAfriVac after the age of two years [6,7].

We used this updated model to assess the impact of a set of new vaccination strategies and compared them to the current strategy followed by African countries, since 2015.

Assuming that the duration of protection is at most 10 years, model results suggest that meningococcal disease cannot be eliminated within the first 50 years after the initial vaccination by the current or new strategies explored. On the contrary, provided that high antibody levels persist for an average of 20 years, all strategies, including the current, result in a possible elimination of NmA cases since there are no predicted cases until at least 2060.

As a long-term strategy, in the absence of any catch-up campaigns, routine vaccination of 10 year olds would lead to the smallest average number of cases. However, including the campaigns, in the case of determining which strategy leads to the least number of cases, assuming that the duration of protection is the same across all ages, no single-dose strategy is superior to the rest as there was considerable overlap in the results. This is due to the mini catch-up campaign, which is part of only the current strategy (EPI@12m) and not the other two (EPI@5y and EPI@10y). The main difference in the results comparing the strategies is in the age distribution of cases. Reductions in the number of cases in one age group results in a rise of cases in another age group. Routine vaccination at 12 months offers better protection in young children, whereas vaccination at older ages reduces disease burden in adolescents and young adults. The risk of developing at least one major sequelae after meningococcal meningitis is higher in children under the age of 5 years [15]. In this study, we do not calculate Disability Adjusted Life Years (DALYs), where an age-specific weight may be appropriate. Assuming that vaccine protection is short-lived in children under the age of 5 years, the model suggests that it would be wiser to change the target age of routine immunization from 12 months to 5 years provided that coverage is at least 50%.

Routine immunization at 10 years of age (EPI@10y) is consistently the most effective strategy across all different assumptions about the duration of vaccine protection in terms of the number of people needed to vaccinate (NNV). This is due to the small number of doses administered, calculated based on Chad's population demography. The high annual growth rate of the country results in a triangle-shaped age distribution with the number of individuals declining with age. The strategy associated with the highest NNV is the strategy with the additional booster dose since the number of doses is almost double that of the rest of the strategies. NNV is widely used in the scientific literature. The nature of the disease (endemic, epidemic, high/low R_0) as well as the way NNV is calculated can produce biased results [12] and, thus, caution should be taken when interpreting results or comparing NNVs with other diseases in the scientific literature. However, the highest NNV value of 485 produced by the simulations for the Booster strategy is far superior to NNV 2800–3700 estimated by Trotter et al. [16] when evaluating the response thresholds for reactive vaccination campaigns.

This is the first model to explore the potential benefits of targeting schoolchildren for routine immunization with MenAfriVac. As in all mathematical models, there is uncertainty around the model structure and certain key model parameters. The results from this work were generated using demographic data from Chad, a country lying entirely in the meningitis belt and which suffered from epidemics every 8 to 12 years before the introduction of MenAfriVac in 2011 [9]. The same structure is used to model different countries across the belt; here, we chose Chad as a typical example, but given that country-specific demography is not substantially different, we believe the results are more broadly generalisable to other meningitis belt countries. A number of key parameters, such as the transmission rate and the duration of natural immunity remain unknown; therefore, were kept the same as in the original study, allowing for a more direct comparison. Mixing parameters are also important in age-structured models. The carriage prevalence produced by our model is consistent with contact studies in Africa, in which the highest intensity of contacts is observed in 5–15 year olds [17].

We used several metrics to compare the different strategies qualitatively and quantitatively. None of the strategies explored in this work is superior in all respects. This is especially true when vaccine induced protection is the same regardless of the age at vaccination. Immunising infants (EPI@12m) offers protection to young children and raise the median age of infection. However, the NNV to prevent one case is higher than the NNV to prevent one case when EPI targets 10-year-olds (EPI@10y). Leaving children up to the age of 10 years unprotected, however, results in more cases in younger ages and less in older age groups. The Booster strategy may result in the least number of cases but it is the most costly intervention since it needs two doses and therefore we assume approximately double the cost of the others. A possible change in the current immunization schedule would have to be based in the prioritization of all the above factors.

The uncertainty around the assumptions regarding the duration of protection was also explored in another mathematical model forecasting the impact of MenAfriVac vaccination by Jackson et al. [18]. In their study, they mainly focused on updating and validating their previous model in light of newly data [19]. In contrast to our work, Jackson et al. assumed that routine vaccination solely targets 9 months old children. They also explored the benefits of adding a booster dose at 10 years of age in a sensitivity analysis. Despite their structural differences, both models highlight the critical need for a long-term immunization strategy to sustain low levels of infection as well as the importance of continuous updating of models when new data become available.

Since the start of immunization with MenAfriVac, there has been an increased disease incidence caused by serogroups other than serogroup A. A new pentavalent vaccine is being currently developed with the expectation of licensure by end of 2022 [20]. In order to estimate the impact of introducing this new pentavalent vaccine in an already vaccinated population, a more robust study, including a multi-serogroup model, should be performed. This will involve a number of new unknown parameters and further increase the complexity of the model structure. Yaesoubi et al. [21] developed a transmission dynamic model to investigate the cost-effectiveness of alternative vaccination strategies using the novel multivalent vaccine. They concluded that the inclusion of a catch-up campaign with the novel vaccine would be a cost-effective way to further reduce the meningococcal disease burden.

Despite the limitations of this work, and the uncertainty surrounding the introduction of the pentavalent vaccine in the countries of the African meningitis belt, this analysis and the conclusions drawn can be used in the future by policymakers to understand the importance of the duration of vaccine protection and support decision making around vaccine scheduling, such as a shift to routine immunization at an older age or the addition of a booster dose. This change can either be the addition of a booster dose at a later age or simply the age of the primary dose. The aim of this study is to identify the optimal way to maintain the success of MenAfriVac in reducing the number of MenA cases in the long-term. Additional work on the feasibility and cost-effectiveness of policy changes is also essential. In the future, with the advent and rollout of affordable multivalent vaccines, protection against NmA and other serogroups will be enhanced.

5. Conclusions

Models can be useful in investigating a range of assumptions and a variety of vaccine strategies. Further empirical studies of the duration of protection (or the duration of the immune response) following MenAfriVac will help to decrease uncertainty about the optimal vaccination policy.

Author Contributions: Conceptualization, A.K.P. and C.L.T.; methodology, A.K.P., A.J.K.C., C.L.T.; model coding, A.K.P.; formal analysis, A.K.P.; writing—original draft preparation, A.K.P.; writing—review and editing, A.K.P., A.J.K.C., C.L.T.; visualization, A.K.P.; supervision, A.J.K.C., C.L.T.; funding acquisition, C.L.T. All authors have read and agreed to the published version of the manuscript.

Funding: This research was funded by the Vaccine Impact Modelling Consortium (www.vaccineimpact.org). VIMC is jointly funded by Gavi, the Vaccine Alliance, and by the Bill & Melinda Gates Foundation. The views expressed are those of the authors and not necessarily those of the Consortium or its funders.

Acknowledgments: We thank Antoine Durupt, James Stuart, Andre Bita, and Marie-Pierre Preziosi for comments that helped to shape our modelling work.

Conflicts of Interest: Caroline L. Trotter reports receiving a consulting payment from GlaxoSmithKline in 2018, outside the submitted work. Other authors declare no conflict of interest. The funders had no role in the design of the study; in the collection, analyses, or interpretation of data; in the writing of the manuscript, or in the decision to publish the results.

Appendix A

Table A1. List of the model compartments and their definitions.

Compartment Name	Definition
S	Susceptible individuals not vaccinated
C	Carriers of NmA not vaccinated
I	Individuals with invasive disease not vaccinated
R	Immune after colonization or disease not vaccinated
SE	Susceptible individuals vaccinated before the age of 5 years
CE	Carriers of NmA vaccinated before the age of 5 years
IE	Diseased individuals vaccinated before the age of 5 years
RE	Immune after colonization or disease vaccinated before the age of 5 years
SV	Susceptible individuals vaccinated after the age of 5 years
CV	Carriers of NmA vaccinated after the age of 5 years
IV	Diseased individuals vaccinated after the age of 5 years
RV	Immune after colonization or disease vaccinated after the age of 5 years

People are born in the susceptible compartment (S). Children vaccinated up to the age of 5 years are transferred to the SE, CE, IE, and RE compartments while individuals who are targeted at 5 years or older are moved to the SV, CV, IV, RV compartments accordingly. For example, during the initial mass campaigns, children in the age groups 1–2 years, 2–3 years, 3–4 years, and 4–5 years are transferred to the early vaccination compartments (SE, CE, IE, RE) while individuals between 5 and 29 years of age are moved to the vaccinated compartments SV, CV, IV, and RV. Note that there is no movement to the IE or IV compartments upon vaccination as we assume that individuals with meningitis do not receive a vaccine dose. Individuals in the vaccinated states (SE, SV, CE, CV, IE, IV, RE, RV) revert to the equivalent unvaccinated S, C, I, R states at the age-specific rates w_1 and w_2, depending on the strategy implemented (Figure A1). When duration of protection is the same for all ages, then $w_1 = w_2$.

With the addition of the extra compartments, the force of infection for age group j becomes

$$\lambda_j = \theta \sum_{k=1}^{100} \beta(z_j, z_k)(I_k + C_k + IV_k + CV_k + IE_k + CE_k) \quad \text{(A1)}$$

where θ is the stochastic term, which changes annually, and was previously described [4] and $\beta(z_j, z_k)$ is the transmission rate between age groups j and k.

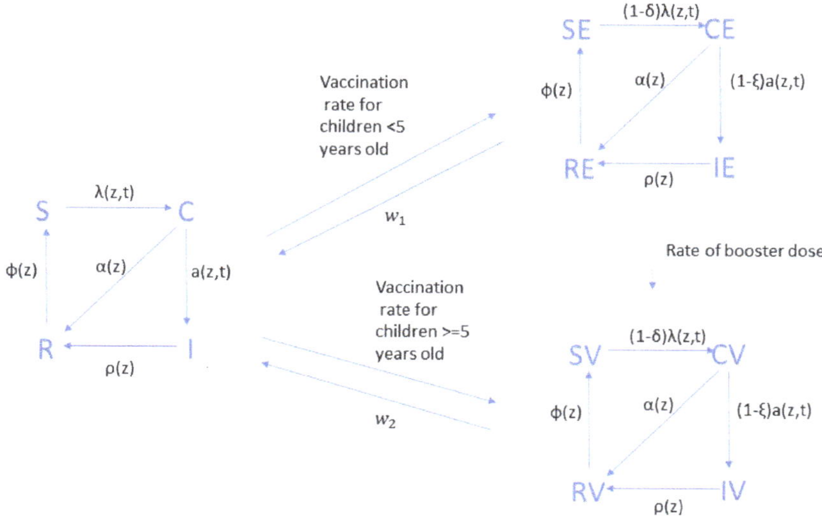

Figure A1. Flow diagram of the model with vaccination. Susceptible individuals become carriers with age and time dependent force of infection (λ(z,t)), which is reduced by the vaccine efficacy against carriage (δ) for vaccinated people. Similarly, the age and time dependent rate at which carriers develop disease (a(z,t)) is reduced by the vaccine efficacy against disease (ξ). Carriers and diseased individuals recover at a rate α and ρ, respectively. Temporary immunity wanes at a rate φ, while vaccine induced protection wanes at a rate w_1 for children vaccinated before the age of 5 years and w_2 for people vaccinated after their 5th birthday. People die at an age-specific natural mortality rate not shown here.

Appendix B

Table 2. Numerical results for the different vaccination scenarios for the time period 2010–2060. Duration of protection is 10 years and vaccine uptake for children routinely immunized over the age of 12 months is assumed to be 80%. Each value presented is the mean and 95% confidence interval is given inside the brackets.

Outcome	No Vaccination	EPI@12m	EPI@5y	EPI@10y	Switch	Booster
Total number of cases (in thousands)	260.7 (258–263.4)	86.3 (84.6–88.1)	87.6 (85.8–89.3)	87.1 (85.1–89.2)	77.3 (75.1–79.5)	54.1 (52.5–55.6)
Cases averted (in thousands)	-	174.4 (171.8–176.9)	173.1 (170.7–175.5)	173.5 (171.1–176)	183.4 (181–185.8)	206.6 (204.3–209)
% of cases averted	-	66.8 (66.2–67.4)	66.4 (65.8–67)	66.6 (65.9–67.2)	70.4 (69.7–71.1)	79.3 (78.7–79.8)
Year of resurgence	-	2032	2030	2027	2034	2031
Total number of doses given (in millions)	-	42.03	33.87	30.02	39.96	63.88
NNV	-	241	196	173	218	309

Table 3. Summary statistics showing the year disease incidence exceeds the threshold of 1 case per 100,000 population from 200 simulation runs for a range of values for the duration of protection. The scenario simulated to generate these results is the EPI@12m.

Duration of Protection	11 Years	12 Years	13 Years	14 Years	15 Years	16 Years	17 Years	18 Years *
Minimum	2033	2035	2037	2040	2044	2047	2056	-
1st Quartile	2038	2041	2044	2048	2053	2053	2057	-
Median	2040	2043	2047	2051	2056	2057	2058	-
Mean	2040	2043	2047	2051	2055	2056	2058	-
3rd Quartile	2042	2045	2049	2054	2058	2059	2058	-
Maximum	2049	2054	2059	2060	2060	2060	2060	-
# of runs leading to elimination	0	0	0	3	58	139	195	200
% of runs leading to elimination	0	0	0	1.5	29	69.5	97.5	100

* All 200 simulations resulted in elimination.

References

1. Greenwood, B. Meningococcal meningitis in Africa. *Trans. R. Soc. Trop. Med. Hyg.* **1999**, *93*, 341–353. [CrossRef]
2. Lingani, C.; Bergeron-Caron, C.; Stuart, J.M.; Fernandez, K.; Djingarey, M.H.; Ronveaux, O.; Schnitzler, J.C.; Perea, W.A. Meningococcal Meningitis Surveillance in the African Meningitis Belt, 2004–2013. *Clin. Infect. Dis.* **2015**, *61*, S410–S415. [CrossRef] [PubMed]
3. Trotter, C.L.; Lingani, C.; Fernandez, K.; Cooper, L.V.; Bita, A.; Tevi-Benissan, C.; Ronveaux, O.; Marie-Pierre, P.; Stuart, J.M. Impact of MenAfriVac in nine countries of the African meningitis belt, 2010–2015: An analysis of surveillance data. *Lancet Infect. Dis.* **2017**, *17*, 867–872. [CrossRef]
4. Karachaliou, A.; Conlan, A.J.K.; Preziosi, M.-P.; Trotter, C.L. Modeling Long-term Vaccination Strategies with MenAfriVac in the African Meningitis Belt. *Clin. Infect. Dis.* **2015**, *61* (Suppl. 5), S594–S600. [CrossRef] [PubMed]
5. World Health Organization (WHO). Meningococcal A conjugate vaccine: Updated guidance, February 2015. *Wkly. Epidemiol. Rec.* **2015**, *90*, 57–62.
6. Frasch, C.E.; Borrow, R.; Donnelly, J. Bactericidal antibody is the immunologic surrogate of protection against meningococcal disease. *Vaccine* **2009**, *27*, B112–B116. [CrossRef] [PubMed]
7. White, M.; Idoko, O.; Sow, S.; Diallo, A.; Kampmann, B.; Borrow, R.; Trotter, C. Antibody kinetics following vaccination with MenAfriVac: An analysis of serological data from randomised trials. *Lancet Infect. Dis.* **2019**, *19*, 327–336. [CrossRef]
8. Yaro, S.; Njanpop Lafourcade, B.-M.; Ouangraoua, S.; Ouoba, A.; Kpoda, H.; Findlow, H.; Tall, H.; Seanehia, J.; Martin, C.; Ouedraogo, J.-B.; et al. Antibody Persistence at the Population Level 5 Years after Mass Vaccination with Meningococcal Serogroup A Conjugate Vaccine (PsA-TT) in Burkina Faso: Need for a Booster Campaign? *Clin. Infect. Dis.* **2019**, *68*, 435–443. [CrossRef] [PubMed]
9. Daugla, D.M.; Gami, J.P.; Gamougam, K.; Naibei, N.; Mbainadji, L.; Narbé, M.; Toralta, J.; Kodbesse, B.; Ngadoua, C.; Coldiron, M.E.; et al. Effect of a serogroup A meningococcal conjugate vaccine (PsA–TT) on serogroup A meningococcal meningitis and carriage in Chad: A community study. *Lancet* **2014**, *383*, 40–47. [CrossRef]
10. Meningitis Vaccine Project: Timeline. Available online: https://www.meningvax.org/timeline.html (accessed on 30 December 2020).
11. t.test Function | R Documentation. Available online: https://www.rdocumentation.org/packages/stats/versions/3.6.2/topics/t.test (accessed on 30 December 2020).
12. Hashim, A.; Dang, V.; Bolotin, S.; Crowcroft, N.S. How and why researchers use the number needed to vaccinate to inform decision making—A systematic review. *Vaccine* **2015**, *33*, 753–758. [CrossRef]
13. Soeters, H.M.; Diallo, A.O.; Bicaba, B.W.; Kadadé, G.; Dembélé, A.Y.; Acyl, M.A.; Nikiema, C.; Sadji, A.Y.; Poy, A.N.; Lingani, C.; et al. Bacterial Meningitis Epidemiology in Five Countries in the Meningitis Belt of Sub-Saharan Africa, 2015–2017. *J. Infect. Dis.* **2019**, *220*, S165–S174. [CrossRef] [PubMed]
14. Cooper, L.V.; Kristiansen, P.A.; Christensen, H.; Karachaliou, A.; Trotter, C.L. Meningococcal carriage by age in the African meningitis belt: A systematic review and meta-analysis. *Epidemiol. Infect.* **2019**, *147*. [CrossRef] [PubMed]
15. Edmond, K.; Clark, A.; Korczak, V.S.; Sanderson, C.; Griffiths, U.K.; Rudan, I. Global and regional risk of disabling sequelae from bacterial meningitis: A systematic review and meta-analysis. *Lancet Infect. Dis.* **2010**, *10*, 317–328. [CrossRef]
16. Trotter, C.L.; Cibrelus, L.; Fernandez, K.; Lingani, C.; Ronveaux, O.; Stuart, J.M. Response thresholds for epidemic meningitis in sub-Saharan Africa following the introduction of MenAfriVac®. *Vaccine* **2015**, *33*, 6212–6217. [CrossRef] [PubMed]
17. le Polain de Waroux, O.; Cohuet, S.; Ndazima, D.; Kucharski, A.J.; Juan-Giner, A.; Flasche, S.; Tumwesigye, E.; Arinaitwe, R.; Mwanga-Amumpaire, J.; Boum, Y., II; et al. Characteristics of human encounters and social mixing patterns relevant to infectious diseases spread by close contact: A survey in Southwest Uganda. *BMC Infect. Dis.* **2018**, *18*, 172. [CrossRef]

18. Jackson, M.L.; Diallo, A.O.; Médah, I.; Bicaba, B.W.; Yaméogo, I.; Koussoubé, D.; Ouédraogo, R.; Sangaré, L.; Mbaeyi, S.A. Initial validation of a simulation model for estimating the impact of serogroup A Neisseria meningitidis vaccination in the African meningitis belt. *PLoS ONE* **2018**, *13*, e0206117. [CrossRef]
19. Tartof, S.; Cohn, A.; Tarbangdo, F.; Djingarey, M.H.; Messonnier, N.; Clark, T.A.; Kambou, J.L.; Novak, R.T.; Diomande, F.; Medah, I.; et al. Identifying Optimal Vaccination Strategies for Serogroup A Neisseria meningitidis Conjugate Vaccine in the African Meningitis Belt. *PLoS ONE* **2013**, *8*, e63605. [CrossRef]
20. Borrow, R.; Caugant, D.A.; Ceyhan, M.; Christensen, H.; Dinleyici, E.C.; Findlow, J.; Glennie, L.; von Gottberg, A.; Kechrid, A.; Moreno, J.V.; et al. Meningococcal disease in the Middle East and Africa: Findings and updates from the Global Meningococcal Initiative. *J. Infect.* **2017**, *75*, 1–11. [CrossRef] [PubMed]
21. Yaesoubi, R.; Trotter, C.; Colijn, C.; Yaesoubi, M.; Colombini, A.; Resch, S.; Kristiansen, P.A.; LaForce, F.M.; Cohen, T. The cost-effectiveness of alternative vaccination strategies for polyvalent meningococcal vaccines in Burkina Faso: A transmission dynamic modeling study. *PLoS Med.* **2018**, *15*, e1002495. [CrossRef] [PubMed]

microorganisms

MDPI

Review

A Narrative Review of the W, X, Y, E, and NG of Meningococcal Disease: Emerging Capsular Groups, Pathotypes, and Global Control

Yih-Ling Tzeng [1] and David S. Stephens [1,2,*]

[1] Division of Infectious Diseases, Department of Medicine, Emory University School of Medicine, Atlanta, GA 30322, USA; ytzeng@emory.edu

[2] Department of Microbiology and Immunology, Emory University School of Medicine, Atlanta, GA 30322, USA

* Correspondence: dstep01@emory.edu; Tel.: +404-727-8357

Abstract: *Neisseria meningitidis*, carried in the human nasopharynx asymptomatically by ~10% of the population, remains a leading cause of meningitis and rapidly fatal sepsis, usually in otherwise healthy individuals. The epidemiology of invasive meningococcal disease (IMD) varies substantially by geography and over time and is now influenced by meningococcal vaccines and in 2020–2021 by COVID-19 pandemic containment measures. While 12 capsular groups, defined by capsular polysaccharide structures, can be expressed by *N. meningitidis*, groups A, B, and C historically caused most IMD. However, the use of mono-, bi-, and quadrivalent-polysaccharide-conjugate vaccines, the introduction of protein-based vaccines for group B, natural disease fluctuations, new drugs (e.g., eculizumab) that increase meningococcal susceptibility, changing transmission dynamics and meningococcal evolution are impacting the incidence of the capsular groups causing IMD. While the ability to spread and cause illness vary considerably, capsular groups W, X, and Y now cause significant IMD. In addition, group E and nongroupable meningococci have appeared as a cause of invasive disease, and a nongroupable *N. meningitidis* pathotype of the hypervirulent clonal complex 11 is causing sexually transmitted urethritis cases and outbreaks. Carriage and IMD of the previously "minor" *N. meningitidis* are reviewed and the need for polyvalent meningococcal vaccines emphasized.

Keywords: *Neisseria meningitidis*; capsule; meningococcal group; nongroupable; meningococcal carriage; invasive meningococcal disease; meningococcal urethritis

Citation: Tzeng, Y.-L.; Stephens, D.S. A Narrative Review of the W, X, Y, E, and NG of Meningococcal Disease: Emerging Capsular Groups, Pathotypes, and Global Control. *Microorganisms* **2021**, *9*, 519. https://doi.org/10.3390/microorganisms9030519

Academic Editor: James Stuart

Received: 5 February 2021
Accepted: 26 February 2021
Published: 3 March 2021

Publisher's Note: MDPI stays neutral with regard to jurisdictional claims in published maps and institutional affiliations.

1. Introduction

Neisseria meningitidis (the meningococcus), a Gram-negative pathogen of humans, causes epidemic meningitis and rapidly fatal sepsis in many parts of the world. *N. meningitidis* is usually a commensal inhabitant of the human respiratory tract, is isolated from the nasopharynx of 3–20% of healthy individuals in the absence of outbreaks or crowding [1,2] and is transmitted from person to person by close contact of large aerosolized droplets or with oral or nasal secretions. Recent studies have suggested a global decline in overall meningococcal carriage.

For both children and adults, carriage of *N. meningitidis* can be an immunizing event, resulting in systemic protective immune response. While in most instances, the acquisition of meningococci in the upper respiratory tract does not result in invasive disease, invasive meningococcal disease (IMD), even with antibiotic therapy and supportive care, has a mortality rate that remains at 10–15%. Factors determining the establishment of carriage versus the development of invasive meningococcal disease following acquisition include expression of capsule and other bacterial virulence determinants (reflected in virulent genotypes, such as clonal complex (cc) 5 and cc7 (group A), cc41/44, cc32, cc18, cc269, cc8, and cc35 (group B), and cc11 (group C)) and host susceptibility.

The meningococcus often produces a capsular polysaccharide (CPS) in which structural differences form the basis of the historic serogroup typing system. Although there are 12 capsular groups expressed by *N. meningitidis* and defined genetically [3], three groups, A, B, and C, have been associated with significant invasive disease. Group A (MenA) *N. meningitidis* expressing a homopolymeric ($\alpha1 \rightarrow 6$) *N*-acetylmannosamine 1-phosphate capsule caused large pandemic outbreaks globally through much of the 20th century that persisted especially in the meningitis belt of sub-Saharan Africa until introduction of the MenA conjugate vaccine, MenAfriVac, in 2010. Meningococci of groups B and C (MenB and MenC), which are ($\alpha2 \rightarrow 8$)- and ($\alpha2 \rightarrow 9$)-linked homopolymers of sialic acid (*N*-acetylneuraminic acid), respectively, cause clusters or local outbreaks (MenC) or localized longer outbreaks and hyperendemic disease (MenB) throughout the world and have been responsible for most sporadic meningococcal disease in developed countries. The "minor" groups, X, Y, and Z were first identified by Slaterus in 1961 and groups E (29E) and W (W135) were identified in 1968. The group Y (MenY) capsular polymer is composed of alternating D-glucose and partially O-acetylated sialic acid, while the group W (MenW) capsular polysaccharide is an alternating D-galactose and partially O-acetylated sialic acid. MenY and MenW have emerged as groups causing epidemic outbreaks and global disease since 1995. Group X meningococci (MenX) expressing homopolymer of ($\alpha1 \rightarrow 4$) *N*-acetylglucosamine 1-phosphate [4] also now cause outbreaks and endemic disease in parts of sub-Saharan Africa. Meningococcal group E (MenE) expresses a capsule consisting of alternating D-galactosamine and 2-keto-3-deoxyoctulosonate (KDO) residues [5] and nongroupable (MenNG) strains, either with capsule null locus (*cnl*) or unencapsulated due to inactivation of capsule synthesis, rarely but are now identified as a cause of invasive disease especially in immunocompromised individuals. Recently an unencapsulated cc11 meningococcal pathotype causing sexually transmitted urethritis cases, outbreaks, and disease clusters has also been recognized [6–8].

Meningococcal disease incidence has historically been cyclical in nature [9]. Epidemiological studies of *N. meningitidis* have clearly shown that IMD varies over time, influenced by the circulating meningococcal groups and clonal complex genotypes, by geographic locations and by the host populations affected [10]. In this report, we provide an overview of carriage and invasive disease caused by "minor" *N. meningitidis* groups W, X, Y, and E as well as nongroupable meningococci causing invasive disease and sexually transmitted infections. During the COVID-19 pandemic, many countries have experienced a significant, sustained reduction in invasive diseases due to *N meningitidis* [11] and other invasive respiratory pathogens (*Hemophilus influenzae* and *Streptococcus pneumoniae*, but not *S. agalactiae*), coinciding with the introduction of COVID-19 containment measures (medRxiv preprint, https://doi.org/10.1101/2020.11.18.20225029, access on 20 November 2020).

2. Minor *N. meningitidis* Capsular Groups

2.1. Group W

Group W meningococci (MenW) prior to 2000 was an infrequent cause of meningococcal disease. The first global epidemic caused by MenW was detected in 2000, after the Hajj pilgrimage to Mecca, Saudi Arabia [12] and was attributed to a specific MenW:cc11 lineage, referred to as the Hajj lineage [13]. The MenW attack rate was high among the pilgrims and household contacts of returning pilgrims. The emergence of MenW in a background of the hypervirulent cc11 lineage was likely related to capsule switching events [14]. Subsequently, MenW strains have continued to evolve and cause global disease. The global spread of diverse cc11 lineages expressing capsular groups C, B, and W, resolved by genomic typing, divided the invasive MenW:cc11 lineage into the Hajj and the South American sublineages [14]. The South American sublineage, with the spread in Brazil, Argentina, and Chile, emerged in the mid-2000s [15] and has been further divided into the original U.K. 2009 lineage [14] and the newly emerging novel U.K. 2013 lineage [16].

Surveillance data from 13 European countries revealed an increase in MenW IMD in the period of 2013–2017 [17]. While the annual incidence of IMD remained stable during that time, the incidence and the proportion of MenW IMD among all IMD increased significantly. Average annual percentage increase in MenW incidence during this period was significant for the Netherlands (133%), Norway (86%), Spain (62%), Sweden (58%), Switzerland (44%), Germany (35%), and England (23%). The proportion of MenW among all IMD cases varied considerably between countries. The proportion of MenW was lowest in Portugal, Greece, and Poland (2–3%), while it was highest in Switzerland, the Netherlands, and England (22–24%) [17]. Of the MenW IMD isolates analyzed by multilocus sequence typing (MLST), 80% belonged to cc11 but cc22, cc174, and cc865 also caused disease. The proportion of MenW cc11 increased from 64% in the year 2013 to 86% in the year 2016. The increase in MenW IMD in England and Wales since 2009 has been mainly due to the novel U.K. 2013 lineage [16]. MenW IMD incidence, with an associated case fatality rate of 28.6%, increased from 0.02/100,000 in 2013 to 0.29/100,000 in 2017 in the Republic of Ireland and the Ireland MenW isolates clustered among both the original UK 2009 and the novel U.K. 2013 lineages [18]. Sweden had a low incidence of MenW IMD with an average incidence of 0.03 case/100,000 population from 1995 to 2014; however, the incidence of MenW increased 5-fold in 2015. This increase in MenW IMD was due to isolates belonging to the novel U.K. 2013 lineage that were introduced into Sweden in 2013 and have since been the dominant lineage of MenW [19]. The increases seen in Europe follows an increasing incidence of MenW in South America since 2004 [15], in Australia since 2013 [20], and in Canada since 2015 [21].

MenW became the predominant meningococcal capsule group in Australia in 2016 [20]. In 2017, an unprecedented outbreak of MenW infection occurred among the Indigenous pediatric population of Central Australia. Among these cases were atypical manifestations, including meningococcal pneumonia, septic arthritis, and conjunctivitis [22]. The Canadian MenW:cc11 isolates have been shown to be distinct from the traditional MenW:cc22. Both the Hajj-related and non-Hajj MenW:cc11 strains were associated with IMD in Canada [21].

A review of IMD in the Asia–Pacific region conducted by Global Meningococcal Initiative (GMI) recently reported that the predominant capsular groups were B, W, and Y in Australia, New Zealand, and China [23]. MenW circulation is significant across the Asia–Pacific region. The Philippines reported that 16.7% of sterile specimens collected in 2018 were MenW. As noted, a higher percentage (28%) of MenW was reported in Australia and a similar proportion (30%) of MenW cases was reported in New Zealand during the same period. An update on the global spread of cc11 provided during the GMI meeting highlighted the presence of the MenW:cc11 Hajj strain sublineage in Russia and Bangladesh; the MenW:cc11 South American strain sublineage in Russia, Japan, and New Zealand; the MenW:cc11 Chinese strain sublineage in China and Japan; and a further distinct MenW:cc11 strain in Bangladesh [23].

The introduction of the MenAfriVac vaccine in 2010 dramatically reduced MenA cases in 26 countries of the meningitis belt but magnified other groups as significant problems in the region, in particular groups C, W, and X. While MenW:cc11 cases have been reported in the African meningitis belt since the late 1990s and no epidemics have occurred since 2001, MenW:cc11 seems to have reemerged after 2010 [24]. In 2016, Togo experienced its second largest epidemic of bacterial meningitis since 1997, where 91.5% were due to MenW:cc11 [25]. The MenW:cc11 isolates collected in Burkina Faso during 2011–2012, Mali during 2012, and Niger during 2015 have been shown to descend from the strain identified during the Hajj-related outbreak of 2000 [26–28]. On the other hand, the MenW:cc11 isolates from Central African Republic in 2015–2016 grouped together in a genetic cluster separated from the Anglo-French Hajj sublineage and the South American/UK sublineage. These data appear to support a multifocal emergence of MenW:cc11 strains. The epidemiology of IMD in South Africa over 14 years [29] shows that MenW accounts for 49.5% IMD. Patients with MenW were 3 times more likely to present with severe disease than those with MenB, and HIV was associated with an increased risk of IMD, especially for MenW and MenY diseases.

2.2. Group X

Sporadic cases of IMD caused by *N. meningitidis* group X (MenX) have been reported in industrialized countries since 1980s [30–32] but since the late 1990s, MenX has emerged as a cause of IMD outbreaks in sub-Saharan African countries [33–39]. The PubMLST database (>75,000 isolate records, updated on 01/03/2021) contains a collection of 636 MenX isolates, 1961–2019. MenX is the latest group to cause large localized outbreaks in Kenya [36,37], Niger [34,35], Ghana [33], Mali, and Burkina Faso [38]. A study examining MenX burden and epidemiological patterns during 2006–2010 [38] showed that in Togo during 2006–2009, MenX accounted for 16% of the bacterial meningitis cases; while in Burkina Faso during 2007–2010, MenX accounted for 7% of meningitis cases, with a significant increase from 2009 to 2010 (4–35% of all confirmed cases, respectively) [38]. With the successful vaccination campaign of the MenA conjugate vaccine starting in 2010, the significance of MenX in the African Meningitis belt has become more evident. In Burkina Faso, a few months after the introduction of MenAfriVac in 2011, among the 258 confirmed meningococcal cases, only 1.6% were MenA, whereas 59% were MenX [39]. Thus, MenX, along with MenC and MenW disease in the meningitis belt, is a major driver for a new pentavalent conjugate (ACXYW) vaccine in clinical trials for sub-Saharan Africa [40]. Of note, IMD due to MenX (cc750) has also been seen in the United Kingdom (Scotland) and elsewhere in Europe. MenX has also been identified rarely in the hypervirulent cc5, cc11, and cc41/44 backgrounds.

MenX expressing genotypes (X:4) can be efficiently transmitted and colonize the nasopharynx as was seen in military recruits in the United Kingdom [41]. In a longitudinal carriage study investigating the dynamics of meningococcal carriage during an inter-epidemic period in Ghana, the disappearance of MenA was accompanied by a sharp increase in carriage of MenX, reaching 17% and coincided with an outbreak of MenX disease [33,42]. During the peak of the MenX wave, the ratio of MenX cases to carriers was found to be between 0.1 and 0.3 per 1000 cases; while the ratio of MenA during the outbreak was between 16.8 and 42.3 per 1000 cases in the respective dry seasons [42]. These studies suggest that MenX has a disease-to-carriage ratio significantly lower than MenA and that MenX have a lower invasive potential. Like other outbreak-causing meningococci, dominant virulent clones are responsible for the majority of MenX disease. Most MenX carrier and disease isolates recovered in the African meningitis belt belonged to cc181, which has been circulating in Africa since the 1970s [43].

2.3. Group Y

Group Y meningococci (MenY) are frequently recovered from the nasopharynx but have historically considered less invasive than groups A, B, and C [44]. However, in the mid-1990s, the rates of IMD due to MenY increased in the United States [45], and subsequently in several European countries [46–49] as well as Israel, South America, and South Africa. Clonal complexes cc23, cc167, and cc175 have been linked to the majority of MenY IMD, but MenY IMD has also been seen with cc22 (Europe), cc174 (the United Kingdom), cc92 (Europe and South America), and cc103 (Europe).

In the mid-1990s, MenY (cc23 and cc167) emerged as a major cause of significant sporadic and hyperendemic disease in the United States. The proportion of MenY IMD cases in the United States was 2% during 1989–1991 [50], increased to 10.6% in 1992, and increase to 32.6% of reported cases in 1996 [51]. Subsequently, the proportion of MenY cases decreased in the United States [45], although still causing 15% of IMD in 2018. The increase in MenY cases has not been as prominent in neighboring Canada [52]. In 1998 at the peak of MenY incidence in the United States, a carriage study of high school students from counties in the metropolitan area of Atlanta, GA, found the rate of meningococcal carriage to be 7.7% and of these isolates, 48% were MenY [53]. However, in 2006–2007, a similar carriage study in high school students found a much lower carriage rate of <3% and a lower proportion of MenY carriage [54]. Thus, like MenX, high rates of acquisition and carriage were associated with increased disease and lower MenY carriage correlated with the decrease in MenY IMD cases [45].

Meningococcal quadrivalent conjugate vaccines against groups A, C, Y, and W (MenACWY) were licensed in the United States beginning in 2005 and coverage has steadily increased among children aged 13–17 years, from 11.7% in 2006 to 86.6% in 2018. A study comparing group distribution of IMD isolates prior to (2000–2005) and post vaccine introduction (2006–2010) reported that among all age groups, the overall IMD incidence declined over time, but there was no evidence of vaccine-induced capsular group replacement. While the incidence of IMD significantly declined in the United States, the proportion of MenY varied from 33% in 2000–2005 and 37% in 2006–2010 to 27% in 2011–2015 [55,56]. Changes in group and clonal complex were observed in isolates of both vaccine targeted and non-targeted groups. These changing profiles are likely representative of natural variation and fluctuations within meningococcal population structure. As noted, clonal complexes cc23 and cc167 accounted for most of MenY disease in both the United States and Canada [55,57,58], again suggesting that closely related strains circulate at high frequencies in a community causing sporadic disease.

MenY disease has recently emerged in Latin American countries and is characterized by clear differences from country to country [59]. Molecular characterization of MenY IMD isolates during 2000–2006 showed variable trends among 5 countries. While no increase in the frequency of MenY isolates was observed in either Brazil or Chile, the proportion of MenY IMD isolates increased in Argentina from 2002, to a level similar to those of groups C and W by 2006. In Colombia, MenY IMD isolates increased from 4% in 2000 to 50% in 2006 [60]. Venezuela also reported an increase in the proportion of cases due to MenY in 2006, representing 50% of all cases identified [59]. Again, most of the IMD isolates belonged to cc23 and cc167 [61]. Recently, IMD cases caused by penicillin- and ciprofloxacin-resistant cc23 MenY were found in El Salvador. These isolates contained a β-lactamase gene (*blaROB-1*) and a mutated DNA gyrase gene (*gyrA*) [62].

Until the last decade, MenY cases were rare in Europe, accounting for <2% of cases [63]. An emergence of MenY IMD cases was noted in several European countries after 2010. For example, in France, MenY accounted for 3% of IMD cases in 2000 to 2005, but increased to 10% in 2013 [64,65]. In Scotland, MenY IMD cases increase from 2.3% in 2010 to 17% in 2013 [65]. Further, significant increases in the incidence and the relative proportion of MenY IMD cases were found in Scandinavian countries: in Norway, the 4-year trend between 2010 and 2013 for Norway is 31–55–25–26% and in Finland, it was 38–21–24–40% [48,65]. Sweden had the highest relative proportion of MenY IMD in Europe—39% in 2010 and ~50% in the following 3 years [65]. The significantly increased MenY IMD in Sweden is mainly due to the emergence of specific cc23 clusters [47,48]. Whole-genome sequencing data of invasive MenY isolates from 1995 to 2012 in Sweden found at least three related but distinct cc23 clusters causing disease in Sweden. Thus, the increase in MenY IMD cases was not caused by the expansion of a single virulent variant [47], but was linked to increased virulence, host adaptive immunity, and transmission dynamics. Comparison to a collection of MenY isolates from England, Wales, and Northern Ireland during 2010 to 2012, which had relatively low MenY incidence, and MenY from the United States showed that the MenY cc23 clusters have a distribution spanning North America and Europe, including Sweden, over a number of years [47] but different strain types were prevalent in each geographic region.

Several carriage studies conducted in the United Kingdom have detected changes in MenY carriage in young adults over the last three decades [66–69]. MenY constituted approximately 8% of recovered isolates when carriage was assessed during 1997–1998 in first-year university students at Nottingham University, the United Kingdom [66]. During 1999–2001 in >48,000 samplings of 15–17 years old throughout the United Kingdom, MenY strains accounted for ~10% of the carriage isolates [70]. A later 2008 carriage study of first-year students carried out again at Nottingham University, the United Kingdom, found that MenY carriage reached 26% [68]. Core genome analysis of carriage-associated MenY isolates recovered in the United Kingdom during 1997–2010 reveals extensive genetic similarities to disease-associated MenY recovered during 2010–2011 [5]. Again, the majority of these MenY belong to cc23 (58% in carriage and 79% in disease) and a long-term temporal stability of MenY clones was suggested [5]. However, in South Africa, a different clonal complex was responsible for increases in MenY disease. MenY cc175 caused significant IMD and was dominant in South Africa in the early 2000s [58]. MenY is also expressed in cc11, cc32, and cc41/44 clonal complex backgrounds that are more frequently associated with other capsular groups. Interestingly, comparison of IMD cases during a 2-month lockdown period in 2020 and the same periods of 2018 and 2019 in France found significant decrease in all IMD cases from prior years, and seemed to mainly involve IMD cases due to groups B and C and W, but not IMD due to group Y and other groups or nongroupable isolates [11]. The MenY genotypes had not changed in 2020 and were cc23 [11]. The observed IMD decreased mainly in the highly transmissible and hyperinvasive isolates belonging to cc11.

2.4. Group E

IMD due to meningococcal group E (MenE), previously known as 29E and Z′ and first identified in 1968 [71], is infrequent and has been most often associated with immunocompromised patients [71,72]. Query of the PubMLST database shows 1003 MenE isolates in the collection. The vast majority (>75%) were from pharyngeal carriers with cc60 (42%) and cc1157 (29%) dominating. The earliest invasive MenE recorded in PubMLST is in 2000, and are predominantly cc60 (33%), cc1157 (26%), cc254 (8%), and cc178. A recent study reported the molecular characterization of three MenE IMD cases in Queensland, Australia; the emergence of these cases was attributed to a circulating cc1157 clone [73]. Globally as noted, MenE carriage is not uncommon. Historic carriage studies of first-year college students in the United Kingdom in 1997 [66] and young adults in the Czech Republic during 1993 [74] found ~6% and ~5% MenE in the respective isolate collections. A 2008 carriage study of first-year students at Nottingham University, the United Kingdom, found MenE clones highly prevalent (21–32%) in residential halls, indicative of rapid clonal expansion [67]. However, a recent carriage study performed in Australia in 2017 identified a single individual with MenE carriage from 421 first-year university students (0.2%) [75]. In contrast, an ongoing study of meningococcal carriage in participants in an STI clinic, MenE was identified in ~13% (Tzeng et al. unpublished data) of carriage isolates. While no group-specific vaccine is currently available for MenE, the protein-based group B vaccine, MenB-4C (Bexsero), contains outer membrane vesicles with multiple surface antigens that can provide cross-reactive protection. Similar data are available for MenB-FHbp (Trumenba) where bactericidal responses to groups C, W, Y, and X expressing different fHbp peptides have been shown.

2.5. Nongroupable

The meningococcal nongroupable (MenNG) phenotype is a result of elimination or minimal capsule production. Responsible mechanisms include down-regulation of capsule gene expression, phase variation in the capsule synthesis genes, transient or permanent inactivation of genes by insertion element movement into the capsule gene cluster (*cps*), frame-shift point mutations within an otherwise intact biosynthesis genes, or transformation/recombination events resulting in major deletion of the *cps* locus [76–78]. Meningococci with a capsule null locus (*cnl*), similar to *N. gonorrhoeae*/*N. lactamica*-like genetic configuration at the *cps* locus, were first identified in healthy carriers in Germany in 2000 and constituted ~16% of all recovered isolates [78]. Subsequently, additional carriage studies showed that nongroupable *cnl* meningococci are prevalent in carriage [79,80].

MenNG rarely cause invasive disease; however, *cnl* isolates have been described as a cause of IMD in immunocompetent individuals [81–85]. Most invasive *cnl* meningococci belong to cc198 [81–83] and cc192 [84,85], with cc192 being most commonly identified in Africa, but rarely elsewhere in the world [79]. Two cc198 invasive *cnl* isolates from Canada were examined using murine intraperitoneal infection model. Although no mortality was seen upon infection with the non-encapsulated MC58 derivative, 18% succumbed to infection with one *cnl* strain and 50% died after infection with the other *cnl* strain. Thus, although virulence potentials of both *cnl* strains were below that of encapsulated strain MC58, both strains exhibit a virulence phenotype [83].

While case reports in the literature across several decades have indicated that *N. meningitidis* is capable of colonizing the urogenital tract and causing sporadic cases of urethritis, cervicitis, or proctitis, very low overall incidence has been reported [86–89]. In one study of 23 meningococci isolated from the urogenital tract and rectum, two are *cnl* isolates [90]. Another collection of 39 urethritis-associated *N. meningitidis* identified 4 *cnl* isolates and 17 MenNG isolates due to various mutations in the *cps* locus [91]. More recently, a meningococcal clade of cc11.2 lineage (US_NmUC) with a nongroupable phenotype due to deletion of capsule biosynthesis genes has caused unprecedented clusters of meningococcal urethritis in heterosexual men [6,8,92]. As one example, in Columbus, Ohio from 2015 to 2016, ~25% of presumed gonococcal urethral infections were determined to be meningococcal urethritis with clinical presentation mirroring that of gonococcal urethritis [8,92]. Other mucosal infections, e.g., neonatal conjunctivitis, [93] and at least five cases of IMD (meningitis and meningococcemia) were also reported with this clade [6], although it is not known if these patients were immunocompetent.

While urogenital colonization and sporadic cases of urethritis caused by *N. meningitidis* are documented across many genotypes and groups [90,91,94], the US_NmUC is unique in its capability of causing multicity epidemiologically unlinked urethritis clusters and US_NmUC appears to be sexually transmitted, like gonococci [8,92]. These observations suggest that a phylogenetically distinct nongroupable cc11 US_NmUC has emerged as a new urotropic pathotype to cause meningococcal urethritis. Specific signatures universal to the US_NmUC include (1) an IS1301-mediated specific deletion of the group C capsule biosynthesis genes, (2) expression of a unique FHbp ID896 protein [95], and (3) the acquisition of gonococcal NorB-AniA denitrification apparatus [7]. These unique features differ from other urogenital meningococcal isolates, many of which express capsule, encode a frame-shifted *fHbp* allele, and have a meningococcal denitrification pathway [91,94]. Loss of capsule has been demonstrated to enhance colonization at the mucosal surface, confer increased invasion into epithelial cells [96], and facilitate biofilm formation [97,98]. The acquisition of gonococcal denitrification pathway likely contributes to the success of this clade in better adapting to the male urethra.

Enhanced national surveillance and whole genome sequencing analysis of invasive, urogenital and rectal isolates at CDC has identified ~300 isolates from over 13 states to be members of US_NmUC [6]. This is certainly an underestimate due to possible misidentification by *N. gonorrhoeae* diagnostic assay [99]. In 2019, two US_NmUC isolates were reported in MSM in the United Kingdom [100]. One of the UK isolates had acquired a frameshifted gonococcal maltose phosphorylase gene, resulting in a carbohydrate utilization profile more typically associated with gonococci [100]. Ecological separation within the human host is proposed as an explanation for the lower frequency of interspecies recombination noted between naturally competent *N. meningitidis* and *N. gonorrhoeae* [101]. However, among the members of the US_NmUC, the genome content of total length of DNA sequences inferred to have originated from *N. gonorrhoeae* varied substantially from ~5 to ~30 kb [6]. Further, one 2015 isolate had a gonococcal-like *mtrR* allele that is associated with elevated azithromycin MICs [6] and 7 out of 10 isolates recovered during 2018–2019 in St. Louis, MO, are non-susceptible to azithromycin [99]. In addition, intermediate penicillin resistance was seen in the clade, and one UK isolate, having acquired part of a gonococcal DNA gyrase (*gyrA*) gene, was resistant to ciprofloxacin [100]. These multiple recombination events demonstrated widespread acquisition of gonococcal DNA by US_NmUC and suggested that co-colonization of these two species had facilitated genetic exchanges and raised the prospect of further acquisition of gonococcal antibiotic resistance determinants [100].

3. Population Structure of Invasive "Minor" Capsular Groups and Nongroupable *N. meningitidis*

The genetic relationships of W, X, Y, E, and NG capsular groups associated with disease are shown in Figure 1. Genome allelic profile comparisons were made based on the core genome MLST (cgMLST v1.0) scheme with a set of 1605 loci present in ≥95% *N. meningitidis* isolates [102]. The analysis examines 1158 disease-causing isolates: 29 MenE, 575 MenW, 39 MenX, 453 MenY, and 62 MenNG (48 *cnl* and 14 urethritis clade) isolates, compiled by selecting a representative isolate that had a unique country/year/ST combination. Isolates not assigned to a clonal complex or without records of year and country origin were excluded. The resultant minimum spanning tree was visualized by GrapeTree [103], which is integrated into the BIGSdb functionality [104]. As shown, the geographically and temporally diverse collection of disease-causing X, E, *cnl*, and urethritis isolates displayed two distinct major groupings that are dominated by the W:cc11 and Y:cc23 isolates. However, the emergence of multiple distinct clonal lineages—cc11, cc22, cc174, and cc865 in MenW; cc181 and cc750 in MenX; cc60 and cc1157 for MenE; and the distinct cc11 urethritis clade—is evident.

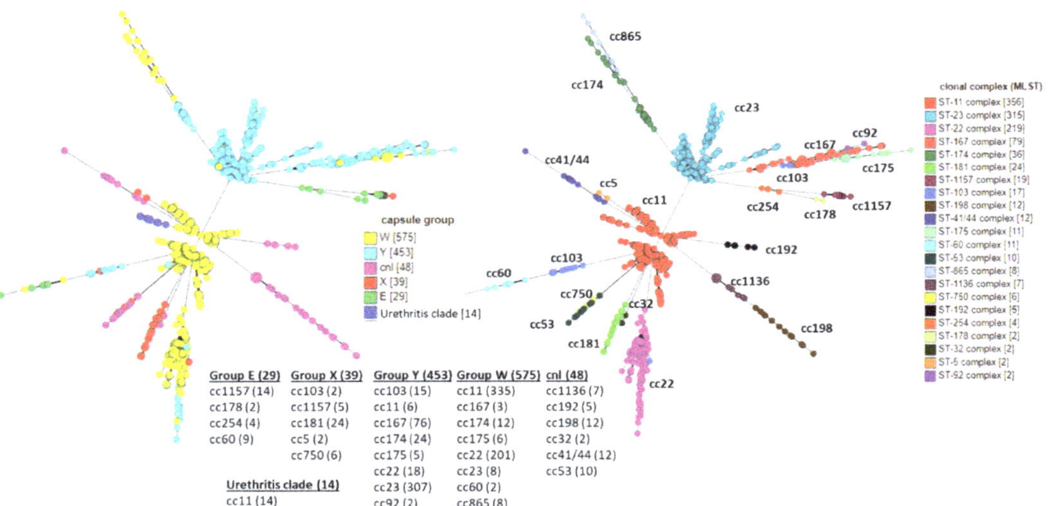

Figure 1. Meningococcal invasive isolates of capsule groups E, W, X, Y, nongroupable *cnl*, and urethritis clade isolates that have whole genome sequencing data and "disease" record entries of "invasive," "meningitis," "septicemia," or "meningitis and septicemia" were retrieved from PubMLST database (http://pubmlst.org/neisseria/, access on 2 March 2021). Isolates without records of "year" and "clonal complex" were excluded. A single isolate that has a unique combination of country/year/sequence type (ST) definition was selected and the clonal complexes with at least two isolates were included in the analysis. The minimum spanning trees were generated and visualized with GrapeTree [103], a plug-in analysis tool at PubMLST, with the scheme of *N. meningitidis* core genome MLST (cgMLST) v 1.0 and default parameters. The trees are colored by capsular groups (**left**) or clonal complexes (**right**). The clonal complex breakdowns of each capsule groups are also listed.

4. Meningococcal Vaccines

Capsular polysaccharides have historically been the targets for group-specific meningococcal vaccine development. The first capsular polysaccharide vaccines were developed in the early 1970s. While a major step forward, they were generally not effective for children less than 2 years and failed to induce long-term memory responses. The subsequent development and introduction of polysaccharide–protein conjugate vaccines in the late 1990s markedly accelerated the prevention of meningococcal disease. Meningococcal polysaccharide–protein conjugate vaccines against groups A, C, Y, and W, developed as monovalent, bivalent, or quadrivalent products, have considerably greater effectiveness than the polysaccharide vaccines and induce immune memory responses [105]. The polysaccharide conjugate vaccines reduce transmission by prevention of meningococcal acquisition, resulting in significant herd or community protection at quite modest levels of vaccine coverage. The introduction of a monovalent MenC conjugate vaccine in 2000 virtually eliminated the incidence of MenC disease in United Kingdom, an effect that has persisted for well over two decades, demonstrating 90% effectiveness at 3 years in 11–18-year-olds [106]. Vaccination against MenC induced herd protection and reduced the rates of MenC carriage and disease in non-vaccinated individuals by more than 50% [107]. A MenACWY conjugate vaccines was first licensed in the United States in early 2005. Additional quadrivalent MenACWY conjugate vaccines were subsequently introduced and are now in use globally (Table 1). The group B capsule has not been developed as a vaccine target given its structural similarity with human polysaccharide antigens. However, two outer membrane protein-based vaccines targeting MenB (Table 1), also with activity against non-MenB *N. meningitidis*, have been licensed and are now in use in meningococcal disease prevention strategies.

Table 1. Meningococcal protein andpolysaccharide conjugate vaccines *.

Vaccine Product	Trade Name	Age Group	Year Licensed
	Polysaccharide Conjugate (Groups A, C, W, and Y)		
MenACWY-D	Menactra	9 months–55 years	2005
MenACWY-CRM	Menveo	≥2 months	2010
MenACWY-TT	MenQuadfi # /Nimenrix	≥1 year/≥6 weeks	2020/2012
MenA-TT	MenAfriVac	3 months–29 years	2010
	Protein based (directed at group B)		
MenB-FHbp	Trumenba	10–25 years	2014
MenB-4C #	Bexsero	≥2 months	2015

Abbreviations: MenACWY-CRM = meningococcal groups A, C, W, and Y capsular polysaccharide-diphtheria CRM_{197} conjugate; -D = diphtheria toxoid conjugate; -TT = tetanus toxoid conjugate vaccine; MenB-4C = four-component meningococcal group B vaccine; MenB-FHbp = meningococcal group B bivalent factor H binding protein vaccine. # MenQuadfi is indicated for ≥2 years in the U.S. Bexsero is licensed for 10–25 years in the U.S. * Monovalent C conjugate vaccines, including Meningitec (MenC-CRM_{197}), Menjugate (MenC-CRM_{197}), NeisVac-C (MenC-TT), and Menitroix (MenC-TT+Hib), are still in use in some countries. Additional conjugate vaccines directed against MenAC and MenC are also available in China. Pentavalent meningococcal conjugate vaccines: MenABCWY (MenACWY-CRM-197 combined with MenB multicomponent recombinant proteins, GlaxoSmithKline), MenABCWY (bivalent FHbp-containing pentavalent vaccine, Pfizer), and MenACXWY (NmCV-5, Serum Institute of India) are in phase 3 clinical trials.

However, meningococcal vaccination strategies with limited capsular group coverage will eventually select for or uncover previously "minor" capsular groups, now causing significant endemic and epidemic meningococcal disease in the last two decades (e.g., groups W, X, and Y). Examples are the Hajj MenW outbreaks and the emergence of MenW disease in South America and Europe. In addition, the dramatic control of MenA disease in the African meningitis belt achieved by the introduction of MenAfriVac in 2010 uncovered outbreaks of MenX and MenW [39]. Due to horizontal gene transfer and recombination *N. meningitidis*, like *Streptococcus pneumoniae*, can undergo capsule structural change, e.g., "capsule switching" [55,108–111] lessening herd immunity. Transformation and homologous recombination of capsule genes with the appearance of otherwise identical MenC strains was first noted during a prolonged MenB outbreak in the 1990s [108]. The MenW outbreaks associated with the Hajj in 2000 may have been the result of a historic capsule switching event from cc11 MenC strains. In large meningococcal isolate collections, capsule switching is detected in ~3% of isolates [112].

Pentavalent meningococcal conjugate vaccines, i.e., MenABCWY (MenACWY-Oligosaccharide diphtheria CRM_{197} conjugate, combined with MenB multicomponent recombinant, GlaxoSmithKline), MenABCWY (bivalent FHbp-containing pentavalent vaccine, Pfizer), and MenACXWY (NmCV-5, Serum Institute of India), are in phase 3 clinical trials and are a next step, if widely implemented, for global control of meningococcal disease. The broad capsule focused coverage together with the MenB protein component(s) can potentially provide protection [113] against other minor disease-causing groups and nongroupable strains [95,114,115].

5. Conclusions

Capsular groups W, X, and Y now cause significant IMD as reflected in the higher numbers of invasive isolates deposited into PubMLST since 2000 (Figure 2) as well as the country and global surveillance data noted above. In addition, group E and nongroupable meningococci have appeared as a cause of invasive disease, and a nongroupable *N. meningitidis* pathotype of the hypervirulent cc11 is causing sexually transmitted urethritis cases and outbreaks. *N. meningitidis* is a human microbe circulating within populations. Due to factors including the introduction of highly effective meningococcal vaccines of limited coverage, the capsular groups causing IMD has changed over time and across geographic regions. Pentavalent meningococcal conjugate vaccines in phase 3 clinical trials appear to be an important next step for enhanced global control. However, the capacity of meningococci to continue to evolve is significant. Genetic transformation and recombination, including transfer of genes between meningococci, gonococci, and commensal *Neisseria* spp. [6,116] and immune selection can all result in the rise, diversification, and disappearance of virulent meningococcal clones. Continued surveillance including molecular characterization is key to recognizing the changing epidemiology of meningococcal disease.

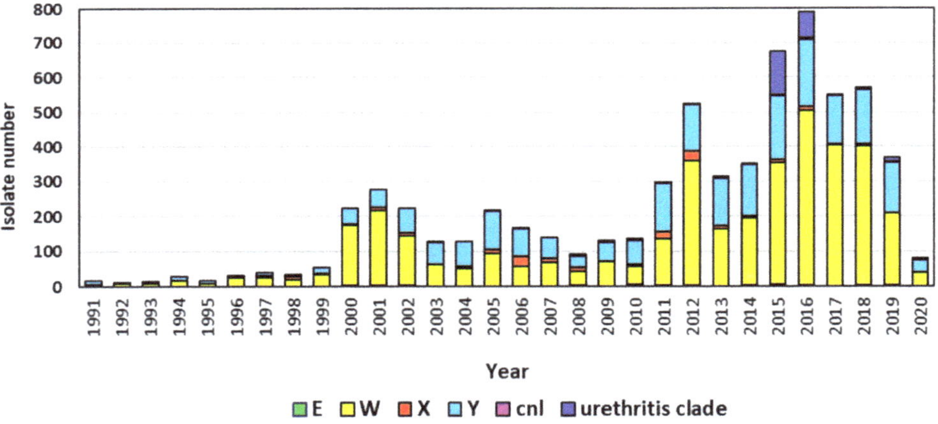

Figure 2. *N. meningitidis* capsular groups E, W, X, Y, and cnl from invasive meningococcal disease and urotropic meningococci submitted to PubMLST database, 1991–2020.

Author Contributions: Y.-L.T. and D.S.S. wrote and reviewed the manuscript. All authors have read and agreed to the published version of the manuscript.

Funding: This work was supported in part by the NIH Grants R01AI127863 and R21AI128313.

Institutional Review Board Statement: Not applicable.

Informed Consent Statement: Not applicable.

Data Availability Statement: Not applicable.

Acknowledgments: This publication made use of the *Neisseria* Multi Locus Sequence Typing and analysis tools hosted on Bacterial Isolate Genome Sequence Database (BIGSdb) (http://pubmlst.org/neisseria/ accessed on 20 November 2020).

Conflicts of Interest: The authors declare no conflict of interest.

References

1. Gold, R.; Goldschneider, I.; Lepow, M.L.; Draper, T.F.; Randolph, M. Carriage of *Neisseria meningitidis* and *Neisseria lactamica* in infants and children. *J. Infect. Dis.* **1978**, *137*, 112–121. [CrossRef]
2. Cartwright, K.A.; Stuart, J.M.; Jones, D.M.; Noah, N.D. The Stonehouse survey: Nasopharyngeal carriage of meningococci and *Neisseria lactamica*. *Epidemiol. Infect.* **1987**, *99*, 591–601. [CrossRef]
3. Harrison, O.B.; Claus, H.; Jiang, Y.; Bennett, J.S.; Bratcher, H.B.; Jolley, K.A.; Corton, C.; Care, R.; Poolman, J.T.; Zollinger, W.D.; et al. Description and nomenclature of *Neisseria meningitidis* capsule locus. *Emerg. Infect. Dis.* **2013**, *19*, 566–573. [CrossRef] [PubMed]
4. Bundle, D.R.; Jennings, H.J.; Kenny, C.P. Studies on the group-specific polysaccharide of *Neisseria meningitidis* serogroup X and an improved procedure for its isolation. *J. Biol. Chem.* **1974**, *249*, 4797–4801. [CrossRef]
5. Bhattacharjee, A.K.; Jennings, H.J.; Kenny, C.P. Structural elucidation of the 3-deoxy-D-manno-octulosonic acid containing meningococcal 29-e capsular polysaccharide antigen using carbon-13 nuclear magnetic resonance. *Biochemistry* **1978**, *17*, 645–651. [CrossRef] [PubMed]
6. Retchless, A.C.; Kretz, C.B.; Chang, H.Y.; Bazan, J.A.; Abrams, A.J.; Norris Turner, A.; Jenkins, L.T.; Trees, D.L.; Tzeng, Y.L.; Stephens, D.S.; et al. Expansion of a urethritis-associated *Neisseria meningitidis* clade in the United States with concurrent acquisition of *N. gonorrhoeae* alleles. *BMC Genom.* **2018**, *19*, 176. [CrossRef]
7. Tzeng, Y.L.; Bazan, J.A.; Turner, A.N.; Wang, X.; Retchless, A.C.; Read, T.D.; Toh, E.; Nelson, D.E.; Del Rio, C.; Stephens, D.S. Emergence of a new *Neisseria meningitidis* clonal complex 11 lineage 11.2 clade as an effective urogenital pathogen. *Proc. Natl. Acad. Sci. USA* **2017**, *114*, 4237–4242. [CrossRef] [PubMed]
8. Bazan, J.A.; Turner, A.N.; Kirkcaldy, R.D.; Retchless, A.C.; Kretz, C.B.; Briere, E.; Tzeng, Y.L.; Stephens, D.S.; Maierhofer, C.; Del Rio, C.; et al. Large Cluster of *Neisseria meningitidis* Urethritis in Columbus, Ohio, 2015. *Clin. Infect. Dis.* **2017**, *65*, 92–99. [CrossRef]
9. Chang, Q.; Tzeng, Y.L.; Stephens, D.S. Meningococcal disease: Changes in epidemiology and prevention. *Clin. Epidemiol.* **2012**, *4*, 237–245. [CrossRef] [PubMed]
10. Bai, X.; Borrow, R.; Bukovski, S.; Caugant, D.A.; Culic, D.; Delic, S.; Dinleyici, E.C.; Eloshvili, M.; Erdosi, T.; Galajeva, J.; et al. Prevention and control of meningococcal disease: Updates from the Global Meningococcal Initiative in Eastern Europe. *J. Infect.* **2019**, *79*, 528–541. [CrossRef]
11. Taha, M.K.; Deghmane, A.E. Impact of COVID-19 pandemic and the lockdown on invasive meningococcal disease. *BMC Res. Notes* **2020**, *13*, 399. [CrossRef] [PubMed]
12. Taha, M.K.; Achtman, M.; Alonso, J.M.; Greenwood, B.; Ramsay, M.; Fox, A.; Gray, S.; Kaczmarski, E. Serogroup W135 meningococcal disease in Hajj pilgrims. *Lancet* **2000**, *356*, 2159. [CrossRef]
13. Mayer, L.W.; Reeves, M.W.; Al-Hamdan, N.; Sacchi, C.T.; Taha, M.K.; Ajello, G.W.; Schmink, S.E.; Noble, C.A.; Tondella, M.L.; Whitney, A.M.; et al. Outbreak of W135 meningococcal disease in 2000: Not emergence of a new W135 strain but clonal expansion within the electophoretic type-37 complex. *J. Infect. Dis.* **2002**, *185*, 1596–1605. [CrossRef] [PubMed]
14. Lucidarme, J.; Hill, D.M.; Bratcher, H.B.; Gray, S.J.; du Plessis, M.; Tsang, R.S.; Vazquez, J.A.; Taha, M.K.; Ceyhan, M.; Efron, A.M.; et al. Genomic resolution of an aggressive, widespread, diverse and expanding meningococcal serogroup B, C and W lineage. *J. Infect.* **2015**, *71*, 544–552. [CrossRef]
15. Abad, R.; Lopez, E.L.; Debbag, R.; Vazquez, J.A. Serogroup W meningococcal disease: Global spread and current affect on the Southern Cone in Latin America. *Epidemiol. Infect.* **2014**, *142*, 2461–2470. [CrossRef] [PubMed]
16. Lucidarme, J.; Scott, K.J.; Ure, R.; Smith, A.; Lindsay, D.; Stenmark, B.; Jacobsson, S.; Fredlund, H.; Cameron, J.C.; Smith-Palmer, A.; et al. An international invasive meningococcal disease outbreak due to a novel and rapidly expanding serogroup W strain, Scotland and Sweden, July to August 2015. *Eurosurveillance* **2016**, *21*. [CrossRef]
17. Krone, M.; Gray, S.; Abad, R.; Skoczynska, A.; Stefanelli, P.; van der Ende, A.; Tzanakaki, G.; Molling, P.; Joao Simoes, M.; Krizova, P.; et al. Increase of invasive meningococcal serogroup W disease in Europe, 2013 to 2017. *Eurosurveillance* **2019**, *24*. [CrossRef]
18. Mulhall, R.M.; Bennett, D.E.; Bratcher, H.B.; Jolley, K.A.; Bray, J.E.; O'Lorcain, P.P.; Cotter, S.M.; Maiden, M.C.J.; Cunney, R.J. cgMLST characterisation of invasive *Neisseria meningitidis* serogroup C and W strains associated with increasing disease incidence in the Republic of Ireland. *PLoS ONE* **2019**, *14*, e0216771. [CrossRef] [PubMed]
19. Eriksson, L.; Hedberg, S.T.; Jacobsson, S.; Fredlund, H.; Molling, P.; Stenmark, B. Whole-Genome Sequencing of Emerging Invasive *Neisseria meningitidis* Serogroup W in Sweden. *J. Clin. Microbiol.* **2018**, *56*. [CrossRef]
20. Martin, N.V.; Ong, K.S.; Howden, B.P.; Lahra, M.M.; Lambert, S.B.; Beard, F.H.; Dowse, G.K.; Saul, N.; Communicable Diseases Network Australia Men, W.W.G. Rise in invasive serogroup W meningococcal disease in Australia 2013–2015. *Commun. Dis. Intell. Q. Rep.* **2016**, *40*, E454–E459. [PubMed]
21. Tsang, R.S.W.; Ahmad, T.; Tyler, S.; Lefebvre, B.; Deeks, S.L.; Gilca, R.; Hoang, L.; Tyrrell, G.; Van Caeseele, P.; Van Domselaar, G.; et al. Whole genome typing of the recently emerged Canadian serogroup W *Neisseria meningitidis* sequence type 11 clonal complex isolates associated with invasive meningococcal disease. *Int. J. Infect. Dis.* **2018**, *69*, 55–62. [CrossRef] [PubMed]
22. Sudbury, E.L.; O'Sullivan, S.; Lister, D.; Varghese, D.; Satharasinghe, K. Case Manifestations and Public Health Response for Outbreak of Meningococcal W Disease, Central Australia, 2017. *Emerg. Infect. Dis.* **2020**, *26*, 1355–1363. [CrossRef] [PubMed]

23. Aye, A.M.M.; Bai, X.; Borrow, R.; Bory, S.; Carlos, J.; Caugant, D.A.; Chiou, C.S.; Dai, V.T.T.; Dinleyici, E.C.; Ghimire, P.; et al. Meningococcal disease surveillance in the Asia-Pacific region (2020): The global meningococcal initiative. *J. Infect.* **2020**, *81*, 698–711. [CrossRef]
24. Collard, J.M.; Maman, Z.; Yacouba, H.; Djibo, S.; Nicolas, P.; Jusot, J.F.; Rocourt, J.; Maitournam, R. Increase in *Neisseria meningitidis* serogroup W135, Niger, 2010. *Emerg. Infect. Dis.* **2010**, *16*, 1496–1498. [CrossRef] [PubMed]
25. Mounkoro, D.; Nikiema, C.S.; Maman, I.; Sakande, S.; Bozio, C.H.; Tall, H.; Sadji, A.Y.; Njanpop-Lafourcade, B.M.; Sibabe, A.; Landoh, D.E.; et al. *Neisseria meningitidis* Serogroup W Meningitis Epidemic in Togo, 2016. *J. Infect. Dis.* **2019**, *220*, S216–S224. [CrossRef] [PubMed]
26. Retchless, A.C.; Hu, F.; Ouedraogo, A.S.; Diarra, S.; Knipe, K.; Sheth, M.; Rowe, L.A.; Sangare, L.; Ky Ba, A.; Ouangraoua, S.; et al. The Establishment and Diversification of Epidemic-Associated Serogroup W Meningococcus in the African Meningitis Belt, 1994 to 2012. *mSphere* **2016**, *1*. [CrossRef] [PubMed]
27. Kretz, C.B.; Retchless, A.C.; Sidikou, F.; Issaka, B.; Ousmane, S.; Schwartz, S.; Tate, A.H.; Pana, A.; Njanpop-Lafourcade, B.M.; Nzeyimana, I.; et al. Whole-Genome Characterization of Epidemic *Neisseria meningitidis* Serogroup C and Resurgence of Serogroup W, Niger, 2015. *Emerg. Infect. Dis.* **2016**, *22*, 1762–1768. [CrossRef] [PubMed]
28. Retchless, A.C.; Congo-Ouedraogo, M.; Kambire, D.; Vuong, J.; Chen, A.; Hu, F.; Ba, A.K.; Ouedraogo, A.S.; Hema-Ouangraoua, S.; Patel, J.C.; et al. Molecular characterization of invasive meningococcal isolates in Burkina Faso as the relative importance of serogroups X and W increases, 2008–2012. *BMC Infect. Dis.* **2018**, *18*, 337. [CrossRef]
29. Meiring, S.; Cohen, C.; de Gouveia, L.; du Plessis, M.; Kularatne, R.; Hoosen, A.; Lekalakala, R.; Lengana, S.; Seetharam, S.; Naicker, P.; et al. Declining Incidence of Invasive Meningococcal Disease in South Africa: 2003–2016. *Clin. Infect. Dis.* **2019**, *69*, 495–504. [CrossRef] [PubMed]
30. Hansman, D. Meningococcal disease in South Australia: Incidence and serogroup distribution 1971–1980. *J. Hyg.* **1983**, *90*, 49–54. [CrossRef]
31. Ryan, N.J.; Hogan, G.R. Severe meningococcal disease caused by serogroups X and Z. *Am. J. Dis. Child.* **1980**, *134*, 1173. [CrossRef] [PubMed]
32. Pastor, J.M.; Fe, A.; Gomis, M.; Gil, D. Meningococcal meningitis caused by *Neisseria meningitidis* of the X serogroup. *Med. Clin.* **1985**, *85*, 208–209.
33. Gagneux, S.P.; Hodgson, A.; Smith, T.A.; Wirth, T.; Ehrhard, I.; Morelli, G.; Genton, B.; Binka, F.N.; Achtman, M.; Pluschke, G. Prospective study of a serogroup X *Neisseria meningitidis* outbreak in northern Ghana. *J. Infect. Dis.* **2002**, *185*, 618–626. [CrossRef]
34. Djibo, S.; Nicolas, P.; Alonso, J.M.; Djibo, A.; Couret, D.; Riou, J.Y.; Chippaux, J.P. Outbreaks of serogroup X meningococcal meningitis in Niger 1995–2000. *Trop. Med. Int. Health* **2003**, *8*, 1118–1123. [CrossRef]
35. Boisier, P.; Nicolas, P.; Djibo, S.; Taha, M.K.; Jeanne, I.; Mainassara, H.B.; Tenebray, B.; Kairo, K.K.; Giorgini, D.; Chanteau, S. Meningococcal meningitis: Unprecedented incidence of serogroup X-related cases in 2006 in Niger. *Clin. Infect. Dis.* **2007**, *44*, 657–663. [CrossRef] [PubMed]
36. Materu, S.; Cox, H.S.; Isaakidis, P.; Baruani, B.; Ogaro, T.; Caugant, D.A. Serogroup X in meningococcal disease, Western Kenya. *Emerg. Infect. Dis.* **2007**, *13*, 944–945. [CrossRef]
37. Mutonga, D.M.; Pimentel, G.; Muindi, J.; Nzioka, C.; Mutiso, J.; Klena, J.D.; Morcos, M.; Ogaro, T.; Materu, S.; Tetteh, C.; et al. Epidemiology and risk factors for serogroup X meningococcal meningitis during an outbreak in western Kenya, 2005–2006. *Am. J. Trop. Med. Hyg.* **2009**, *80*, 619–624. [CrossRef]
38. Delrieu, I.; Yaro, S.; Tamekloe, T.A.; Njanpop-Lafourcade, B.M.; Tall, H.; Jaillard, P.; Ouedraogo, M.S.; Badziklou, K.; Sanou, O.; Drabo, A.; et al. Emergence of epidemic *Neisseria meningitidis* serogroup X meningitis in Togo and Burkina Faso. *PLoS ONE* **2011**, *6*, e19513. [CrossRef]
39. Xie, O.; Pollard, A.J.; Mueller, J.E.; Norheim, G. Emergence of serogroup X meningococcal disease in Africa: Need for a vaccine. *Vaccine* **2013**, *31*, 2852–2861. [CrossRef]
40. Chen, W.H.; Neuzil, K.M.; Boyce, C.R.; Pasetti, M.F.; Reymann, M.K.; Martellet, L.; Hosken, N.; LaForce, F.M.; Dhere, R.M.; Pisal, S.S.; et al. Safety and immunogenicity of a pentavalent meningococcal conjugate vaccine containing serogroups A, C, Y, W, and X in healthy adults: A phase 1, single-centre, double-blind, randomised, controlled study. *Lancet Infect. Dis.* **2018**, *18*, 1088–1096. [CrossRef]
41. Jones, G.R.; Christodoulides, M.; Brooks, J.L.; Miller, A.R.; Cartwright, K.A.; Heckels, J.E. Dynamics of carriage of *Neisseria meningitidis* in a group of military recruits: Subtype stability and specificity of the immune response following colonization. *J. Infect. Dis.* **1998**, *178*, 451–459. [CrossRef]
42. Leimkugel, J.; Hodgson, A.; Forgor, A.A.; Pfluger, V.; Dangy, J.P.; Smith, T.; Achtman, M.; Gagneux, S.; Pluschke, G. Clonal waves of Neisseria colonisation and disease in the African meningitis belt: Eight year longitudinal study in northern Ghana. *PLoS Med.* **2007**, *4*, e101. [CrossRef]
43. Gagneux, S.; Wirth, T.; Hodgson, A.; Ehrhard, I.; Morelli, G.; Kriz, P.; Genton, B.; Smith, T.; Binka, F.; Pluschke, G.; et al. Clonal groupings in serogroup X *Neisseria meningitidis*. *Emerg. Infect. Dis.* **2002**, *8*, 462–466. [CrossRef]
44. Stephens, D.S.; Greenwood, B.; Brandtzaeg, P. Epidemic meningitis, meningococcaemia, and *Neisseria meningitidis*. *Lancet* **2007**, *369*, 2196–2210. [CrossRef]

45. Cohn, A.C.; MacNeil, J.R.; Harrison, L.H.; Hatcher, C.; Theodore, J.; Schmidt, M.; Pondo, T.; Arnold, K.E.; Baumbach, J.; Bennett, N.; et al. Changes in *Neisseria meningitidis* disease epidemiology in the United States, 1998–2007: Implications for prevention of meningococcal disease. *Clin. Infect. Dis.* **2010**, *50*, 184–191. [CrossRef]
46. Broker, M.; Bukovski, S.; Culic, D.; Jacobsson, S.; Koliou, M.; Kuusi, M.; Simoes, M.J.; Skoczynska, A.; Toropainen, M.; Taha, M.K.; et al. Meningococcal serogroup Y emergence in Europe: High importance in some European regions in 2012. *Hum. Vaccines Immunother.* **2014**, *10*, 1725–1728. [CrossRef]
47. Toros, B.; Hedberg, S.T.; Unemo, M.; Jacobsson, S.; Hill, D.M.; Olcen, P.; Fredlund, H.; Bratcher, H.B.; Jolley, K.A.; Maiden, M.C.; et al. Genome-Based Characterization of Emergent Invasive *Neisseria meningitidis* Serogroup Y Isolates in Sweden from 1995 to 2012. *J. Clin. Microbiol.* **2015**, *53*, 2154–2162. [CrossRef] [PubMed]
48. Hedberg, S.T.; Toros, B.; Fredlund, H.; Olcen, P.; Molling, P. Genetic characterisation of the emerging invasive *Neisseria meningitidis* serogroup Y in Sweden, 2000 to 2010. *Eurosurveillance* **2011**, *16*, 19885. [PubMed]
49. Ladhani, S.N.; Lucidarme, J.; Newbold, L.S.; Gray, S.J.; Carr, A.D.; Findlow, J.; Ramsay, M.E.; Kaczmarski, E.B.; Borrow, R. Invasive meningococcal capsular group Y disease, England and Wales, 2007–2009. *Emerg. Infect. Dis.* **2012**, *18*, 63–70. [CrossRef]
50. Jackson, L.A.; Wenger, J.D. Laboratory-based surveillance for meningococcal disease in selected areas, United States, 1989–1991. *Morb. Mortal. Wkly. Rep. CDC Surveill. Summ.* **1993**, *42*, 21–30.
51. Rosenstein, N.E.; Perkins, B.A.; Stephens, D.S.; Lefkowitz, L.; Cartter, M.L.; Danila, R.; Cieslak, P.; Shutt, K.A.; Popovic, T.; Schuchat, A.; et al. The changing epidemiology of meningococcal disease in the United States, 1992–1996. *J. Infect. Dis.* **1999**, *180*, 1894–1901. [CrossRef] [PubMed]
52. Baccarini, C.; Ternouth, A.; Wieffer, H.; Vyse, A. The changing epidemiology of meningococcal disease in North America 1945–2010. *Hum. Vaccines Immunother.* **2013**, *9*, 162–171. [CrossRef]
53. Kellerman, S.E.; McCombs, K.; Ray, M.; Baughman, W.; Reeves, M.W.; Popovic, T.; Rosenstein, N.E.; Farley, M.M.; Blake, P.; Stephens, D.S.; et al. Genotype-specific carriage of *Neisseria meningitidis* in Georgia counties with hyper- and hyposporadic rates of meningococcal disease. *J. Infect. Dis.* **2002**, *186*, 40–48. [CrossRef] [PubMed]
54. Harrison, L.H.; Shutt, K.A.; Arnold, K.E.; Stern, E.J.; Pondo, T.; Kiehlbauch, J.A.; Myers, R.A.; Hollick, R.A.; Schmink, S.; Vello, M.; et al. Meningococcal Carriage Among Georgia and Maryland High School Students. *J. Infect. Dis.* **2014**. [CrossRef]
55. Wang, X.; Shutt, K.A.; Vuong, J.T.; Cohn, A.; MacNeil, J.; Schmink, S.; Plikaytis, B.; Messonnier, N.E.; Harrison, L.H.; Clark, T.A.; et al. Changes in the Population Structure of Invasive *Neisseria meningitidis* in the United States After Quadrivalent Meningococcal Conjugate Vaccine Licensure. *J. Infect. Dis.* **2015**. [CrossRef] [PubMed]
56. Potts, C.C.; Joseph, S.J.; Chang, H.Y.; Chen, A.; Vuong, J.; Hu, F.; Jenkins, L.T.; Schmink, S.; Blain, A.; MacNeil, J.R.; et al. Population structure of invasive *Neisseria meningitidis* in the United States, 2011–2015. *J. Infect.* **2018**, *77*, 427–434. [CrossRef]
57. Tsang, R.S.; Henderson, A.M.; Cameron, M.L.; Tyler, S.D.; Tyson, S.; Law, D.K.; Stoltz, J.; Zollinger, W.D. Genetic and antigenic analysis of invasive serogroup Y *Neisseria meningitidis* isolates collected from 1999 to 2003 in Canada. *J. Clin. Microbiol.* **2007**, *45*, 1753–1758. [CrossRef]
58. Whitney, A.M.; Coulson, G.B.; von Gottberg, A.; Block, C.; Keller, N.; Mayer, L.W.; Messonnier, N.E.; Klugman, K.P. Genotypic comparison of invasive *Neisseria meningitidis* serogroup Y isolates from the United States, South Africa, and Israel, isolated from 1999 through 2002. *J. Clin. Microbiol.* **2009**, *47*, 2787–2793. [CrossRef]
59. Safadi, M.A.; Cintra, O.A. Epidemiology of meningococcal disease in Latin America: Current situation and opportunities for prevention. *Neurol. Res.* **2010**, *32*, 263–271. [CrossRef]
60. Ines Agudelo, C.; Sanabria, O.M.; Ovalle, M.V. Serogroup Y meningococcal disease, Colombia. *Emerg. Infect. Dis.* **2008**, *14*, 990–991. [CrossRef] [PubMed]
61. Abad, R.; Agudelo, C.I.; Brandileone, M.C.; Chanto, G.; Gabastou, J.M.; Hormazabal, J.C.; MC, O.G.; Maldonado, A.; Moreno, J.; Muros-Le Rouzic, E.; et al. Molecular characterization of invasive serogroup Y *Neisseria meningitidis* strains isolated in the Latin America region. *J. Infect.* **2009**, *59*, 104–114. [CrossRef] [PubMed]
62. Marin, J.E.O.; Villatoro, E.; Luna, M.J.; Barrientos, A.M.; Mendoza, E.; Lemos, A.P.S.; Camargo, C.H.; Sacchi, C.T.; Cunha, M.P.V.; Galas, M.; et al. Emergence of MDR invasive *Neisseria meningitidis* in El Salvador, 2017–2019. *J. Antimicrob. Chemother.* **2021**. [CrossRef] [PubMed]
63. Sullivan, C.B.; Diggle, M.A.; Davies, R.L.; Clarke, S.C. Clonal analysis of meningococci during a 26 year period prior to the introduction of meningococcal serogroup C vaccines. *PLoS ONE* **2015**, *10*, e115741. [CrossRef]
64. Broker, M.; Jacobsson, S.; DeTora, L.; Pace, D.; Taha, M.K. Increase of meningococcal serogroup Y cases in Europe: A reason for concern? *Hum. Vaccines Immunother.* **2012**, *8*, 685–688. [CrossRef] [PubMed]
65. Broker, M.; Emonet, S.; Fazio, C.; Jacobsson, S.; Koliou, M.; Kuusi, M.; Pace, D.; Paragi, M.; Pysik, A.; Simoes, M.J.; et al. Meningococcal serogroup Y disease in Europe: Continuation of high importance in some European regions in 2013. *Hum. Vaccines Immunother.* **2015**, *11*, 2281–2286. [CrossRef] [PubMed]
66. Ala'Aldeen, D.A.; Neal, K.R.; Ait-Tahar, K.; Nguyen-Van-Tam, J.S.; English, A.; Falla, T.J.; Hawkey, P.M.; Slack, R.C. Dynamics of meningococcal long-term carriage among university students and their implications for mass vaccination. *J. Clin. Microbiol.* **2000**, *38*, 2311–2316. [CrossRef]
67. Bidmos, F.A.; Neal, K.R.; Oldfield, N.J.; Turner, D.P.; Ala'Aldeen, D.A.; Bayliss, C.D. Persistence, replacement, and rapid clonal expansion of meningococcal carriage isolates in a 2008 university student cohort. *J. Clin. Microbiol.* **2011**, *49*, 506–512. [CrossRef] [PubMed]

68. Ala'aldeen, D.A.; Oldfield, N.J.; Bidmos, F.A.; Abouseada, N.M.; Ahmed, N.W.; Turner, D.P.; Neal, K.R.; Bayliss, C.D. Carriage of meningococci by university students, United Kingdom. *Emerg. Infect. Dis.* **2011**, *17*, 1762–1763. [CrossRef]
69. Jeppesen, C.A.; Snape, M.D.; Robinson, H.; Gossger, N.; John, T.M.; Voysey, M.; Ladhani, S.; Okike, I.O.; Oeser, C.; Kent, A.; et al. Meningococcal carriage in adolescents in the United Kingdom to inform timing of an adolescent vaccination strategy. *J. Infect.* **2015**, *71*, 43–52. [CrossRef]
70. Maiden, M.C.; Ibarz-Pavon, A.B.; Urwin, R.; Gray, S.J.; Andrews, N.J.; Clarke, S.C.; Walker, A.M.; Evans, M.R.; Kroll, J.S.; Neal, K.R.; et al. Impact of meningococcal serogroup C conjugate vaccines on carriage and herd immunity. *J. Infect. Dis.* **2008**, *197*, 737–743. [CrossRef]
71. Evans, J.R.; Artenstein, M.S.; Hunter, D.H. Prevalence of meningococcal serogroups and description of three new groups. *Am. J. Epidemiol.* **1968**, *87*, 643–646. [CrossRef]
72. Wachter, E.; Brown, A.E.; Kiehn, T.E.; Lee, B.J.; Armstrong, D. *Neisseria meningitidis* serogroup 29E (Z') septicemia in a patient with far advanced multiple myeloma (plasma cell leukemia). *J. Clin. Microbiol.* **1985**, *21*, 464–466. [CrossRef] [PubMed]
73. Thangarajah, D.; Guglielmino, C.J.D.; Lambert, S.B.; Khandaker, G.; Vasant, B.R.; Malo, J.A.; Smith, H.V.; Jennison, A.V. Genomic Characterization of Recent and Historic Meningococcal Serogroup E Invasive Disease in Australia: A Case Series. *Clin. Infect. Dis.* **2020**, *70*, 1761–1763. [CrossRef]
74. Jolley, K.A.; Kalmusova, J.; Feil, E.J.; Gupta, S.; Musilek, M.; Kriz, P.; Maiden, M.C. Carried meningococci in the Czech Republic: A diverse recombining population. *J. Clin. Microbiol.* **2000**, *38*, 4492–4498. [CrossRef] [PubMed]
75. McMillan, M.; Walters, L.; Mark, T.; Lawrence, A.; Leong, L.E.X.; Sullivan, T.; Rogers, G.B.; Andrews, R.M.; Marshall, H.S. B Part of It study: A longitudinal study to assess carriage of *Neisseria meningitidis* in first year university students in South Australia. *Hum. Vaccines Immunother.* **2019**, *15*, 987–994. [CrossRef] [PubMed]
76. Tzeng, Y.L.; Thomas, J.; Stephens, D.S. Regulation of capsule in Neisseria meningitidis. *Crit. Rev. Microbiol.* **2016**, *42*, 759–772. [CrossRef]
77. Dolan-Livengood, J.M.; Miller, Y.K.; Martin, L.E.; Urwin, R.; Stephens, D.S. Genetic basis for nongroupable *Neisseria meningitidis*. *J. Infect. Dis.* **2003**, *187*, 1616–1628. [CrossRef]
78. Claus, H.; Maiden, M.C.J.; Maag, R.; Frosch, M.; Vogel, U. Many carried meningococci lack the genes required for capsule synthesis and transport. *Microbiology* **2002**, *148*, 1813–1819. [CrossRef]
79. Barnes, G.K.; Kristiansen, P.A.; Beyene, D.; Workalemahu, B.; Fissiha, P.; Merdekios, B.; Bohlin, J.; Preziosi, M.P.; Aseffa, A.; Caugant, D.A. Prevalence and epidemiology of meningococcal carriage in Southern Ethiopia prior to implementation of MenAfriVac, a conjugate vaccine. *BMC Infect. Dis.* **2016**, *16*, 639. [CrossRef]
80. Neri, A.; Fazio, C.; Ambrosio, L.; Vacca, P.; Barbui, A.; Daprai, L.; Vocale, C.; Santino, I.; Conte, M.; Rossi, L.; et al. Carriage meningococcal isolates with capsule null locus dominate among high school students in a non-endemic period, Italy, 2012–2013. *Int. J. Med. Microbiol.* **2019**, *309*, 182–188. [CrossRef]
81. Hoang, L.M.; Thomas, E.; Tyler, S.; Pollard, A.J.; Stephens, G.; Gustafson, L.; McNabb, A.; Pocock, I.; Tsang, R.; Tan, R. Rapid and fatal meningococcal disease due to a strain of *Neisseria meningitidis* containing the capsule null locus. *Clin. Infect. Dis.* **2005**, *40*, e38–42. [CrossRef] [PubMed]
82. Xu, Z.; Zhu, B.; Xu, L.; Gao, Y.; Shao, Z. First case of *Neisseria meningitidis* capsule null locus infection in China. *Infect. Dis.* **2015**, *47*, 591–592. [CrossRef]
83. Johswich, K.O.; Zhou, J.; Law, D.K.; St Michael, F.; McCaw, S.E.; Jamieson, F.B.; Cox, A.D.; Tsang, R.S.; Gray-Owen, S.D. Invasive potential of nonencapsulated disease isolates of *Neisseria meningitidis*. *Infect. Immun.* **2012**, *80*, 2346–2353. [CrossRef] [PubMed]
84. Findlow, H.; Vogel, U.; Mueller, J.E.; Curry, A.; Njanpop-Lafourcade, B.M.; Claus, H.; Gray, S.J.; Yaro, S.; Traore, Y.; Sangare, L.; et al. Three cases of invasive meningococcal disease caused by a capsule null locus strain circulating among healthy carriers in Burkina Faso. *J. Infect. Dis.* **2007**, *195*, 1071–1077. [CrossRef] [PubMed]
85. Ganesh, K.; Allam, M.; Wolter, N.; Bratcher, H.B.; Harrison, O.B.; Lucidarme, J.; Borrow, R.; de Gouveia, L.; Meiring, S.; Birkhead, M.; et al. Molecular characterization of invasive capsule null *Neisseria meningitidis* in South Africa. *BMC Microbiol.* **2017**, *17*, 40. [CrossRef] [PubMed]
86. Hayakawa, K.; Itoda, I.; Shimuta, K.; Takahashi, H.; Ohnishi, M. Urethritis caused by novel *Neisseria meningitidis* serogroup W in man who has sex with men, Japan. *Emerg. Infect. Dis.* **2014**, *20*, 1585–1587. [CrossRef]
87. Khemees, T.A.; Porshinsky, B.S.; Patel, A.P.; McClung, C.D. Fournier's Gangrene in a Heterosexual Man: A Complication of *Neisseria meningitidis* Urethritis. *Case Rep. Urol.* **2012**, *2012*, 312365. [CrossRef]
88. Maini, M.; French, P.; Prince, M.; Bingham, J.S. Urethritis due to *Neisseria meningitidis* in a London genitourinary medicine clinic population. *Int. J. STD AIDS* **1992**, *3*, 423–425. [CrossRef]
89. Leoni, A.F.; Littvik, A.; Moreno, S.B. Pelvic inflammatory disease associated with *Neisseria meningitis* bacteremia. *Rev. Fac. Cienc. Med.* **2003**, *60*, 77–81.
90. Harrison, O.B.; Cole, K.; Peters, J.; Cresswell, F.; Dean, G.; Eyre, D.W.; Paul, J.; Maiden, M.C. Genomic analysis of urogenital and rectal *Neisseria meningitidis* isolates reveals encapsulated hyperinvasive meningococci and coincident multidrug-resistant gonococci. *Sex. Transm. Infect.* **2017**, *93*, 445–451. [CrossRef]
91. Ma, K.C.; Unemo, M.; Jeverica, S.; Kirkcaldy, R.D.; Takahashi, H.; Ohnishi, M.; Grad, Y.H. Genomic Characterization of Urethritis-Associated *Neisseria meningitidis* Shows that a Wide Range of *N. meningitidis* Strains Can Cause Urethritis. *J. Clin. Microbiol.* **2017**, *55*, 3374–3383. [CrossRef]

92. Bazan, J.A.; Peterson, A.S.; Kirkcaldy, R.D.; Briere, E.C.; Maierhofer, C.; Turner, A.N.; Licon, D.B.; Parker, N.; Dennison, A.; Ervin, M.; et al. Notes from the Field: Increase in *Neisseria meningitidis*-Associated Urethritis Among Men at Two Sentinel Clinics—Columbus, Ohio, and Oakland County, Michigan, 2015. *MMWR Morb. Mortal. Wkly. Rep.* **2016**, *65*, 550–552. [CrossRef] [PubMed]
93. Kretz, C.B.; Bergeron, G.; Aldrich, M.; Bloch, D.; Del Rosso, P.E.; Halse, T.A.; Ostrowsky, B.; Liu, Q.; Gonzalez, E.; Omoregie, E.; et al. Neonatal Conjunctivitis Caused by *Neisseria meningitidis* US Urethritis Clade, New York, USA, August 2017. *Emerg. Infect. Dis.* **2019**, *25*, 972–975. [CrossRef]
94. Taha, M.K.; Claus, H.; Lappann, M.; Veyrier, F.J.; Otto, A.; Becher, D.; Deghmane, A.E.; Frosch, M.; Hellenbrand, W.; Hong, E.; et al. Evolutionary Events Associated with an Outbreak of Meningococcal Disease in Men Who Have Sex with Men. *PLoS ONE* **2016**, *11*, e0154047. [CrossRef]
95. Tzeng, Y.L.; Giuntini, S.; Berman, Z.; Sannigrahi, S.; Granoff, D.M.; Stephens, D.S. *Neisseria meningitidis* Urethritis Outbreak Isolates Express a Novel Factor H Binding Protein Variant That Is a Potential Target of Group B-Directed Meningococcal (MenB) Vaccines. *Infect. Immun.* **2020**, *88*. [CrossRef]
96. Bartley, S.N.; Tzeng, Y.L.; Heel, K.; Lee, C.W.; Mowlaboccus, S.; Seemann, T.; Lu, W.; Lin, Y.H.; Ryan, C.S.; Peacock, C.; et al. Attachment and Invasion of *Neisseria meningitidis* to Host Cells Is Related to Surface Hydrophobicity, Bacterial Cell Size and Capsule. *PLoS ONE* **2013**, *8*, e55798. [CrossRef] [PubMed]
97. Yi, K.; Rasmussen, A.W.; Gudlavalleti, S.K.; Stephens, D.S.; Stojiljkovic, I. Biofilm Formation by *Neisseria meningitidis*. *Infect. Immun.* **2004**, *72*, 6132–6138. [CrossRef] [PubMed]
98. Neil, R.B.; Shao, J.Q.; Apicella, M.A. Biofilm formation on human airway epithelia by encapsulated *Neisseria meningitidis* serogroup B. *Microbes Infect.* **2009**, *11*, 281–287. [CrossRef] [PubMed]
99. Sukhum, K.V.; Jean, S.; Wallace, M.; Anderson, N.; Burnham, C.A.; Dantas, G. Genomic characterization of emerging bacterial uropathogen Neisseria meningitidis misidentified as Neisseria gonorrhoeae by nucleic acid amplification testing. *J. Clin. Microbiol.* **2020**. [CrossRef] [PubMed]
100. Brooks, A.; Lucidarme, J.; Campbell, H.; Campbell, L.; Fifer, H.; Gray, S.; Hughes, G.; Lekshmi, A.; Schembri, G.; Rayment, M.; et al. Detection of the United States *Neisseria meningitidis* urethritis clade in the United Kingdom, August and December 2019—Emergence of multiple antibiotic resistance calls for vigilance. *Eurosurveillance* **2020**, *25*. [CrossRef] [PubMed]
101. Vazquez, J.A.; de la Fuente, L.; Berron, S.; O'Rourke, M.; Smith, N.H.; Zhou, J.; Spratt, B.G. Ecological separation and genetic isolation of *Neisseria gonorrhoeae* and *Neisseria meningitidis*. *Curr. Biol.* **1993**, *3*, 567–572. [CrossRef]
102. Bratcher, H.B.; Corton, C.; Jolley, K.A.; Parkhill, J.; Maiden, M.C. A gene-by-gene population Genom. platform: De novo assembly, annotation and genealogical analysis of 108 representative *Neisseria meningitidis* genomes. *BMC Genom.* **2014**, *15*, 1138. [CrossRef] [PubMed]
103. Zhou, Z.; Alikhan, N.F.; Sergeant, M.J.; Luhmann, N.; Vaz, C.; Francisco, A.P.; Carrico, J.A.; Achtman, M. GrapeTree: Visualization of core genomic relationships among 100,000 bacterial pathogens. *Genome Res.* **2018**, *28*, 1395–1404. [CrossRef] [PubMed]
104. Jolley, K.A.; Maiden, M.C. BIGSdb: Scalable analysis of bacterial genome variation at the population level. *BMC Bioinform.* **2010**, *11*, 595. [CrossRef] [PubMed]
105. Baxter, R.; Keshavan, P.; Welsch, J.A.; Han, L.; Smolenov, I. Persistence of the immune response after MenACWY-CRM vaccination and response to a booster dose, in adolescents, children and infants. *Hum. Vaccines Immunother.* **2016**, *12*, 1300–1310. [CrossRef]
106. Trotter, C.L.; Andrews, N.J.; Kaczmarski, E.B.; Miller, E.; Ramsay, M.E. Effectiveness of meningococcal serogroup C conjugate vaccine 4 years after introduction. *Lancet* **2004**, *364*, 365–367. [CrossRef]
107. Ramsay, M.E.; Andrews, N.J.; Trotter, C.L.; Kaczmarski, E.B.; Miller, E. Herd immunity from meningococcal serogroup C conjugate vaccination in England: Database analysis. *BMJ* **2003**, *326*, 365–366. [CrossRef]
108. Swartley, J.S.; Marfin, A.A.; Edupuganti, S.; Liu, L.J.; Cieslak, P.; Perkins, B.; Wenger, J.D.; Stephens, D.S. Capsule switching of *Neisseria meningitidis*. *Proc. Natl. Acad. Sci. USA* **1997**, *94*, 271–276. [CrossRef]
109. Tsang, R.S.; Law, D.K.; Tyler, S.D.; Stephens, G.S.; Bigham, M.; Zollinger, W.D. Potential capsule switching from serogroup Y to B: The characterization of three such *Neisseria meningitidis* isolates causing invasive meningococcal disease in Canada. *Can. J. Infect. Dis. Med. Microbiol.* **2005**, *16*, 171–174. [CrossRef] [PubMed]
110. Lancellotti, M.; Guiyoule, A.; Ruckly, C.; Hong, E.; Alonso, J.M.; Taha, M.K. Conserved virulence of C to B capsule switched *Neisseria meningitidis* clinical isolates belonging to ET-37/ST-11 clonal complex. *Microbes Infect.* **2006**, *8*, 191–196. [CrossRef]
111. Beddek, A.J.; Li, M.S.; Kroll, J.S.; Jordan, T.W.; Martin, D.R. Evidence for capsule switching between carried and disease-causing *Neisseria meningitidis* strains. *Infect. Immun.* **2009**, *77*, 2989–2994. [CrossRef]
112. Harrison, L.H.; Shutt, K.A.; Schmink, S.E.; Marsh, J.W.; Harcourt, B.H.; Wang, X.; Whitney, A.M.; Stephens, D.S.; Cohn, A.A.; Messonnier, N.E.; et al. Population structure and capsular switching of invasive *Neisseria meningitidis* isolates in the pre-meningococcal conjugate vaccine era—United States, 2000–2005. *J. Infect. Dis.* **2010**, *201*, 1208–1224. [CrossRef]
113. Harris, S.L.; Tan, C.; Andrew, L.; Hao, L.; Liberator, P.A.; Absalon, J.; Anderson, A.S.; Jones, T.R. The bivalent factor H binding protein meningococcal serogroup B vaccine elicits bactericidal antibodies against representative non-serogroup B meningococci. *Vaccine* **2018**, *36*, 6867–6874. [CrossRef] [PubMed]
114. Muzzi, A.; Brozzi, A.; Serino, L.; Bodini, M.; Abad, R.; Caugant, D.; Comanducci, M.; Lemos, A.P.; Gorla, M.C.; Krizova, P.; et al. Genetic Meningococcal Antigen Typing System (gMATS): A genotyping tool that predicts 4CMenB strain coverage worldwide. *Vaccine* **2019**, *37*, 991–1000. [CrossRef] [PubMed]

115. Biolchi, A.; De Angelis, G.; Moschioni, M.; Tomei, S.; Brunelli, B.; Giuliani, M.; Bambini, S.; Borrow, R.; Claus, H.; Gorla, M.C.O.; et al. Multicomponent meningococcal serogroup B vaccination elicits cross-reactive immunity in infants against genetically diverse serogroup C, W and Y invasive disease isolates. *Vaccine* **2020**, *38*, 7542–7550. [CrossRef] [PubMed]
116. Linz, B.; Schenker, M.; Zhu, P.; Achtman, M. Frequent interspecific genetic exchange between commensal *Neisseriae* and *Neisseria meningitidis*. *Mol. Microbiol.* **2000**, *36*, 1049–1058. [CrossRef]

MDPI

Review

Long Term Impact of Conjugate Vaccines on *Haemophilus influenzae* Meningitis: Narrative Review

Mary Paulina Elizabeth Slack

Gold Coast Campus, School of Medicine & Dentistry, Griffith University, Southport, QLD 4222, Australia; m.slack@griffith.edu.au

Abstract: *H. influenzae* serotype b (Hib) used to be the commonest cause of bacterial meningitis in young children. The widespread use of Hib conjugate vaccine has profoundly altered the epidemiology of *H. influenzae* meningitis. This short review reports on the spectrum of *H. influenzae* meningitis thirty years after Hib conjugate vaccine was first introduced into a National Immunization Program (NIP). Hib meningitis is now uncommon, but meningitis caused by other capsulated serotypes of *H. influenzae* and non-typeable strains (NTHi) should be considered. *H. influenzae* serotype a (Hia) has emerged as a significant cause of meningitis in Indigenous children in North America, which may necessitate a Hia conjugate vaccine. Cases of Hie, Hif, and NTHi meningitis are predominantly seen in young children and less common in older age groups. This short review reports on the spectrum of *H. influenzae* meningitis thirty years after Hib conjugate vaccine was first introduced into a NIP.

Keywords: *Haemophilus influenzae*; Hib; impact of Hib conjugate vaccine; Hia; NTHi

Citation: Slack, M.P.E. Long Term Impact of Conjugate Vaccines on *Haemophilus influenzae* Meningitis: Narrative Review. *Microorganisms* **2021**, *9*, 886. https://doi.org/10.3390/microorganisms9050886

Academic Editor: James Stuart

Received: 17 March 2021
Accepted: 19 April 2021
Published: 21 April 2021

1. Introduction

Haemophilus influenzae is a small, pleiomorphic Gram-negative coccobacillus, which is restricted to humans. It is fastidious in its growth requirement, only growing in culture media supplemented with both X factor (hemin) and V factor (nicotinamide adenine dinucleotide, NAD), for example chocolate agar. *H. influenzae* strains can be differentiated into two major groups: capsulated and non-capsulated strains (generally referred to as non-typeable strains, NTHi). The capsulated strains are further divided into six groups (a to f) based on the chemical structure of their polysaccharide capsules [1]. The most virulent type of *H. influenzae* is type b (Hib) and the major virulence determinant of Hib is its polysaccharide capsule, composed of polyribosyl ribitol phosphate (PRP).

H. influenzae colonizes the nasopharynx [2] and to a lesser extent the conjunctivae [3] and genital tract [4–6]. The respiratory tract is mainly colonized by *H. influenzae* and to a lesser extent *H. parainfluenzae* [2]. Approximately 80% of individuals carry NTHi strains in the nasopharynx, while 3–5% carry capsulated strains in the upper respiratory tract [7,8]. Spread from one person to another occurs via respiratory droplets or by direct contact with secretions [4].

Before the introduction of Hib conjugate vaccines, Hib was the commonest cause of bacterial meningitis in young children in the United States [9,10], Sweden [11], Iceland [12], the Netherlands [13], and England and Wales [14]. Seventy five percent of Hib meningitis cases occurred in children between the ages of three months and three years [15,16]. The case fatality ratio of Hib meningitis was ~5 to 10% in high-income countries [17].

In 1933, Fothergill and Wright [18] reported that blood from children aged less than two years lacked bactericidal activity against Hib, whereas blood from older children and adults demonstrated bactericidal activity. They speculated that naturally acquired antibodies to Hib were protective and as the mean level of Hib antibodies increased through exposure to the organism, so Hib meningitis incidence declined. The paucity of cases of Hib meningitis in infants aged <two months correlates with the presence of maternal Hib antibodies. This was confirmed by Peltola et al. [19] who demonstrated the incidence of

Hib meningitis declined as the mean level of anti-Hib antibodies increased. Studies on un-immunized individuals established a putative short-term correlate of protection against Hib infection of ≥0.15 μg/mL anti-PRP antibodies [20]. Later studies established that an anti-PRP antibody titer of ≥1.0 μg/mL was required for long-term protection [21].

It is now more than three decades since Hib conjugate vaccines were first developed and a variety of vaccine formulations, with a Hib component, are now included in the NIP of almost all countries in the world. Wherever Hib conjugate vaccine has been used the epidemiology of *H. influenzae* meningitis has changed, with Hib meningitis now infrequently seen in young children [22]. However, *H. influenzae* serotype a (Hia) has emerged as a significant cause of meningitis in Indigenous children in North America [23], and non-typeable strains of *H. influenzae* (NTHi) are associated with invasive infections, including meningitis, in neonates, older adults, and other vulnerable patient groups [24]. In 2020, the World Health Organization (WHO) published the document" Defeating meningitis by 2030: a global road map" [25]. The aims of the road map include the reduction of cases and deaths from vaccine-preventable meningitis; introduction of new vaccines; increasing vaccine coverage; and improving surveillance and advocacy. This short review will review the current epidemiology of *H. influenzae* meningitis in the second decade of the twenty first century to assess the progress made to date in achieving the goals set out in this document.

2. Method

A PubMed search was performed to identify published papers on the epidemiology of *H. influenzae* meningitis, before and after the introduction of Hib conjugate vaccines, using the terms: (((invasive) AND haemophilus) AND influenzae) AND ("meningitis" OR "nontypable" OR "NTHi" OR "serotype a" OR "serotype b" OR "serotype c" OR "serotype d" OR "serotype e" OR "serotype f" OR "non-b" OR "Hib") AND (epidemiology OR "burden" OR "risk factor" OR "impact" OR "Hib vaccine" OR "Hib conjugate vaccine" OR "surveillance" OR "review" OR "clinical" OR "outcome" OR "neonate" OR "adult" OR "children") for papers published between 1985 and 2020. Relevant papers on *H. influenzae* meningitis were reviewed.

3. Global Burden of Hib Meningitis before the Introduction of Routine Hib Immunization

Acute meningitis was the most serious presentation of Hib infection, following invasion of the blood stream. Often the child initially developed upper respiratory tract symptoms or otitis media before signs of meningeal involvement [4]. Before the introduction of Hib conjugate vaccines, *H. influenzae* serotype b (Hib) was the commonest cause of bacterial meningitis in young children in the US [9], Sweden [11], Iceland [12], the Netherlands [13], and the UK [14]. The mean annual incidence of Hib meningitis in the US was 54/100,000 (range 19–69/100,000) in children aged < five years and ~120 to 130/100,000 in infants aged < 12 months [26]. Annual rates in Europe, Australia (non-Indigenous children) and South America ranged from <20 to 50/100,000 children aged <5 years [17]. A much higher rate was reported from The Gambia (60/100,000 < 5 years and 297/100,000 < one year of age) [27]. Incidence rates of 282/100,000, 254/100,000, 152/100,000, and 450/100,000 in children aged < five years were reported in Alaska Native [28], White Mountain Apache [29], Navajo Indian [30], and Indigenous Australian [31] children, respectively. A rate of 530/100,000 in children aged <five years was reported in the Keewatin District of Northern Canada, mostly afflicting Inuit children [32].

The majority of cases of Hib meningitis cases occurred in children aged between three months and three years [15,16]. The proportion varied in different parts of the world, with approximately 50%, 40%, and 80% of cases of Hib meningitis occurring in infants aged < 12 months in the US, Europe, and Africa, respectively [17]. In the US and Europe, the peak incidence occurred at eight to nine months of age with less than 10% of cases occurring before the age of six months, and approximately 40% of all cases of Hib meningitis occurred in the first year of life [33]. In Indigenous communities in North America and Australia, and in low and middle income countries (LMICs), the proportion of cases of Hib meningitis occurring in the first six months of life was higher than in industrialized communities [33]. In Australia, the median age of onset of Hib meningitis (and the proportion of cases in the first 12 months of life) in Indigenous and non-Indigenous children was six months (60%) and 15 months (17%), respectively [34]. In The Gambia, 44% and 84% of cases occurred in the first six and twelve months of life respectively [33]. In Alaska Native children, 34% and 67% of cases of Hib meningitis occurred in the first six and twelve respectively [27] (Table 1).

Table 1. Incidence of Hib meningitis before the introduction of routine Hib conjugate vaccination.

Region	Hib Meningitis (Cases/100,000 Children < 5 Years of Age)
USA	54
North America (Indigenous)	152–530
Europe	23–31
Israel	18
The Gambia	60
Australia and New Zealand (non-Indigenous)	25–34
Australia and New Zealand (Indigenous)	450
Latin America	35
Asia	25
Mongolia	28

Data derived from: USA [9,10]; North American (Indigenous) [28–30,35]; Europe [36–40]; Israel [41]; The Gambia (Reference [27]; Australia and New Zealand (non-Indigenous) [34,42]; Australia and New Zealand [31,43]; Latin America [44]; Asia [45]; and Mongolia [46].

The mean case fatality ratio (CFR) of Hib meningitis ranged from approximately five to ten % in high-income countries to 28% in Africa [17]. Fifteen to 30% of survivors had long-term sequelae, including sensorineural hearing loss, intellectual impairment, epilepsy, cerebral palsy, or hydrocephalus [26,47–50]. Thirty eight percent of children who survived an episode of Hib meningitis in The Gambia had long-term sequelae [51].

4. Hib Vaccines

4.1. Plain PRP Vaccine

The first Hib vaccine was a plain polysaccharide vaccine consisting of PRP. It was used in a large field trial in Finland involving 100,000 children aged three months to five years [52]. Although efficacious in children >18 months, it did not induce protective levels of anti-PRP antibodies in children aged <18 months, i.e., those most at risk of Hib meningitis [20,53]. It also failed to have any impact on nasopharyngeal carriage of Hib and so had no impact on transmission [52]. Plain polysaccharide vaccines activate B cells via a T-cell independent pathway, which is poorly developed in children <18 months of age [54]. The antibody response is short-lived, mainly IgM with little isotype switching and no induction of immune memory [55].

4.2. Hib Protein-Conjugate Vaccines

In the late 1980s conjugate Hib vaccines were developed in which PRP was covalently linked to a protein carrier. The PRP-protein conjugate induces a T-cell dependent response, which develops at a much younger age in infants, who are able to respond to conjugate vaccines from the age of six to eight weeks [56]. The protein antigen encourages class switching from IgM to IgG via T-helper cells [55]. The IgG generated is predominantly IgG1, which in vitro induces complement-mediated opsonization and bacteriolysis. The antibodies produced are of a higher avidity than those produced by a plain polysaccharide vaccine [55]. Furthermore, PRP-conjugate vaccines have a marked impact on nasopharyngeal carriage of Hib [57]. By reducing nasopharyngeal carriage, transmission of Hib to other susceptible children and adults is interrupted, thereby reducing infection in other non-immunized groups. This is called "herd effect" or "herd protection".

Four different protein carriers were initially used for Hib conjugate vaccines: tetanus toxoid (PRP-TT), diphtheria toxoid (PRP-D), a non-toxic mutant *Corynebacterium diphtheriae* protein CRM 197 (PRP-CRM) and an outer membrane complex of *Neisseria meningitidis* (PRP-OMP) [58]. The different Hib vaccines were equally immunogenic in adults but elicited different responses in infants <18 months of age. PRP-D was the least immunogenic, generating antibody titers of $\geq$1.0 μg/mL in approximately 30% of infants after two or three doses [59]. This vaccine was subsequently withdrawn. PRP-OMP vaccine generated antibody titers $\geq$1.0 μg/mL in 70–80% of infants at two months of age [60] and was the preferred vaccine for use in Indigenous populations in North America and Australia, where there was a very high burden of disease in very young infants [60]. The PRP-TT and PRP-CRM vaccines were similar in their immunogenicity eliciting antibody titers $\geq$ 1.0 μg/mL after three priming doses [61]. Over time monovalent Hib conjugate vaccines have largely been replaced by combination vaccines, including a bivalent Hib + meningococcus serogroup C vaccine (Hib-MenC), and pentavalent and hexavalent vaccines, where Hib is combined with diphtheria toxoid (D), tetanus toxoid (T), pertussis whole cell (wP) or acellular (aP), and/or hepatitis B (HepB), and/ or inactivated polio vaccine (IPV).

5. Introduction of Hib Conjugate Vaccine in National Immunization Programs (NIPs)

Hib vaccine was introduced into the NIP of Finland in 1986 [52], followed by the US in 1987 [62]. In the early 1990s Hib vaccine was added to the NIP in many Western European countries. By 2004, Hib vaccine had been included in the NIP of all European countries and $\geq$90% high-income countries. The introduction of Hib vaccine into the NIP of LMICs has taken longer, because of several factors. These include a lack of local data on the burden of Hib disease as a result of the difficulties in culturing this fastidious organism, widespread use of antibiotics before collection of blood and cerebro-spinal fluid (CSF) samples for culture and the relatively high cost of the vaccine. In 2004 WHO and the Global Alliance for Vaccines and Immunization (GAVI) sought to address this. Vaccine probe studies [63], in which a randomized controlled trial assesses the difference in incidence of meningitis between children immunized with Hib vaccine and unimmunized children, and the Hib Rapid Assessment Tool (HibRAT) [64] provided data on the burden of Hib meningitis for many LMICs. In 2005, GAVI established the Hib Initiative to accelerate the introduction of Hib vaccine in GAVI-eligible countries [65]. In 2006, WHO recommended the use of Hib conjugate vaccines in all countries [66], thereby allowing GAVI-eligible countries to apply for Hib vaccine without the need to have local data on Hib disease burden. With these measures, the number of countries using Hib vaccine increased from 89/193 (46%) in 2004 to 158/193 (82%) in 2009 [67]. Hib vaccine has now been added to the NIP of all countries in the world, except China, where it is available in the private market and in the Russian Federation, where it is recommended for certain groups of children [68].

6. Impact of Hib Conjugate Vaccine on Hib Meningitis

Wherever Hib vaccine has been introduced there has been a significant and sustained decline in Hib meningitis [69–71]. In 2000, the global incidence of Hib meningitis was estimated to be 31 (uncertainty range (UR) 16–39) cases/100,000 children aged < five years [72]. The estimated incidence varied considerably by region (Table 2). At that time, the only regions that had widespread use of Hib vaccine were the Americas and Europe. A further analysis of the burden of Hib meningitis in 2000–2015 [73] estimated the global incidence of Hib meningitis had declined to five (UR 2–8) cases/100,000 children aged < five years. There were still regional variations, with the highest estimated incidences in the South East Asian and Western Pacific Regions, which may reflect the lack of introduction of Hib vaccine into some countries in these regions at that time.

Table 2. Estimated incidence and case fatality ratio of Hib meningitis (with uncertainty estimates) by WHO region in 2000 and 2015.

	Global	African Region	Region of the Americas	Eastern Mediterranean Region	European Region	South East Asia Region	Western Pacific Region
2000 Estimates							
Incidence	31 (16–39)	46 (31–52)	25 (16–30)	24 (14–35)	16 (12–22)	27 (11–38)	34 (12–48)
CFR	43% (23–55%)	67% (44–75%)	28% (15–36%)	44% (26–62%)	27% (17–41%)	44% (17–62%)	22% (8–34%)
2015 Estimates [73]							
Incidence	5 (2–8)	2 (1–3)	0 (0–0)	1 (0–1)	3 (1–5)	8 (3–12)	11 (6–18)
CFR	19% (7–29%)	61% (20–98%)	30% (7–51%)	54% (16–89%)	5% (2–9%)	32% (12–49%)	5% (2–8%)

Data are estimates (uncertainty range) Incidence is /100,000 children aged <5 years. CFR: case fatality ratio. Data derived from: Watt et al. [61] and Wahl et al. [62].

By 2015, the burden of Hib meningitis was limited to a small number of countries that had not yet or only recently introduced Hib vaccine in their NIP. In the six years since this study almost all countries have now introduced Hib vaccine and the global burden will have been further reduced. This excellent control depends on maintaining high coverage of Hib vaccine combined with on-going surveillance of all cases of Hib meningitis in all ages of patients.

The estimated global CFR of Hib meningitis in 2000 was 43% (UR 23–55%), ranging from 22% (8–34%) in the Western Pacific Region to 67% (44–75%) in the African Region [72]. By 2015, the global CFR had declined to 19% (7–29%), ranging from 5% (2–8) in Europe and the Western Pacific Region to 61% (20–98%) in the African Region [73].

In 2013, a systematic review of the impact of Hib conjugate vaccine on childhood meningitis mortality, estimated the dose-specific impact (one dose: relative risk, RR = 0.64, 95% CI 0.38–1.06; two doses; RR = 0.09, 95% CI 0.03–0.27; three doses: RR = 0.06, 95% CI 0.02–0.22) [74]. The relative risk (RR) or risk ratio is the ratio of the probability of meningitis in children vaccinated with Hib vaccine to the probability of meningitis in unvaccinated children. This review estimated that three doses of Hib vaccine would prevent 38–43% of childhood meningitis mortality [74].

After the introduction of Hib immunization into several NIPs in the 1990s, the incidence of Hib meningitis declined rapidly [26]. Hib conjugate vaccines have proved to be highly effective in all countries, where there is sustained high coverage of the vaccine [75]. In the US active surveillance of invasive *H. influenzae* disease is undertaken in the Active Bacterial Core Surveillance (ABC) sites, coordinated by the Centers for Disease Control and Prevention (CDC). This surveillance system covers a population of over 42 million in five states and five metropolitan areas across the US [76]. In the 1990s, the rate of bacterial meningitis declined by 55% in the USA following Hib vaccine introduction [77]. Between 1998 and 2007, there were 187 cases of *H. influenzae* meningitis cases identified in the CDC ABC surveillance sites, 9.4% of cases were due to Hib. The overall incidence of *H. influenzae* meningitis declined between 1998–1999 and 2006–2007, from 0.12/100,000 population (95% CI, 0.09 to 0.17) to 0.08/100,000 (95% CI, 0.05 to 0.11) [77]. In 2018, only 38 cases of invasive Hib infection in children aged <five years (incidence 0.19/100,000) were notified throughout the US [78].The number of cases of Hib meningitis was not specified.

In a population-based observational study in Finland, where Hib conjugate vaccine was introduced in 1986, there were 1361 reported cases of bacterial meningitis between 1995 and 2014. Four percent of cases were caused by *H. influenzae* (incidence 0.06/100,000 population) and 92% of the isolates were non-b [79].The median age of *H. influenzae* meningitis was 29 years. From 2004 to 2014 two of 26 *H. influenzae* isolates were Hib [79].

Hib meningitis incidence declined by 72–83% at sentinel hospitals in Pakistan and Bangladesh, respectively, within two years of implementing nationwide Hib conjugate vaccination [80]. In a hospital-based multi-center prospective survey of bacterial meningitis in Turkey from 2015 to 2018, 994 cases of suspected bacterial meningitis in children, aged one month to 18 years, were identified [81]. Three (2.4%) of the 125 culture-positive cases were caused by Hib. Hib conjugate vaccine was introduced in the Japanese NIP in 2013, although Hib vaccine had been available on a voluntary basis since 2008. A nationwide population-based surveillance of invasive *H. influenzae* diseases in children in Japan [82] identified 336 cases of *H. influenzae* meningitis between 2008 and 2017. Between 2008–2012 and 2013–2017 there were 336 and 6 cases of *H. influenzae* meningitis, respectively. No cases of invasive Hib meningitis have been identified since 2014.

Although Hib meningitis has been virtually eliminated in almost all countries with established immunization programs and high vaccine coverage, there have been a few examples of countries that have experienced a re-emergence of invasive Hib infections, including Hib meningitis.

7. Resurgence of Hib Meningitis in Some Countries

7.1. Resurgence of Hib in the UK

In the UK there was a resurgence in cases in the late 1990s. The UK introduced Hib vaccine in 1992 as a three-dose infant schedule of PRP-TT (at two, three, and four months) with no booster dose in the second year of life, together with a catch-up campaign for all children <five years of age. Hib infections declined rapidly in all age groups through direct and indirect (herd) protection. The incidence of invasive Hib disease in England and Wales declined from 22.9/100,000 children < five years in 1990 to 0.65/ 100,000 in 1998 [83]. From 1999 Hib infections began to increase, especially among toddlers, most of whom were fully immunized. After 1999, the incidence of Hib disease increased to 4.6/100,000 in children aged <five years [84], with many of the infections, including meningitis, occurring in toddlers [55]. Studies established that there was a greater than expected decline in Hib antibodies after primary immunization, which had been initially masked by the catch-up campaign [85–87]. The catch-up campaign also contributed to indirect protection by reducing nasopharyngeal carriage. By 1998, all children aged <five years had received three priming doses of Hib vaccine in infancy. A single dose of Hib vaccine administered at the age of 12 months was more immunogenic than three doses given in infancy. Another factor was the use of a less immunogenic Hib combination vaccine with diphtheria, tetanus, and acellular pertussis (DTaP-Hib) in 2000–2001 [84,88]. The resurgence was controlled by the re-introduction of a whole-cell pertussis-containing Hib vaccine (DTwP-Hib) in 2002, an Hib booster campaign for toddlers in 2003, and the introduction of a routine 12-month Hib booster in 2006 [89,90].

Since that time Hib infections, including meningitis, have remained at a very low level in the UK, A review of invasive Hib infections in England and Wales, between 2009 and 2012, identified only 14 cases in 2012 [22]. Hib incidence was 0.06/100,000 (two cases) in children aged <five years [22]. Most of the cases that occurred over those four years were in adults (73%), many of whom had underlying comorbidities and presented with pneumonia (56%) [22]. The Hib-associated case fatality rate was 9.4% (10/106 cases) [22]. There were 20 cases (18.9%) of meningitis: ten in children aged < one year; five in children aged one to five years, two in adults aged 20 to 44 years, two in adults aged 45 to 64 years and one case in an older adult aged ≥65 years [22]. There was only one death in the vaccine-eligible age cohort: a child with Hib meningitis who was partially vaccinated and had a complement deficiency [22]. Hib meningitis is now uncommon in the UK.

The current Hib vaccination program in the UK is hexavalent vaccine (DTaP-Hib-HepB-IPV) administered at two, three, and four months, with a 12-month booster dose of Hib-MenC vaccine [91]. The number of cases of invasive Hib infection is at a very low level, with only five cases of invasive Hib disease (cases that were meningitis not specified) in the vaccine eligible population in 2017–2018 [92].

7.2. Resurgence of Hib in South Africa

South Africa introduced Hib conjugate vaccine (PRP-TT) in 1999 as an early accelerated schedule of three doses at six, ten, and fourteen weeks without a booster dose in the second year of life [93]. The number of cases of invasive Hib infection initially declined, but from 2005 increasing number of cases in fully vaccinated children were detected [94]. Despite high vaccination coverage the detection rate of invasive Hib infection in children aged < five years increased from 0.7/100,000 in 2003 to 1.3/100,000 in 2009 ($p < 0.001$), and 135/263 (51%) of cases in children with known vaccination status were Hib vaccine failures [93]. From 2003 to 2009 the surveillance program (GERMS) identified 349 cases of invasive Hib infection in children aged <five years, of which 211 (60%) presented as meningitis [94] with a CFR of 19%. Fifty-five% of the children, where HIV status was documented, were HIV negative. Following the addition of a booster dose of Hib vaccine in 2009, as a pentavalent vaccine (DTaP-Hib-IPV) the incidence of invasive Hib declined [92]. In 2018, GERMS identified 327 cases of invasive *H. influenzae* infection, of which 201 were available for typing. Seventeen percent (34/201) were Hib, of which eight cases presented with meningitis [95].

7.3. Resurgence of Hib in the Gambia

The Gambia introduced Hib vaccine in 1997. Before the Gambia introduced routine Hib vaccination, Hib meningitis incidence was 297/100,000 in infants <one year of age and 60/100,000 in children aged <five years [27]. The Gambia used a three-dose primary series of PRP-TT Hib vaccine, administered at two, three, and four months without a booster dose. For 14 years invasive Hib disease was well controlled in this country with consistently high coverage, low carriage rates and high levels of protective antibodies [96]. On-going surveillance in eastern Gambia identified an increase in Hib infections between 2011 and 2013, with 17 cases of invasive Hib infection, including 14 cases of Hib meningitis [97]. Although the reason for this re-emergence is not entirely clear, it does emphasize the importance of on-going surveillance.

7.4. Is a Booster Dose of Hib Vaccine Needed?

Although these instances where invasive Hib infections have emerged were in countries using a three dose primary series of Hib vaccine without a booster dose, Kenya and most LMICs use this schedule with no evidence of a resurgence of invasive Hib cases [98]. A three dose primary series of Hib vaccine without a booster dose is recommended by WHO [66]. A meta-analysis of 20 RCTs, conducted in 15 countries comparing different Hib vaccination schedules (3 + 0, 3 + 1, and 2 + 1) and different intervals between the primary, and the primary and booster doses, concluded that there was no difference between the schedules in terms of preventing invasive Hib disease, clinical effectiveness or immunologic response. All of the schedules protected against Hib infection and local epidemiology should determine the schedule, with three doses in the first six months of life being more appropriate where the greatest burden of Hib infection is in the first year of life, as in sub-Saharan Africa. Where the burden of infection occurs at a later age, the third dose could be given in the second year of life. In countries like the UK, where Hib infections resurged with a 3 + 0 schedule, a booster in the second year of life may be required [99]. Children who are HIV infected may require a booster dose of vaccine [100].

8. Current Burden of *H. influenzae* Meningitis

When Hib vaccine was first introduced there were concerns that Hib meningitis might be replaced by infections caused by other serotypes of *H. influenzae*. This has generally not happened, except in the Indigenous communities of North America, where *H. influenzae* serotype a (Hia) has emerged as a significant pathogen [23,101]. There has also been a slight increase in infections, including meningitis, caused by Hie and Hif in Europe [102,103]. Invasive infections caused by non-typeable strains of *H. influenzae* (NTHi) have increased significantly in many regions of the world [103,104].

9. Meningitis Due to Non-b Serotypes of *H. influenzae*

9.1. Meningitis Due to Serotype a (Hia)

Before the introduction of Hib vaccine, invasive *H. influenzae* serotype a (Hia) disease was very uncommon, although Hia was responsible for 12% of cases of bacterial meningitis in young children in Papua New Guinea before the introduction of Hib immunization [105,106]. Over the last two decades Hia has emerged as a significant pathogen, particularly in Indigenous populations in North America [23]. High incidences of Hia infection have been reported in Alaska Native, American Indian, and Canadian Inuit children [29,107–110]. In 2011, a population-based study in 12 Canadian pediatric tertiary care centers reported an Hia incidence of 418.8/100,000 in Inuit children aged < five years in the Keewatin region [111]. Hia is the second most virulent capsular serotype of *H. influenzae* [112] and can cause meningitis, pneumonia, septic arthritis, and bacteremia [23]. Most Hia infections occur in children aged six months to two years [23]. Between 1998 and 2003 38/76 (50%) of cases of Hia infection identified in Navajo and White Mountain Apache children presented with meningitis [107]. Hia meningitis was the commonest presentation in Indigenous children in the North American Arctic and Northern Canada [108,110]. Hia has also emerged as a significant pathogen in Utah and North and South Dakota [113–117]. In a study in Utah from 1998 to 2008, 28% of all invasive disease in children aged < five years was due to Hia, and 18% due to Hib. Fifty percent of the Hia cases presented as meningitis [115]. Hia infections in these states were not exclusively in American Indian children. Hia infections have also been reported from Brazil [118–120]. The case fatality rate of Hia meningitis was 14% in Brazil [120], 16% in Northern Canada [110], and 6% in the North American Arctic [112]. Hia has also been reported in Italy [121] and England [122] but there were no cases of meningitis in these reports of infections, which predominantly occurred in adults. Hia meningitis in a 10 month old infant and a 3 year old child was reported from Saudi Arabia [123]. The emergence of Hia as a significant cause of invasive infections in certain populations has prompted the development of an Hia conjugate vaccine [124].

9.2. Meningitis Due to Serotypes e and f (Hie and Hif)

There has also been increasing recognition of cases of meningitis caused by Hie and Hif [102]. Between 2001 and 2010 the year on year incidence of Hie and Hif infections in England and Wales increased by 7.4% and 11.0% respectively [100]. In 2009–2010, the incidences of Hie and Hif infections were 0.03/100,000 persons and 0.09/100,000 persons, respectively, with the highest rates being seen in infants and older adults [102]. Nine of 10 cases occurring in infants aged <one year presented with meningitis (three Hie, six Hif). All of the infants with Hif meningitis survived, but one child with Hie meningitis died, one had severe bilateral sensorineural deafness and one developed seizures [102]. Meningitis was a less common presentation in older children and adults, with three cases of Hie meningitis (one child aged one to four years, one child aged five to 14 years, one adult aged 15–64 years) and four cases of Hif meningitis (one in a child aged one to four years, three in adults aged 15–64 years). The case fatality rates of Hie and Hif meningitis were 14.3% and 0%, respectively. In this study Hie meningitis was associated with more complications and a higher case fatality rate.

Whittaker et al. [125] analyzed reports of invasive *H. influenzae* infection reported by 12 European countries to the European Centre for Disease Prevention and Control (ECDC) between 2007 and 2014. Five hundred and ninety-six cases of meningitis were reported, representing 9% of all infections. Sixty percent and 40% of infants aged <one year with Hie or Hif infection were reported to have meningitis [125]. National surveillance in Germany between 2001 and 2016 identified 351 cases of capsulated *H. influenzae* invasive infection: 241 cases of Hif, 45 cases of Hie, seven cases of Hia, and 58 cases of Hib(126). Forty cases of Hif infection were in children aged < four years with 40% of these cases presenting as meningitis. There were 185 cases of Hif infection in adults aged $\geq$ 40 years with meningitis accounting for 15% [126]. Hif meningitis has also been reported in the United States [127–129] and in Sweden [130].

9.3. Meningitis Due to Non-Typeable H.influenzae (NTHi)

Since the introduction of Hib vaccine, NTHi infections have emerged as the most common cause of invasive *H. infuenzae* infection in many parts of the world, where surveillance has been undertaken [104,113,125–128,130–139]. The highest burden of NTHi infections is seen in neonates, children aged <one year, pregnant/post-partum women, and in older adults (≥65 years) [104]. The clinical presentation varies by age, with meningitis more commonly seen in older infants and children and pneumonia more common in older adults [104].

Over a five year period (2009–2013), there were 115 cases of neonatal invasive NTHi infection in England and Wales (incidence 4.1/100,000; 95% CI 3.4–5.0) [24]. The incidence was significantly higher in premature babies (28.4/100,000; 95% CI 22.8–35.0) compared to those born at term (0.9/100,000; 95% CI 0.6–1.4) and increased exponentially with increasing prematurity. For infants born at <28 weeks' gestation the incidence was 342/100,000 (95% CI, 234–483). Most cases (110/115, 96%) presented within 48 h of birth. Although most of the infants developed a bacteremia, 11 (10%) presented with meningitis. One infant with meningitis died and five (50%) developed long-term sequelae [24].

Active surveillance for invasive *H. influenzae* disease in the US ABC surveillance sites from 2009 to 2015, reported that invasive NTHi infections had the highest incidence (1.22/100,000) [113]. Among 317 cases of invasive *H. influenzae* infection in children aged <one year, 25.1% presented with meningitis. One hundred and ninety six of 294 (66.7%) invasive infections (where the serotype was known) in this age group were due to NTHi. Although the serotyping of the meningitis cases was not reported it is probable that they included cases of NTHi meningitis.

Between 2001 and 2008, there were 396 cases of invasive NTHi infection documented by the Netherlands Reference Laboratory for Bacterial Meningitis [134]. Overall, the most common presenting clinical syndrome was invasive pneumonia (190/396, 48%) followed by bacteremia (75/396, 19%). Fifty-seven (14%) of the cases presented with meningitis. Among children aged seven weeks to <five years 28/60 (47%) of cases were meningitis. Nationwide active surveillance in Germany between 1998 and 2005 identified 70 cases of invasive NTHi infection. The median age of presentation was 26 months (0–73 months) and 34% presented with meningitis [135]. Thirty eight percent of children with NTHi meningitis had predisposing conditions, including prematurity, immunodeficiency, and Down's syndrome [135]. In a study from England [131] 26% of children who survived NTHi meningitis suffered long-term sequelae, including deafness, seizures, and hydrocephalus [131]. The case fatality rate of NTHi meningitis is similar to that of Hib meningitis [131].

10. Conclusions

Hib conjugate vaccine has been a remarkable success story, reducing the incidence of Hib meningitis to a very low level in countries with a well-established Hib immunization program and sustained high vaccine coverage [140]. There has been considerable progress in achieving the elimination of *H. influenzae* meningitis, but more still needs to be done. Cases of Hib meningitis do still occur, in unimmunized or partially vaccinated children, and as rare instances of true Hib vaccine failures. In 2015, Wahl et al. [73] estimated that there were still 12,900 cases (UR 6400 to 21,500) of Hib meningitis globally. Since then, Hib vaccine has been introduced into the NIP of almost all countries, including India and Thailand, except for China and the Russian Federation (where Hib vaccine is recommended for certain risk groups). Every child in the world should be offered Hib vaccine and vaccine coverage needs to be maintained at a high level in all countries. Hia has emerged as a significant cause of meningitis in Indigenous populations of North America, potentially requiring the use of Hia conjugate vaccine in these high-risk populations. Hie, Hif, and NTHi have also been associated with cases of meningitis. The changing epidemiology of *H.*

influenzae meningitis emphasizes the importance of on-going surveillance. Epidemiologic and microbiologic surveillance should be comprehensive, covering all ages and all types of *H. influenzae*. Accurate typing of strains, using molecular methods combined with clinical ascertainment of clinical presentation, underlying risk factors and outcome should be undertaken to fully document these changes. Considerable progress in achieving the elimination of *H. influenzae* meningitis has been made, but more still needs to be done.

Funding: This research received no external funding.

Institutional Review Board Statement: Not applicable.

Informed Consent Statement: Not applicable.

Data Availability Statement: Data available in a publicly available repository (PubMed).

Conflicts of Interest: MPES has received personal fees from GSK, Pfizer, AstraZeneca, and Sanofi Pasteur as a speaker at international meetings and as a member of advisory boards (unrelated to the submitted work).

References

1. Pittman, M. Variation and type specificity in the bacterial species *Haemophilus influenzae*. *J. Exp. Med.* **1931**, *53*, 471–492. [CrossRef] [PubMed]
2. Kilian, M.; Frederiksen, W. Ecology of *Haemophilus, Pasteurella* and *Actinobacillus*. In *Haemophilus, Pasteurella and Actinobacillus*; Kilian, M., Frederiksen, W., Biberstein, E.L., Eds.; Academic Press: London, UK, 1981; pp. 11–38.
3. Ingham, H.R.; Turk, D.C. Haemophili from eyes. *J. Clin. Pathol.* **1969**, *22*, 258–262. [CrossRef]
4. Moxon, E.R. Haemophilus influenzae. In *Mandell, Douglas abnd Bennett's Principles and Practice of Infectious Disease*, 4th ed.; Mandell, G.L., Bennett, J., Dolin, R., Eds.; Churchill Livingstone: New York, NY, USA, 1995; Volume 2, pp. 2039–2050.
5. Hammerschlag, M.R.; Alpert, S.; Rosner, I.; Thurston, P.; Semine, D.; McComb, D.; McCormack, W.M. Microbiology of the vagina in children: Normal and potentially pathogenic organisms. *Pediatrics* **1978**, *62*, 57–62. [PubMed]
6. Cox, R.A.; Slack, M.P. Clinical and microbiological features of *Haemophilus influenzae* vulvovaginitis in young girls. *J. Clin. Pathol.* **2002**, *55*, 961–964. [CrossRef] [PubMed]
7. Moxon, E.R. The carrier state: Haemophilus influenzae. *J. Antimicrob. Chemother.* **1986**, *18*, 17–24. [CrossRef] [PubMed]
8. Kilian, M.; Frederiksen, W.; Biberstein, E.L. *Haemophilus, Pasteurella and Actinobacillus*; Academic Press Inc.: London, UK, 1981; 294p.
9. Wenger, J.D.; Hightower, A.W.; Facklam, R.R.; Gaventa, S.; Broome, C.V. Bacterial meningitis in the United States, 1986: Report of a multistate surveillance study. The Bacterial Meningitis Study Group. *J. Infect. Dis.* **1990**, *162*, 1316–1323. [CrossRef]
10. Broome, C.V. Epidemiology of *Haemophilus influenzae* type b infections in the United States. *Pediatr. Infect. Dis. J.* **1987**, *6*, 779–782. [CrossRef]
11. Salwén, K.M.; Vikerfors, T.; Olcén, P. Increased incidence of childhood bacterial meningitis. A 25-year study in a defined population in Sweden. *Scand. J. Infect. Dis.* **1987**, *19*, 1–11. [CrossRef]
12. Snaebjarnardóttir, K.; Erlendsdóttir, H.; Reynisson, I.K.; Kristinsson, K.; Halldórsdóttir, S.; Hardardóttir, H.; Gudnason, T.; Gottfredsson, M.; Haraldsson, Á. Bacterial meningitis in children in Iceland, 1975–2010: A nationwide epidemiological study. *Scand. J. Infect. Dis.* **2013**, *45*, 819–824. [CrossRef]
13. Spanjaard, L.; Bol, P.; Ekker, W.; Zanen, H.C. The incidence of bacterial meningitis in the Netherlands–a comparison of three registration systems, 1977–1982. *J. Infect.* **1985**, *11*, 259–268. [CrossRef]
14. Davison, K.L.; Ramsay, M.E. The epidemiology of acute meningitis in children in England and Wales. *Arch. Dis. Child.* **2003**, *88*, 662–664. [CrossRef]
15. Måkelå, P.H.; Takala, A.K.; Peltola, H.; Eskola, J. Epidemiology of Invasive *Haemophilus influenzae* Type b Disease. *J. Infect. Dis.* **1992**, *165*, S2–S6. [CrossRef]
16. Wilfert, C.M. Epidemiology of *Haemophilus influenzae* type b infections. *Pediatrics* **1990**, *85*, 631–635.
17. World Health Organization. *Haemophilus influenzae Type b (Hib) Meningitis in the Pre-Vaccine Era: A Global Review of Incidence, Age Distributions, and Case-Fatality Rates*; World Health Organization: Geneva, Switzerland, 2002; Available online: https://apps.who.int/iris/handle/10665/67572 (accessed on 6 April 2021).
18. Fothergill, L.D.; Wright, J. Influenzal Meningitis. *J. Immunol.* **1933**, *24*, 273.
19. Peltola, H.; Käyhty, H.; Sivonen, A.; Mäkelä, H. *Haemophilus influenzae* type b capsular polysaccharide vaccine in children: A double-blind field study of 100,000 vaccinees 3 months to 5 years of age in Finland. *Pediatrics* **1977**, *60*, 730–737.
20. Käyhty, H.; Peltola, H.; Karanko, V.; Mäkelä, P.H. The protective level of serum antibodies to the capsular polysaccharide of *Haemophilus influenzae* type b. *J. Infect. Dis.* **1983**, *147*, 1100. [CrossRef]
21. Anderson, P. The protective level of serum antibodies to the capsular polysaccharide of *Haemophilus influenzae* type b. *J. Infect. Dis.* **1984**, *149*, 1034–1035. [CrossRef] [PubMed]

22. Collins, S.; Ramsay, M.; Campbell, H.; Slack, M.P.; Ladhani, S.N. Invasive *Haemophilus influenzae* type b disease in England and Wales: Who is at risk after 2 decades of routine childhood vaccination? *Clin. Infect. Dis. Off. Publ. Infect. Dis. Soc. Am.* **2013**, *57*, 1715–1721. [CrossRef] [PubMed]
23. Ulanova, M.; Tsang, R.S.W. *Haemophilus influenzae* serotype a as a cause of serious invasive infections. *Lancet Infect. Dis.* **2014**, *14*, 70–82. [CrossRef]
24. Collins, S.; Litt, D.J.; Flynn, S.; Ramsay, M.E.; Slack, M.P.; Ladhani, S.N. Neonatal invasive *Haemophilus influenzae* disease in England and Wales: Epidemiology, clinical characteristics, and outcome. *Clin. Infect. Dis. Off. Publ. Infect. Dis. Soc. Am.* **2015**, *60*, 1786–1792. [CrossRef] [PubMed]
25. World Health Organization. Defeating Meningitis by 2030. Available online: https://www.who.int/initiatives/defeating-meningitis-by-2030 (accessed on 6 April 2021).
26. Peltola, H. Worldwide *Haemophilus influenzae* type b disease at the beginning of the 21st century: Global analysis of the disease burden 25 years after the use of the polysaccharide vaccine and a decade after the advent of conjugates. *Clin. Microbiol. Rev.* **2000**, *13*, 302–317. [CrossRef]
27. Bijlmer, H.A.; van Alphen, L.; Greenwood, B.M.; Brown, J.; Schneider, G.; Hughes, A.; Menon, A.; Zanen, H.C.; Valkenburg, H.A. The epidemiology of *Haemophilus influenzae* meningitis in children under five years of age in The Gambia, West Africa. *J. Infect. Dis.* **1990**, *161*, 1210–1215. [CrossRef]
28. Ward, J.I.; Margolis, H.S.; Lum, M.K.; Fraser, D.W.; Bender, T.R.; Anderson, P. *Haemophilus influenzae* disease in Alaskan Eskimos: Characteristics of a population with an unusual incidence of invasive disease. *Lancet* **1981**, *1*, 1281–1285. [CrossRef]
29. Losonsky, G.A.; Santosham, M.; Sehgal, V.M.; Zwahlen, A.; Moxon, E.R. *Haemophilus influenzae* disease in the White Mountain Apaches: Molecular epidemiology of a high risk population. *Pediatr. Infect. Dis.* **1984**, *3*, 539–547. [CrossRef] [PubMed]
30. Coulehan, J.L.; Michaels, R.H.; Hallowell, C.; Schults, R.; Welty, T.K.; Kuo, J.S. Epidemiology of *Haemophilus influenzae* type B disease among Navajo Indians. *Public Health Rep.* **1984**, *99*, 404–409.
31. Hanna, J.N.; Wild, B.E. Bacterial meningitis in children under five years of age in Western Australia. *Med. J. Aust.* **1991**, *155*, 160–164. [CrossRef] [PubMed]
32. Hammond, G.W.; Rutherford, B.E.; Malazdrewicz, R.; MacFarlane, N.; Pillay, N.; Tate, R.B.; Nicolle, L.E.; Postl, B.D.; Stiver, H.G. *Haemophilus influenzae* meningitis in Manitoba and the Keewatin District, NWT: Potential for mass vaccination. *Can. Med. Assoc. J.* **1988**, *139*, 743–747.
33. Bijlmer, H.A. World-wide epidemiology of *Haemophilus influenzae* meningitis; industrialized versus non-industrialized countries. *Vaccine* **1991**, *9*, S5–S9. [CrossRef]
34. Gilbert, G.L. Epidemiology of *Haemophilus influenzae* type b disease in Australia and New Zealand. *Vaccine* **1991**, *9*, S10–S13, discussion S25. [CrossRef]
35. Ward, J.I.; Lum, M.K.; Hall, D.B.; Silimperi, D.R.; Bender, T.R. Invasive *Haemophilus influenzae* type b disease in Alaska: Background epidemiology for a vaccine efficacy trial. *J. Infect. Dis.* **1986**, *153*, 17–26. [CrossRef] [PubMed]
36. Takala, A.K.; Eskola, J.; Peltola, H.; Måkelå, P.H. Epidemiology of invasive *Haemophilus influenzae* type b disease among children in Finland before vaccination with Haemophilus influenzae type b conjugate vaccine. *Pediatr. Infect. Dis. J.* **1989**, *8*, 297–302. [CrossRef] [PubMed]
37. Anderson, E.C.; Begg, N.T.; Crawshaw, S.C.; Hargreaves, R.M.; Howard, A.J.; Slack, M.P. Epidemiology of invasive *Haemophilus influenzae* infections in England and Wales in the pre-vaccination era (1990-2). *Epidemiol. Infect.* **1995**, *115*, 89–100. [CrossRef] [PubMed]
38. Clements, D.A.; Booy, R.; Dagan, R.; Gilbert, G.L.; Moxon, E.R.; Slack, M.P.; Takala, A.; Zimmermann, H.P.; Zuber, P.L.; Eskola, J. Comparison of the epidemiology and cost of *Haemophilus influenzae* type b disease in five western countries. *Pediatr. Infect. Dis. J.* **1993**, *12*, 362–367. [CrossRef]
39. Gervaix, A.; Suter, S. Epidemiology of invasive *Haemophilus influenzae* type b infections in Geneva, Switzerland, 1976 to 1989. *Pediatr. Infect. Dis. J.* **1991**, *10*, 370–374. [CrossRef] [PubMed]
40. Trollfors, B.; Claesson, B.A.; Strangert, K.; Taranger, J. *Haemophilus influenzae* meningitis in Sweden 1981–1983. *Arch. Dis. Child.* **1987**, *62*, 1220–1223. [CrossRef] [PubMed]
41. Dagan, R. Epidemiology of invasive *Haemophilus influenzae* type b (Hib) disease in Israel. *Vaccine* **1993**, *11* (Suppl. 1), S43–S45. [CrossRef]
42. Gilbert, G.L.; Clements, D.A.; Broughton, S.J. *Haemophilus influenzae* type b infections in Victoria, Australia, 1985 to 1987. *Pediatr. Infect. Dis. J.* **1990**, *9*, 252–257. [CrossRef]
43. Hansman, D.; Hanna, J.; Morey, F. High prevalence of invasive *Haemophilus influenzae* disease in central Australia, 1986. *Lancet* **1986**, *2*, 927. [CrossRef]
44. Peltola, H. *Haemophilus influenzae* type b disease and vaccination in Latin America and the Caribbean. *Pediatr. Infect. Dis. J.* **1997**, *16*, 780–787. [CrossRef]
45. Peltola, H. Spectrum and burden of severe *Haemophilus influenzae* type b diseases in Asia. *Bull. World Health Organ.* **1999**, *77*, 878–887.
46. Scott, S.; Altanseseg, D.; Sodbayer, D.; Nymadawa, P.; Bulgan, D.; Mendsaikhan, J.; Watt, J.P.; Slack, M.P.; Carvalho, M.G.; Hajjeh, R.; et al. Impact of *Haemophilus influenzae* Type b conjugate vaccine in Mongolia: Prospective population-based surveillance, 2002–2010. *J. Pediatr.* **2013**, *163*, S8–S11. [CrossRef] [PubMed]

47. Grimwood, K.; Anderson, V.A.; Bond, L.; Catroppa, C.; Hore, R.L.; Keir, E.H.; Nolan, T.; Roberton, D.M. Adverse outcomes of bacterial meningitis in school-age survivors. *Pediatrics* **1995**, *95*, 646–656.
48. Baraff, L.J.; Lee, S.I.; Schriger, D.L. Outcomes of bacterial meningitis in children: A meta-analysis. *Pediatr. Infect. Dis. J.* **1993**, *12*, 389–394. [CrossRef] [PubMed]
49. Letson, G.W.; Gellin, B.G.; Bulkow, L.R.; Parks, D.J.; Ward, J.I. Severity and frequency of sequelae of bacterial meningitis in Alaska Native infants. Correlation with a scoring system for severity of sequelae. *Am. J. Dis. Child.* **1992**, *146*, 560–566. [CrossRef] [PubMed]
50. D'Angio, C.T.; Froehlke, R.G.; Plank, G.A.; Meehan, D.J.; Aguilar, C.M.; Lande, M.B.; Hugar, L. Long-term Outcome of *Haemophilus influenzae* Meningitis in Navajo Indian Children. *Arch. Pediatr. Adolesc. Med.* **1995**, *149*, 1001–1008. [CrossRef]
51. Goetghebuer, T.; West, T.E.; Wermenbol, V.; Cadbury, A.L.; Milligan, P.; Lloyd-Evans, N.; Adegbola, R.A.; Mulholland, E.K.; Greenwood, B.M.; Weber, M.W. Outcome of meningitis caused by *Streptococcus pneumoniae* and *Haemophilus influenzae* type b in children in The Gambia. *Trop. Med. Int. Health* **2000**, *5*, 207–213. [CrossRef]
52. Eskola, J.; Peltola, H.; Takala, A.K.; Käyhty, H.; Hakulinen, M.; Karanko, V.; Kela, E.; Rekola, P.; Ronnberg, P.R.; Samuelson, J.S.; et al. Efficacy of *Haemophilus influenzae* type b polysaccharide-diphtheria toxoid conjugate vaccine in infancy. *N. Engl. J. Med.* **1987**, *317*, 717–722. [CrossRef]
53. Käyhty, H.; Karanko, V.; Peltola, H.; Mäkelä, P.H. Serum antibodies after vaccination with *Haemophilus influenzae* type b capsular polysaccharide and responses to reimmunization: No evidence of immunologic tolerance or memory. *Pediatrics* **1984**, *74*, 857–865.
54. O'Brien, K.L.; Steinhoff, M.C.; Edwards, K.; Keyserling, H.; Thoms, M.L.; Madore, D. Immunologic priming of young children by pneumococcal glycoprotein conjugate, but not polysaccharide, vaccines. *Pediatr. Infect. Dis. J.* **1996**, *15*, 425–430. [CrossRef]
55. Ladhani, S.N. Two decades of experience with the *Haemophilus influenzae* serotype b conjugate vaccine in the United Kingdom. *Clin. Ther.* **2012**, *34*, 385–399. [CrossRef]
56. Anderson, P.; Pichichero, M.E.; Insel, R.A. Immunization of 2-month-old infants with protein-coupled oligosaccharides derived from the capsule of *Haemophilus influenzae* type b. *J. Pediatr.* **1985**, *107*, 346–351. [CrossRef]
57. Barbour, M.L.; Booy, R.; Crook, D.W.; Griffiths, H.; Chapel, H.M.; Moxon, E.R.; Mayon-White, D. *Haemophilus influenzae* type b carriage and immunity four years after receiving the *Haemophilus influenzae* oligosaccharide-CRM197 (HbOC) conjugate vaccine. *Pediatr. Infect. Dis. J.* **1993**, *12*, 478–484. [CrossRef]
58. Kelly, D.F.; Moxon, E.R.; Pollard, A.J. *Haemophilus influenzae* type b conjugate vaccines. *Immunology* **2004**, *113*, 163–174. [CrossRef] [PubMed]
59. Ward, J.; Brenneman, G.; Letson, G.W.; Heyward, W.L. Limited efficacy of a *Haemophilus influenzae* type b conjugate vaccine in Alaska Native infants. The Alaska *H. influenzae* Vaccine Study Group. *N. Engl. J. Med.* **1990**, *323*, 1393–1401. [CrossRef] [PubMed]
60. Bulkow, L.R.; Wainwright, R.B.; Letson, G.W.; Chang, S.J.; Ward, J.I. Comparative immunogenicity of four *Haemophilus influenzae* type b conjugate vaccines in Alaska Native infants. *Pediatr. Infect. Dis. J.* **1993**, *12*, 484–492. [CrossRef] [PubMed]
61. Decker, M.D.; Edwards, K.M.; Bradley, R.; Palmer, P. Comparative trial in infants of four conjugate *Haemophilus influenzae* type b vaccines. *J. Pediatr.* **1992**, *120*, 184–189. [CrossRef]
62. Centers for Disease Control. Haemophilus b Conjugate Vaccines for Prevention of Haemophilus influenzae Type b Disease among Infants and Children Two Months of Age and Older: Recommendations of the Immunization Practices Advisory Committee (ACIP). Available online: https://www.cdc.gov/mmwr/preview/mmwrhtml/00041736.htm (accessed on 6 April 2021).
63. Gessner, B.D.; Sutanto, A.; Linehan, M.; Djelantik, I.G.; Fletcher, T.; Gerudug, I.K.; Mercer, D.; Moniaga, V.; Moulton, L.H.; Mulholland, K.; et al. Incidences of vaccine-preventable *Haemophilus influenzae* type b pneumonia and meningitis in Indonesian children: Hamlet-randomised vaccine-probe trial. *Lancet* **2005**, *365*, 43–52. [CrossRef]
64. Feikin, D.R.; Nelson, C.B.; Watt, J.P.; Mohsni, E.; Wenger, J.D.; Levine, O.S. Rapid assessment tool for *Haemophilus influenzae* type b disease in developing countries. *Emerg. Infect. Dis.* **2004**, *10*, 1270–1276. [CrossRef]
65. Hajjeh, R.A.; Privor-Dumm, L.; Edmond, K.; O'Loughlin, R.; Shetty, S.; Griffiths, U.K.; Bear, A.P.; Cohen, A.L.; Chandran, A.; Schuchat, A.; et al. Supporting new vaccine introduction decisions: Lessons learned from the Hib Initiative experience. *Vaccine* **2010**, *28*, 7123–7129. [CrossRef]
66. World Health Organization. WHO Position Paper on *Haemophilus influenzae* type b conjugate vaccines. *Wkly. Epidemiol. Rec.* **2006**, *81*, 445–452.
67. Ojo, L.R.; O'Loughlin, R.E.; Cohen, A.L.; Loo, J.D.; Edmond, K.M.; Shetty, S.S.; Bear, A.P.; Privor-Dumm, L.; Griffiths, U.K.; Hajjeh, R. Global use of *Haemophilus influenzae* type b conjugate vaccine. *Vaccine* **2010**, *28*, 7117–7122. [CrossRef]
68. International Vaccine Access Center. VIEW-Hub Report:Global Vaccine Introduction and Implementation. Available online: https://www.jhsph.edu/ivac/wp-content/uploads/2021/02/VIEW-hubReport_Dec2020.pdf (accessed on 6 April 2021).
69. Swingler, G.; Fransman, D.; Hussey, G. Conjugate vaccines for preventing *Haemophilus influenzae* type b infections. *Cochrane Database Syst. Rev.* **2007**, Cd001729. [CrossRef]
70. Watt, J.P.; Levine, O.S.; Santosham, M. Global reduction of Hib disease: What are the next steps? Proceedings of the meeting Scottsdale, Arizona, 22–25 September 2002. *J. Pediatr.* **2003**, *143*, S163–S187. [CrossRef]
71. McIntyre, P.B.; O'Brien, K.L.; Greenwood, B.; van de Beek, D. Effect of vaccines on bacterial meningitis worldwide. *Lancet* **2012**, *380*, 1703–1711. [CrossRef]

72. Watt, J.P.; Wolfson, L.J.; O'Brien, K.L.; Henkle, E.; Deloria-Knoll, M.; McCall, N.; Lee, E.; Levine, O.S.; Hajjeh, R.; Mulholland, K.; et al. Burden of disease caused by *Haemophilus influenzae* type b in children younger than 5 years: Global estimates. *Lancet* **2009**, *374*, 903–911. [CrossRef]
73. Wahl, B.; O'Brien, K.L.; Greenbaum, A.; Majumder, A.; Liu, L.; Chu, Y.; Luksic, I.; Nair, H.; McAllister, D.A.; Campbell, H.; et al. Burden of *Streptococcus pneumoniae* and *Haemophilus influenzae* type b disease in children in the era of conjugate vaccines: Global, regional, and national estimates for 2000-15. *Lancet. Glob. Health* **2018**, *6*, e744–e757. [CrossRef]
74. Davis, S.; Feikin, D.; Johnson, H.L. The effect of *Haemophilus influenzae* type b and pneumococcal conjugate vaccines on childhood meningitis mortality: A systematic review. *BMC Public Health* **2013**, *13* (Suppl. 3), S21. [CrossRef]
75. Morris, S.K.; Moss, W.J.; Halsey, N. *Haemophilus influenzae* type b conjugate vaccine use and effectiveness. *Lancet Infect. Dis.* **2008**, *8*, 435–443. [CrossRef]
76. Schuchat, A.; Hilger, T.; Zell, E.; Farley, M.M.; Reingold, A.; Harrison, L.; Lefkowitz, L.; Danila, R.; Stefonek, K.; Barrett, N.; et al. Active bacterial core surveillance of the emerging infections program network. *Emerg. Infect. Dis.* **2001**, *7*, 92–99. [CrossRef] [PubMed]
77. Thigpen, M.C.; Whitney, C.G.; Messonnier, N.E.; Zell, E.R.; Lynfield, R.; Hadler, J.L.; Harrison, L.H.; Farley, M.M.; Reingold, A.; Bennett, N.M.; et al. Bacterial meningitis in the United States, 1998–2007. *N. Engl. J. Med.* **2011**, *364*, 2016–2025. [CrossRef] [PubMed]
78. Centers for Disease Control and Prevention. Nationally Notifiable Infectious Diseases and Conditions, United States: Annual Tables. Available online: https://wonder.cdc.gov/nndss/static/2018/annual/2018-table1.html (accessed on 6 April 2021).
79. Polkowska, A.; Toropainen, M.; Ollgren, J.; Lyytikäinen, O.; Nuorti, J.P. Bacterial meningitis in Finland, 1995–2014: A population-based observational study. *BMJ Open* **2017**, *7*, e015080. [CrossRef]
80. Ali, M.; Chang, B.A.; Johnson, K.W.; Morris, S.K. Incidence and aetiology of bacterial meningitis among children aged 1–59 months in South Asia: Systematic review and meta-analysis. *Vaccine* **2018**, *36*, 5846–5857. [CrossRef]
81. Ceyhan, M.; Ozsurekci, Y.; Tanır Basaranoglu, S.; Gurler, N.; Sali, E.; Keser Emiroglu, M.; Oz, F.N.; Belet, N.; Duman, M.; Ulusoy, E.; et al. Multicenter Hospital-Based Prospective Surveillance Study of Bacterial Agents Causing Meningitis and Seroprevalence of Different Serogroups of *Neisseria meningitidis, Haemophilus influenzae* type b, and *Streptococcus pneumoniae* during 2015 to 2018 in Turkey. *mSphere* **2020**, *5*. [CrossRef] [PubMed]
82. Suga, S.; Ishiwada, N.; Sasaki, Y.; Akeda, H.; Nishi, J.; Okada, K.; Fujieda, M.; Oda, M.; Asada, K.; Nakano, T.; et al. A nationwide population-based surveillance of invasive *Haemophilus influenzae* diseases in children after the introduction of the *Haemophilus influenzae* type b vaccine in Japan. *Vaccine* **2018**, *36*, 5678–5684. [CrossRef] [PubMed]
83. Ladhani, S.N.; Ramsay, M.; Slack, M.P. The impact of *Haemophilus influenzae* serotype b resurgence on the epidemiology of childhood invasive *Haemophilus influenzae* disease in England and Wales. *Pediatr. Infect. Dis. J.* **2011**, *30*, 893–895. [CrossRef] [PubMed]
84. McVernon, J.; Andrews, N.; Slack, M.P.; Ramsay, M.E. Risk of vaccine failure after *Haemophilus influenzae* type b (Hib) combination vaccines with acellular pertussis. *Lancet* **2003**, *361*, 1521–1523. [CrossRef]
85. McVernon, J.; Johnson, P.D.; Pollard, A.J.; Slack, M.P.; Moxon, E.R. Immunologic memory in *Haemophilus influenzae* type b conjugate vaccine failure. *Arch. Dis. Child.* **2003**, *88*, 379–383. [CrossRef] [PubMed]
86. Ramsay, M.E.; McVernon, J.; Andrews, N.J.; Heath, P.T.; Slack, M.P. Estimating *Haemophilus influenzae* type b vaccine effectiveness in England and Wales by use of the screening method. *J. Infect. Dis.* **2003**, *188*, 481–485. [CrossRef]
87. McVernon, J.; Howard, A.J.; Slack, M.P.; Ramsay, M.E. Long-term impact of vaccination on *Haemophilus influenzae* type b (Hib) carriage in the United Kingdom. *Epidemiol. Infect.* **2004**, *132*, 765–767. [CrossRef]
88. Johnson, N.G.; Ruggeberg, J.U.; Balfour, G.F.; Lee, Y.C.; Liddy, H.; Irving, D.; Sheldon, J.; Slack, M.P.; Pollard, A.J.; Heath, P.T. *Haemophilus influenzae* type b reemergence after combination immunization. *Emerg. Infect. Dis.* **2006**, *12*, 937–941. [CrossRef]
89. Trotter, C.L.; McVernon, J.; Andrews, N.J.; Burrage, M.; Ramsay, M.E. Antibody to *Haemophilus influenzae* type b after routine and catch-up vaccination. *Lancet* **2003**, *361*, 1523–1524. [CrossRef]
90. Ladhani, S.; Slack, M.P.; Heys, M.; White, J.; Ramsay, M.E. Fall in *Haemophilus influenzae* serotype b (Hib) disease following implementation of a booster campaign. *Arch. Dis. Child.* **2008**, *93*, 665–669. [CrossRef] [PubMed]
91. Department of Health. Immunisation against Infectious Diseases Chapter 16: Haemophilus influenzae Type b (Hib). Available online: https://www.gov.uk/government/publications/haemophilus-influenzae-type-hib-the-green-book-chapter-16 (accessed on 6 April 2021).
92. Litt, D.; Fry, N.K.; Ladhani, S.N. Epidemiology of invasive *Haemophilus influenzae* disease in England: National surveillance, 2017–2018 PO-019-(EMGM2019-13268). In Proceedings of the 15th EMGM Congress, Lisbon, Portugal, 27–30 May 2019; Available online: https://www.gov.uk/government/publications/haemophilus-influenzae-type-hib-the-green-book-chapter-16 (accessed on 6 April 2021).
93. Von Gottberg, A.; de Gouveia, L.; Madhi, S.A.; du Plessis, M.; Quan, V.; Soma, K.; Huebner, R.; Flannery, B.; Schuchat, A.; Klugman, K. Impact of conjugate *Haemophilus influenzae* type b (Hib) vaccine introduction in South Africa. *Bull. World Health Organ.* **2006**, *84*, 811–818. [CrossRef]
94. Von Gottberg, A.; Cohen, C.; Whitelaw, A.; Chhagan, M.; Flannery, B.; Cohen, A.L.; de Gouveia, L.; Plessis, M.; Madhi, S.A.; Klugman, K.P. Invasive disease due to *Haemophilus influenzae* serotype b ten years after routine vaccination, South Africa, 2003–2009. *Vaccine* **2012**, *30*, 565–571. [CrossRef] [PubMed]

95. National Institute for Communicable Diseases. GERMS South Africa Annual Surveillance Review 2018. Available online: https://www.nicd.ac.za/wp-content/uploads/2019/11/GERMS-SA-AR-2018-Final.pdf (accessed on 6 April 2021).
96. Howie, S.R.; Oluwalana, C.; Secka, O.; Scott, S.; Ideh, R.C.; Ebruke, B.E.; Balloch, A.; Sambou, S.; Erskine, J.; Lowe, Y.; et al. The effectiveness of conjugate *Haemophilus influenzae* type b vaccine in The Gambia 14 years after introduction. *Clin. Infect. Dis. Off. Publ. Infect. Dis. Soc. Am.* **2013**, *57*, 1527–1534. [CrossRef] [PubMed]
97. Mackenzie, G.A.; Ikumapayi, U.N.; Scott, S.; Idoko, O.; Odutola, A.; Ndiaye, M.; Sahito, S.M.; Osuorah, C.D.; Manjang, A.; Jarju, S.; et al. Increased disease due to *Haemophilus influenzae* type b: Population-based surveillance in eastern Gambia, 2008–2013. *Pediatr. Infect. Dis. J.* **2015**, *34*, e107–e112. [CrossRef]
98. Hammitt, L.L.; Crane, R.J.; Karani, A.; Mutuku, A.; Morpeth, S.C.; Burbidge, P.; Goldblatt, D.; Kamau, T.; Sharif, S.; Mturi, N.; et al. Effect of *Haemophilus influenzae* type b vaccination without a booster dose on invasive *H influenzae* type b disease, nasopharyngeal carriage, and population immunity in Kilifi, Kenya: A 15-year regional surveillance study. *Lancet. Glob. Health* **2016**, *4*, e185–e194. [CrossRef]
99. Low, N.; Redmond, S.M.; Rutjes, A.W.; Martínez-González, N.A.; Egger, M.; di Nisio, M.; Scott, P. Comparing *Haemophilus influenzae* type b conjugate vaccine schedules: A systematic review and meta-analysis of vaccine trials. *Pediatr. Infect. Dis. J.* **2013**, *32*, 1245–1256. [CrossRef] [PubMed]
100. Mangtani, P.; Mulholland, K.; Madhi, S.A.; Edmond, K.; O'Loughlin, R.; Hajjeh, R. *Haemophilus influenzae* type b disease in HIV-infected children: A review of the disease epidemiology and effectiveness of Hib conjugate vaccines. *Vaccine* **2010**, *28*, 1677–1683. [CrossRef]
101. Tsang, R.S.W.; Ulanova, M. The changing epidemiology of invasive *Haemophilus influenzae* disease: Emergence and global presence of serotype a strains that may require a new vaccine for control. *Vaccine* **2017**, *35*, 4270–4275. [CrossRef]
102. Ladhani, S.N.; Collins, S.; Vickers, A.; Litt, D.J.; Crawford, C.; Ramsay, M.E.; Slack, M.P. Invasive *Haemophilus influenzae* serotype e and f disease, England and Wales. *Emerg. Infect. Dis.* **2012**, *18*, 725–732. [CrossRef] [PubMed]
103. Ladhani, S.; Slack, M.P.; Heath, P.T.; von Gottberg, A.; Chandra, M.; Ramsay, M.E. Invasive *Haemophilus influenzae* Disease, Europe, 1996–2006. *Emerg. Infect. Dis.* **2010**, *16*, 455–463. [CrossRef]
104. Van Eldere, J.; Slack, M.P.; Ladhani, S.; Cripps, A.W. Non-typeable *Haemophilus influenzae*, an under-recognised pathogen. *Lancet Infect. Dis.* **2014**, *14*, 1281–1292. [CrossRef]
105. Gratten, M.; Barker, J.; Shann, F.; Gerega, G.; Montgomery, J.; Kajoi, M.; Lupiwa, T. Non-type b *Haemophilus influenzae* meningitis. *Lancet* **1985**, *1*, 1343–1344. [CrossRef]
106. Gratten, M.; Montgomery, J. The bacteriology of acute pneumonia and meningitis in children in Papua New Guinea: Assumptions, facts and technical strategies. *Papua New Guin. Med. J.* **2005**, *48*, 73–86.
107. Millar, E.V.; O'Brien, K.L.; Watt, J.P.; Lingappa, J.; Pallipamu, R.; Rosenstein, N.; Hu, D.; Reid, R.; Santosham, M. Epidemiology of invasive *Haemophilus influenzae* type A disease among Navajo and White Mountain Apache children, 1988–2003. *Clin. Infect. Dis. Off. Publ. Infect. Dis. Soc. Am.* **2005**, *40*, 823–830. [CrossRef] [PubMed]
108. Tsang, R.S.; Mubareka, S.; Sill, M.L.; Wylie, J.; Skinner, S.; Law, D.K. Invasive *Haemophilus influenzae* in Manitoba, Canada, in the postvaccination era. *J. Clin. Microbiol.* **2006**, *44*, 1530–1535. [CrossRef]
109. Tsang, R.S.; Bruce, M.G.; Lem, M.; Barreto, L.; Ulanova, M. A review of invasive *Haemophilus influenzae* disease in the Indigenous populations of North America. *Epidemiol. Infect.* **2014**, *142*, 1344–1354. [CrossRef]
110. Bruce, M.G.; Deeks, S.L.; Zulz, T.; Navarro, C.; Palacios, C.; Case, C.; Hemsley, C.; Hennessy, T.; Corriveau, A.; Larke, B.; et al. Epidemiology of *Haemophilus influenzae* serotype a, North American Arctic, 2000–2005. *Emerg. Infect. Dis.* **2008**, *14*, 48–55. [CrossRef]
111. McConnell, A.; Tan, B.; Scheifele, D.; Halperin, S.; Vaudry, W.; Law, B.; Embree, J. Invasive infections caused by *Haemophilus influenzae* serotypes in twelve Canadian IMPACT centers, 1996–2001. *Pediatr. Infect. Dis. J.* **2007**, *26*, 1025–1031. [CrossRef] [PubMed]
112. Zwahlen, A.; Kroll, J.S.; Rubin, L.G.; Moxon, E.R. The molecular basis of pathogenicity in *Haemophilus influenzae*: Comparative virulence of genetically-related capsular transformants and correlation with changes at the capsulation locus cap. *Microb. Pathog.* **1989**, *7*, 225–235. [CrossRef]
113. Soeters, H.M.; Blain, A.; Pondo, T.; Doman, B.; Farley, M.M.; Harrison, L.H.; Lynfield, R.; Miller, L.; Petit, S.; Reingold, A.; et al. Current Epidemiology and Trends in Invasive *Haemophilus influenzae* Disease-United States, 2009–2015. *Clin. Infect. Dis. Off. Publ. Infect. Dis. Soc. Am.* **2018**, *67*, 881–889. [CrossRef] [PubMed]
114. Soeters, H.M.; Oliver, S.E.; Plumb, I.D.; Blain, A.E.; Zulz, T.; Simons, B.C.; Barnes, M.; Farley, M.M.; Harrison, L.H.; Lynfield, R.; et al. Epidemiology of Invasive *Haemophilus influenzae* Serotype a Disease-United States, 2008–2017. *Clin. Infect. Dis. Off. Publ. Infect. Dis. Soc. Am.* **2020**. [CrossRef]
115. Bender, J.M.; Cox, C.M.; Mottice, S.; She, R.C.; Korgenski, K.; Daly, J.A.; Pavia, A.T. Invasive *Haemophilus influenzae* disease in Utah children: An 11-year population-based study in the era of conjugate vaccine. *Clin. Infect. Dis. Off. Publ. Infect. Dis. Soc. Am.* **2010**, *50*, e41–e46. [CrossRef]
116. Crandall, H.; Christiansen, J.; Varghese, A.A.; Russon, A.; Korgenski, E.K.; Bengtson, E.K.; Dickey, M.; Killpack, J.; Knackstedt, E.D.; Daly, J.A.; et al. Clinical and Molecular Epidemiology of Invasive *Haemophilus influenzae* Serotype a Infections in Utah Children. *J. Pediatr. Infect. Dis.* **2019**, *9*, 650–655. [CrossRef] [PubMed]

117. Antony, S.; Kaushik, A.; Mauriello, C.; Chatterjee, A. Non-Type b *Haemophilus influenzae* Invasive Infections in North Dakota and South Dakota, 2013–2015. *J. Pediatr. Infect. Dis.* **2017**, *6*, 281–284.
118. Ribeiro, G.S.; Lima, J.B.; Reis, J.N.; Gouveia, E.L.; Cordeiro, S.M.; Lobo, T.S.; Pinheiro, R.M.; Ribeiro, C.T.; Neves, A.B.; Salgado, K.; et al. *Haemophilus influenzae* meningitis 5 years after introduction of the *Haemophilus influenzae* type b conjugate vaccine in Brazil. *Vaccine* **2007**, *25*, 4420–4428. [CrossRef]
119. Zanella, R.C.; Bokermann, S.; Andrade, A.L.; Flannery, B.; Brandileone, M.C. Changes in serotype distribution of *Haemophilus influenzae* meningitis isolates identified through laboratory-based surveillance following routine childhood vaccination against *H. influenzae* type b in Brazil. *Vaccine* **2011**, *29*, 8937–8942. [CrossRef] [PubMed]
120. Lima, J.B.; Ribeiro, G.S.; Cordeiro, S.M.; Gouveia, E.L.; Salgado, K.; Spratt, B.G.; Godoy, D.; Reis, M.G.; Ko, A.I.; Reis, J.N. Poor clinical outcome for meningitis caused by Haemophilus influenzae serotype a strains containing the IS1016-bexA deletion. *J. Infect. Dis.* **2010**, *202*, 1577–1584. [CrossRef] [PubMed]
121. Giufrè, M.; Cardines, R.; Brigante, G.; Orecchioni, F.; Cerquetti, M. Emergence of Invasive *Haemophilus influenzae* type a Disease in Italy. *Clin. Infect. Dis. Off. Publ. Infect. Dis. Soc. Am.* **2017**, *64*, 1626–1628. [CrossRef]
122. Collins, S.; Fry, N.K.; Ladhani, S.N.; Litt, D. The rise (and fall) of invasive *Haemophilus influenzae* serotype a (Hia) disease in England)C-(EMGM2019-13311). In Proceedings of the 15th EMGM Congress, Lisbon, Portugal, 27–30 May 2019; Available online: https://emgm.eu/meetings/emgm2019/emgm2019_abstracts.pdf (accessed on 6 April 2021).
123. Roaa, Z.; Abdulsalam, A.; Shahid, G.; Kamaldeen, B.; Tariq, A.F. Pediatric invasive disease due to *Haemophilus influenzae* serogroup a in Riyadh, Saudi Arabia: Case series. *J. Infect. Dev. Ctries* **2016**, *10*, 528–532. [CrossRef] [PubMed]
124. Cox, A.D.; Barreto, L.; Ulanova, M.; Bruce, M.G.; Tsang, R. Developing a vaccine for *Haemophilus influenzae* serotype a: Proceedings of a workshop. *Can. Commun. Dis. Rep.* **2017**, *43*, 89–95. [CrossRef] [PubMed]
125. Whittaker, R.; Economopoulou, A.; Dias, J.G.; Bancroft, E.; Ramliden, M.; Celentano, L.P. Epidemiology of Invasive *Haemophilus influenzae* Disease, Europe, 2007–2014. *Emerg. Infect. Dis.* **2017**, *23*, 396–404. [CrossRef]
126. Takla, A.; Schönfeld, V.; Claus, H.; Krone, M.; An der Heiden, M.; Koch, J.; Vogel, U.; Wichmann, O.; Lâm, T.-T. Invasive *Haemophilus influenzae* Infections in Germany After the Introduction of Routine Childhood Immunization, 2001–2016. *Open Forum Infect. Dis.* **2020**, *7*, ofaa444. [CrossRef]
127. Urwin, G.; Krohn, J.A.; Deaver-Robinson, K.; Wenger, J.D.; Farley, M.M. Invasive disease due to *Haemophilus influenzae* serotype f: Clinical and epidemiologic characteristics in the *H. influenzae* serotype b vaccine era. The *Haemophilus influenzae* Study Group. *Clin. Infect. Dis. Off. Publ. Infect. Dis. Soc. Am.* **1996**, *22*, 1069–1076. [CrossRef]
128. MacNeil, J.R.; Cohn, A.C.; Farley, M.; Mair, R.; Baumbach, J.; Bennett, N.; Gershman, K.; Harrison, L.H.; Lynfield, R.; Petit, S.; et al. Current epidemiology and trends in invasive *Haemophilus influenzae* disease–United States, 1989–2008. *Clin. Infect. Dis. Off. Publ. Infect. Dis. Soc. Am.* **2011**, *53*, 1230–1236. [CrossRef]
129. Abou-Hanna, J.; Panning, K.; Mehta, H. *Haemophilus influenzae* Type f Meningitis Complicated by Bilateral Subdural Empyema, Central Venous Thrombosis, and Bilateral Sensorineural Hearing Loss in an Immunocompetent 12-month-old. *Cureus* **2019**, *11*, e4850. [CrossRef] [PubMed]
130. Resman, F.; Ristovski, M.; Ahl, J.; Forsgren, A.; Gilsdorf, J.R.; Jasir, A.; Kaijser, B.; Kronvall, G.; Riesbeck, K. Invasive disease caused by *Haemophilus influenzae* in Sweden 1997–2009; evidence of increasing incidence and clinical burden of non-type b strains. *Clin. Microbiol. Infect. Off. Publ. Eur. Soc. Clin. Microbiol. Infect. Dis.* **2011**, *17*, 1638–1645. [CrossRef] [PubMed]
131. Collins, S.; Vickers, A.; Ladhani, S.N.; Flynn, S.; Platt, S.; Ramsay, M.E.; Litt, D.J.; Slack, M.P. Clinical and Molecular Epidemiology of Childhood Invasive Nontypeable *Haemophilus influenzae* Disease in England and Wales. *Pediatr. Infect. Dis. J.* **2016**, *35*, e76–e84. [CrossRef] [PubMed]
132. Ulu-Kilic, A.; Altay, F.A.; Gürbüz, Y.; Otgun, S.N.; Sencan, I. Haemophilus influenzae serotype e meningitis in an adult. *J. Infect. Dev. Ctries* **2010**, *4*, 253–255. [CrossRef]
133. Gkentzi, D.; Slack, M.P.; Ladhani, S.N. The burden of nonencapsulated *Haemophilus influenzae* in children and potential for prevention. *Curr. Opin. Infect. Dis.* **2012**, *25*, 266–272. [CrossRef]
134. Van Wessel, K.; Rodenburg, G.D.; Veenhoven, R.H.; Spanjaard, L.; van der Ende, A.; Sanders, E.A. Nontypeable *Haemophilus influenzae* invasive disease in The Netherlands: A retrospective surveillance study 2001–2008. *Clin. Infect. Dis. Off. Publ. Infect. Dis. Soc. Am.* **2011**, *53*, e1–e7. [CrossRef] [PubMed]
135. Cardines, R.; Giufre, M.; Mastrantonio, P.; Ciofi degli Atti, M.L.; Cerquetti, M. Nontypeable *Haemophilus influenzae* meningitis in children: Phenotypic and genotypic characterization of isolates. *Pediatr. Infect. Dis. J.* **2007**, *26*, 577–582. [CrossRef]
136. Kalies, H.; Siedler, A.; Grondahl, B.; Grote, V.; Milde-Busch, A.; von Kries, R. Invasive *Haemophilus influenzae* infections in Germany: Impact of non-type b serotypes in the post-vaccine era. *BMC Infect. Dis.* **2009**, *9*, 45. [CrossRef] [PubMed]
137. Sarangi, J.; Cartwright, K.; Stuart, J.; Brookes, S.; Morris, R.; Slack, M. Invasive *Haemophilus influenzae* disease in adults. *Epidemiol. Infect.* **2000**, *124*, 441–447. [CrossRef] [PubMed]
138. Dworkin, M.S.; Park, L.; Borchardt, S.M. The changing epidemiology of invasive *Haemophilus influenzae* disease, especially in persons $\geq$ 65 years old. *Clin. Infect. Dis. Off. Publ. Infect. Dis. Soc. Am.* **2007**, *44*, 810–816. [CrossRef]
139. Desai, S.; Jamieson, F.B.; Patel, S.N.; Seo, C.Y.; Dang, V.; Fediurek, J.; Navaranjan, D.; Deeks, S.L. The Epidemiology of Invasive *Haemophilus influenzae* Non-Serotype b Disease in Ontario, Canada from 2004 to 2013. *PLoS ONE* **2015**, *10*, e0142179. [CrossRef]
140. Slack, M.; Esposito, S.; Haas, H.; Mihalyi, A.; Nissen, M.; Mukherjee, P.; Harrington, L. *Haemophilus influenzae* type b disease in the era of conjugate vaccines: Critical factors for successful eradication. *Exp. Rev. Vaccines* **2020**, *19*, 903–917. [CrossRef]

Article

Global Landscape Review of Serotype-Specific Invasive Pneumococcal Disease Surveillance among Countries Using PCV10/13: The Pneumococcal Serotype Replacement and Distribution Estimation (PSERENADE) Project

Maria Deloria Knoll [1,*], Julia C. Bennett [1], Maria Garcia Quesada [1], Eunice W. Kagucia [2], Meagan E. Peterson [1], Daniel R. Feikin [3], Adam L. Cohen [4,‡], Marissa K. Hetrich [1], Yangyupei Yang [1], Jenna N. Sinkevitch [1], Krow Ampofo [5], Laurie Aukes [6], Sabrina Bacci [7], Godfrey Bigogo [8], Maria-Cristina C. Brandileone [9], Michael G. Bruce [10], Romina Camilli [11], Jesús Castilla [12,13], Guanhao Chan [14], Grettel Chanto Chacón [15], Pilar Ciruela [12,16], Heather Cook [17], Mary Corcoran [18], Ron Dagan [19], Kostas Danis [20], Sara de Miguel [21], Philippe De Wals [22], Stefanie Desmet [23,24], Yvonne Galloway [25], Theano Georgakopoulou [26], Laura L. Hammitt [1,2], Markus Hilty [27], Pak-Leung Ho [28], Sanjay Jayasinghe [29], James D. Kellner [30], Jackie Kleynhans [31,32], Mirjam J. Knol [33], Jana Kozakova [34], Karl Gústaf Kristinsson [35], Shamez N. Ladhani [36], Claudia S. Lara [37], Maria Eugenia León [38], Tiia Lepp [39], Grant A. Mackenzie [40,41,42], Lucia Mad'arová [43], Allison McGeer [44], Tuya Mungun [45], Jason M. Mwenda [46], J. Pekka Nuorti [47,48], Néhémie Nzoyikorera [49,50], Kazunori Oishi [51], Lucia Helena De Oliveira [52], Metka Paragi [53], Tamara Pilishvili [54], Rodrigo Puentes [55], Eric Rafai [56], Samir K. Saha [57], Larisa Savrasova [58,59], Camelia Savulescu [60], J. Anthony Scott [2], Kevin J. Scott [61], Fatima Serhan [4], Lena Petrova Setchanova [62], Nadja Sinkovec Zorko [63], Anna Skoczyńska [64], Todd D. Swarthout [65,66], Palle Valentiner-Branth [67], Mark van der Linden [68], Didrik F. Vestrheim [69], Anne von Gottberg [31,70], Inci Yildirim [71], Kyla Hayford [1,§] and the PSERENADE Team [†]

Citation: Deloria Knoll, M.; Bennett, J.C.; Garcia Quesada, M.; Kagucia, E.W.; Peterson, M.E.; Feikin, D.R.; Cohen, A.L.; Hetrich, M.K.; Yang, Y.; Sinkevitch, J.N.; et al. Global Landscape Review of Serotype-Specific Invasive Pneumococcal Disease Surveillance among Countries Using PCV10/13: The Pneumococcal Serotype Replacement and Distribution Estimation (PSERENADE) Project. *Microorganisms* **2021**, *9*, 742. https://doi.org/10.3390/microorganisms9040742

Academic Editor: James Stuart

Received: 3 March 2021
Accepted: 26 March 2021
Published: 2 April 2021

1 Johns Hopkins Bloomberg School of Public Health, Baltimore, MD 21205, USA; jbenne63@jhu.edu (J.C.B.); mgarci64@jhmi.edu (M.G.Q.); meaganepeterson@gmail.com (M.E.P.); mhetric2@jhmi.edu (M.K.H.); yyang165@jhmi.edu (Y.Y.); jsinkev1@jhu.edu (J.N.S.); lhammitt@jhu.edu (L.L.H.); kylahayford@jhu.edu (K.H.)
2 Epidemiology and Demography Department, KEMRI-Wellcome Trust Research Programme, Centre for Geographic Medicine-Coast, P.O. Box 230-80108, Kilifi, Kenya; EKagucia@kemri-wellcome.org (E.W.K.); anthony.scott@lshtm.ac.uk (J.A.S.)
3 Independent Consultant, 1296 Coppet, Switzerland; drf3217@gmail.com
4 World Health Organization, 1202 Geneva, Switzerland; dvj1@cdc.gov (A.L.C.); serhanfa@who.int (F.S.)
5 Division of Pediatric Infectious Diseases, Department of Pediatrics, University of Utah Health Sciences Center, Salt Lake City, UT 84132, USA; Krow.Ampofo@hsc.utah.edu
6 Vaccine Study Center, Kaiser Permanente, Oakland, CA 94612, USA; Laurie.A.Aukes@kp.org
7 European Centre for Disease Prevention and Control, 169 73 Solna, Sweden; Sabrina.Bacci@ecdc.europa.eu
8 Centre for Global Health Research, Kenya Medical Research Institute, P.O. Box 1578-40100, Kisumu, Kenya; GBigogo@kemricdc.org
9 National Laboratory for Meningitis and Pneumococcal Infections, Center of Bacteriology, Institute Adolfo Lutz (IAL), São Paulo 01246-902, Brazil; maria.brandileone@ial.sp.gov.br
10 Arctic Investigations Program, Division of Preparedness and Emerging Infections, National Center for Emerging and Zoonotic Infectious Diseases, Centers for Disease Control and Prevention, Anchorage, AK 99508, USA; zwa8@cdc.gov
11 Department of Infectious Diseases, Italian National Institute of Health (Istituto Superiore di Sanità, ISS), 00161 Rome, Italy; romina.camilli@iss.it
12 CIBER Epidemiología y Salud Pública, (CIBERESP), 28029 Madrid, Spain; jcastilc@navarra.es (J.C.); pilar.ciruela@gencat.cat (P.C.)
13 Instituto de Salud Pública de Navarra-IdiSNA, 31003 Pamplona, Spain
14 Singapore Ministry of Health, Communicable Diseases Division, Singapore 308442, Singapore; CHAN_Guanhao@moh.gov.sg
15 Instituto Costarricense de Investigación y Enseñanza en Nutrición y Salud, Tres Ríos, 30301 Cartago, Costa Rica; gchanto@inciensa.sa.cr
16 Surveillance and Public Health Emergency Response, Public Health Agency of Catalonia, 08005 Barcelona, Spain
17 Centre for Disease Control, Department of Health and Community Services, Darwin City, NT 8000, Australia; hcdarwin@gmail.com
18 Irish Meningitis and Sepsis Reference Laboratory, Children's Health Ireland at Temple Street, Temple Street, D01 YC76 Dublin 1, Ireland; mary.corcoran@cuh.ie

[19] The Faculty of Health Sciences, Ben-Gurion University of the Negev, 8410501 Beer-Sheva, Israel; rdagan@bgu.ac.il
[20] Santé Publique France, the French National Public Health Agency, FR-94410 Saint Maurice, France; Costas.DANIS@santepubliquefrance.fr
[21] Epidemiology Department, Dirección General de Salud Pública, 28009 Madrid, Spain; sarade.miguel@salud.madrid.org
[22] Department of Social and Preventive Medicine, Laval University, Québec, QC G1V 0A6, Canada; philippe.dewals@criucpq.ulaval.ca
[23] Department of Microbiology, Immunology and Transplantation, KU Leuven, 3000 Leuven, Belgium; stefanie.desmet@uzleuven.be
[24] National Reference Centre for Streptococcus Pneumoniae, University Hospitals Leuven, 3000 Leuven, Belgium
[25] Epidemiology Team, Institute of Environmental Science and Research, Porirua, Wellington 5022, New Zealand; Yvonne.Galloway@esr.cri.nz
[26] National Public Health Organisation, 15123 Athens, Greece; t.georgakopoulou@eody.gov.gr
[27] Swiss National Reference Centre for invasive Pneumococci, Institute for Infectious Diseases, University of Bern, 3012 Bern, Switzerland; Markus.Hilty@ifik.unibe.ch
[28] Department of Microbiology and Carol Yu Centre for Infection, Queen Mary Hospital, The University of Hong Kong, Hong Kong, China; plho@hku.hk
[29] National Centre for Immunisation Research and Surveillance and Discipline of Child and Adolescent Health, Children's Hospital Westmead Clinical School, Faculty of Medicine and Health, University of Sydney, Westmead, NSW 2145, Australia; sanjay.jayasinghe@health.nsw.gov.au
[30] Department of Pediatrics, University of Calgary, and Alberta Health Services, Calgary, AB T3B 6A8, Canada; kellner@ucalgary.ca
[31] Centre for Respiratory Diseases and Meningitis, National Institute for Communicable Diseases of the National Health Laboratory Service, Sandringham, Johannesburg 2192, South Africa; JackieL@nicd.ac.za (J.K.); annev@nicd.ac.za (A.v.G.)
[32] School of Public Health, Faculty of Health Sciences, University of the Witwatersrand, Braamfontein, Johannesburg 2000, South Africa
[33] National Institute for Public Health and the Environment, 3721 MA Bilthoven, The Netherlands; mirjam.knol@rivm.nl
[34] National Institute of Public Health (NIPH), 100 42 Praha, Czech Republic; jana.kozakova@szu.cz
[35] Department of Clinical Microbiology, Landspitali-The National University Hospital, Hringbraut, 101 Reykjavik, Iceland; karl@landspitali.is
[36] Immunisation and Countermeasures Division, Public Health England, London NW9 5EQ, UK; shamez.ladhani@phe.gov.uk
[37] Servicio de Bacteriología Clínica, Departamento de Bacteriología, INEI-ANLIS "Dr. Carlos G. Malbrán", Buenos Aires C1282 AFF, Argentina; cslara@anlis.gob.ar
[38] Laboratorio Central de Salud Pública, Asunción, Paraguay (Central Laboratory of Public Health, Asunción, Paraguay), Asunción, Paraguay; bacteriologia.lcsp@mspbs.gov.py
[39] Department of Communicable Disease and Control and Health Protection, Public Health Agency of Sweden, 171 82 Solna, Sweden; tiia.lepp@folkhalsomyndigheten.se
[40] Faculty of Infectious and Tropical Diseases, London School of Hygiene & Tropical Medicine, Keppel St, London WC1E 7HT, UK; gmackenzie@mrc.gm
[41] Medical Research Council Unit the Gambia at London School of Hygiene & Tropical Medicine, P.O. Box 273, Banjul, The Gambia
[42] New Vaccines Group, Murdoch Children's Research Institute, Parkville, Melbourne, VIC 3052, Australia
[43] National Reference Centre for Pneumococcal and Haemophilus Diseases, Regional Authority of Public Health, 975 56 Banská Bystrica, Slovakia; madarova@vzbb.sk
[44] Toronto Invasive Bacterial Diseases Network, and Department of Laboratory Medicine and Pathobiology, University of Toronto, Toronto, ON M5S 1A8, Canada; Allison.McGeer@sinaihealth.ca
[45] National Center of Communicable Diseases (NCCD), Ministry of Health, Bayanzurkh District, Ulaanbaatar 13336, Mongolia; tuya_mungun@yahoo.com
[46] World Health Organization Regional Office for Africa, P.O. Box 06, Brazzaville, Congo; mwendaj@who.int
[47] Department of Health Security, Finnish Institute for Health and Welfare, 00271 Helsinki, Finland; pekka.nuorti@tuni.fi
[48] Health Sciences Unit, Faculty of Social Sciences, Tampere University, 33100 Tampere, Finland
[49] Bacteriology-Virology and Hospital Hygiene Laboratory, Ibn Rochd University Hospital Centre, Casablanca 20250, Morocco; nzoyikorera@yahoo.fr
[50] Department of Microbiology, Faculty of Medicine and Pharmacy, Hassan II University of Casablanca, Casablanca 20000, Morocco
[51] Toyama Institute of Health, Imizu, Toyama 939-0363, Japan; toyamaeiken1@chic.ocn.ne.jp

[52] Pan American Health Organization, World Health Organization, Washington, DC 20037, USA; oliveirl@paho.org
[53] Centre for Medical Microbiology, National Laboratory of Health, Environment and Food, 2000 Maribor, Slovenia; metka.paragi@nlzoh.si
[54] National Center for Immunizations and Respiratory Diseases, Centers for Disease Control and Prevention, Atlanta, GA 30333, USA; tpilishvili@cdc.gov
[55] Instituto de Salud Pública de Chile, Santiago 7780050, Santiago Metropolitan, Chile; rpuentes@ispch.cl
[56] Ministry of Health and Medical Services, Suva, Fiji; eric.rafai@govnet.gov.fj
[57] Child Health Research Foundation, Dhaka 1207, Bangladesh; samirk.sks@gmail.com
[58] Centre for Disease Prevention and Control of Latvia, 1005 Riga, Latvia; larisa.savrasova@spkc.gov.lv
[59] Doctoral Studies Department, Riga Stradinš University, 1007 Riga, Latvia
[60] Epidemiology Department, Epiconcept, 75012 Paris, France; c.savulescu@epiconcept.fr
[61] Bacterial Respiratory Infection Service, Scottish Microbiology Reference Laboratory, NHS GG&C, Glasgow, G31 2ER, UK; kevin.scott@ggc.scot.nhs.uk
[62] Department of Medical Microbiology, Medical University of Sofia, Faculty of Medicine, 1431 Sofia, Bulgaria; lenasetchanova@hotmail.com
[63] Communicable Diseases Centre, National Institute of Public Health, 1000 Ljubljana, Slovenia; Nadja.Sinkovec-Zorko@nijz.si
[64] National Reference Centre for Bacterial Meningitis, National Medicines Institute, 00-725 Warsaw, Poland; a.skoczynska@nil.gov.pl
[65] Malawi-Liverpool-Wellcome Trust Clinical Research Programme, P.O. Box 30096, Chichiri, Blantyre, Malawi; tswarthout@mlw.mw
[66] NIHR Global Health Research Unit on Mucosal Pathogens, Division of Infection and Immunity, UCL, Bloomsbury, London WC1E 6BT, UK
[67] Infectious Disease Epidemiology and Prevention, Statens Serum Institut, DK-2300 Copenhagen, Denmark; pvb@ssi.dk
[68] National Reference Center for Streptococci, Department of Medical Microbiology, University Hospital RWTH Aachen, 52074 Aachen, Germany; mlinden@ukaachen.de
[69] Department of Infection Control and Vaccine, Norwegian Institute of Public Health, 0456 Oslo, Norway; didrik.frimann.vestrheim@fhi.no
[70] School of Pathology, Faculty of Health Sciences, University of the Witwatersrand, Braamfontein, Johannesburg 2000, South Africa
[71] Department of Pediatrics, Yale New Haven Children's Hospital, New Haven, CT 06504, USA; inci.yildirim@yale.edu
* Correspondence: mknoll2@jhu.edu
† Members are listed in Appendix A.
‡ Present address: Centers for Disease Control and Prevention, Atlanta, GA, USA. Affiliation at the time this work was completed was World Health Organization.
§ Present address: Pfizer, Inc., Collegeville, PA, USA. Affiliation at the time this work was completed was Johns Hopkins Bloomberg School of Public Health.

Abstract: Serotype-specific surveillance for invasive pneumococcal disease (IPD) is essential for assessing the impact of 10- and 13-valent pneumococcal conjugate vaccines (PCV10/13). The Pneumococcal Serotype Replacement and Distribution Estimation (PSERENADE) project aimed to evaluate the global evidence to estimate the impact of PCV10/13 by age, product, schedule, and syndrome. Here we systematically characterize and summarize the global landscape of routine serotype-specific IPD surveillance in PCV10/13-using countries and describe the subset that are included in PSERENADE. Of 138 countries using PCV10/13 as of 2018, we identified 109 with IPD surveillance systems, 76 of which met PSERENADE data collection eligibility criteria. PSERENADE received data from most (n = 63, 82.9%), yielding 240,639 post-PCV10/13 introduction IPD cases. Pediatric and adult surveillance was represented from all geographic regions but was limited from lower income and high-burden countries. In PSERENADE, 18 sites evaluated PCV10, 42 PCV13, and 17 both; 17 sites used a 3 + 0 schedule, 38 used 2 + 1, 13 used 3 + 1, and 9 used mixed schedules. With such a sizeable and generally representative dataset, PSERENADE will be able to conduct robust analyses to estimate PCV impact and inform policy at national and global levels regarding adult immunization, schedule, and product choice, including for higher valency PCVs on the horizon.

Keywords: global; invasive pneumococcal disease; pneumococcal meningitis; surveillance; pneumococcal conjugate vaccines

1. Introduction

Streptococcus pneumoniae is an important cause of morbidity and mortality globally, in both children and adults [1,2]. In 2007, the World Health Organization (WHO) first recommended including pneumococcal conjugate vaccines (PCV) in childhood immunization programs worldwide to prevent pneumococcal disease. WHO encouraged countries to implement surveillance of invasive pneumococcal disease (IPD) to establish a baseline rate of disease for evaluating vaccine impact [3]. In 2019, WHO expanded IPD surveillance recommendations to encourage high-quality sentinel surveillance to monitor the distribution of serotypes causing IPD and ideally population-based surveillance for evaluating PCV impact on IPD incidence and serotype replacement disease [4]. By 2020, 145 countries, including countries from all regions of the world, had introduced PCV into infant immunization programs [5], many of which have IPD surveillance systems [6–10]. However, an individual country's ability to assess vaccine impact and inform policy can be limited by small sample size, limited years of available data either pre- or post-vaccine introduction, limited serotyping capacity, lack of a population catchment area for estimating incidence rates, changes in surveillance systems over time that bias inferences on vaccine impact, or insufficient characterization of cases or evaluation of the detection system to enable assessment of potential bias [11]. Further, unrelated events and temporal changes that influence health or access to care and natural fluctuations in pneumococcal serotypes over time may obscure PCV impact. Even sites not affected by these issues cannot assess the long-term relative merits across PCV products or schedules among both vaccinated and unvaccinated individuals, and their results may not be generalizable to other settings without robust data. Multi-site analyses that include data from many surveillance sites representing a variety of settings and PCV regimens can overcome these limitations. Multisite analyses also lead to greater understanding of pneumococcal epidemiology and PCV impact around the world, and where there is heterogeneity, to greater understanding of the factors driving it, e.g., differences in local epidemiology versus PCV use.

WHO's Strategic Advisory Group of Experts (SAGE) on Immunization previously commissioned an analysis of PCV7 (Prevenar/Prevnar, Pfizer) impact [11] and several global and regional systematic reviews of IPD serotype distribution have also been conducted [12–15]. However, these reviews do not reflect the current setting of PCV10 (Synflorix, GlaxoSmithKline) and PCV13 (Prevenar13/Prevnar13, Pfizer) use, evaluate only published data, do not evaluate effects of PCV10 and PCV13 separately, or do not account for duration of PCV use. An updated, more comprehensive global analysis of the long-term effects of PCV10/13 on serotype-specific IPD incidence and serotype distribution is needed to inform policy related to pneumococcal epidemiology in PCV10/13-using countries, the potential value of future higher-valency PCVs, and global and national vaccination policy around product choice and schedule for children and immunization recommendations for adults.

WHO commissioned the Pneumococcal Serotype Replacement and Distribution Estimation (PSERENADE) project to summarize and estimate the impact of PCV10/13 programs on IPD incidence and serotype distribution among children and adults. Here we aimed to describe the landscape of available published and unpublished serotype-specific IPD surveillance data globally that can be used for evaluating vaccine impact, to identify limitations and gaps in the availability of IPD surveillance data globally, and to describe the surveillance sites included in PSERENADE to provide greater clarity in how the data used in PSERENADE analyses were gathered and processed.

2. Materials and Methods

2.1. Identification of Surveillance Sites

We aimed to systematically identify sites conducting serotype-specific IPD surveillance in countries where PCV10 or PCV13 was universally recommended for all infants by 1 January 2017 to ensure at least one full year of post-PCV10/13 surveillance data. Countries using PCV10/13 and their year of introduction were identified using View-Hub, a publicly available database with current information on PCV use worldwide [5]. IPD surveillance sites were identified using multiple approaches. First, we contacted the following surveillance networks: WHO-coordinated Global Invasive Bacterial Vaccine Preventable Disease (IB-VPD) Surveillance Network, the Pan American Health Organization (PAHO) Sistema de Redes de Vigilancia de los Agentes responsables de Neumonias y Meningitis (SIREVA) Network, the European Centre for Disease Prevention and Control *Streptococcus pneumoniae* Invasive Disease Network (SpIDnet), The European Surveillance System (ECDC), and the U.S. Centers for Disease Control and Prevention (CDC) Active Bacterial Core Surveillance (ABCs) system. Second, we conducted a systematic literature review including articles published in any language with publication dates between 1 January 2011 and 20 December 2018 to identify additional sites where serotype-specific IPD surveillance was conducted for at least a full year following PCV10/13 introduction. Seven databases (Embase (with Medline), PubMed, Web of Science (all databases), Global Index Medicus (including regional databases), Africa Wide Information, Global Health Database, and PASCAL) were searched using search terms modified for each database that were reviewed by a specialist librarian (Supplementary Materials C). Third, results from the PCV Review of Impact Evidence (PRIME) literature review [16] and the View-Hub PCV10/13 impact study module database [5] were used to identify other sites and to validate the search terms to ensure relevant studies were captured. Two reviewers fluent in the language of the written report independently screened all studies and a third reviewer adjudicated disagreements. Fourth, we reviewed citations from a prior literature search on changes in IPD incidence after PCV7 introduction, which included studies published in 1994–2010 [11]. Fifth, International Symposium on Pneumococci and Pneumococcal Diseases (ISPPD) abstracts from 2012–2018 were reviewed. Finally, experts on pneumococcal disease surveillance suggested additional countries or sites not yet identified.

2.2. Data Collection

Site investigators of identified surveillance sites and corresponding authors of studies identified in the literature review were contacted by email. Surveillance data were evaluated for suitability for inclusion in analyses of IPD serotype distribution and PCV impact on IPD incidence over time using standardized criteria intended to ensure comparability of methods and PCV uptake across sites (Table 1). Sites with suitable data were invited to participate in the PSERENADE project and contribute IPD surveillance data. IPD was defined as the isolation or detection of pneumococcus from a normally sterile site or detection of pneumococcus in cerebral spinal fluid (CSF) or pleural fluid using *lytA*-based PCR or antigen testing; pneumococcus detected in blood by PCR was not considered IPD given its low specificity [17]. Datasets provided by sites were preferentially used over data abstracted from literature in order to include the most up-to-date and comprehensive data available and to optimize the level of detail needed for planned analyses. Characterization of PSERENADE-eligible sites that chose not to participate in PSERENADE is based on descriptions in the published literature.

Table 1. Data collection inclusion criteria.

Data Collection Inclusion Criterion	Rationale
1. Site reports annual serotype-specific and age-specific IPD case counts obtained from normally sterile sites	Data must meet a standardized case definition of IPD to ensure comparability across sites, be stratified by age in order to evaluate direct and indirect effects, and be characterized by serotype in order to estimate serotype distributions and evaluate vaccine-type and serotype-specific changes in disease rates over time.
2. At least 50% of isolates serotyped per year in at least one age stratum	A minimum proportion of isolates must be serotyped to limit risks of non-representativeness of serotyped isolates, i.e., to ensure the serotype distribution based only on serotyped cases is not biased and to minimize chance from selective testing.
3. At least one complete year of data post-PCV10/13, excluding the year of introduction	Twelve continuous months are required to ensure data are not limited to an outbreak period and to control for seasonal fluctuations in disease or serotype-specific distribution.
4. At least 50% uptake for the primary PCV series at 12 months of age in at least one year post-PCV10/13	The goal is to evaluate PCV, not the immunization program. Therefore, vaccine uptake must be high enough to be able to affect serotype distribution/IPD incidence rates at the population level [18,19] and to represent the experience of countries with high coverage. It also serves to help eliminate heterogeneous results due to coverage to enable focus on effects of product and schedule.
5. Testing or reporting not limited to immunocompromised individuals or other specialized populations	PCV impact and serotype distribution may be different in specialized populations (e.g., HIV-positive populations) and may not be representative of the wider population [20].
6. No major changes or biases in surveillance that would affect estimates of serotype-specific proportions or rates	Changes in the surveillance system over the analysis period, such as a change in indication for blood culturing, introduction of new serotyping methods, or change in the population under surveillance, may bias interpretations of changes in incidence rates making it difficult to distinguish PCV effects from a change in the system. If changes are correlated with vaccine introduction, results may be incorrectly attributed to vaccine program impact.

Surveillance sites shared annual serotype-specific IPD case data by age in either an individual case-level or aggregate format using a standardized template. Population-based denominators were provided where available. Prior to sharing, data were de-identified and anonymized per The US Health Insurance Portability and Accountability Act (HIPAA) and The European Union (EU) General Data Protection Regulation 2016/679 (GDPR). Data were stored on a secure database at Johns Hopkins University. Where possible, the following additional case characteristics were provided: hospitalized vs. outpatient status (for children under five years of age), HIV status, specimen type, and clinical syndrome (meningitis vs. pneumonia). For meningitis, two case definitions were used: confirmed positive CSF (CSF+) and site-defined clinical meningitis syndrome. Pneumonia cases were defined based on site-specific definitions. Characterization of non-pneumonia/non-meningitis IPD cases was not requested given limitations in data availability.

Site investigators also completed a questionnaire describing the site's surveillance system and laboratory methods for detection of pneumococcus and serotyping of cases. The questionnaire requested information on the country's pneumococcal immunization program, including annual immunization uptake estimates representative of the population under surveillance, PCV schedule, year of PCV introduction and product used (including use of PCV7 prior to introduction of PCV10/13), catch-up campaigns, and adult pneumococcal vaccination programs. We also abstracted WHO and UNICEF Estimates of National Immunization Coverage (WUENIC) for national uptake with three doses of PCV for all years of available surveillance data [21]. In the absence of evidence to the contrary, we assumed countries receiving funding from Gavi, the Vaccine Alliance to support PCV implementation did not have an adult pneumococcal vaccine program.

For eligible PAHO countries participating in the SIREVA II surveillance network, the WHO-coordinated Global IB-VPD network facilitated data transfer for children under five years of age. For countries with additional data reported in SIREVA II reports beyond what was available in the WHO Global IB-VPD database, data for patients of all ages were abstracted from 2006–2016 (the last year of available data at the time of abstraction) by year, age group, and serotype [22]. Discrepancies in abstraction were adjudicated by a third reviewer (MGQ) fluent in Spanish. Colombia's SIREVA II data were abstracted from a separate report published by the country, which included annual data through 2018 [23]. SIREVA II diagnostic and laboratory methods were abstracted from a standardized laboratory manual [24].

A standard data quality review was conducted independently for each site by two PSERENADE team members. Descriptive figures of the data with respect to each of the data quality check elements in Table 2 were shared with investigators with expertise in IPD surveillance at each site to assess the quality of the data. These characterizations and discussions with investigators at each site were used to define eligibility by year, age group, and syndrome for the various subsequent primary and secondary analyses of the study.

PCV-using countries that had IPD surveillance data were summarized by data collection eligibility criteria and participation in PSERENADE. Sites were characterized by UN region [25], World Bank income level [26], under five mortality rate [27], childhood pneumococcal disease burden prior to PCV introduction [5], Gavi-eligibility status, PCV product, and PCV schedule. The surveillance systems and PCV programs were also described and summarized for sites included in PSERENADE.

Table 2. PSERENADE standard data quality review.

Data Quality Check	Rationale
A. Are there dramatic changes in overall IPD incidence rates (IR) from year to year that might not be explained by vaccine introduction?	Stable surveillance system, population structure and clinical practices should not exhibit dramatic unexplained changes.
B. Are vaccine-serotype IRs decreasing in the target age groups after vaccine introduction as expected?	Vaccine-serotype IRs should be decreasing in target age groups after vaccine introduction, given sufficient vaccine uptake.
C. Are there dramatic changes in overall case counts from year to year that might not be explained by vaccine introduction?	Dramatic unexplained changes in case counts could indicate changes in the surveillance system or clinical practices.
D. Are vaccine-serotype case counts decreasing in the target age groups after vaccine introduction as expected?	Vaccine-serotype case counts should be decreasing in target age groups after vaccine introduction, given sufficient vaccine uptake.
E. Have the number of cases due to serotype 14 and 6B among children < 5 years been eliminated or greatly reduced in the post-PCV era?	Serotype 14 and 6B should be decreasing after vaccine introduction. Persistent serotype 14 or 6B cases may indicate low immunization coverage or surveillance system changes.
F. Do the denominators used to calculate IRs in each age group change over time?	Population-based denominators should vary slightly but not substantially over time. If annual population denominators are not available (i.e., denominator only available in some years) rates may be an under- or over-estimate.
G. Do the denominators in each age group make sense relative to each other?	Based on conventional population age structures, we expect the number of children aged < 5 years to be less than adults aged $\geq$ 18 years. The number of adults aged $\geq$ 65 years would be expected to be less than that of adults aged 18–64 years.
H. Do all IPD IRs in each age group make sense relative to each other and the setting?	Expect IPD IRs to be highest in young children and older adults who are most vulnerable, but there can be exceptions in some settings where other age groups have age-associated excess risk [28].
I. Do at least 50% of cases for each age group/surveillance year stratum have a known serotype?	Ensures that the serotype distribution of serotyped cases is not biased or different from the serotype distribution of cases that were not serotyped or not fully serotyped. An exception can be made if the specimens were randomly selected for serotyping, when costs may prohibit all serotyped.
J. Does the site distinguish between: Serotype 6A and serotype 6C cases? Serotype 6B and serotype 6D cases?	In 2007 researchers discovered that pneumococci classified as serotype 6A on the basis of phenotype could be further distinguished chemically, resulting in identification of a new serotype, 6C [29]. Similarly, in 2009 serotype 6D was discovered as a chemically distinct serotype from 6B [30]. Pneumococci previously classified as serotype 6A or 6B would have to be retrospectively reevaluated to distinguish serotypes 6C and 6D, respectively.
K. Are undistinguished PCV13-type serotypes identifiable (e.g., '6A/6C' instead of '6A')?	Because undistinguished PCV13-type cases (e.g., 6A/6C) will need to be reapportioned based on the distribution of fully serotyped PCV13-type cases, confirmed '6A' cases need to be differentiated from unconfirmed (i.e., might be 6C). Dates of changes in serotyping methods or documentation of retrospective reclassification efforts are required.

3. Results

Pediatric and adult IPD surveillance data were available in every UN region of the world, representing countries from all World Bank income levels, under five mortality rate strata, levels of IPD disease burden, PCV products, and infant PCV schedules (Table 3). Of 138 countries with a universal infant PCV10/13 program operational for one or more years by January 1, 2018, we identified 109 conducting IPD surveillance (Table 3, Figure 1). Of these, 76 (69.7%) had surveillance that met PSERENADE eligibility criteria for data collection (Table 1) and 62 (81.6%) of those eligible participated. Surveillance sites in 14 countries that met data collection eligibility criteria did not contribute data to PSERENADE because they either did not respond or declined to participate. Characteristics associated with participation were not evaluated, but the proportion of participating eligible sites are detailed for each region (Table 3). The resulting dataset contained incidence rate data from 38 countries for evaluating PCV impact and case count data only from 24 additional countries for estimating serotype distribution.

Eligibility of IPD surveillance data varied by region, income level, and epidemiological setting (Table 3). In Asia and Africa, where most pneumococcal deaths occur, fewer than half (48.3%) of the 58 countries conducting IPD surveillance met PSERENADE inclusion criteria, compared to 75.0–100% of countries elsewhere, and only 57.1% of the 28 that were eligible participated in PSERENADE. Although most (90.5%) PCV-using low-income countries (LICs) had IPD surveillance, the surveillance was less likely to meet eligibility criteria than that in upper-middle-(UMICs) or high-income countries (HICs) (47.4% for LICs vs. 78.9 for UMICs and 88.9% for HICs). Among those countries with surveillance meeting eligibility criteria, LICs were also less likely to contribute to PSERENADE (44.4% vs. 82.5–93.3%). Similarly, countries with high or medium under-5 mortality rates were less likely to have surveillance systems meeting eligibility criteria (38.5% and 44.0%, respectively) than low mortality countries (84.5%), and of the 13 high-mortality countries with IPD surveillance, only 5 (38.5%) were eligible for PSERENADE and only 2 participated, neither of which had population-based denominators to estimate incidence rates. There were 19 Gavi-eligible PCV-using countries with IPD surveillance eligible for PSERENADE, 13 (68.4%) of which participated, including 5 with incidence data. Of the 61 countries using a schedule with three primary doses and no booster (3 + 0), only 22 (36.1%) had eligible data, compared to 56 (70.0%) of 80 countries using an infant PCV schedule with a booster dose (3 + 1 or 2 + 1). Although the proportion of countries with surveillance systems meeting eligibility criteria was similar by PCV product (PCV13: 64.7%; PCV10: 71.4%), there were more PCV13-using countries eligible for PSERENADE analyses (n = 44 vs. 15).

Seventy-seven sites from 62 countries participated in PSERENADE (Tables 3 and 4). All surveillance sites contributing data to PSERENADE collected pediatric data; although 88.0% overall also collected adult IPD data, those that did not were disproportionately from Sub-Saharan Africa and Asia where only 54.5% and 60.0% of sites, respectively, collected adult IPD data (Table 4). Data from the period prior to PCV introduction was available from 58 (77.3%) of surveillance sites. Although 51 (68.0%) sites conducted population-based surveillance with population denominators enabling incidence estimation, few of these were from the regions of Latin America and the Caribbean (three sites from two countries), Sub-Saharan Africa (six sites from four countries), and Northwestern Africa and Western Asia (two sites from two countries) (Table 4 and Supplementary Table S2).

All surveillance sites collected both blood and CSF except those in Sub-Saharan Africa, of which two (18.2%) collected blood only (Table 4); 68.9% of surveillance sites collected pleural fluid, with this proportion also lowest in Sub-Saharan Africa (2/11; 18.2%). Cases were characterized by clinical syndrome at 77.3% of sites overall, but those that did not characterize cases by clinical syndrome were disproportionately from the Latin America and the Caribbean region (11 of 19). Most surveillance sites (77.3%) used detection methods on CSF beyond culture (42.7% used antigen detection and 72.0% used nucleic acid detection). To identify the serotype, most (85.1%) sites used Quellung reaction and 73.0%

used another method, primarily PCR (62.2%) and latex agglutination (29.7%) (Table 4 and Supplementary Table S3).

In total, PSERENADE collected data on over 240,000 post-PCV10/13 IPD cases, with the majority from Europe (n = 142,586) and North America (n = 37,187), but with a substantial number also from Latin America and the Caribbean (n = 20,609), Sub-Saharan Africa (n = 19,734), and Oceania (n = 13,038) (Table 3). The average number of annual cases post-PCV10/13 was lowest among Sub-Saharan Africa (median across sites = 10) and Latin America and the Caribbean (median = 50) compared to other regions (median range: 124–548). The number of cases per site in total was generally lower for sites without surveillance among all ages, those with smaller population catchment areas, and those with fewer years since PCV10/13 introduction (data not shown). The median number of surveillance years post-PCV10/13 across regions ranged from 4 (Asia) to 7 (North America, Europe and Northern Africa/Western Asia; Table 3).

Most (54.5%) PSERENADE sites used PCV13, 23.4% used PCV10 and 22.1% used both products concurrently or switched between products (Tables 3 and 5). The majority of sites introduced PCV10/13 without a catch-up program (69.9%) and have a booster dose schedule (77.9%). PCV10/13 immunization coverage across the post-PCV10/13 period was high in most sites (mean uptake 87.9%, range 55–98%). The majority of sites have an adult pneumococcal vaccine program for polysaccharide vaccine (PPV23) and/or PCV13. Among these, 62.3% and 63.6% of sites recommend PPV23 for older adults and individuals at high risk for IPD, respectively, and 35.1% and 55.8%, respectively, recommend PCV13. Data on adult PPV23 and PCV13 uptake were available from 24 sites; 45.8% had >50% uptake (data not shown).

Table 3. Availability of invasive pneumococcal disease (IPD) surveillance data globally in PCV10/13-using countries.

Strata	Category	All PCV-Using Countries [1]				Data in PSERENADE				
		A. Countries Using PCV, N (% of Countries [2])	B. PCV-Using Countries with IPD Surveillance, N (% of A [2])	C. Countries Eligible for PSERENADE, N (% of B [2,3])	D. Countries in PSERENADE, N (% of C [2])	E. Countries with Incidence Data [4], N (% of D [2])	F. Number of Surveillance Sites	G. Total Number of Cases in Post-PCV10/13 Years [5]	H. Annual Number of Cases Averaged across Post-PCV10/13 Years, Median (IQR) [5,6]	I. Number of Years Post-PCV10/13 with Data, Median (IQR) [5]
Total	Total	138 (70.4%)	109 (79.0%)	76 (69.7%)	62 (81.6%)	38 (61.3%)	77	241,442	117 (26, 513)	7 (5, 7)
Region [7]	North America	2 (100.0%)	2 (100.0%)	2 (100.0%)	2 (100.0%)	2 (100.0%)	10	37,187	124 (55, 269)	7 (7, 8)
	Latin America and the Caribbean	22 (66.7%)	19 (86.4%)	19 (100.0%)	18 (94.7%)	2 (11.1%)	19	20,609	50 (21, 227)	5 (4, 6)
	Europe	31 (73.8%)	26 (83.9%)	24 (92.3%)	23 (95.8%)	20 (87.0%)	26	142,586	548 (115, 918)	7 (6, 8)
	Sub-Saharan Africa	39 (81.2%)	31 (79.5%)	14 (45.2%)	9 (64.3%)	4 (44.4%)	11	19,734	10 (7, 23)	6 (5, 6)
	Northern Africa and Western Asia	17 (73.9%)	14 (82.4%)	7 (50.0%)	2 (28.6%)	2 (100.0%)	2	4,380	313 (171, 454)	7 (7, 7)
	Asia	17 (53.1%)	13 (76.5%)	7 (53.8%)	5 (71.4%)	5 (100.0%)	5	3,908	126 (77, 179)	4 (3, 7)
	Oceania	10 (62.5%)	4 (40.0%)	3 (75.0%)	3 (100.0%)	3 (100.0%)	4	13,038	274 (48, 748)	6 (6, 7)
World Bank Income level [8]	High income	52 (83.9%)	45 (86.5%)	40 (88.9%)	33 (82.5%)	29 (87.9%)	46	206,562	377 (99, 824)	7 (6, 8)
	Upper middle income	27 (50.0%)	19 (70.4%)	15 (78.9%)	14 (93.3%)	3 (21.4%)	14	33,085	55 (21, 272)	6 (4, 7)
	Lower middle income	38 (74.5%)	26 (68.4%)	12 (46.2%)	11 (91.7%)	4 (36.4%)	12	968	12 (8, 19)	5 (4, 6)
	Low income	21 (72.4%)	19 (90.5%)	9 (47.4%)	4 (44.4%)	2 (50.0%)	5	827	10 (9, 29)	6 (5, 6)
Under 5 mortality rate (2018) [9]	Low	87 (66.4%)	71 (81.6%)	60 (84.5%)	52 (86.7%)	33 (63.5%)	65	221,478	179 (42, 587)	7 (5, 7)
	Medium	35 (79.5%)	25 (71.4%)	11 (44.0%)	8 (72.7%)	5 (62.5%)	10	19,948	14 (9, 65)	6 (5, 7)
	High	16 (76.2%)	13 (81.2%)	5 (38.5%)	2 (40.0%)	0 (0.0%)	2	16	3 (3, 3)	6 (6, 6)
Pre-PCV under 5 Spn disease burden (2000) [10]	Low burden	42 (77.8%)	38 (90.5%)	33 (86.8%)	29 (87.9%)	27 (93.1%)	41	200,066	469 (117, 878)	7 (7, 8)
	Medium burden	34 (60.7%)	23 (67.6%)	20 (87.0%)	17 (85.0%)	3 (17.6%)	18	20,356	64 (24, 275)	6 (5, 7)
	High burden	62 (74.7%)	48 (77.4%)	23 (47.9%)	16 (69.6%)	8 (50.0%)	18	21,020	17 (9, 30)	5 (4, 6)

Table 3. *Cont.*

Strata	Category	All PCV-Using Countries [1]				Data in PSERENADE				
		A. Countries Using PCV, N (% of Countries [2])	B. PCV-Using Countries with IPD Surveillance, N (% of A [2])	C. Countries Eligible for PSERENADE, N (% of B [2,3])	D. Countries in PSERENADE, N (% of C [2])	E. Countries with Incidence Data [4], N (% of D [2])	F. Number of Surveillance Sites	G. Total Number of Cases in Post-PCV10/13 Years [5]	H. Annual Number of Cases Averaged across Post-PCV10/13 Years, Median (IQR) [5,6]	I. Number of Years Post-PCV10/13 with Data, Median (IQR) [5]
Gavi status [11]	Gavi	57 (78.1%)	44 (77.2%)	19 (43.2%)	13 (68.4%)	5 (38.5%)	15	1455	10 (7, 15)	5 (4, 6)
	Non-Gavi	81 (65.9%)	65 (80.2%)	57 (87.7%)	49 (86.0%)	33 (67.3%)	62	239,987	262 (57, 625)	7 (6, 8)
Product	PCV10	22 (15.9%) [15]	21 (95.5%)	15 (71.4%)	14 (93.3%)	8 (57.1%)	18	23,967	49 (14, 416)	6 (5, 7)
	PCV13	93 (67.4%) [15]	68 (73.1%)	44 (64.7%)	34 (77.3%)	19 (55.9%)	42	183,610	123 (31, 594)	7 (6, 7)
	PCV10 and PCV13 [12]	23 (16.7%) [15]	20 (87.0%)	17 (85.0%)	14 (82.4%)	11 (78.6%)	17	33,865	209 (56, 386)	7 (6, 7)
Schedule [13]	3 + 0	58 (42.0%) [16]	44 (75.9%)	20 (45.5%)	14 (70.0%)	5 (35.7%)	17	10,825	12 (8, 29)	6 (5, 6)
	2 + 1	48 (34.8%) [16]	40 (83.3%)	36 (90.0%)	33 (91.7%)	20 (60.6%)	38	151,942	308 (70, 594)	7 (5, 7)
	3 + 1	19 (13.8%) [16]	13 (68.4%)	10 (76.9%)	6 (60.0%)	5 (83.3%)	13	32,716	92 (42, 247)	7 (7, 8)
	3 + 0 and 2 + 1/3 + 1 [14]	3 (2.2%) [16]	2 (66.7%)	2 (100.0%)	1 (50.0%)	1 (100.0%)	0[14]	0	0 (0, 0)	0 (0, 0)
	3 + 1 and 2 + 1 [15]	10 (7.2%) [16]	10 (100.0%)	8 (80.0%)	8 (100.0%)	7 (87.5%)	9	45,959	634 (276, 932)	7 (7, 8)

[1] Countries with a full year of a PCV10/13 immunization program for infants by 2018 (i.e., introduced by 1 January 2017). Countries with only a risk immunization program rather than universal are also not counted as PCV-using countries. Data from View-Hub [5]. Taiwan and Hong Kong are not merged with China in this table given differences in PCV use and availability of IPD surveillance data compared to the rest of China. [2] Percentage by category unless otherwise specified. [3] To be eligible for PSERENADE, a surveillance site must have had at least one full year of post-PCV10/13 IPD incidence or four years of post-PCV10/13 IPD case counts, over 50% vaccination uptake, and over 50% of cases serotyped by age/year group (Table 1). [4] Incidence data are only available for pneumococcal meningitis cases in Brazil and Greece, as opposed to all IPD in all other countries. [5] Post-PCV10/13 years exclude the year of introduction. [6] The average number of cases in post-PCV10/13 years was calculated for each surveillance site and used to estimate the median (IQR) across strata categories. [7] United Nations (UN) regions adapted from UN Statistics Division [25]. [8] World Bank Income level as of November 2020 [26]. [9] Under 5-year mortality rate data from United Nations Interagency Group for Child Mortality Estimation (2020), 2018 estimate by country. Low: <30 deaths per 1000 livebirths, medium: 30 to <75 deaths, high: 75 to <150 deaths [27]. [10] Pre-PCV pneumococcal disease burden estimates for children <5 years calculated as the sum of estimated pneumonia, meningitis, and invasive non-pneumonia, non-meningitis incidence rates in 2000 [5]. Strata were defined as fewer than 300 cases per 100,000 children (low burden), 300 to fewer than 2000 cases per 100,000 children (medium burden), or 2000 or more cases per 100,000 children (high burden). Countries missing any or all incidence rates were categorized as "Unknown". [11] Gavi countries are those that are eligible or have graduated. [12] Countries that either used both products concurrently or switched between PCV10 and PCV13. [13] 3 + 0: three primary doses and no booster; 2 + 1: two primary doses and a booster; 3 + 1: three primary doses and a booster. [14] Countries that used PCV10/13 schedules with and without a booster dose. Australia, included in PSERENADE, uses 3 + 1 among indigenous populations and used 3 + 0 among non-indigenous populations until 2018, when non-indigenous changed to 2 + 1. Because Australia (non-indigenous) predominantly used 3 + 0 during the years described here, that surveillance site was categorized as 3 + 0 in columns FI, and Australia (indigenous) was categorized as a 3 + 1 surveillance site in columns F–I. Not included in PSERENADE were Trinidad and Tobago (switched from 3 + 0 to 3 + 1) and Libyan Arab Jamahiriya (switched from 3 + 0 to 2 + 1). [15] Countries that used 3 + 1 and 2 + 1 PCV10/13 schedules. All switched from 3 + 1 to 2 + 1 except for Poland, which uses 2 + 1 in the National Immunization Program (NIP) and 3 + 1 in the private market, and Canada, which uses 3 + 1 and/or 2 + 1 in different provinces. Canadian surveillance sites for individual provinces are categorized accordingly in columns F-I. [16] Percentage is of the 138 PCV-using countries.

Table 4. Summary of PSERENADE surveillance sites by region [1,2].

	North America N = 9	Latin America and the Caribbean N = 19	Europe N = 26	Sub-Saharan Africa N = 11	N. Africa and W. Asia N = 2	Asia N = 5	Oceania N = 3	Total N = 75
Availability of data, N (%)								
0–17 years	9 (100%)	19 (100%)	26 (100%)	11 (100%)	2 (100%)	5 (100%)	3 (100%)	75 (100%)
≥18 years	7 (77.8%)	19 (100%)	26 (100%)	6 (54.5%)	2 (100%)	3 (60.0%)	3 (100%)	66 (88.0%)
Pre-PCV period	7 (77.8%)	19 (100%)	17 (65.4%)	6 (54.5%)	2 (100%)	4 (80.0%)	3 (100%)	58 (77.3%)
PCV7 period [3]	8 (88.9%)	8 (100%)	16 (84.2%)	2 (100%)	1 (100%)	3 (75.0%)	2 (100%)	40 (88.9%)
PCV10/13 period	9 (100%)	19 (100%)	26 (100%)	11 (100%)	2 (100%)	5 (100%)	3 (100%)	75 (100%)
Incidence data [4]	9 (100%)	3 (15.8%)	23 (88.5%)	6 (54.5%)	2 (100%)	5 (100%)	3 (100%)	51 (68.0%)
Clinical syndrome data	8 (88.9%)	11 (57.9%)	20 (76.9%)	10 (90.9%)	1 (50.0%)	5 (100%)	3 (100%)	58 (77.3%)
Specimens collected, N (%) [5]								
Blood	9 (100%)	19 (100%)	25 (100%)	11 (100%)	2 (100%)	5 (100%)	3 (100%)	74 (100%)
CSF	9 (100%)	19 (100%)	25 (100%)	9 (81.8%)	2 (100%)	5 (100%)	3 (100%)	72 (97.3%)
Pleural fluid	7 (77.8%)	18 (94.7%)	17 (68.0%)	2 (18.2%)	1 (50.0%)	4 (80.0%)	2 (66.7%)	51 (68.9%)
Additional detection methods, N (%)								
Nucleic acid	2 (22.2%)	16 (84.2%)	23 (88.5%)	6 (54.5%)	1 (50.0%)	4 (80.0%)	2 (66.7%)	54 (72.0%)
Antigen detection	0 (0.0%)	13 (68.4%)	14 (53.8%)	0 (0.0%)	0 (0.0%)	3 (60.0%)	2 (66.7%)	32 (42.7%)
Serotyping methods, N (%) [5]								
Quellung	9 (100%)	19 (100%)	23 (92.0%)	4 (36.4%)	2 (100%)	3 (60.0%)	3 (100%)	63 (85.1%)
Non-Quellung	2 (22.2%)	12 (63.2%)	22 (88.0%)	11 (100%)	1 (50.0%)	4 (80.0%)	2 (66.7%)	54 (73.0%)
Latex agglutination	1 (11.1%)	2 (10.5%)	15 (60.0%)	3 (27.3%)	1 (50.0%)	0 (0.0%)	0 (0.0%)	22 (29.7%)
Any PCR method [6]	2 (22.2%)	12 (63.2%)	14 (56.0%)	11 (100%)	1 (50.0%)	4 (80.0%)	2 (66.7%)	46 (62.2%)
PCR35/37/38 [7,8]	0 (0.0%)	10 (52.6%)	2 (8.0%)	2 (18.2%)	0 (0.0%)	1 (20.0%)	0 (0.0%)	15 (20.3%)
PCR70/76 [7,8]	2 (22.2%)	5 (26.3%)	8 (32.0%)	0 (0.0%)	1 (50.0%)	3 (60.0%)	1 (33.3%)	20 (27.0%)
Other method [9]	1 (11.1%)	0 (0.0%)	6 (24.0%)	0 (0.0%)	0 (0.0%)	0 (0.0%)	0 (0.0%)	7 (9.5%)

[1] Subpopulations (e.g., indigenous and non-indigenous) from the same surveillance system were presented as one site. Countries with more than one surveillance site are represented more than once. Data for individual surveillance sites are in Supplementary Table S3. [2] United Nations (UN) regions adapted from UN Statistics Division [25]. N. Africa and W. Asia: Northern Africa and Western Asia. [3] Sites that did not use PCV7 were excluded from calculations of PCV7 period data availability (not applicable). Total calculations are out of the 45 sites that used PCV7. [4] Incidence data from Brazil and Greece are for pneumococcal meningitis only. [5] One site (Lithuania) with unknown specimen type and serotyping data was excluded from calculations. Total calculations are out of 74 sites. [6] Comprised of sites that use PCR at any capacity—including those with unknown or custom PCR schemes that do not fall into PCR35/37/38 or PCR70/76 categories. [7] The number following "PCR" indicates the number of serotypes able to be identified by PCR. Similar serotyping capacities were grouped together. [8] Argentina, Mexico, and Paraguay use both PCR37 and PCR70 and are counted in both of those categories. [9] Includes sites that reported other serotyping methods: Whole genome sequencing (WGS), Next generation sequencing (NGS), Capsular sequence typing (CST), or Gel diffusion (GD).

A. Children < 18 years of age

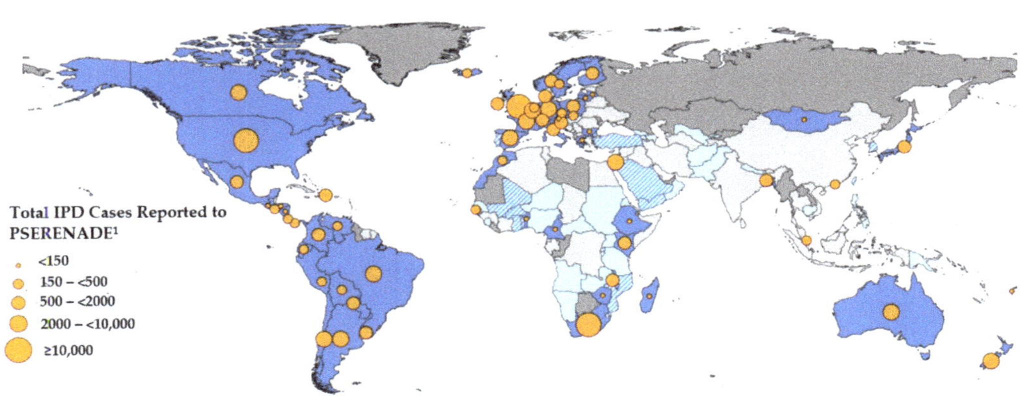

B. Adults ≥ 18 years of age

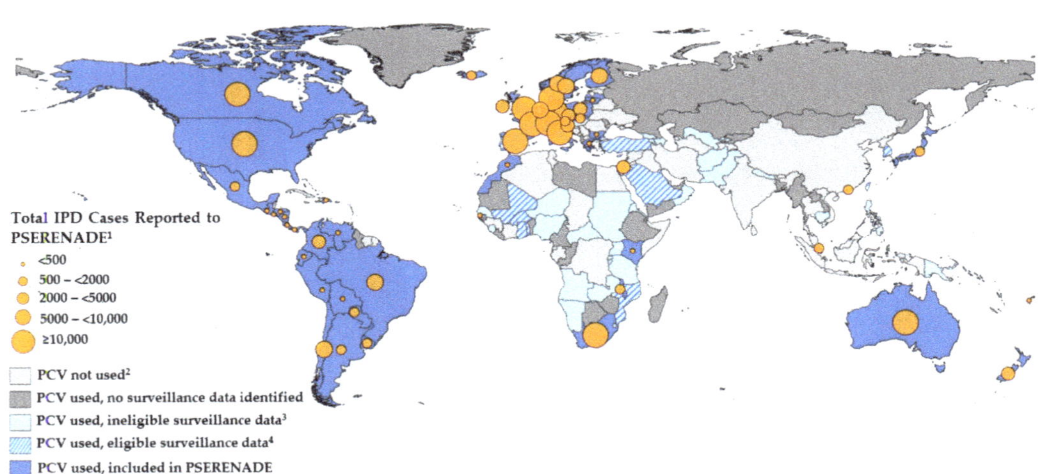

Figure 1. Availability of IPD surveillance data for countries with universal recommendations for PCV in the infant immunization program. [1] Cases from multiple surveillance sites within the same country were aggregated.[2] PCV not universally introduced into the routine infant immunization program by 2018 (includes India which began sub-national introduction in 2017). [3] IPD surveillance data did not meet PSERENADE data collection eligibility criteria (Box 1). [4] IPD surveillance data met data collection eligibility criteria but did not participate in PSERENADE.

Table 5. Description of infant and adult pneumococcal immunization use for PSERENADE sites.

Infant PCV Product [1]	PCV10/13 Schedule [1]	Region [2]	Site [3]	PCV7 Period	PCV10 Period	PCV13 Period	PCV10/13 Catch-Up	Mean PCV10/13 Uptake (%)		Other PCV/PPV Recommendations [7]	
								Primary Series [5]	WUENIC PCV3 [6]	Adult	High Risk
PCV10	3 + 0	LA and C	Ecuador (SIREVA, WHO)	2010–2011	2010–	–	N	–	85	–	–
		Sub-Saharan Africa	Ethiopia (WHO)	– [4]	2011–	–	N	–	58	–	–
			Kenya, Asembo	–	2011–	–	Y	86	81	–	–
			Kenya, Kibera	–	2011–	–	N	87	81	–	–
			Kenya, Kilifi	–	2011–	–	Y	83	78	–	–
			Madagascar (WHO)	–	2012–	–	N	–	74	–	–
		Asia	Bangladesh	–	2015–	–	N	–	97	–	–
		Oceania	Fiji	–	2012–	–	N	90	99	–	–
		LA and C	Colombia (SIREVA)	–	2011–	–	N	–	82	–	PPV
	2 + 1	Europe	Austria (ECDC) [10]	–	2012–	–	N	–	–	PPV, PCV	PPV
			Finland	–	2010–	–	N	95	90	PPV, PCV	PPV, PCV
			Iceland	–	2011–	–	N	–	89	PPV	PPV, PCV
			Latvia	2010–2011	2012–	–	N	91	85	–	–
			Lithuania (ECDC)	–	2014–	–	Unk	–	82	–	–
			Slovenia	–	2015–2019	2019–	N	55 [12]	55	PPV, PCV	PPV, PCV
	3 + 1	Europe	Bulgaria	–	2010–	–	N	–	91	PPV, PCV	–
	3 + 1/2 + 1	LA and C	Brazil	–	2010–	–	Y	91 [8]	88	–	PPV, PCV
		Europe	Netherlands	2006–2011	2011–	–	N	95	94	–	PPV, PCV
PCV13	3 + 0	LA and C	Bolivia (SIREVA)	–	–	2014–	N	–	–	–	–
			Honduras (SIREVA, WHO)	–	–	2011–	N	–	99	–	–
			Nicaragua (SIREVA, WHO)	–	–	2010–	N	–	98	PPV	PPV
		Sub Saharan Africa	Benin (WHO)	–	–	2011–	N	–	73	–	–
			Cameroon (WHO)	–	–	2011–	N	–	72	–	–
			Malawi, Blantyre District	–	–	2011–	Y	92	88	–	–
			The Gambia, Basse	2009–2011	–	2011–	N	77	95	–	–
			Zimbabwe (WHO)	–	–	2012–	N	–	90	–	–

Table 5. *Cont.*

Infant PCV Product [1]	PCV10/13 Schedule [1]	Region [2]	Site [3]	PCV7 Period	PCV10 Period	PCV13 Period	PCV10/13 Catch-Up	Mean PCV10/13 Uptake (%)		Other PCV/PPV Recommendations [7]	
								Primary Series [5]	WUENIC PCV3 [6]	Adult	High Risk
	2 + 1	N. Am.	Canada, Alberta	2002–2010	–	2010–	N	88 [8]	77	PPV	PPV, PCV
		LA and C	Argentina	–	–	2012–	Y	–	84	PPV, PCV	PPV, PCV
			Costa Rica	–	–	2011–	N	94	92	–	PPV, PCV
			Dominican Republic (SIREVA)	–	–	2013–	N	–	–	–	–
			Guatemala (SIREVA)	–	–	2012–	N	–	81	–	–
			Mexico (SIREVA)	2009–2013	–	2011–	Y	–	90	PPV	PPV
			Panama (SIREVA)	2010–2011	–	2011–	Unk	–	93	PPV, PCV	PPV, PCV
			Uruguay (SIREVA)	2008–2010	–	2010–	Y	–	94	PPV	–
		Europe	Denmark	2007–2010	–	2010–	N	91 [8]	93	PPV, PCV	PPV, PCV
			France	2006–2010	–	2010–	N	93	90	–	PPV, PCV
			Ireland	2008–2010	–	2010–	N	91	91	PPV	PPV, PCV
			Italy	2006–2009	–	2010–	N	86 [8]	87	PPV, PCV	–
			Norway	2006–2011	–	2011–	N	93	93	PPV	PPV, PCV
			Spain, Madrid	2006–2010	–	2010–	N	98	93	PPV, PCV	PPV, PCV
			Switzerland	2005–2010	–	2010–	Y	79 [8]	77	–	PCV
			UK, England	2006–2009	–	2010–	N	94	92	PPV	PPV
			UK, Scotland	2006–2010	–	2010–	N	97	92	PPV	PPV, PCV
		N. Africa and W. Asia	Israel	2009–2010	–	2010–	N	95	93	PPV, PCV	PPV, PCV
		Sub-Saharan Africa	South Africa	2009–2011	–	2011–	Y	77 [8]	77	PPV, PCV	PPV, PCV
		Asia	Mongolia	–	–	2016–	Y	93	20	–	–
			Singapore	2009–2011	–	2011–	Y	83	74	PPV, PCV	PPV, PCV
	3 + 1	N. Am.	USA, ABCs	2000–2009	–	2010–	Y	88	93	PPV, PCV	PPV, PCV
			USA, Alaska	2001–2009	–	2010–	Y	83	93	PPV, PCV	PPV, PCV
			USA, California	2000–2009	–	2010–	Y	96	93	PPV, PCV	PPV, PCV
			USA, Massachusetts	2000–2009	–	2010–	Y	94	93	PPV, PCV	PPV, PCV
			USA, Southwest (Indigenous)	2000–2009	–	2010–	Y	82	93	PPV, PCV	PPV, PCV
			USA, Utah	2000–2009	–	2010–	Y	88	93	PPV, PCV	PPV, PCV
		Europe	Greece [10]	2006–2009	–	2010–	N	82	75	PPV, PCV	PPV, PCV
		Asia	Japan	2010–2013	–	2013–	N	94 [8]	98	PPV	PPV

Table 5. *Cont.*

Infant PCV Product [1]	PCV10/13 Schedule [1]	Region [2]	Site [3]	PCV7 Period	PCV10 Period	PCV13 Period	PCV10/13 Catch-Up	Mean PCV10/13 Uptake (%)		Other PCV/PPV Recommendations [7]	
								Primary Series [5]	WUENIC PCV3 [6]	Adult	High Risk
	3 + 0/2 + 1	Oceania	Australia (Non-Indigenous) [13]	2005–2011	–	2011–	Y	92	92	PPV, PCV	PPV, PCV
	3 + 1/2 + 1	LA and C	Venezuela (SIREVA)	–	–	2014–	Unk	–	7	–	PPV, PCV
		Europe	Germany [10]	2006–2009	–	2009–	N	85	84	PPV	PPV, PCV
			Spain, Catalonia	2001–2010 [9]	–	2010–2015 [9] 2016–	N	70 [9]	93	PPV	PPV, PCV
			Spain, Navarra	2004–2009 [9]	–	2010–2015 [9] 2016–	N	71 [9]	93	PPV	PPV, PCV
PCV10/13	2 + 1	N. Am.	Canada, Quebec (excluding Nunavik)	2004–2009	2009–2010 2018–	2011–2018	N	97	75	PPV	PPV, PCV
		LA and C	Chile, Metropolitan Region	2009–2010	2011–2015	2016–	N	97	89	PPV	PPV
			Chile, Non-Metropolitan Regions	–	2011–2017	2017–	N	97	89	PPV	PPV
			El Salvador (SIREVA, WHO)	2010–2011	2018–	2011–2018	Unk	–	87	PPV	PPV, PCV
			Paraguay	–	2012–2017	2017–	Y	78	91	PPV	–
			Peru (SIREVA, WHO)	2009–2011	2011–2015	2015–	N	–	–	PPV	PPV
		Europe	Belgium	2007–2011	2015–2019	2011–2015 2019–	N	93 [8]	94	PPV, PCV	PPV, PCV
			Slovakia	2009–2010	2011–	2011–	Y	97	97	PCV	PCV
			Sweden	2009–2010	2010–	2010–2019	N	97 [8]	97	PPV	PPV, PCV
		N. Africa and W. Asia	Morocco, Grand Casablanca	–	2012–	2010–2012	N	91	90	–	–

Table 5. *Cont.*

Infant PCV Product [1]	PCV10/13 Schedule [1]	Region [2]	Site [3]	PCV7 Period	PCV10 Period	PCV13 Period	PCV10/13 Catch-Up	Mean PCV10/13 Uptake (%)		Other PCV/PPV Recommendations [7]	
								Primary Series [5]	WUENIC PCV3 [6]	Adult	High Risk
	3 + 1	N. Am.	Canada, Quebec-Nunavik	2002–2009	2009–2010	2011–	N	97	75	PPV	PPV, PCV
			Canada, Ontario	2005–2009	2009–2010	2010–	Y	72 [8]	77	PPV	PPV, PCV
		Asia	Hong Kong	2009–2010	2010–2011	2011–	N	98	–	PPV, PCV	PPV, PCV
		Oceania	New Zealand	2008–2011	2011–2014, 2017–	2014-2017	N	93	93	PPV, PCV	PPV, PCV
			Australia, Northern Territory	2001–2009	2009–2011	2011–	Y	88	92	PPV, PCV	PPV, PCV
	3 + 1/2 + 1	Europe	Czech Republic	–	2010–	2010–	N	74 [8]	–	PPV, PCV	PPV, PCV
			Poland [11]	–	2017–	2017–	N	94	60	PPV, PCV	PPV, PCV

[1] Product and schedule classifications intend to represent what was widely used in the population as of 2018, which occasionally differ from the national universal recommendation. [2] United Nations (UN) regions adapted from UN Statistics Division [25]. [3] (WHO): WHO Global Invasive Bacterial Vaccine-Preventable Diseases (IB-VPD) Surveillance Network; (SIREVA): Pan American Health Organization Sistema de Redes de Vigilancia de los Agentes Responsables de Neumonias y Meningitis Bacterianas (SIREVA); (ECDC): The European Surveillance System (ECDC). [4] "—" represents PCV not universally used. [5] Annual PCV uptake estimates provided by the surveillance site for the primary series of PCV by 12 months of age (if available, for some sites up to 15 months of age), excluding year of vaccine rollout; '—' represents no coverage information provided to PSERENADE project. [6] WUENIC PCV3 uptake, excluding the year of vaccine rollout (PCV3 represents the third dose whether given before 12 months or at or after 12 months, but in some cases uptake estimates may reflect the percentage of surviving infants who received two doses of PCV prior to the 1st birthday); '—' represents no WUENIC coverage information available. [7] Pneumococcal vaccine recommendation for other age groups or risk conditions. Adult recommendations are for all adults aged 50 years and above, aged 60 years and above, or aged 65 years and above. High-risk population age recommendations and populations included varies across sites. '–' represents no product is recommended. [8] Annual PCV uptake estimates provided by the surveillance site for the primary series plus the booster dose by 23 months of age, excluding year of vaccine rollout. [9] Recommended for high-risk populations only but had substantial (≥50% annually) private market uptake among the general population. [10] Although both PCV10 and PCV13 were recommended in the guidelines, the country was classified according to the product that was in substantially wide use. [11] PCV10 and PCV13 became available in the private market in 2009 and 2010, respectively. Widespread use began in 2017 when PCV was introduced into the Polish National Immunization Program (NIP). Private market use of PCV uses a different schedule (3 + 1) than the NIP (2 + 1). To date, PCV10 has been chosen for the NIP and private market PCV13 use is approximately 30%. [12] Range of vaccine uptake is 49–60%. [13] Australia (non-indigenous) switched from 3 + 0 to 2 + 1 in 2018. Abbreviations: N. Am.: North America; LA and C: Latin America and the Caribbean; N. Africa and W. Asia: Northern Africa and Western Asia; Y: Yes; N: No; Unk: unknown; PCV: pneumococcal conjugate vaccine; PPV: pneumococcal polysaccharide vaccine; 3 + 1: 3 primary doses plus booster; 2 + 1: 2 primary doses plus booster; 3 + 0: 3 primary doses and no booster.

4. Discussion

As part of the PSERENADE project, the largest and most comprehensive global serotype-specific IPD database was compiled though a comprehensive and systematic search. All available serotype-specific IPD surveillance data in countries using PCV10/13 were identified and characterized to evaluate the global evidence to estimate the impact of PCV10/13 by age, product, schedule, and syndrome. IPD surveillance is recommended by the WHO [4] and nearly 80% of countries using PCV in 2018 had an active IPD surveillance system. Seventy percent of these met PSERENADE eligibility criteria for potential to evaluate PCV impact or post-PCV-era serotype distribution, and over half of the eligible countries had annual IPD incidence rate data. Eligible IPD surveillance data existed for both children and adults, in all regions of the world, from both PCV10- and PCV13-using countries, from countries with and without a booster dose schedule, and from all income and infant mortality rate strata. The majority of countries that met PSERENADE eligibility criteria were from HICs and used a booster dose. Although there were eligible data from at least 15 countries representing low- or lower-middle income countries (LMICs), 3 + 0 schedules, and the African or Asian regions, when restricted to analyses of incidence or stratified by product, data become very sparse for addressing some questions. While there are challenges in drawing inferences from observational surveillance data and some gaps remain, the breadth and depth of the data compiled by PSERENADE increase our capacity to address many questions.

This multisite database overcomes common IPD data limitations, including having too few cases or years of data available, temporal confounding, and changes of surveillance systems over time. It can facilitate more robust results with greater accuracy by observing trends at many sites, thus increasing confidence in interpretation and improving PCV policy relevance at both national and global levels. The PSERENADE collaboration will enable robust analyses to answer policy-relevant epidemiologic questions. These questions relate to how well PCVs performed in reducing vaccine-type disease, the magnitude of indirect effects of PCVs on vaccine-type disease in unimmunized older children and adults, the degree and heterogeneity of serotype replacement (i.e., where non-vaccine type disease increases as a result of reduction of vaccine-types) in all age groups, how these may differ by product or schedule, what optimizes the impact of available pneumococcal vaccines, and what the potential impacts of future higher valency PCVs in addressing residual IPD will be. Furthermore, serotype-specific IPD incidence rate data will allow for estimation of the effectiveness of PCV10/13 against important vaccine-type or -related serotypes, in particular serotypes 3, 6A, 6C, and 19A, and the magnitude of replacement disease due to specific non-vaccine serotypes.

Although there was representativeness across a wide array of settings, the quantity and depth of data at some sites were limited, reducing its usefulness for analyses. Resource-poor settings, which are often those with the highest disease burden, face many challenges. As a result, these were more likely to have the smallest IPD sample sizes and less well characterized cases. For example, resource-poor settings are more likely to have frequent use of antibiotics that limit detection of *S. pneumoniae* by bacterial culture, an inability to identify cases meeting the IPD case definition, a lack of capacity to perform more sensitive PCR-based tests of the CSF or using PCR serotyping methods that are limited in the number of serotypes they can identify (both of which may bias the serotype distribution [31,32]), or an inability to link laboratory results with clinical data. In addition, outbreak-prone settings can be overwhelmed during peak seasons with case management, and may not be able to keep up with specimen collection, testing, and reporting to national surveillance systems [33]. These challenges are reflected by a greater proportion of surveillance sites from LMICs not meeting PSERENADE data collection eligibility criteria. Further, some sites with eligible data were unable to contribute to the project due to a lack of data management resources. However, many of these surveillance sites are still able to serve local purposes, such as serving as sentinels to identify pockets of vaccine-type disease where immunization uptake may be suboptimal. Eligible sites from LMICs that participated in PSERENADE were also more likely to have small sample size compared to those from UMICs or HICs. Further, although most (88%) surveillance included adults in addition to children, surveillance of adult disease was less common in Africa (54%) and Asia (60%). Pneumococcal pneumonia and meningitis are common in adults and surveillance in this age group is important for assessing indirect effects of infant immunization, including replacement disease, particularly in the meningitis belt of Africa, where serotype 1 IPD outbreaks are common [33–39]. Improving the quality of existing systems in key high-burden settings could help address remaining questions and increase representation from all settings in global analyses.

The majority of PSERENADE sites were able to classify IPD cases by specimen type, thus enabling identification of pneumococcal meningitis cases (cases with detection of pneumococcus in CSF), which is important for understanding syndrome-specific vaccine impact. This is important in regions with a history of meningitis outbreaks, such as the African meningitis belt. However, only a small proportion of surveillance sites have laboratory data systematically linked to clinical data to allow characterization of cases by clinical diagnoses. Therefore, few PSERENADE sites were able to identify bacteremic pneumonia cases because blood cultures are also obtained for other non-pneumonia IPD syndromes. As a result, few surveillance sites can directly assess the relative PCV impact on meningitis versus bacteremic pneumonia. Countries within the African meningitis belt also tended to have less comprehensive surveillance systems, resulting in IPD being reported predominantly from meningitis cases [36,40,41] or having incomplete data on age or serotype [42]. Improving the availability and characterization of cases has direct relevance to WHO's global roadmap for defeating meningitis by 2030, which set targets for vaccine-preventable meningitis surveillance, including for pneumococcal meningitis. In particular, the roadmap calls for strengthening of surveillance systems where a lack of laboratory capacity and resources for conducting surveillance hinder meningitis outbreak responses and provide data of limited quality to inform vaccine use and evaluate vaccine impact [43].

Despite investment in IPD surveillance globally, important gaps remain in the availability of IPD data needed for some assessments of PCV impact across diverse settings. The vast majority of IPD surveillance data are from HICs in Europe and North America. In these settings, pneumococcal epidemiology, serotype distribution, and disease burden differ from LMICs, which makes it difficult to be confident that global analyses are fully representative. Africa and Asia are the most under-represented. Most sites in these regions report lower case counts despite having a higher disease burden. Furthermore, stable population-based surveillance over time, important for estimating incidence rates, is particularly sparse in LMICs using a 3 + 0 schedule; only 5 countries (Bangladesh, Fiji, The Gambia, Kenya, and Malawi) provided data, with Bangladesh providing data for children only. A limitation of PSERENADE was not being able to get all key data that are available, including from two important countries using 3 + 0 schedules and conducting surveillance in the meningitis belt, Ghana and Burkina Faso. Population-based surveillance data over time for estimating impact were particularly limited from Latin America and the Caribbean, where only Chile had population-based surveillance data for all IPD, and Brazil had population-based surveillance data for pneumococcal meningitis only. Currently, the ability to answer important sub-group questions, including the indirect effects and some serotype-specific effects of 3 + 0 schedules in high-burden settings (particularly in meningitis outbreak prone settings), is limited. Additional high-quality and well-characterized data for all ages in these settings will be needed to answer these questions.

5. Conclusions

A large amount of IPD data is available globally for both children and adults. PSERENADE's systematic assessment and combined database of the available serotype-specific IPD surveillance data in countries using PCV10/13 will facilitate answering important questions as well as highlight the gaps needed to be filled to address remaining questions.

These data have the potential to inform policy around pneumococcal vaccine use, for both PCV and PPV, at national and global levels, including recommendations concerning product choice, schedule, and adult immunization. The PSERENADE project has informed WHO SAGE recommendations around pneumococcal vaccine use in adults and in community outbreak settings [33] and will contribute important evidence for other pneumococcal vaccine policy decisions. The ongoing collection of serotype-specific IPD surveillance data in countries that have introduced or plan to introduce PCV, as recommended by WHO, will provide data needed to understand PCV impact and inform pneumococcal vaccine policy decisions, particularly if efforts are made to support and enhance surveillance capacity in key areas underrepresented in global analyses.

Supplementary Materials: The following are available online at https://www.mdpi.com/article/10.3390/microorganisms9040742/s1, Search Terms, Table S1: Surveillance data sources for sites in PSERENADE, Table S2: PCV-using countries with IPD surveillance, whether included in PSERENADE and if so whether incidence data were provided, Table S3: Characteristics of data in PSERENADE by surveillance site.

Author Contributions: Conceptualization, M.D.K., K.H., J.C.B., M.G.Q., E.W.K., M.E.P.; Methodology, M.D.K., K.H., J.C.B., M.G.Q., M.E.P., E.W.K.; Software and Analysis, J.C.B., M.E.P., M.G.Q.; Writing—Original Draft Preparation, M.D.K., J.C.B., K.H., M.E.P.; Writing—Review & Editing, M.D.K., K.H., M.G.Q., E.W.K., D.R.F., M.K.H., Y.Y., J.N.S.; Visualization, J.N.S., M.G.Q., M.K.H.; Supervision, K.H., M.D.K., A.L.C.; Project Administration, J.C.B., M.G.Q.; Funding Acquisition, A.L.C., M.D.K., K.H. All other authors contributed data and critically reviewed results. All authors have read and agreed to the published version of the manuscript.

Funding: The PSERENADE project is funded by the Bill and Melinda Gates Foundation as part of the World Health Organization Pneumococcal Vaccines Technical Coordination Project, grant number INV-010429/OPP1189065.

Institutional Review Board Statement: This study was determined to not quality as human subjects research as defined by DHHS regulations 45 CFR 46.102 by the Johns Hopkins Bloomberg School of Public Health Institutional Review Board, due to the use of existing, de-identified data. Therefore, Institutional Review Board oversight was not required, and ethical review and approval were waived for this study.

Informed Consent Statement: Not applicable.

Data Availability Statement: Restrictions apply to the availability of these data. Data were obtained under data sharing agreements from contributing surveillance sites and can only be shared by contributing organizations with their permission.

Acknowledgments: This work was only possible through the contributions of many other individuals and organizations. The list of names of key individuals is provided in the Supplementary Acknowledgement List. We thank all individuals involved in the surveillance activities at the sites contributing to the PSERENADE project for trusting us with their data and for the assistance along the way to ensure this work provides value. This includes the surveillance network coordinators at Epi-Concept, European Centre for Disease Prevention and Control, Pan American Health Organization and WHO who facilitated access to and understanding of the data. We thank the Johns Hopkins University librarians and research assistants for providing literature review and data collection support. We thank the technical and strategic advisors from the WHO for providing guidance on the objectives and for facilitating introductions to the many surveillance sites globally. Finally, we are greatly indebted to the members of our PSERENADE Technical Advisory Group for providing us with expert technical review and strategic advice on all aspects of the methods, presentation and interpretation throughout the project. The PSERENADE project is funded by the Bill and Melinda Gates Foundation as part of the World Health Organization Pneumococcal Vaccines Technical Coordination Project.

Conflicts of Interest: KH conducted the study and analyses while working at the Johns Hopkins School of Public Health but is an employee at Pfizer, Inc. as of 26 October 2020. JCB reports funding from Pfizer in the past year, unrelated to the submitted work. MDK reports grants from Merck, personal fees from Merck, and grants from Pfizer, outside the submitted work. JAS reports grants from the Bill & Melinda Gates Foundation, the Wellcome Trust, the UK MRC, National Institute of

Health Research, outside the submitted work. MCB reports lectures fee from MSD outside from submitted work. AS reports grants and personal fees from Pfizer and personal fees from MSD and Sanofi Pasteur, outside the submitted work. ML has been a member of advisory boards and has received speakers' honoraria from Pfizer and Merck. German pneumococcal surveillance has been supported by Pfizer and Merck. SD reports grant from Pfizer, outside the submitted work. KA reports a grant from Merck, outside the submitted work. MC has previously received a professional fee from Pfizer (Ireland), an unrestricted research grant from Pfizer Ireland (2007–2016) and an Investigator Initiated Reward from Pfizer Ireland in 2018 (W1243730). AvG has received researching funding from Pfizer (last year 2017, Pfizer Investigator-Initiated Research [IIR] Program IIR WI 194379); attended advisory board meetings for Pfizer and Merck. AM received research support to her institution from Pfizer and Merck; honoraria for advisory board membership from GlaxoSmithKline, Merck and Pfizer. SNL performs contract research for GSK, Pfizer, Sanofi Pasteur on behalf of St. George's University of London, but receives no personal remuneration. IY stated she was a member of mRNA-1273 study group and has received funding to her institution to conduct clinical research from BioFire, MedImmune, Regeneron, PaxVax, Pfizer, GSK, Merck, Novavax, Sanofi-Pasteur, and Micron. RD has received grants/research support from Pfizer, Merck Sharp & Dohme and Medimmune; has been a scientific consultant for Pfizer, MeMed, Merck Sharp & Dohme, and Biondvax; had served on advisory boards of Pfizer, Merck Sharp & Dohme and Biondvax and has been a speaker for Pfizer. MH received an educational grant from Pfizer AG for partial support of this project. However, Pfizer AG had no role in the data analysis and content of the manuscript. CLB, MD has intellectual property in BioFire Diagnostics and receives royalties through the University of Utah. CLB is an advisor to IDbyDNA. AK reports personal fees from Pfizer, outside the submitted work. MT reports grants from GlaxoSmithKline and grants from Pfizer Inc. to the Finnish Institute for Health and Welfare for research projects outside the submitted work, in which she has been a co-investigator. JCS reports had received assistance from Pfizer for attending scientific meetings outside the submitted work. NPK has received research support from Pfizer for studies on pneumococcal serotyping and pneumococcal conjugate vaccine effectiveness. NPK has also received research support for unrelated studies from GlaxoSmithKline, Sanofi Pasteur, Merck, and Protein Sciences (now Sanofi Pasteur). SCGA received a travel grant from Pfizer. CMA reports grants and personal fees from Pfizer, Qiagen, and BioMerieux and grants from Genomica SAU, outside the submitted work. BL had two research grants from Pfizer on *Streptococcus pneumoniae*. EV reports grants from the French public health agency, during the conduct of the study; grants from Pfizer, grants from Merck, outside the submitted work. LLH reports research grants to her institution from GSK, Pfizer, and Merck. JDK has received an unrestricted grant-in-aid from Pfizer Canada that supports, in part, the CASPER invasive pneumococcal disease surveillance project. CGS reports grant funding from Pfizer, Merck, and AstraZeneca in the past three years. NMvS reports grants and fee for service from Pfizer, fee for service from MSD and GSK, outside the submitted work; In addition, NMvS has a patent WO 2013/020090 A3 with royalties paid to University of California San Diego (inventors: Nina van Sorge/Victor Nizet). All other authors did not declare any conflicts of interest. The funders had no role in the design of the study; in the collection, analyses, or interpretation of data; in the writing of the manuscript, or in the decision to publish the results.

Disclaimer: The findings and conclusions in this report are those of the authors and do not necessarily represent the official position of the World Health Organization (WHO) or of the Centers for Disease Control and Prevention (CDC). The authors Lucia Helena de Oliveira and Gloria Rey-Benito are staff members of the Pan American Health Organization. The authors alone are responsible for the views expressed in this publication, and they do not necessarily represent the decisions or policies of the Pan American Health Organization.

Appendix A

Table A1. PSERENADE team.

Name	Affiliation
Pedro Alarcon	Instituto de Salud Pública de Chile, 7780050 Santiago, Santiago Metropolitan, Chile
Samanta C. G. Almeida	National Laboratory for Meningitis and Pneumococcal Infections, Center of Bacteriology, Institute Adolfo Lutz (IAL), São Paulo, 01246-902, Brazil
Zahin Amin-Chowdhury	Immunisation and Countermeasures Division, Public Health England, NW9 5EQ, London, UK
Michelle Ang	National Public Health Laboratory, National Centre for Infectious Diseases, Singapore 308442, Singapore
Mária Avdicová	National Reference Centre for Pneumococcal and Haemophilus Diseases, Regional Authority of Public Health, 975 56 Banská Bystrica, Slovak Republic
Sithembile Bilima	Malawi-Liverpool-Wellcome Trust Clinical Research Programme, P.O. Box 30096, Chichiri, Blantyre 3, Malawi
Rita Born	Federal Office of Public Health, 3097 Liebefeld, Switzerland
Dana Bruden	Arctic Investigations Program, Division of Preparedness and Emerging Infections, National Center for Emerging and Zoonotic Infectious Diseases, Centers for Disease Control and Prevention, Anchorage, AK 99508, USA
Carrie L. Byington	University of Utah Department of Pediatrics (emeritus), Salt Lake City, UT 84108, United States; University of California Health System, Oakland, CA 94607, USA
Gustavo Chamorro	National Program on Immunopreventable Diseases and Expanded Program on Immunizations, Asunción, Paraguay
Kin-Hung Chow	Department of Microbiology and Carol Yu Centre for Infection, Queen Mary Hospital, The University of Hong Kong, Hong Kong, China
Urtnasan Chuluunbat	National Center of Communicable Diseases (NCCD), Ministry of Health, Bayanzurkh district, 13336 Ulaanbaatar, Mongolia
Cheryl Cohen	Centre for Respiratory Diseases and Meningitis, National Institute for Communicable Diseases of the National Health Laboratory Service, Sandringham, 2192, Johannesburg, South Africa; School of Public Health, Faculty of Health Sciences, University of the Witwatersrand, 2000, Johannesburg, South Africa
Edoardo Colzani	European Centre for Disease Prevention and Control, 169 73 Solna, Sweden
Tine Dalby	Bacteria, Parasites and Fungi, Statens Serum Institut, DK-2300 Copenhagen S, Denmark
Geneviève Deceuninck	Quebec University Hospital Research Centre, Québec, QC G1V 4G2, Canada
Elina Dimina	Centre for disease prevention and control of Latvia, Ziemelu District, Riga, LV-1005, Latvia
Mignon du Plessis	Centre for Respiratory Diseases and Meningitis, National Institute for Communicable Diseases of the National Health Laboratory Service, Sandringham, 2192, Johannesburg, South Africa
Helga Erlendsdottir	Department of Clinical Microbiology, Landspitali-The National University Hospital, Hringbraut, 101 Reykjavik, Iceland
Ryan Gierke	National Center for Immunizations and Respiratory Diseases, Centers for Disease Control and Prevention, Atlanta, GA 30333, USA
Charlotte Gilkison	Epidemiology Team, Institute of Environmental Science and Research, Porirua, 5022 Wellington, New Zealand
Noga Givon-Lavi	Pediatric Infectious Disease Unit and Clinical Microbiology Laboratory, Soroka University Medical Center, Ben-Gurion University of the Negev, Beer-Sheva, 8410501, Israel
Marcela Guevara	Instituto de Salud Pública de Navarra-IdiSNA, 31003 Pamplona, Navarra, Spain; CIBER Epidemiología y Salud Pública, (CIBERESP), 28029 Madrid, Spain
Md. Hasanuzzaman	Child Health Research Foundation, Dhaka 1207, Bangladesh
Ilias Hossain	Medical Research Council Unit The Gambia at London School of Hygiene & Tropical Medicine, P.O. Box 273, Banjul, The Gambia

Table A1. *Cont.*

Name	Affiliation
Aníbal Kawabata	Laboratorio Central de Salud Pública, Asunción, Paraguay (Central Laboratory of Public Health, Asunción, Paraguay), Asunción, Paraguay
Nicola P Klein	Vaccine Study Center, Kaiser Permanente, Oakland, CA 94612, USA
Vicki Krause	Centre for Disease Control, Department of Health and Community Services, Darwin City NT 8000, Australia
Pavla Krizova	National Institute of Public Health (NIPH), 100 42, Praha, Czech Republic
Alicja Kuch	National Reference Centre for Bacterial Meningitis, National Medicines Institute, 00-725 Warsaw, Poland
Brigitte Lefebvre	Laboratoire de Santé Publique du Québec, Sainte-Anne-de-Bellevue, Quebec H9X 3R5, Canada
Laura MacDonald	Public Health Scotland, G2 6QE Glasgow, UK
Ioanna Magaziotou	National Public Health Organisation, 15123 Athens, Greece
Jolita Mereckiene	HSE Health Protection Surveillance Centre, Mountjoy, Dublin, D01 A4A3, Ireland
Eva Morfeldt	Department of Microbiology Public Health Agency of Sweden, 171 82 Solna, Sweden
Carmen Muñoz-Almagro	Molecular Microbiology Department, Hospital Sant Joan de Déu Research Institute, 08950 Esplugues de Llobregat, Barcelona, Spain; Medicine Department, Universitat Internacional de Catalunya, 08017 Barcelona, Spain; CIBER Epidemiología y Salud Pública, (CIBERESP), 28029 Madrid, Spain
Tomoka Nakamura	World Health Organization, 1202 Geneva, Switzerland
Stephen I. Pelton	Boston University Schools of Medicine and Public Health, Boston, MA 02118, USA
Kate Pennington	Communicable Disease Epidemiology and Surveillance Section, Office of Health Protection, Australian Government Department of Health, 2606 Canberra ACT, Australia
Marie-Cecile Ploy	University Hospital Centre Limoges, Regional Observatories for Pneumococci, 87000 Limoges, France
Gloria Rey-Benito	Pan American Health Organization, World Health Organization, Washington, DC 20037, United States
Flavia Riccardo	Department of Infectious Diseases, Italian National Institute of Health (Istituto Superiore di Sanità, ISS), 00161 Rome, Italy
Leah J. Ricketson	Department of Pediatrics, University of Calgary, Calgary AB T3B 6A8, Canada
Fiona Russell	Murdoch Children's Research Institute, Parkville VIC 3052, Australia
Nahuel M. Sánchez Eluchans	Servicio de Bacteriología Clínica, Departamento de Bacteriología, INEI-ANLIS "Dr. Carlos G. Malbrán", C1282 AFF, Buenos Aires, Argentina
Juan Carlos Sanz	Laboratorio Regional de Salud Pública, Dirección General de Salud Pública, Comunidad de Madrid, 28053 Madrid, Spain
Shigeru Suga	Infectious Disease Center and Department of Clinical Research, National Hospital Organization Mie Hospital, Tsu, Mie 514-0125, Japan
Catherine G. Sutcliffe	Johns Hopkins Bloomberg School of Public Health, Baltimore, MD 21205, United States
Koh Cheng Thoon	KK Women's and Children's Hospital, 229899, Singapore
Maija Toropainen	Department of Health Security, Finnish Institute for Health and Welfare, 00271 Helsinki, Finland
Georgina Tzanakaki	National Meningitis Reference Laboratory, National School of Public Health Athens, Athens, Greece
Maria Teresa Valenzuela	Universidad de los Andes, Santiago, Las Condes, Metropolitan Region, Chile
Nina M. van Sorge	Medical Microbiology and Infection Prevention, Netherlands Reference Laboratory for Bacterial Meningitis, Amsterdam University Medical Centers, location AMC, University of Amsterdam, 1105 AZ Amsterdam, The Netherlands
Emmanuelle Varon	National Reference Centre for Pneumococci, Centre Hospitalier Intercommunal de Créteil, 94000 Créteil, France
Jennifer R. Verani	National Center for Immunizations and Respiratory Diseases, Centers for Disease Control and Prevention, Atlanta, GA 30333, USA; Centers for Disease Control and Prevention (CDC), Center for Global Health (CGH), Division of Global Health Protection (DGHP), Nairobi, Kenya

Table A1. *Cont.*

Name	Affiliation
Brita A. Winje	Department of Infection Control and Vaccine, Norwegian Institute of Public Health, 0456 Oslo, Norway
Khalid Zerouali	Department of Microbiology, Faculty of Medicine and Pharmacy, Hassan II University of Casablanca, Casablanca 20000, Morocco; Bacteriology-Virology and Hospital Hygiene Laboratory, Ibn Rochd University Hospital Centre, Casablanca 20250, Morocco
Toronto Invasive Bacterial Diseases Network	
WHO Invasive Bacterial Vaccine-Preventable Diseases (IB-VPD) Network	

Table A2. Acknowledgement list.

PSERENADE Technical Advisory Group
Thomas Cherian
William P. Hausdorff
Marc Lipsitch
Shabir A. Madhi
Elizabeth Miller
Catherine Satzke
Cynthia G. Whitney
World Health Organization
Katherine L. O'Brien
Jenny A. Walldorf
Johns Hopkins University
Yunfeng Cao
Peggy Gross
Donna Hesson
Ananya Kumar
Kate Perepezko
Francesca Schiaffino Salazar
Daniel Stephens
Epidemiology Department, EpiConcept
Germaine Hanquet
Dirección General de Salud Pública, Comunidad de Madrid, Spain
Luis García Comas
Maria Ordobás Gavín
Department of Infectious Diseases, Italian National Institute of Health (Istituto Superiore di Sanità, ISS), Rome, Italy
Martina Del Manso
Faculty of Sciences and Health Techniques, Mohammed VI University of Health Sciences (UM6SS) of Casablanca, Casablanca, Morocco; National Reference Laboratory, Mohammed VI University of Health Sciences (UM6SS), Casablanca, Morocco
Idrissa Diawara

Table A2. *Cont.*

Surveillance and Public Health Emergency Response, Public Health Agency of Catalonia, Barcelona, Spain
Sonia Broner
Conchita Izquierdo
Australian National Notifiable Diseases Surveillance data were provided by the Office of Health Protection, Australian Government Department of Health, on behalf of the Communicable Diseases Network Australia and the Enhanced Invasive Pneumococcal Disease Surveillance Working Group.

References

1. Wahl, B.; O'Brien, K.L.; Greenbaum, A.; Majumder, A.; Liu, L.; Chu, Y.; Lukšić, I.; Nair, H.; McAllister, D.A.; Campbell, H.; et al. Burden of Streptococcus Pneumoniae and Haemophilus Influenzae Type b Disease in Children in the Era of Conjugate Vaccines: Global, Regional, and National Estimates for 2000–15. *Lancet Glob. Health* **2018**, *6*, e744–e757. [CrossRef]
2. World Health Organization. Pneumococcal Disease. Available online: https://www.who.int/ith/diseases/pneumococcal/en/ (accessed on 4 January 2021).
3. World Health Organization. Pneumococcal Conjugate Vaccine for Childhood Immunization—WHO Position Paper. *Wkly. Epidemiol. Rec.* **2007**, *82*, 93–104.
4. World Health Organization. Pneumococcal Conjugate Vaccines in Infants and Children under 5 Years of Age: WHO Position Paper. *Wkly. Epidemiol. Rec.* **2019**, *94*, 85–104.
5. International Vaccine Access Center (IVAC), Johns Hopkins Bloomberg School of Public Health VIEW-Hub. Available online: https://view-hub.org (accessed on 29 December 2020).
6. World Health Organization. Invasive Bacterial Vaccine Preventable Diseases Laboratory Network. Available online: http://www.who.int/immunization/monitoring_surveillance/burden/laboratory/IBVPD/en/ (accessed on 5 January 2021).
7. Johnson, H.L.; Deloria-Knoll, M.; Levine, O.S.; Stoszek, S.K.; Hance, L.F.; Reithinger, R.; Muenz, L.R.; O'Brien, K.L. Systematic Evaluation of Serotypes Causing Invasive Pneumococcal Disease among Children Under Five: The Pneumococcal Global Serotype Project. *PLoS Med.* **2010**, *7*, e1000348. [CrossRef] [PubMed]
8. Hausdorff, W.P.; Bryant, J.; Paradiso, P.R.; Siber, G.R. Which Pneumococcal Serogroups Cause the Most Invasive Disease: Implications for Conjugate Vaccine Formulation and Use, Part I. *Clin. Infect. Dis.* **2000**, *30*, 100–121. [CrossRef]
9. Deloria-Knoll, M.; Nonyane, B.A.; Garcia, C.; Levine, O.S.; O'Brien, K.L.; Johnson, H.L.; AGEDD Project Team. Global Serotype Distribution of Invasive Pneumococcal Disease (IPD) in Older Children and Adults: AGEDD Study. In Proceedings of the 9th International Symposium on Pneumococci and Pneumococcal Diseases (ISPPD), Hyderabad, India, 9–13 March 2014.
10. World Health Organization. Meeting of the Strategic Advisory Group of Experts on Immunization, November 2013—Conclusions and Recommendations. *Wkly. Epidemiol. Rep.* **2014**, *89*, 1–20.
11. Feikin, D.R.; Kagucia, E.W.; Loo, J.D.; Link-Gelles, R.; Puhan, M.A.; Cherian, T.; Levine, O.S.; Whitney, C.G.; O'Brien, K.L.; Moore, M.R.; et al. Serotype-Specific Changes in Invasive Pneumococcal Disease after Pneumococcal Conjugate Vaccine Introduction: A Pooled Analysis of Multiple Surveillance Sites. *PLoS Med.* **2013**, *10*, e1001517. [CrossRef]
12. Savulescu, C.; Krizova, P.; Lepoutre, A.; Mereckiene, J.; Vestrheim, D.F.; Ciruela, P.; Ordobas, M.; Guevara, M.; McDonald, E.; Morfeldt, E.; et al. Effect of High-Valency Pneumococcal Conjugate Vaccines on Invasive Pneumococcal Disease in Children in SpIDnet Countries: An Observational Multicentre Study. *Lancet Respir. Med.* **2017**, *5*, 648–656. [CrossRef]
13. Iroh Tam, P.-Y.; Thielen, B.K.; Obaro, S.K.; Brearley, A.M.; Kaizer, A.M.; Chu, H.; Janoff, E.N. Childhood Pneumococcal Disease in Africa—A Systematic Review and Meta-Analysis of Incidence, Serotype Distribution, and Antimicrobial Susceptibility. *Vaccine* **2017**, *35*, 1817–1827. [CrossRef] [PubMed]
14. Cui, Y.A.; Patel, H.; O'Neil, W.M.; Li, S.; Saddier, P. Pneumococcal Serotype Distribution: A Snapshot of Recent Data in Pediatric and Adult Populations around the World. *Hum. Vaccines Immunother.* **2017**, *13*, 1229–1241. [CrossRef]
15. Balsells, E.; Guillot, L.; Nair, H.; Kyaw, M.H. Serotype Distribution of Streptococcus Pneumoniae Causing Invasive Disease in Children in the Post-PCV Era: A Systematic Review and Meta-Analysis. *PLoS ONE* **2017**, *12*, e0177113. [CrossRef]
16. Cherian, T.; Cohen, M.; de Oliveira, L.; Farrar, J.L.; Goldblatt, D.; Knoll, M.; Moisi, J.C.; O'Brien, K.L.; Pilishvili, T.; Ramakrishnan, M.; et al. *Pneumococcal Conjugate Vaccine (PCV) Review of Impact Evidence (PRIME): Summary of Findings from Systematic Review*; Report to Strategic Advisory Group of Experts on Immunization (SAGE) of the World Health Organization; WHO: Geneva, Switzerland, October 2017; Volume 1, pp. 1–215.
17. Morpeth, S.C.; Deloria Knoll, M.; Scott, J.A.G.; Park, D.E.; Watson, N.L.; Baggett, H.C.; Brooks, W.A.; Feikin, D.R.; Hammitt, L.L.; Howie, S.R.C.; et al. Detection of Pneumococcal DNA in Blood by Polymerase Chain Reaction for Diagnosing Pneumococcal Pneumonia in Young Children From Low- and Middle-Income Countries. *Clin. Infect. Dis.* **2017**, *64*, S347–S356. [CrossRef] [PubMed]

18. Grant, L.R.; Hammitt, L.L.; O'Brien, S.E.; Jacobs, M.R.; Donaldson, C.; Weatherholtz, R.C.; Reid, R.; Santosham, M.; O'Brien, K.L. Impact of the 13-Valent Pneumococcal Conjugate Vaccine on Pneumococcal Carriage Among American Indians. *Pediatr. Infect. Dis. J.* **2016**, *35*, 907–914. [CrossRef]
19. Loughlin, A.M.; Hsu, K.; Silverio, A.L.; Marchant, C.D.; Pelton, S.I. Direct and Indirect Effects of PCV13 on Nasopharyngeal Carriage of PCV13 Unique Pneumococcal Serotypes in Massachusetts' Children. *Pediatr. Infect. Dis. J.* **2014**, *33*, 504–510. [CrossRef]
20. Von Gottberg, A.; de Gouveia, L.; Tempia, S.; Quan, V.; Meiring, S.; von Mollendorf, C.; Madhi, S.A.; Zell, E.R.; Verani, J.R.; O'Brien, K.L.; et al. Effects of Vaccination on Invasive Pneumococcal Disease in South Africa. *N. Engl. J. Med.* **2014**, *371*, 1889–1899. [CrossRef]
21. World Health Organization; UNICEF. WHO-UNICEF Estimates of PCV3 Coverage. Available online: https://apps.who.int/immunization_monitoring/globalsummary/timeseries/tswucoveragepcv3.html (accessed on 30 December 2020).
22. Pan American Health Organization; World Health Organization. Sistema Regional de Vacunas (SIREVA) II. Available online: https://www.paho.org/es/sireva (accessed on 31 August 2020).
23. Dirección Redes en Salud Pública; Subdirección Laboratorio Nacional De Referencia; Grupo de Microbiología. *Vigilancia por Laboratorio de aislamientos invasores de Streptococcus pneumoniae Colombia 2006-2018: SIREVA II*; Instituto Nacional de Salud: Bogotá, Colombia, 2019; pp. 1–16.
24. Organización Panamericana de la Salud; Instituto Nacional de Salud Colombia. *Procedimientos Para El Diagnóstico de Neumonías y Meningitis Bacterianas y La Caracterización de Cepas de Streptococcus Pneumoniae y Haemophilus Influenzae, SIREVA II*; Grupo Microbiología, Instituto Nacional de Salud Bogotá—Colombia, Organización Panamericana dela Salud: Colombia, 2012; pp. 1–121. Available online: http://antimicrobianos.com.ar/ATB/wp-content/uploads/2013/01/PAHO-Manual-Neumo-Haemophilus-SIREVA-2012.pdf (accessed on 29 March 2021).
25. United Nations. Methodology: Geographic Regions. Available online: https://unstats.un.org/unsd/methodology/m49/#geo-regions (accessed on 22 December 2020).
26. The World Bank. World Bank Country and Lending Groups. Available online: https://datahelpdesk.worldbank.org/knowledgebase/articles/906519-world-bank-country-and-lending-groups (accessed on 5 November 2020).
27. United Nations. Levels and Trends in Child Mortality: 2020 Report. Available online: https://www.un.org/development/desa/pd/news/levels-and-trends-child-mortality-2020-report (accessed on 30 January 2021).
28. Cohen, C.; Walaza, S.; Moyes, J.; Groome, M.; Tempia, S.; Pretorius, M.; Hellferscee, O.; Dawood, H.; Haffejee, S.; Variava, E.; et al. Epidemiology of Severe Acute Respiratory Illness (SARI) among Adults and Children Aged $\geq$ 5 Years in a High HIV-Prevalence Setting, 2009–2012. *PLoS ONE* **2015**, *10*, e0117716. [CrossRef]
29. Park, I.H.; Pritchard, D.G.; Cartee, R.; Brandao, A.; Brandileone, M.C.C.; Nahm, M.H. Discovery of a New Capsular Serotype (6C) within Serogroup 6 of Streptococcus Pneumoniae. *J. Clin. Microbiol.* **2007**, *45*, 1225–1233. [CrossRef]
30. Jin, P.; Kong, F.; Xiao, M.; Oftadeh, S.; Zhou, F.; Liu, C.; Russell, F.; Gilbert, G.L. First Report of Putative Streptococcus Pneumoniae Serotype 6D among Nasopharyngeal Isolates from Fijian Children. *J. Infect. Dis.* **2009**, *200*, 1375–1380. [CrossRef]
31. Swarthout, T.D.; Gori, A.; Bar-Zeev, N.; Kamng'ona, A.W.; Mwalukomo, T.S.; Bonomali, F.; Nyirenda, R.; Brown, C.; Msefula, J.; Everett, D.; et al. Evaluation of Pneumococcal Serotyping of Nasopharyngeal-Carriage Isolates by Latex Agglutination, Whole-Genome Sequencing (PneumoCaT), and DNA Microarray in a High-Pneumococcal-Carriage-Prevalence Population in Malawi. *J. Clin. Microbiol.* **2020**, *59*, e02103-20. [CrossRef]
32. Satzke, C.; Dunne, E.M.; Porter, B.D.; Klugman, K.P.; Mulholland, E.K. The PneuCarriage Project: A Multi-Centre Comparative Study to Identify the Best Serotyping Methods for Examining Pneumococcal Carriage in Vaccine Evaluation Studies. *PLoS Med.* **2015**, *12*, e1001903. [CrossRef]
33. World Health Organization. *Strategic Advisory Group of Experts on Immunization Yellow Book*; World Health Organization Department of Immunization, Vaccines and Biologicals (IVB): Geneva, Switzerland, 2020; p. 8.
34. Moïsi, J.C.; Makawa, M.-S.; Tall, H.; Agbenoko, K.; Njanpop-Lafourcade, B.-M.; Tamekloe, S.; Amidou, M.; Mueller, J.E.; Gessner, B.D. Burden of Pneumococcal Disease in Northern Togo before the Introduction of Pneumococcal Conjugate Vaccine. *PLoS ONE* **2017**, *12*, e0170412. [CrossRef] [PubMed]
35. Tornheim, J.A.; Manya, A.S.; Oyando, N.; Kabaka, S.; Breiman, R.F.; Feikin, D.R. The Epidemiology of Hospitalized Pneumonia in Rural Kenya: The Potential of Surveillance Data in Setting Public Health Priorities. *Int. J. Infect. Dis.* **2007**, *11*, 536–543. [CrossRef]
36. Bozio, C.H.; Abdul-Karim, A.; Abenyeri, J.; Abubakari, B.; Ofosu, W.; Zoya, J.; Ouattara, M.; Srinivasan, V.; Vuong, J.T.; Opare, D.; et al. Continued Occurrence of Serotype 1 Pneumococcal Meningitis in Two Regions Located in the Meningitis Belt in Ghana Five Years after Introduction of 13-Valent Pneumococcal Conjugate Vaccine. *PLoS ONE* **2018**, *13*, e0203205. [CrossRef] [PubMed]
37. Campagne, G.; Schuchat, A.; Djibo, S.; Ousséini, A.; Cissé, L.; Chippaux, J.P. Epidemiology of Bacterial Meningitis in Niamey, Niger, 1981-96. *Bull. World Health Organ.* **1999**, *77*, 499–508. [PubMed]
38. Aku, F.Y. Meningitis Outbreak Caused by Vaccine-Preventable Bacterial Pathogens—Northern Ghana, 2016. *MMWR.* **2017**, *66*, 806–810. [CrossRef]
39. Soeters, H.M.; Diallo, A.O.; Bicaba, B.W.; Kadadé, G.; Dembélé, A.Y.; Acyl, M.A.; Nikiema, C.; Sadji, A.Y.; Poy, A.N.; Lingani, C.; et al. Bacterial Meningitis Epidemiology in Five Countries in the Meningitis Belt of Sub-Saharan Africa, 2015–2017. *J. Infect. Dis.* **2019**, *220*, S165–S174. [CrossRef]

40. Soeters, H.M.; Kambiré, D.; Sawadogo, G.; Ouédraogo-Traoré, R.; Bicaba, B.; Medah, I.; Sangaré, L.; Ouédraogo, A.-S.; Ouangraoua, S.; Yaméogo, I.; et al. Impact of 13-Valent Pneumococcal Conjugate Vaccine on Pneumococcal Meningitis, Burkina Faso, 2016–2017. *J. Infect. Dis.* **2019**, *220*, S253–S262. [CrossRef]
41. Novak, R.T.; Ronveaux, O.; Bita, A.F.; Aké, H.F.; Lessa, F.C.; Wang, X.; Bwaka, A.M.; Fox, L.M. Future Directions for Meningitis Surveillance and Vaccine Evaluation in the Meningitis Belt of Sub-Saharan Africa. *J. Infect. Dis.* **2019**, *220*, S279–S285. [CrossRef]
42. Patel, J.C.; Soeters, H.M.; Diallo, A.O.; Bicaba, B.W.; Kadadé, G.; Dembélé, A.Y.; Acyl, M.A.; Nikiema, C.; Lingani, C.; Hatcher, C.; et al. MenAfriNet: A Network Supporting Case-Based Meningitis Surveillance and Vaccine Evaluation in the Meningitis Belt of Africa. *J. Infect. Dis.* **2019**, *220*, S148–S154. [CrossRef]
43. World Health Organization. *Global Vaccine Action Plan: Defeating Meningitis by 2030 Meningitis Prevention and Control*; World Health Organization Seventy-third World Health Assembly; WHO: Geneva, Switzerland, 2020; pp. 1–4.

 microorganisms

Article

Serotype Distribution of Remaining Pneumococcal Meningitis in the Mature PCV10/13 Period: Findings from the PSERENADE Project

Maria Garcia Quesada [1,*], Yangyupei Yang [1], Julia C. Bennett [1], Kyla Hayford [1,†], Scott L. Zeger [1], Daniel R. Feikin [2], Meagan E. Peterson [1], Adam L. Cohen [3,‡], Samanta C. G. Almeida [4], Krow Ampofo [5], Michelle Ang [6], Naor Bar-Zeev [1,7], Michael G. Bruce [8], Romina Camilli [9], Grettel Chanto Chacón [10], Pilar Ciruela [11,12], Cheryl Cohen [13,14], Mary Corcoran [15], Ron Dagan [16], Philippe De Wals [17], Stefanie Desmet [18,19], Idrissa Diawara [20,21], Ryan Gierke [22], Marcela Guevara [11,23], Laura L. Hammitt [1], Markus Hilty [24], Pak-Leung Ho [25], Sanjay Jayasinghe [26], Jackie Kleynhans [13,14], Karl G. Kristinsson [27], Shamez N. Ladhani [28], Allison McGeer [29], Jason M. Mwenda [30], J. Pekka Nuorti [31,32], Kazunori Oishi [33], Leah J. Ricketson [34] Juan Carlos Sanz [35], Larisa Savrasova [36,37], Lena Petrova Setchanova [38], Andrew Smith [39,40], Palle Valentiner-Branth [41], Maria Teresa Valenzuela [42], Mark van der Linden [43], Nina M. van Sorge [44] Emmanuelle Varon [45], Brita A. Winje [46], Inci Yildirim [47], Jonathan Zintgraff [48], Maria Deloria Knoll [1,*] and the PSERENADE Team [§]

Citation: Garcia Quesada, M.; Yang, Y.; Bennett, J.C.; Hayford, K.; Zeger, S.L.; Feikin, D.R.; Peterson, M.E.; Cohen, A.L.; Almeida, S.C.G.; Ampofo, K.; et al. Serotype Distribution of Remaining Pneumococcal Meningitis in the Mature PCV10/13 Period: Findings from the PSERENADE Project. *Microorganisms* **2021**, *9*, 738. https://doi.org/10.3390/microorganisms9040738

Academic Editor: James Stuart

Received: 4 March 2021
Accepted: 28 March 2021
Published: 1 April 2021

Publisher's Note: MDPI stays neutral with regard to jurisdictional claims in published maps and institutional affiliations.

1. Johns Hopkins Bloomberg School of Public Health, Baltimore, MD 21205, USA; yyang165@jhmi.edu (Y.Y.); jbenne63@jhu.edu (J.C.B.); kylahayford@jhu.edu (K.H.); sz@jhu.edu (S.L.Z.); meaganepeterson@gmail.com (M.E.P.); nbarzee1@jhu.edu (N.B.-Z.); lhammitt@jhu.edu (L.L.H.)
2. Independent Consultant, 1296 Coppet, Switzerland; drf3217@gmail.com
3. World Health Organization, 1202 Geneva, Switzerland; dvj1@cdc.gov
4. Center of Bacteriology, National Laboratory for Meningitis and Pneumococcal Infections, Institute Adolfo Lutz (IAL), São Paulo 01246-902, Brazil; samanta.almeida@ial.sp.gov.br
5. Department of Pediatrics, Division of Pediatric Infectious Diseases, University of Utah Health Sciences Center, Salt Lake City, UT 84132, USA; Krow.Ampofo@hsc.utah.edu
6. National Centre for Infectious Diseases, National Public Health Laboratory, Singapore 308442, Singapore; michelle_lt_ang@ncid.sg
7. Malawi-Liverpool-Wellcome Trust Clinical Research Programme, P.O. Box 30096, Chichiri, Blantyre 3, Malawi
8. National Center for Emerging and Zoonotic Infectious Diseases, Centers for Disease Control and Prevention, Arctic Investigations Program, Division of Preparedness and Emerging Infections, Anchorage, AK 99508, USA; zwa8@cdc.gov
9. Department of Infectious Diseases, Italian National Institute of Health (Istituto Superiore di Sanità, ISS), 00161 Rome, Italy; romina.camilli@iss.it
10. Instituto Costarricense de Investigación y Enseñanza en Nutrición y Salud, Tres Ríos, 30301 Cartago, Costa Rica; gchanto@inciensa.sa.cr
11. CIBER Epidemiología y Salud Pública, (CIBERESP), 28029 Madrid, Spain; pilar.ciruela@gencat.cat (P.C.); mp.guevara.eslava@navarra.es (M.G.)
12. Surveillance and Public Health Emergency Response, Public Health Agency of Catalonia, 08005 Barcelona, Spain
13. Centre for Respiratory Diseases and Meningitis, National Institute for Communicable Diseases of the National Health Laboratory Service, 2192 Johannesburg, South Africa; cherylc@nicd.ac.za (C.C.); JackieL@nicd.ac.za (J.K.)
14. School of Public Health, Faculty of Health Sciences, University of the Witwatersrand, 2000 Johannesburg, South Africa
15. Irish Meningitis and Sepsis Reference Laboratory, Children's Health Ireland at Temple Street, Temple Street, D01 YC76 Dublin 1, Ireland; mary.corcoran@cuh.ie
16. Distinguished Professor of Pediatrics and Infectious Diseases, The Faculty of Health Sciences, Ben-Gurion University of the Negev, Beer-Sheva, Israel; rdagan@bgu.ac.il
17. Department of Social and Preventive Medicine, Laval University, Québec, QC G1V 0A6, Canada; philippe.dewals@criucpq.ulaval.ca
18. Department of Microbiology, Immunology and Transplantation, KU Leuven, 3000 Leuven, Belgium; stefanie.desmet@uzleuven.be
19. National Reference Centre for Streptococcus Pneumoniae, University Hospitals Leuven, 3000 Leuven, Belgium
20. Faculty of Sciences and Health Techniques, Mohammed VI University of Health Sciences (UM6SS) of Casablanca, 20250 Casablanca, Morocco; idiawara@um6ss.ma

[21] National Reference Laboratory, Mohammed VI University of Health Sciences (UM6SS), 82403 Casablanca, Morocco
[22] National Center for Immunizations and Respiratory Diseases, Centers for Disease Control and Prevention, Atlanta, GA 30333, USA; ipe3@cdc.gov
[23] Instituto de Salud Pública de Navarra—IdiSNA, 31003 Pamplona, Spain
[24] Swiss National Reference Centre for Invasive Pneumococci, Institute for Infectious Diseases, University of Bern, 3012 Bern, Switzerland; Markus.Hilty@ifik.unibe.ch
[25] Department of Microbiology and Carol Yu Centre for Infection, Queen Mary Hospital, The University of Hong Kong, Hong Kong, China; plho@hku.hk
[26] National Centre for Immunisation Research and Surveillance and Discipline of Child and Adolescent Health, Faculty of Medicine and Health, Children's Hospital Westmead Clinical School, University of Sydney, Westmead, NSW 2145, Australia; sanjay.jayasinghe@health.nsw.gov.au
[27] Department of Clinical Microbiology, Landspitali—The National University Hospital, Hringbraut, 101 Reykjavik, Iceland; karl@landspitali.is
[28] Immunisation and Countermeasures Division, Public Health England, London NW9 5EQ, UK; shamez.ladhani@phe.gov.uk
[29] Toronto Invasive Bacterial Diseases Network, and Department of Laboratory, Medicine and Pathobiology, University of Toronto, Toronto, ON M5S 1A8, Canada; Allison.McGeer@sinaihealth.ca
[30] World Health Organization Regional Office for Africa, P.O. Box 06, Brazzaville, Congo; mwendaj@who.int
[31] Department of Health Security, Finnish Institute for Health and Welfare, 00271 Helsinki, Finland; pekka.nuorti@tuni.fi
[32] Health Sciences Unit, Faculty of Social Sciences, Tampere University, 33100 Tampere, Finland
[33] Toyama Institute of Health, Imizu, Toyama 939-0363, Japan; toyamaeiken1@chic.ocn.ne.jp
[34] Department of Pediatrics, University of Calgary, Calgary, AB T3B 6A8, Canada; ljricket@ucalgary.ca
[35] Laboratorio Regional de Salud Pública, Dirección General de Salud Pública, Comunidad de Madrid, 28053 Madrid, Spain; juan.sanz@salud.madrid.org
[36] Centre for Disease Prevention and Control of Latvia, 1005 Riga, Latvia; larisa.savrasova@spkc.gov.lv
[37] Doctoral Studies Department, Riga Stradinš University, 1007 Riga, Latvia
[38] Department of Medical Microbiology, Faculty of Medicine, Medical University of Sofia, 1431 Sofia, Bulgaria; lenasetchanova@hotmail.com
[39] Bacterial Respiratory Infection Service, Scottish Microbiology Reference Laboratory, NHS GG&C, Glasgow G4 0SF, UK; andrew.smith@glasgow.ac.uk
[40] College of Medical, Veterinary & Life Sciences, Glasgow Dental Hospital & School, University of Glasgow, Glasgow G2 3JZ, UK
[41] Infectious Disease Epidemiology and Prevention, Statens Serum Institut, DK-2300 Copenhagen S, Denmark; pvb@ssi.dk
[42] Department of Public Health and Epidemiology, Faculty of Medicine, Universidad de Los Andes, 12455 Santiago, Chile; mtvalenzuela@uandes.cl
[43] National Reference Center for Streptococci, Department of Medical Microbiology, University Hospital RWTH Aachen, 52074 Aachen, Germany; mlinden@ukaachen.de
[44] Medical Microbiology and Infection Prevention, Netherlands Reference Laboratory for Bacterial Meningitis, Amsterdam University Medical Centers, Location AMC, University of Amsterdam, 1105 AZ Amsterdam, The Netherlands; n.m.vansorge@amsterdamumc.nl
[45] National Reference Centre for Pneumococci, Centre Hospitalier Intercommunal de Créteil, 94000 Créteil, France; Emmanuelle.Varon@chicreteil.fr
[46] Department of Infection Control and Vaccine, Norwegian Institute of Public Health, 0456 Oslo, Norway; Brita.Askeland.Winje@fhi.no
[47] Department of Pediatrics, Yale New Haven Children's Hospital, New Haven, CT 06504, USA; inci.yildirim@yale.edu
[48] Servicio de Bacteriología Clínica, Departamento de Bacteriología, INEI—ANLIS "Dr. Carlos G. Malbrán", C1282 AFF Buenos Aires, Argentina; jzintgraff@anlis.gob.ar
* Correspondence: mgarci64@jhmi.edu (M.G.Q.); mknoll2@jhu.edu (M.D.K.)
† Present address: Pfizer, Inc. Collegeville, PA 19426, USA. Affiliation at the time this work was completed was Johns Hopkins Bloomberg School of Public Health.
‡ Present address: Centers for Disease Control and Prevention, Atlanta, GA 30333, USA. Affiliation at the time this work was completed was World Health Organization.
§ Members are listed in the Appendix A Table A1.

Abstract: Pneumococcal conjugate vaccine (PCV) introduction has reduced pneumococcal meningitis incidence. The Pneumococcal Serotype Replacement and Distribution Estimation (PSERENADE) project described the serotype distribution of remaining pneumococcal meningitis in countries using PCV10/13 for least 5–7 years with primary series uptake above 70%. The distribution was estimated using a multinomial Dirichlet regression model, stratified by PCV product and age. In PCV10-using sites (*N* = 8; cases = 1141), PCV10 types caused 5% of cases <5 years of age and 15% among ≥5 years; the top serotypes were 19A, 6C, and 3, together causing 42% of cases <5 years and 37% ≥5 years. In PCV13-using sites (*N* = 32; cases = 4503), PCV13 types caused 14% in <5 and 26% in ≥5 years; 4% and 13%, respectively, were serotype 3. Among the top serotypes are five (15BC, 8, 12F, 10A, and 22F) included in higher-valency PCVs under evaluation. Other top serotypes (24F, 23B, and 23A) are not in any known investigational product. In countries with mature vaccination programs, the proportion of pneumococcal meningitis caused by vaccine-in-use serotypes is lower (≤26% across all ages) than pre-PCV (≥70% in children). Higher-valency PCVs under evaluation target over half of remaining pneumococcal meningitis cases, but questions remain regarding generalizability to the African meningitis belt where additional data are needed.

Keywords: pneumococcal meningitis; serotype distribution; PCV impact; global; meta-analysis

1. Introduction

Pneumococcal meningitis is a major cause of childhood morbidity and mortality globally, estimated to have caused 83,900 cases and 37,900 deaths in 2015 [1]. Meningitis is estimated to make up approximately 2% of all severe pneumococcal disease and 12% of pneumococcal deaths [1]. Prior to the introduction of pneumococcal conjugate vaccines (PCV) into routine childhood immunization programs, over 70% of invasive pneumococcal disease (IPD), a serious form of pneumococcal disease that includes bacteremic pneumonia, meningitis, and sepsis, was estimated to have been caused by serotypes targeted by the vaccines currently available [2]. Since then, PCVs have been introduced into infant immunization programs in over 140 countries [3].

Immunizing children with PCV is an effective method for preventing IPD, providing not only direct protection in vaccinated children but also indirect protection (i.e., herd immunity) among unvaccinated individuals by decreasing the circulation of pneumococci of the serotypes included in the vaccines [4–7]. PCVs currently in wide use include a 10-valent vaccine (PCV10; GlaxoSmithKline (GSK), Synflorix) and a 13-valent vaccine (PCV13; Pfizer, Prevnar13/Prevenar13). Another 10-valent vaccine (Serum Institute of India (SII), Pneumosil) became available in 2019. The 23-valent pneumococcal polysaccharide vaccine (PPV23; Merck, Pneumovax23) is recommended in many countries for older adults and those at high risk for pneumococcal disease but is not widely used [8].

Significant reductions in vaccine-type IPD of 41–97%, including pneumococcal meningitis, have been observed in the pediatric population and other age groups after the introduction of PCVs [5,7]. However, these changes are not immediate; indirect protection in unvaccinated individuals takes more time, and several countries in the African meningitis belt had pneumococcal meningitis outbreaks due to vaccine serotypes after the introduction of PCVs [9]. Although low immunization coverage is a possible explanation, it raised questions about the speed and degree of indirect protection in high burden settings without a booster dose, primarily administered in the second year of life. PCV formulations covering 15–24 serotypes have been developed, though they are not yet licensed [10–13]; these may offer a solution to address much of the remaining disease, but the preventable fraction depends on how much of the remaining disease is caused by the added serotypes. A World Health Organization (WHO) roadmap to defeat meningitis by 2030 was recently endorsed by the World Health Assembly and includes a path to address the remaining leading causes of acute bacterial meningitis, including pneumococcus [14].

Since many countries have now used PCV10/13 extensively, it is possible to examine if serotypes covered by these vaccines have been eliminated in all age groups and what proportion of the remaining disease is caused by the serotypes included in higher-valency PCV formulations under development and in PPV23. We aimed to estimate the global serotype distribution of pneumococcal meningitis cases, by PCV product used and age group, in countries with well-established PCV10/13 routine infant immunization programs and high uptake. This is part of a larger effort investigating the impact of PCV on IPD incidence and serotype distribution to inform current global and national pneumococcal vaccination policies.

2. Materials and Methods

2.1. Site Identification and Eligibility

Site identification and data collection methods are described in detail elsewhere [15]. Briefly, various methods were used to identify countries conducting serotype-specific IPD surveillance where PCV10 (referring throughout to the GSK product unless otherwise specified, as SII's vaccine was not in use) or PCV13 was universally recommended for all infants for at least one year by 2018. Known surveillance networks were contacted, and the WHO and experts in the field provided contacts for possible data sources; previous systematic reviews were used to identify potential sites and validate search terms for a literature review that included articles published between 1 January 2011 and 20 December 2018; and International Symposium on Pneumococci and Pneumococcal Diseases (ISPPD) abstracts were reviewed from 2012 to 2018. Individuals at each institution that collected IPD data, including research groups as well as national laboratory testing centers, were invited to participate. All datasets underwent extensive data quality checks to identify any sources of potential biases that could impact the serotype distribution, and these were reviewed with site investigators [15].

Assessing eligibility for the meningitis serotype distribution analysis involved a multi-step process: sites had to first have eligible IPD data (step 1), then those sites had to have eligible meningitis data (step 2), which were then assessed to determine the number of years of PCV10/13 use until the serotype distribution stabilized (step 3), and then sites with data after that threshold were included in analyses (step 4).

Step 1: Data collection eligibility criteria were established to capture years of data where the serotype distribution had likely begun to stabilize, and where there was sufficient serotype data to estimate an unbiased distribution. A site or network had to report serotype-specific IPD case counts, regardless of syndrome. IPD was defined as *Streptococcus pneumoniae* isolated by culture from any normally sterile fluid or using *lytA*-based PCR or antigen-based tests in cerebrospinal fluid (CSF) or pleural fluid. Sites had to have a minimum of four years of post-PCV10/13 introduction surveillance data, including the year of introduction, with a minimum of 12 months of continuous surveillance; have at least 50% of isolates serotyped; have no major changes or biases in surveillance that would affect estimates of serotype-specific percentages; and not be limited to HIV-positive or immunocompromised populations. The year of introduction was defined as the year PCV10/13 was introduced if it was introduced in the first three quarters of the year, or as the following year otherwise. For data submitted in epidemiologic years rather than calendar years, the introduction year was defined accordingly.

Step 2: Sites had to identify which IPD cases were either confirmed positive for pneumococcus in CSF (CSF+) or had meningitis described as the clinical syndrome in a patient for whom pneumococcus was isolated in blood.

Step 3: Data from sites meeting the above criteria were used to assess the number of years after PCV10/13 introduction needed until the serotype distribution stabilized. To determine when this occurred, the change over time in the annual serotype distribution of all IPD was examined at each site, separately for children and adults. The change in percentage due to individual serotypes, both vaccine types and non-vaccine types, were examined. Particular attention was given to sites with robust data and high-quality surveillance systems. "Stabilization" was defined when trends in vaccine type and prevalent non-vaccine type serotype percentages over time were no longer evident, and the period after this was defined as the "mature" PCV10/13 period. For children under 5 years of age, the number of years of continuous and exclusive PCV10 or PCV13 use required to reach the mature PCV10/13 period varied depending on (a) whether and how long PCV7 was used prior to PCV10/13 introduction (or if there was a period of use of the alternate PCV10/13 product), and (b) whether PCV10/13 was introduced with a catch-up program. For sites without catch-up or prior use of another PCV product, time to reach the mature period was seven years of continuous and exclusive PCV10 or PCV13 use, including the year of introduction. For sites with a PCV10/13 catch-up program, time to reach the mature period was six years and for sites that used another PCV product for three or more years prior to PCV10/13, it was five years. For older children and adults, it took seven years of PCV10/13 use in infants to reach the mature PCV10/13 period, regardless of prior PCV7 use or catch-up, as these had no meaningful observed impact on time to stabilization in these age groups. This process and the thresholds defined here were reviewed by the PSERENADE Technical Advisory Group and the site investigators.

Step 4: Sites were included in the analyses if they had serotyped meningitis cases during the defined mature PCV10/13 period for their site. Only mature years where the average proportion of children immunized was greater than 70% in the three years preceding were included. WHO-UNICEF estimates of PCV10/13 uptake [16] were used for sites without local immunization coverage information. Sites with concurrent use of PCV10 and PCV13 or that switched between PCV10 and PCV13 and did not use either one for long enough to meet inclusion criteria were excluded. Cases with unknown age were excluded from analyses. Cases from all eligible mature period years were pooled for each site by age group.

The primary analysis was restricted to cases with *S. pneumoniae* identified from CSF (CSF+), given limited availability of clinical syndrome data and differences in the definitions of clinical meningitis across sites (Table S4). A sensitivity analysis included additional clinically defined meningitis cases (i.e., blood-culture positive, CSF-negative/not tested cases). Additionally, the serotype distribution of CSF+ cases was compared to those with only a clinical meningitis diagnosis within sites with large sample sizes to assess any differences.

2.2. Defining Serotype Categories

Cases were grouped into serotype categories for serotypes included in PCV7 (4, 6B, 9V, 14, 18C, 19F, and 23F), PCV10 (PCV7 plus 1, 5, and 7F), PCV13 (PCV10 plus 3, 6A, and 19A), PCV15 (PCV13 plus 22F and 33F) [17], PCV20 (PCV15 plus 8, 10A, 11A, 12F, and 15BC) [18], PCV24 (PCV20 plus 2, 9N, 17F, and 20) [12,13], and PPV23 (PCV24 minus 6A). Serotypes 15B and 15C were grouped as 15BC because they can switch due to a slipped strand mispairing of a tandem thymine–adenine repeat [19]. Non-vaccine type serotypes were defined as serotypes not in the indicated vaccine.

Serotyping methods for each site are summarized in Table S4. Cases without a specific serotype identified were rare and were grouped into four categories: "not serotyped", "untypeable", "typed, serotype not identified", and "serogrouped only". "Not serotyped" cases, those for whom serotyping was not attempted for any reason, were excluded from analyses after site investigators confirmed these to be missing at random. "Untypeable" cases had a comprehensive serotyping methodology performed but did not identify any serotype, such as non-encapsulated strain-prohibiting serotyping, an isolate that produced less capsule under lab conditions and could not be typed phenotypically, or a new serotype; these were grouped with non-vaccine type serotypes and excluded from serotype-specific analyses. "Typed, serotype not identified" cases had serotyping performed with a method that does not assess all serotypes, such as PCR assessing only 37 serotypes; these were grouped with non-PCV10, non-PCV13, and/or non-PPV23 cases depending on the serotyping method used by the site and were excluded from serotype-specific analyses. "Serogrouped only" cases (e.g., 6A/6B/6C/6D), undistinguished cases (e.g., 6A/6C), cases with two serotypes reported, and Quellung Pool-only cases were grouped into serotype categories where possible and excluded from serotype-specific analyses, though these were few.

2.3. Analytic Model

The predicted probability of pneumococcal meningitis due to serotype categories and specific serotypes was estimated using multinomial Dirichlet regression [20]. When data were insufficient for the model to converge, distributions were estimated by pooling data across sites. This model assumes that each site has an underlying unknown serotype distribution that varies in its deviation from the "site-averaged" distribution. The model estimates the magnitude of this deviation for all sites as a measure of possible heterogeneity among sites, which is then used along with sample size to determine each site's weight. When a high degree of heterogeneity exists across sites, sites are weighted more similarly; otherwise, sites are weighted more proportionally to their sample size, so larger sites are weighted more. The model ensures that all proportions in an estimated distribution sum to 1.0.

The serotype distribution for all observed serotypes could not be modeled because the model cannot estimate distributions when sites have many serotypes with zero counts. To identify which serotypes appeared frequently enough to be modeled, the data were first pooled across sites to estimate the rank of serotypes by product and age group. The top 25 ranking serotypes were selected for each age group and product stratum plus serotype 1, which was added to all strata because it is a key serotype of interest. These serotypes were then used to generate a modeled distribution for each age group and product stratum.

The robustness of the model is affected by the number of categories (i.e., serotypes) estimated, such that the more categories there are, the less robust the estimates are. Therefore, a hierarchal, or stepped, approach was used where an initial distribution was estimated for serotypes grouped into categories (e.g., PCV13-type), and subsequent models were run on further subdivided categories (e.g., PCV10-type, 19A, 6A, and 3) until reaching individual serotypes. This enabled sites that did not test for all serotypes to contribute to higher-order categories, even if not for some individual serotypes.

For each distribution estimated, whether for serotype categories or specific serotypes within a category, the model was first run on all eligible data using 30 iterations to estimate initial coefficients. Then, the model was run within a bootstrap with 100 replicates of resampling with replacement, using the initial coefficients estimated and stratified resampling based on the covariate used (e.g., PCV product). Within the bootstrap, the model was limited to a single iteration for each replicate. The means of the bootstrap replicates were used as the estimated distribution. For serotype categories, 95% confidence intervals (95% CI) around the mean values were calculated using adjusted bootstrap percentile (BCa) intervals. Due to limited sample size when estimating the distribution of specific serotypes, jackknife resampling was used in place of bootstrapping where one site was removed at a time for specific serotype estimates. In this case, confidence intervals were calculated using the estimated standard errors. When distributions were estimated via pooling, binomial confidence intervals were used. All analyses were performed in R (R Core Team, 2019), and the model used the VGAM package [20].

3. Results

3.1. Data Included

Of the 76 sites with IPD data eligible for data collection and that participated in the PSERENADE project, 32 PCV13-using sites and eight PCV10-using sites had serotype-specific pneumococcal meningitis cases in the mature PCV10/13 period eligible for this analysis. Reasons for exclusion included: IPD cases were not disaggregated by CSF+ (N = 23), no serotyped meningitis cases in the mature PCV10/13 period were reported (N = 11), and concurrent PCV10 and PCV13 use (N = 4). The majority of meningitis cases were CSF+ (73.3%, range across sites: 24.1–100%) (Figure S1). Most sites (N = 26, 65.0%) were from Europe or North America, and only nine (22.5%) were from low- and middle-income countries (Figure 1). Most PCV13 sites (87.5%) previously used PCV7 compared to only two (25.0%) for PCV10 sites. All but four sites used a booster dose schedule, two of which only report data for children < 5 years. Additional site details and characteristics are described elsewhere [15].

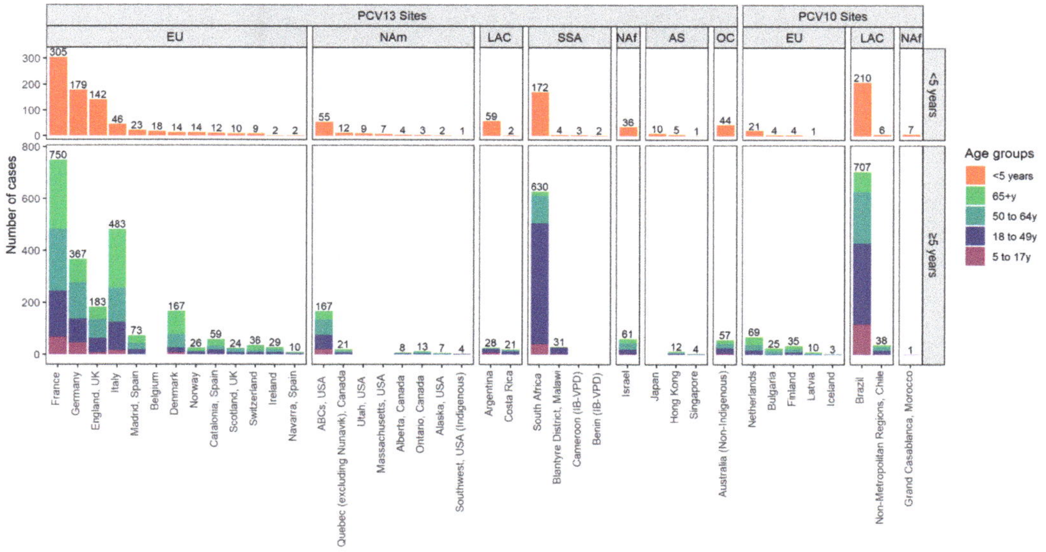

Figure 1. Number of serotyped cerebrospinal fluid positive (CSF+) pneumococcal meningitis cases per site in mature PCV10/13 years by UN region, pneumococcal conjugate vaccine (PCV) product used during years included in the analysis, and age group. Abbreviations: EU = Europe, NAm = North America, LAC = Latin America and Caribbean, SSA = Sub-Saharan Africa, NAf = Northern Africa and Western Asia, AS = Asia, OC = Oceania. PCV13 is Pfizer's Prevnar13/Prevenar13; PCV10 is GSK's Synflorix.

3.2. *Pneumococcal Meningitis Due to Vaccine-Type Serotypes*

In PCV13-using sites, 14.1% (95% CI: 10.4–16.2%) of the remaining CSF+ pneumococcal meningitis during the mature period in children <5 years of age was PCV13-type; among individuals ≥5 years of age, 25.8% (23.6–27.6%) were PCV13-type (Figure 2 and Table S1). Serotype 3 was the most common PCV13-type in PCV13-using sites (4.0% and 13.1% in <5 and ≥5 years, respectively).

For PCV10-using sites, due to the large case counts in Brazil (n = 210 among <5 years) relative to the other PCV10-using sites (n = 43 total among <5 years), data from Brazil were shown separately from all other PCV10-using sites. Data from PCV10-using sites excluding Brazil could not be modeled due to the small sample sizes and were, therefore, pooled. Among children <5 years, the percent PCV10-type was similar in Brazil (4.8%) and the other PCV10 sites (4.9%; Figure 2). The most common PCV10-type serotype was 7F (2.1%). For cases ≥5 years of age, where sample sizes were larger (Brazil: n = 707; other PCV10 sites: n = 181), the percent PCV10-type cases was lower in Brazil (7.6%; 5.8–9.8%) than in the other sites (14.9%; 10.1–21.0%). Modeled results for all PCV10 sites combined (i.e., including Brazil) are shown in Figure S3.

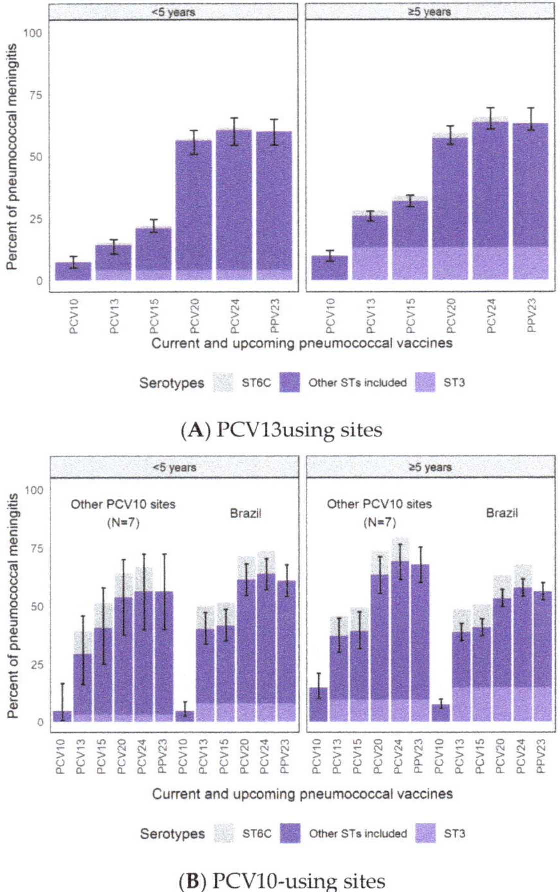

(**A**) PCV13using sites

(**B**) PCV10-using sites

Figure 2. Percentage of CSF+ pneumococcal meningitis cases in the mature PCV10/13 period due to serotypes included in current and upcoming products. PCV13 is Pfizer's Prevnar13/Prevenar13; PCV10 is GSK's Synflorix. PCV13 results are modeled output. PCV10 results are a pooled distribution of 210 cases in Brazil and 43 cases in other PCV10 sites for <5 years of age and of 707 cases in Brazil and 181 cases in other PCV10 sites for ≥5 years of age. ST3 is illustrated separately in lighter purple in the bars corresponding to products that include ST3 due to the uncertain effectiveness against ST3 in current products [21–23]. ST6C is illustrated in grey above the bars where ST6A is included. Although ST6C is not included in PCV10 or PCV13, PCV13 offers cross-protection through ST6A [24]. ST6A also benefits from cross-protection with ST6B, included in both PCV10 and PCV13. Therefore, ST6A causes a very small fraction of disease in both settings and age groups, and it is not shown. Confidence intervals do not include ST6C, as this serotype is not included in PCV10/13.

The percentage of PCV13-type cases was greater in PCV10-using sites than in PCV13-using sites (29.3–40.0% vs. 14.1% for <5 years and 37.0–38.5% vs. 25.8% for ≥5 years). If serotype 6C is considered a PCV13-type serotype because it has possible cross-protection from 6A [24], the difference in the proportion of PCV13-type increases even more. Differences between PCV10- and PCV13-using sites persist for PCV15-type cases but diminish for PCV20-, PCV24-, and PPV23-type cases, which ranged from 54–69% across all age and

PCV-use groups. Results restricted to adults aged ≥50 years and ≥65 years showed no meaningful differences compared to results for those aged ≥5 years (Table S3).

Across both PCV products and age groups, there was wide heterogeneity in the site-specific percentages for the various vaccine-type groups, with the lowest and highest percentages from sites with very small sample sizes (Figure 3). This heterogeneity was evident within regions and overlapped across regions; no clear regional differences were apparent. For the PCV13 sites where modeled estimates were possible, they were consistently within +/-5% of the median of the site-specific estimates.

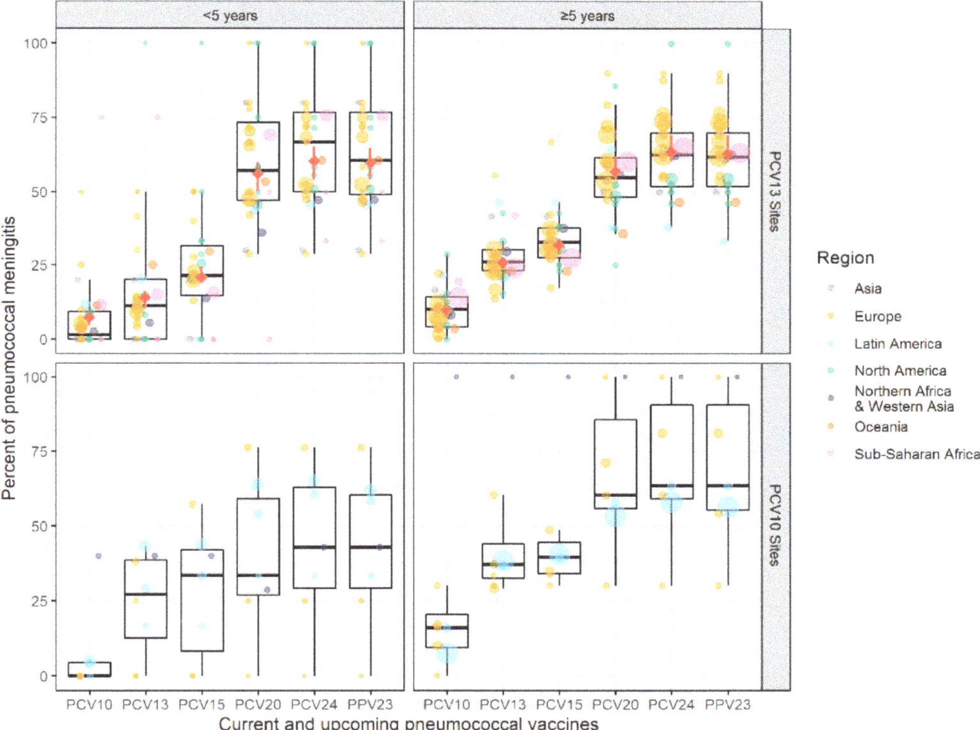

Figure 3. Site-specific percentages of CSF+ pneumococcal meningitis in the mature PCV10/13 period due to serotypes included in current and upcoming products. Each site is represented by a dot, which are colored by region and sized proportionally to the number of cases contributed by that site. The black boxes illustrate the IQR for the site-specific percentages. For PCV13 sites, the modeled results shown in Figure 2 are shown here in red diamonds. Data from Singapore are not shown for confidentiality but contributed to the PCV13 modeled results. PCV13 is Pfizer's Prevnar13/Prevenar13; PCV10 is GSK's Synflorix.

3.3. Serotype Distribution

Serotype-specific percentages are described for all PCV10-using sites combined (i.e., Brazil plus other sites) because the most common serotypes found in Brazil were similar to those in the other PCV10 sites for both age groups (Figure S4); their data were pooled rather than modeled due to small sample size. Among children <5 years of age, the most common serotypes were those not covered by the vaccines in use (Figure 4 and Table S2). The three PCV13-related serotypes (19A, 6C, and 3) in cases <5 years at PCV10-using sites totaled 42.1% compared to 7.5% at PCV13-using sites. Otherwise, the next most common serotypes at PCV10-using sites were similar to the most common at PCV13-using sites.

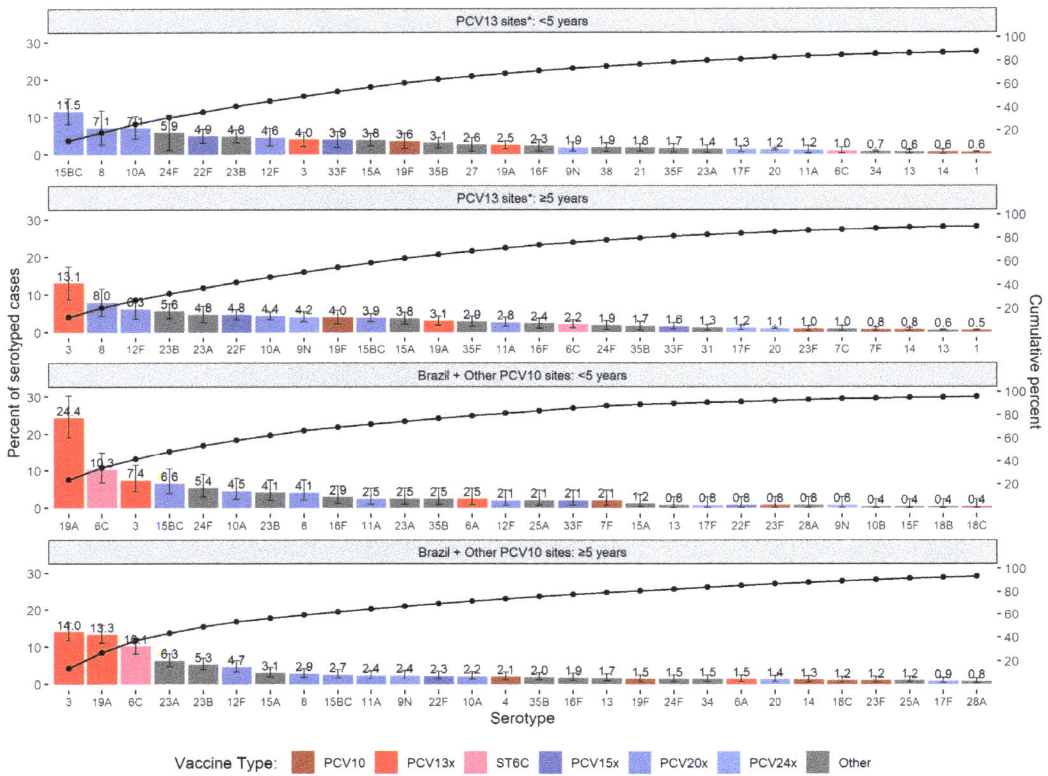

Figure 4. Serotype-specific distribution of CSF+ pneumococcal meningitis in the mature PCV10/13 period. Serotypes are colored by the lowest valency PCV product they are included in. The "x" in the PCV legend represents the extra serotypes included in that product relative to the next lower product (i.e., PCV13x includes serotypes 3, 6A, and 19A not in PCV10). PCV13 is Pfizer's Prevnar13/Prevenar13; PCV10 is GSK's Synflorix. Serotype 6C is colored separately because, although it is not included in any product, it is covered through cross-protection with PCV13-type serotype 6A [24]. Morocco and Bulgaria were not included in the PCV10 distribution due to serotyping limitations. * Serotype distribution for PCV13 sites is modeled output for the top 25 serotypes based on a pooled ranking plus serotype 1. Serotype distribution for PCV10 sites is from pooling cases across sites.

Serotype 19A, a PCV13-type, was the top serotype at PCV10-using sites among children <5 years, causing approximately 24% (95% CI: 18.6–30.7%) of cases, and was also common in cases ≥5 years (14.4% in Brazil and 7.8% in other PCV10-using sites). In contrast, at PCV13-using sites, it was uncommon (2.5% for <5 years of age and 3.1% for ≥5 years).

The second most common serotype at PCV10-using sites among cases <5 years was serotype 6C (10.3%; 95% CI: 6.8–14.9%), another serotype thought to be preventable by PCV13 because of cross-protection from 6A (6C was 1.0% at PCV13-using sites; Figure 2). The proportion of serotype 6C cases was also greater in cases ≥5 years at PCV10-using sites (approximately 10%) than at PCV13-using sites (2.2%).

Serotype 3 was the top serotype among cases ≥5 years at both PCV13-using sites (13.1%; 8.8–17.4%) and PCV10-using sites (13.9%; 11.7–16.4) and was consistent across most site specific distributions (Table S2 and Figure S5) with the exception of South Africa (Site 55, n = 623), where 12F and 8 were the dominant serotypes (approximately 13% each), and serotype 3 was ranked third (5.9%). For children <5 years, the percentage was lower

(4.0 and 7.4% at PCV13- and PCV10-using sites, respectively) and was ranked higher at PCV10-using sites (rank = 3) than PCV13-using sites (rank = 8).

Several non-PCV13-type serotypes that are included in higher-valency PCV products in development (15BC, 8, 12F, 10A, and 22F) were common in both age groups and PCV settings, cumulatively causing between 15 and 36% of pneumococcal meningitis (Figure 4 and Figure S5). Serotype 22F, which is included in PCV15 was generally ranked lower than the others. Serotype 15BC was a leading non-PCV13-type serotype among children <5 years in both PCV10- and PCV13-using sites (6.6 and 11.5%, respectively) but was less common in cases ≥5 years (2.7 and 3.9%, respectively). An exception for children <5 years was South Africa (Site 55, n = 172), where serotype 8 was the dominant serotype (38.4% compared to <6% for all other serotypes). Serotypes in the top 5 of at least one age or PCV group that are not included in PCV15, PCV20, or PCV24 include serotypes 24F, 23B, and 23A, which cumulatively caused between 10 and 12% of pneumococcal meningitis across both PCV settings and age groups.

3.4. Sensitivity Analyses

A sensitivity analysis including all meningitis IPD cases (CSF+ cases and non-CSF+ clinically defined cases) was conducted to ensure restriction to CSF+ only did not bias the selection of sites or cases. These showed no meaningful differences from the primary results restricted to CSF+ cases only (Table S1 and Figure S2). A review of the site-specific serotype distributions comparing CSF+ to clinically-defined meningitis cases also showed no meaningful differences for children <5 years. For cases ≥5 years, serotype 3 was more commonly a top serotype for CSF+ cases than for clinically defined meningitis cases, and serotype 19F was sometimes more prevalent in the clinically defined meningitis cases than in the CSF+ cases. Other sensitivity analyses that excluded sites with fewer than five cases in an age group or restricted to sites with data for both age groups did not result in any meaningful differences (data not shown).

4. Discussion

We found that in settings where PCV10 or PCV13 have been used for about seven years with primary series uptake above 70%, the percentage of remaining pneumococcal meningitis due to serotypes covered by the vaccines in use was low: 5.3% in PCV10 sites and 14.1% in PCV13 sites in children <5 years and 15.3 and 25.8%, respectively, in older children and adults. This is a substantial reduction compared to the era before PCVs when 70–88% of IPD cases and 62–72% of meningitis cases were caused by PCV10/13-type serotypes in children <5 years of age, depending on the vaccine and region [2,25]. Serotype 19A, a PCV13-type, was rare (≤3%) at PCV13-using sites but caused almost a quarter of cases ≤5 years at PCV10-using sites. A large fraction of the remaining disease was due to serotypes found in higher-valency PCV products in development. PCV15 covered an additional 36% of cases <5 years excluding PCV10-types at PCV10-using sites, although only an additional 7% at PCV13-using sites excluding PCV13-types. PCV20 and PCV24 covered an additional 49–59% in PCV10-using sites and 43–47% in PCV13-using sites. In older children and adults, the percent of pneumococcal meningitis covered by PPV23 was greater than 62%, suggesting there is much vaccine-preventable pneumococcal meningitis still remaining in older age groups. Our results may not represent the local experience of any one country, and while there was heterogeneity observed in vaccine-type distributions, large deviations from the average estimates were predominantly only at small sites. Heterogeneity for some specific non-vaccine serotypes highlights the importance of continued monitoring of serotypes by national surveillance systems.

Serotype 3 was uncommon in children <5 years of age, as it was in the era before PCVs (estimated 1.4%) [2], causing 3–8% of cases at PCV10-using sites and 4% at PCV13-using sites. However, serotype 3 was the top ranked serotype among those aged ≥5 years in both PCV10- and PCV13-using sites, causing approximately 13–14% of cases. Although population-level direct effects of PCV13 against serotype 3 are not well understood, this suggests that if PCV13 has some direct effects on serotype 3 disease, the indirect effects are likely limited given the high percentage of cases caused by serotype 3 in adults. This has been previously suggested by De Wals in a review of immunologic and effectiveness evidence who concluded that PCV13 may confer some protection in vaccinated children but that it is likely to be lower than for other vaccine serotypes and short term [26]. A subsequent meta-analysis estimated PCV13 effectiveness against serotype 3 IPD in children to be 51–69% [21]. However, none of the studies estimated effectiveness against serotype 3 meningitis, half the studies used case-control methods to assess effectiveness, results only included data from 12 European and North American sites, and results included data with only 4–6 years of PCV13 use [27–33]. Studies in the US and UK not included in the meta-analysis showed contrasting results after 7 years of PCV13 use, finding no meaningful change in serotype 3 IPD incidence [22,23]. Another small study from Italy that included only six serotype 3 meningitis cases suggests that PCV13 in children may be effective against serotype 3 sepsis and meningitis but not pneumonia [34].

The remaining pneumococcal meningitis in PCV10 sites was largely driven by serotypes 19A and 6C, two PCV13-related serotypes, which together caused 34.7 and 23.4% of disease among those <5 years and ≥5 years, respectively. The potential for 19F in PCV10 to provide cross protection to 19A is not supported by our results, as 19A was found to be the dominant serotype of cases <5 years of age at PCV10-using sites, causing nearly a quarter of meningitis in that age group. Although our results for PCV10-using countries other than Brazil are based on sparse data, 19A was commonly seen across PCV10-using sites. Further evidence will be provided by upcoming serotype distribution analyses of all IPD that will increase the number of PCV10-using sites and number of cases, and by analyses of the change in 19A incidence from the pre-PCV to post-PCV periods. Earlier reviews of incidence-based and other studies of serotype 19A were inconclusive [5] and more recent studies have not found evidence of cross-protection from serotype 19F [35–37]. Prior evidence of cross-protection by PCV13 from serotype 6A to 6C is stronger [5,24] and consistent with the small (1–2%) percentage of serotype 6C cases at PCV13-using sites compared to approximately 10% at PCV10-using sites. This suggests that by also protecting against serotype 6C, PCV13 could potentially address up to a quarter of remaining pneumococcal meningitis in PCV10-using countries, if the corresponding replacement disease is small.

Another PCV10 product from SII (Pneumosil) has recently been licensed and is important in low- and middle-income countries for its affordability [38]. SII's PCV10 includes most of the same serotypes as GSK's PCV10, but replaces serotypes 4 and 18C with 19A and 6A. Using pre-PCV distribution data from Africa and Asia, we expect SII's PCV10 to cover roughly the same percentage of disease (72–73%) as GSK's (72–74%), assuming cross-protection from 6B to 6A, but not from 6A to 6C, which was not estimated in the pre-PCV era. SII's PCV10 covers slightly less than PCV13 if serotype 3 is excluded (76–77%) [2]. The percentage of pneumococcal meningitis in mature PCV10/13 settings covered by SII's PCV10 could not be estimated as we could not account for serotypes 4 and 18C that are covered by PCV10/13 but not SII's PCV10. We can speculate that these may expand to their pre-PCV incidence, and possibly greater with replacement disease. However, assessments of the relative impact of PCV products can only be based on comparisons of the change in incidence over time and should evaluate all pneumococcal syndromes, not just the small portion that are meningitis. Further, vaccine product choice involves many factors beyond epidemiologic settings, including programmatic and financial considerations.

Although higher-valency PCV products in development may not further reduce the remaining burden of serotype 3 disease, they target important non-PCV13-type serotypes responsible for much remaining pneumococcal meningitis. PCV20 (PCV13 + 8, 10A, 11A, 12F, 15BC, 22F, 33F) and PCV24 (PCV20 + 2, 9N, 17F, 20) covered more than half (53–69%) of the remaining pneumococcal meningitis across mature PCV10/13 settings and age groups. PCV15 (PCV13 + 22F, 33F) covers less, 9 and 6% of pneumococcal meningitis cases <5 years and ≥5 years, respectively, in PCV13 sites, and 3% in PCV10 sites among both age groups, but is nearest to being available for children. Merck has submitted applications to the U.S. Food and Drug Administration (FDA) and European Medicines Agency (EMA) for licensure of their PCV15 for all ages [10]. Pfizer's Biologics License Application (BLA) for use of PCV20 in adults 18 years and older has been accepted by the FDA for priority review with a decision expected in June 2021 [39] and Phase 3 trials in children have begun [40]. There are several PCV24 products in the pipeline, including products from Merck [12] and Affinivax, which are currently in Phase 1 testing [13]. If indirect effects for these products are similar to what has been observed for PCV10/13, their use in children <5 years may have more impact than direct immunization with PPV23 in older age groups.

An important limitation of this analysis is the lack of robust data from PCV10-using sites, with the exception of Brazil, which contributed 86% of those cases. Only one other country (The Netherlands) had more than 10 cases for children <5 years in the mature PCV10 period. This was due to a lack of data rather than insufficient years using PCV10 or lack of participation in PSERENADE. Of 23 countries using PCV10 exclusively, most (15 of 16) of those eligible for PSERENADE contributed data [15]; only eight met analytic eligibility criteria. A future update of this analysis adding data from countries excluded because they had not yet reached the mature PCV10 period will not greatly improve the data paucity problem because they generally had low annual numbers of cases. However, a forthcoming analysis of all IPD cases (not restricting to meningitis) will have more robust results due to larger sample sizes.

Another important limitation is the paucity of data from high burden settings using a PCV schedule without a booster dose, particularly the African meningitis belt where pneumococcal meningitis outbreaks occur in all age groups and where serotype 1 is a dominant serotype [41–43]. PSERENADE received data for only three meningitis belt countries, all using PCV13: Benin and Cameroon contributed two and three cases, respectively, for children <5 years of age, and The Gambia had no meningitis cases in the mature PCV13 period so could not contribute to analyses. Our findings from non-meningitis belt countries that primarily used a booster dose schedule showed serotype 1 consistently caused less than 1% of disease after 7 years of use, compared to 8% in the pre-PCV era among children <5 years [2]. In other PSERENADE analyses published separately in this issue, serotype 1 IPD incidence declined in all ages by 95% after 6 years of PCV10/13 use in non-meningitis belt countries [44]. The persistence of serotype 1 outbreaks in unvaccinated older children and adults in the meningitis belt despite 3–4 years of PCV10/13 use [41–43] may suggest that indirect protection may be lower than for other regions, although these results are from the "early" PCV use period so the full effects of PCV13 may not have occurred. In Burkina Faso and Niger, the percentage of pneumococcal meningitis that was PCV13-type was high, approximately 29–45% for children <5 years, with 4–30% due to serotype 1 [41,42]. For cases $\geq$5 years in Burkina Faso, Niger, and Ghana, 53–74% were PCV13-type and 30–64% were serotype 1 [41–43]. These data include cases occurring during meningitis outbreaks, which are commonly caused by serotype 1. Continued monitoring of the serotype distribution in meningitis belt countries using a 3+0 schedule could provide data to understand whether the speed or degree of PCV impact is lower than for countries using a booster dose, or if results are in fact similar after 7 years of PCV13 use.

An important consideration when interpreting the percentages and serotype distribution results is that a similar sized percentage in the mature PCV10/13 period and in the pre-PCV period represent largely different disease burdens, as PCV has greatly reduced the overall disease burden (i.e., 5% of 100 cases is a much smaller disease burden than 5% of 1000 cases). Consequently, inferences about disease burden cannot be made by comparing percentage sizes alone across age or PCV product strata. In addition, the percentage of any given serotype is affected by the incidence of the other serotypes, such that even if a serotype's incidence remains stable, the percentage will increase when another serotype's decreases. Therefore, an analysis of the change in incidence is needed to assess impact, which will be forthcoming in another PSERENADE analysis for sites with incidence data over time.

5. Conclusions

In countries that have used PCV10 or PCV13 for at least 5–7 years and with high uptake, the percentage of pneumococcal meningitis that was vaccine type was less than 15% among children <5 years of age, which is small when compared to percentages above 70% observed prior to PCV introduction, suggesting that PCV10/13 use greatly reduces the proportion of pneumococcal meningitis due to vaccine-type serotypes. Serotype 19A, a PCV13-type, was the most common serotype found in children at PCV10-using sites but rare at PCV13-using sites. Among older children and adults, the percentage vaccine type was <26%, but over 62% was PPV23-type despite common recommendations for PPV23 use among older adults and those at high risk for pneumococcal disease. Higher-valency PCVs currently under evaluation, particularly PCV20 and PCV24, target over half of the remaining pneumococcal meningitis.

Supplementary Materials: The following are available online at https://www.mdpi.com/article/10.3390/microorganisms9040738/s1, Table S1: Percentage of pneumococcal meningitis in the mature PCV10/13 period due to serotypes included in current and upcoming PCV products, Table S2: Serotype-specific distribution of CSF+ pneumococcal meningitis in the mature PCV10/13 period, by age group; Figure S1: Distribution of CSF+ meningitis cases and non-CSF clinically defined meningitis cases in the mature PCV10/13 period by age group; Figure S2: Serotype-specific distribution of pneumococcal meningitis in the mature PCV10/13 period, including all meningitis cases (CSF+ and non-CSF+ clinically defined meningitis cases); Figure S3: Modeled distribution for the percentage of CSF+ pneumococcal meningitis due to serotypes included in current and upcoming products, in both PCV10- and PCV13-using sites; Table S3: Modeled percentage of CSF+ pneumococcal meningitis due to serotypes included in current and upcoming pneumococcal vaccine products, by age group; Figure S4: Serotype-specific distribution of CSF+ pneumococcal meningitis in the mature PCV10/13 period, comparing Brazil to the other PCV10-using sites; Figure S5: Distribution of top 10 ranking serotypes causing CSF+ pneumococcal meningitis in each PCV13-using site; Table S4: Characteristics of sites contributing pneumococcal meningitis cases to analyses.

Author Contributions: Conceptualization, M.G.Q., M.D.K. and K.H.; methodology, M.G.Q., S.L.Z., M.D.K., K.H. and M.E.P.; software and analysis, M.G.Q. and S.L.Z.; writing—original draft preparation, M.G.Q., M.D.K. and Y.Y.; writing—review and editing, M.D.K., J.C.B., S.L.Z. and D.R.F.; visualization, M.G.Q. and Y.Y.; supervision, M.D.K. and K.H.; project administration, J.C.B. and M.G.Q.; funding acquisition, A.L.C., M.D.K. and K.H. All other authors contributed data for the analysis, provided input for the analytic methodology, and critically reviewed results. All authors have read and agreed to the published version of the manuscript.

Funding: The PSERENADE project is funded by the Bill and Melinda Gates Foundation as part of the World Health Organization Pneumococcal Vaccines Technical Coordination Project, grant number INV-010429 / OPP1189065.

Data Availability Statement: Restrictions apply to the availability of these data. Data were obtained under data sharing agreements from contributing surveillance sites and can only be shared by contributing organizations with their permission.

Acknowledgments: This work was only possible through the contributions of many other individuals and organizations. The list of names of key individuals is provided in Appendix A Table A2. We thank all individuals involved in the surveillance activities at the sites contributing to the PSERENADE project for trusting us with their data and for the assistance along the way to ensure this work provides value. This includes the surveillance network coordinators at EpiConcept, European Centre for Disease Prevention and Control, Pan American Health Organization, and WHO who facilitated access to and understanding of the data. We thank the Johns Hopkins University librarians and research assistants for providing literature review and data collection support. We thank the technical and strategic advisors from the WHO for providing guidance on the objectives and for facilitating introductions to the many surveillance sites globally. Finally, we are greatly indebted to the members of our PSERENADE Technical Advisory Group for providing us with expert technical review and strategic advice on all aspects of the methods, presentation, and interpretation throughout the project.

Conflicts of Interest: K.H. conducted the study and analyses while working at the Johns Hopkins School of Public Health but is an employee at Pfizer, Inc. as of 26 October 2020. M.D.K. reports grants from Merck, personal fees from Merck, and grants from Pfizer, outside the submitted work. J.C.B. reports funding from Pfizer in the past year, unrelated to the submitted work. M.L. has been a member of advisory boards and has received speaker's honoraria from Pfizer and Merck. German pneumococcal surveillance has been supported by Pfizer and Merck. S.D. reports grant from Pfizer, outside the submitted work. J.C.S. had received assistance from Pfizer for attending scientific meetings outside the submitted work. K.A. reports a grant from Merck, outside the submitted work. M.C. has previously received a professional fee from Pfizer (Ireland), an unrestricted research grant from Pfizer Ireland (2007–2016), and an Investigator Initiated Reward from Pfizer Ireland in 2018 (W1243730). S.C.G.A. received travel grant from Pfizer. A.M. has received research support to their institution from Pfizer and Merck; honoraria for advisory board membership from GlaxoSmithKline, Merck and Pfizer. S.N.L. performs contract research for GSK, Pfizer, Sanofi Pasteur on behalf of St. George's University of London but receives no personal remuneration. I.Y. was a member of mRNA-1273 study group and has received funding to her institution to conduct clinical research from BioFire, MedImmune, Regeneron, PaxVax, Pfizer, GSK, Merck, Novavax, Sanofi-Pasteur, and Micron. R.D. has received grants/research support from Pfizer, Merck Sharp and Dohme, and Medimmune; has been a scientific consultant for Pfizer, MeMed, Merck Sharp and Dohme, and Biondvax; had served on advisory boards of Pfizer, Merck Sharp and Dohme, and Biondvax and has been a speaker for Pfizer. L.L.H. reports research grants to her institution from GSK, Pfizer, and Merck. E.V. reports grants from the French public health agency, during the conduct of the study and grants from Pfizer and grants from Merck outside the submitted work. M.H. received an educational grant from Pfizer AG for partial support of this project. However, Pfizer AG had no role in the data analysis and content of the manuscript. N.B.Z. has received investigator-initiated research grants from GlaxoSmithKline, Takeda Pharmaceuticals, Merck, and the Serum Institute of India, all unrelated to this research. N.Mv.S. reports grants and fees for service from Pfizer, fees for service from MSD and GSK, outside the submitted work; in addition, N.Mv.S. has a patent WO 2013/020090 A3 with royalties paid to University of California San Diego (inventors: Nina van Sorge/Victor Nizet). M.C.B. reports lecture fees from MSD outside from submitted work. C.L.B., M.D., has intellectual property in BioFire Diagnostics and receives royalties through the University of Utah. C.L.B. is an advisor to IDbyDNA. M.T. reports grants from GlaxoSmithKline and grants from Pfizer Inc. to the Finnish Institute for Health and Welfare for research projects outside the submitted work, in which she has been a co-investigator. Av.G. as received researching funding from Pfizer (last year 2017, Pfizer Investigator-Initiated Research [IIR] Program IIR WI 194379); they have attended advisory board meetings for Pfizer and Merck. C.M.A. reports grants and personal fees from Pfizer, Qiagen, and BioMerieux and grants from Genomica SAU, outside the submitted work. BL had two research grants from Pfizer on Streptococcus pneumoniae. J.D.K. has received an unrestricted grant-in-aid from Pfizer Canada that supports, in part, the CASPER invasive pneumococcal disease surveillance project. C.G.S. reports grant funding from Pfizer, Merck, and AstraZeneca in the past 3 years. All other authors did not declare any conflicts of interest. The funders had no role in the design of the study; in the collection, analyses, or interpretation of data; in the writing of the manuscript, or in the decision to publish the results.

Disclaimer: The findings and conclusions in this report are those of the authors and do not necessarily represent the official position of the Centers for Disease Control and Prevention or the World Health Organization (WHO).

Appendix A

Table A1. PSERENADE team.

Name	Affiliation
Zahin Amin-Chowdhury	Immunisation and Countermeasures Division, Public Health England, London NW9 5EQ, UK
Rita Born	Federal Office of Public Health, 3097 Liebefeld, Switzerland

Table A1. *Cont.*

Name	Affiliation
Maria-Cristina C. Brandileone	National Laboratory for Meningitis and Pneumococcal Infections, Center of Bacteriology, Institute Adolfo Lutz (IAL), São Paulo 01246-902, Brazil
Dana Bruden	Arctic Investigations Program, Division of Preparedness and Emerging Infections, National Center for Emerging and Zoonotic Infectious Diseases, Centers for Disease Control and Prevention, Anchorage, AK 99508, USA
Carrie L. Byington	University of Utah Department of Pediatrics (emeritus), Salt Lake City, UT 84108, United States; University of California Health System, Oakland, CA 94607, USA
Claire Cameron	Public Health Scotland, Glasgow G2 6QE, UK
Jesús Castilla	Instituto de Salud Pública de Navarra—IdiSNA, 31003 Pamplona, Navarra, Spain; CIBER Epidemiología y Salud Pública, (CIBERESP), 28029 Madrid, Spain
Guanhao Chan	Singapore Ministry of Health, Communicable Diseases Division, Singapore 308442, Singapore
Kin-Hung Chow	Department of Microbiology and Carol Yu Centre for Infection, Queen Mary Hospital, The University of Hong Kong, Hong Kong SAR, China
Tine Dalby	Bacteria, Parasites and Fungi, Statens Serum Institut, DK-2300 Copenhagen S, Denmark
Kostas Danis	Santé Publique France, the French National Public Health Agency, Saint Maurice CEDEX, 94415 Paris, France
Linda de Gouveia	Centre for Respiratory Diseases and Meningitis, National Institute for Communicable Diseases of the National Health Laboratory Service, Sandringham, 2192 Johannesburg, South Africa
Sara de Miguel	Epidemiology Department, Dirección General de Salud Pública, 28009 Madrid, Spain
Geneviève Deceuninck	Quebec University Hospital Research Centre, Québec, QC G1V 4G2, Canada
Martina Del Manso	Department of Infectious Diseases, Italian National Institute of Health (Istituto Superiore di Sanità, ISS), 00161 Rome, Italy
Janepsy Díaz	Instituto de Salud Pública de Chile, 7780050 Santiago, Santiago Metropolitan, Chile
Elina Dimina	Centre for disease prevention and control of Latvia, 1005 Riga, Latvia
Helga Erlendsdottir	Department of Clinical Microbiology, Landspitali—The National University Hospital, Hringbraut, 101 Reykjavik, Iceland
Noga Givon-Lavi	Pediatric Infectious Disease Unit and Clinical Microbiology Laboratory, Soroka University Medical Center, Ben-Gurion University of the Negev, Beer-Sheva 8410501, Israel
James D. Kellner	Department of Pediatrics, University of Calgary, and Alberta Health Services, Calgary, AB T3B 6A8, Canada
Mirjam J. Knol	National Institute for Public Health and the Environment, 3721 MA Bilthoven, The Netherlands
Brigitte Lefebvre	Laboratoire de Santé Publique du Québec, Sainte-Anne-de-Bellevue, Quebec City, QC H9X 3R5, Canada
Jolita Mereckiene	HSE Health Protection Surveillance Centre, Mountjoy, D01 A4A3 Dublin, Ireland

Table A1. *Cont.*

Name	Affiliation
Carmen Muñoz-Almagro	Molecular Microbiology Department, Hospital Sant Joan de Déu Research Institute, 08950 Esplugues de Llobregat, Barcelona, Spain; Medicine Department, Universitat Internacional de Catalunya, 08017 Barcelona, Spain; CIBER Epidemiología y Salud Pública (CIBERESP), 28029 Madrid, Spain
Daniela Napoli	Servicio de Bacteriología Clínica, Departamento de Bacteriología, INEI—ANLIS "Dr. Carlos G. Malbrán", C1282 AFF Buenos Aires, Argentina
Néhémie Nzoyikorera	Department of Microbiology, Faculty of Medicine and Pharmacy, Hassan II University of Casablanca, Casablanca 20000, Morocco; Bacteriology-Virology and Hospital Hygiene Laboratory, Ibn Rochd University Hospital Centre, Casablanca 20250, Morocco
Stephen I. Pelton	Boston University Schools of Medicine and Public Health, Boston, MA 02118, USA
Kate Pennington	Communicable Disease Epidemiology and Surveillance Section, Office of Health Protection, Australian Government Department of Health, Canberra, ACT 2606, Australia
Tamara Pilishvili	National Center for Immunizations and Respiratory Diseases, Centers for Disease Control and Prevention, Atlanta, GA 30333, USA
Marie-Cecile Ploy	University Hospital Centre Limoges, Regional Observatories for Pneumococci, 87000 Limoges, France
Rodrigo Puentes	Instituto de Salud Pública de Chile, 7780050 Santiago, Santiago Metropolitan, Chile
Shigeru Suga	Infectious Disease Center and Department of Clinical Research, National Hospital Organization Mie Hospital, Tsu, Mie 514-0125, Japan
Catherine G. Sutcliffe	Johns Hopkins Bloomberg School of Public Health, Baltimore, MD 21205, USA
Todd D. Swarthout	NIHR Global Health Research Unit on Mucosal Pathogens, Division of Infection and Immunity, UCL, Bloomsbury, London WC1E 6BT, UK; Malawi-Liverpool-Wellcome Trust Clinical Research Programme, P.O. Box 30096, Chichiri, Blantyre 3, Malawi
Koh Cheng Thoon	KK Women's and Children's Hospital, Singapore 229899, Singapore
Maija Toropainen	Department of Health Security, Finnish Institute for Health and Welfare, 00271 Helsinki, Finland
Didrik F. Vestrheim	Department of Infection Control and Vaccine, Norwegian Institute of Public Health, 0456 Oslo, Norway
Anne von Gottberg	Centre for Respiratory Diseases and Meningitis, National Institute for Communicable Diseases of the National Health Laboratory Service, Sandringham, 2192 Johannesburg, South Africa; School of Pathology, Faculty of Health Sciences, University of the Witwatersrand, Braamfontein 2000, Johannesburg, South Africa
Toronto Invasive Bacterial Diseases Network	
WHO Invasive Bacterial Vaccine-Preventable Diseases (IB-VPD) Network	

Table A2. Acknowledgement List.

PSERENADE Technical Advisory Group
Thomas Cherian
William P. Hausdorff
Marc Lipsitch
Shabir A. Madhi
Elizabeth Miller
Catherine Satzke
Cynthia G. Whitney
World Health Organization
Katherine L. O'Brien
Jenny A. Walldorf
Johns Hopkins University
Yunfeng Cao
Peggy Gross
Donna Hesson
Marissa Hetrich
Ananya Kumar
Kate Perepezko
E. Wangeci Kagucia
Francesca Schiaffino Salazar
Jenna Sinkevitch
Daniel Stephens
Epiconcept
Camelia Savulescu
European Centre for Disease Prevention and Control
Edoardo Colzani
Pan American Health Organization
Lúcia Helena de Oliveira
Epidemiology Department, Dirección General de Salud Pública, Comunidad de Madrid, Madrid, Spain
Luis García Comas
Maria Ordobás Gavín
Department of Infectious Diseases, Italian National Institute of Health (Istituto Superiore di Sanità, ISS), Rome, Italy
Flavia Riccardo
Surveillance and Public Health Emergency Response, Public Health Agency of Catalonia, Barcelona, Spain
Sonia Broner
Conchita Izquierdo
Laboratoire du Centre Mère et Enfant Fondation Chantal Biya, Yaoundé, Cameroun
Angeline Boula
University of Parakou, Alibori in PARAKOU, Benin
Joseph Agossou

Table A2. *Cont.*

WHO Collaborating Centre for New Vaccines Surveillance, Medical Research Council Unit The Gambia at London School of Hygiene and Tropical Medicine, Banjul, The Gambia
Brenda A. Kwambana-Adams
Martin Antonio
Archibald Worwui
Peter S. Ndow
WHO Regional Office for Africa, Inter Country Support Team, Ouagadougou, Burkina Faso
Joseph Biey
Bernard Ntsama
WHO Regional Office for Africa, Inter Country Support Team, Libreville, Gabon
Gilson Paluku
Aboubacar N'DIAYE
WHO Country offices, Benin and Cameroon
Sentinel site surveillance teams and countries part of the African Pediatric Bacterial Meningitis Surveillance Network
Australian National Notifiable Diseases Surveillance data were provided by the Office of Health Protection, Australian Government Department of Health, on behalf of the Communicable Diseases Network Australia and the Enhanced Invasive Pneumococcal Disease Surveillance Working Group.

References

1. Wahl, B.; O'Brien, K.L.; Greenbaum, A.; Majumder, A.; Liu, L.; Chu, Y.; Lukšić, I.; Nair, H.; McAllister, D.A.; Campbell, H.; et al. Burden of Streptococcus Pneumoniae and Haemophilus Influenzae Type b Disease in Children in the Era of Conjugate Vaccines: Global, Regional, and National Estimates for 2000–2015. *Lancet Glob. Health* **2018**, *6*, e744–e757. [CrossRef]
2. Johnson, H.L.; Deloria-Knoll, M.; Levine, O.S.; Stoszek, S.K.; Hance, L.F.; Reithinger, R.; Muenz, L.R.; O'Brien, K.L. Systematic Evaluation of Serotypes Causing Invasive Pneumococcal Disease among Children Under Five: The Pneumococcal Global Serotype Project. *PLoS Med.* **2010**, *7*, e1000348. [CrossRef] [PubMed]
3. International Vaccine Access Center (IVAC), Johns Hopkins Bloomberg School of Public Health VIEW-Hub. Available online: https://view-hub.org (accessed on 29 December 2020).
4. Lucero, M.; Dulalia, V.; Nillos, L.; Williams, G.; Parreño, R.; Nohynek, H.; Riley, I.; Makela, H. Pneumococcal Conjugate Vaccines for Preventing Vaccine-Type Invasive Pneumococcal Disease and Pneumonia with Consolidation on X-ray in Children under Two Years of Age. *Cochrane Database Syst. Rev.* **2004**, *4*, CD004977. [CrossRef]
5. Cherian, T.; Cohen, M.; de Oliveira, L.; Farrar, J.L.; Goldblatt, D.; Knoll, M.; Moisi, J.C.; O'Brien, K.L.; Pilishvili, T.; Ramakrishnan, M.; et al. *Pneumococcal Conjugate Vaccine (PCV) Review of Impact Evidence (PRIME): Summary of Findings from Systematic Review*; Report to Strategic Advisory Group of Experts on Immunization (SAGE) of the World Health Organization; WHO: Geneva, Switzerland, 2017; Volume 1, pp. 1–215.
6. Dagan, R. Relationship between Immune Response to Pneumococcal Conjugate Vaccines in Infants and Indirect Protection after Vaccine Implementation. *Expert Rev. Vaccines* **2019**, *18*, 641–661. [CrossRef] [PubMed]
7. Feikin, D.R.; Kagucia, E.W.; Loo, J.D.; Link-Gelles, R.; Puhan, M.A.; Cherian, T.; Levine, O.S.; Whitney, C.G.; O'Brien, K.L.; Moore, M.R.; et al. Serotype-Specific Changes in Invasive Pneumococcal Disease after Pneumococcal Conjugate Vaccine Introduction: A Pooled Analysis of Multiple Surveillance Sites. *PLoS Med.* **2013**, *10*, e1001517. [CrossRef]
8. Falkenhorst, G.; Remschmidt, C.; Harder, T.; Hummers-Pradier, E.; Wichmann, O.; Bogdan, C. Effectiveness of the 23-Valent Pneumococcal Polysaccharide Vaccine (PPV23) against Pneumococcal Disease in the Elderly: Systematic Review and Meta-Analysis. *PLoS ONE* **2017**, *12*, e0169368. [CrossRef]
9. Franklin, K.; Ronveaux, O.; Fernandez, K.; Kwambana-Adams, B.; Lessa, F.C.; Soeters, H.M.; Copper, L.; Coldiron, M.E.; Mwenda, J.; Antonio, M.; et al. *Pneumococcal Meningitis Outbreaks in Africa, 2000–2018: Systematic Literature Review and Meningitis Surveillance Database Analyses*; Systematic Review Registration (PROSPERO) Number: 102614 (interim number); 2020.
10. Merck Sharp & Dohme Corp. Merck Submits Applications for Licensure of V114, the Company's Investigational 15-Valent Pneumococcal Conjugate Vaccine, for Use in Adults to the U.S. FDA and European Medicines Agency. Available online: https://www.merck.com/news/merck-submits-applications-for-licensure-of-v114-the-companys-investigational-15-valent-pneumococcal-conjugate-vaccine-for-use-in-adults-to-the-u-s-fda-and-european-medicines-agency/ (accessed on 10 December 2020).

11. Pfizer. A Phase 3, Randomized, Open-Label Trial to Evaluate the Safety and Immunogenicity of a 20-Valent Pneumococcal Conjugate Vaccine in Adults 65 Years of Age and Older with Prior Pneumococcal Vaccination. 2020. Available online: https://clinicaltrials.gov/ct2/show/NCT03835975 (accessed on 20 December 2020).
12. Skinner, J.; Kaufhold, R.; McGuinness, D. Immunogenicity of PCV24, a next Generation Pneumococcal Conjugate Vaccine. In Proceedings of the 11th International Symposium on Pneumococci and Pneumococcal Diseases, Melbourne, Australia, 15 April 2018.
13. Affinivax. Affinivax Announces the Presentation of Phase 1 Clinical Data for Its MAPS™ Vaccine for Streptococcus Pneumoniae at IDWeek 2020. Available online: https://affinivax.com/press-release-october-21-2020/ (accessed on 10 December 2020).
14. World Health Organization. *Global Vaccine Action Plan: Defeating Meningitis by 2030 Meningitis Prevention and Control*; World Health Organization Seventy-third World Health Assembly: Geneva, Switzerland, 2020; pp. 1–4.
15. Deloria-Knoll, M.; Bennett, J.C.; Garcia Quesada, M.; Kagucia, E.W.; Peterson, M.E.; Feikin, D.R.; Cohen, A.L.; Hetrich, M.K.; Yangyupei, Y.; Sinkevitch, J.N.; et al. Global Landscape Review of Serotype-Specific Invasive Pneumococcal Disease Surveillance among Countries Using PCV10/13: The Pneumococcal Serotype Replacement and Distribution Estimation (PSERENADE) Project. *Microorganisms* **2021**.
16. World Health Organization. WHO-UNICEF Estimates of PCV3 Coverage. Available online: https://apps.who.int/immunization_monitoring/globalsummary/timeseries/tswucoveragepcv3.html (accessed on 30 December 2020).
17. Merck Sharp & Dohme Corp. A Phase 3, Multicenter, Randomized, Double-Blind, Active Comparator-Controlled Study to Evaluate the Safety, Tolerability, and Immunogenicity of V114 in Healthy Adults 50 Years of Age or Older (PNEU-AGE). 2020. Available online: https://clinicaltrials.gov/ct2/show/NCT03950622 (accessed on 20 December 2020).
18. Pfizer. A Phase 3, Randomized, Double-Blind Trial to Evaluate the Safety and Immunogenicity of a 20-Valent Pneumococcal Conjugate Vaccine in Pneumococcal Vaccine-Naive Adults 18 Years of Age and Older. 2020. Available online: https://clinicaltrials.gov/ct2/show/NCT03760146 (accessed on 20 December 2020).
19. van Selm, S.; vac Cann, L.M.; Kolkman, M.A.B.; van der Zeijst, B.A.M.; van Putten, J.P.M. Genetic Basis for the Structural Difference between Streptococcus Pneumoniae Serotype 15B and 15C Capsular Polysaccharides-PubMed. *Infect. Immun.* **2003**, *71*, 6192–6198. [CrossRef]
20. Yee, T.W. *Vector Generalized Linear and Additive Models*; Springer Series in Statistics; Springer: New York, NY, USA, 2015; ISBN 978-1-4939-2817-0.
21. Sings, H.L.; De Wals, P.; Gessner, B.D.; Isturiz, R.; Laferriere, C.; McLaughlin, J.M.; Pelton, S.; Schmitt, H.-J.; Suaya, J.A.; Jodar, L. Effectiveness of 13-Valent Pneumococcal Conjugate Vaccine Against Invasive Disease Caused by Serotype 3 in Children: A Systematic Review and Meta-Analysis of Observational Studies. *Clin. Infect. Dis.* **2019**, *68*, 2135–2143, Corrigendum to: *Clin. Infect. Dis.* **2021**, *1*, 1–2, doi:10.1093/cid/ciaa1767. [CrossRef]
22. Lapidot, R.; Shea, K.M.; Yildirim, I.; Cabral, H.J.; Pelton, S.I. Characteristics of Serotype 3 Invasive Pneumococcal Disease before and after Universal Childhood Immunization with PCV13 in Massachusetts. *Pathogens* **2020**, *9*, 396. [CrossRef]
23. Ladhani, S.N.; Collins, S.; Djennad, A.; Sheppard, C.L.; Borrow, R.; Fry, N.K.; Andrews, N.J.; Miller, E.; Ramsay, M.E. Rapid Increase in Non-Vaccine Serotypes Causing Invasive Pneumococcal Disease in England and Wales, 2000–2017: A Prospective National Observational Cohort Study. *Lancet Infect. Dis.* **2018**, *18*, 441–451. [CrossRef]
24. Cooper, D.; Yu, X.; Sidhu, M.; Nahm, M.H.; Fernsten, P.; Jansen, K.U. The 13-Valent Pneumococcal Conjugate Vaccine (PCV13) Elicits Cross-Functional Opsonophagocytic Killing Responses in Humans to Streptococcus Pneumoniae Serotypes 6C and 7A. *Vaccine* **2011**, *29*, 7207–7211. [CrossRef] [PubMed]
25. Iroh Tam, P.-Y.; Thielen, B.K.; Obaro, S.K.; Brearley, A.M.; Kaizer, A.M.; Chu, H.; Janoff, E.N. Childhood Pneumococcal Disease in Africa–A Systematic Review and Meta-Analysis of Incidence, Serotype Distribution, and Antimicrobial Susceptibility. *Vaccine* **2017**, *35*, 1817–1827. [CrossRef]
26. De Wals, P. Commentary on Paradoxical Observations Pertaining to the Impact of the 13-Valent Pneumococcal Conjugate Vaccine on Serotype 3 Streptococcus Pneumoniae Infections in Children. *Vaccine* **2018**, *36*, 5495–5496. [CrossRef]
27. Andrews, N.J.; Waight, P.A.; Burbidge, P.; Pearce, E.; Roalfe, L.; Zancolli, M.; Slack, M.; Ladhani, S.N.; Miller, E.; Goldblatt, D. Serotype-Specific Effectiveness and Correlates of Protection for the 13-Valent Pneumococcal Conjugate Vaccine: A Postlicensure Indirect Cohort Study. *Lancet Infect. Dis.* **2014**, *14*, 839–846. [CrossRef]
28. Moore, M.R.; Link-Gelles, R.; Schaffner, W.; Lynfield, R.; Holtzman, C.; Harrison, L.H.; Zansky, S.M.; Rosen, J.B.; Reingold, A.; Scherzinger, K.; et al. Effectiveness of 13-Valent Pneumococcal Conjugate Vaccine for Prevention of Invasive Pneumococcal Disease in Children in the USA: A Matched Case-Control Study. *Lancet Respir. Med.* **2016**, *4*, 399–406. [CrossRef]
29. Harboe, Z.B.; Dalby, T.; Weinberger, D.M.; Benfield, T.; Mølbak, K.; Slotved, H.C.; Suppli, C.H.; Konradsen, H.B.; Valentiner-Branth, P. Impact of 13-Valent Pneumococcal Conjugate Vaccination in Invasive Pneumococcal Disease Incidence and Mortality. *Clin. Infect. Dis.* **2014**, *59*, 1066–1073. [CrossRef] [PubMed]
30. Domínguez, Á.; Ciruela, P.; Hernández, S.; García-García, J.J.; Soldevila, N.; Izquierdo, C.; Moraga-Llop, F.; Díaz, A.; de Sevilla, M.F.; González-Peris, S.; et al. Effectiveness of the 13-Valent Pneumococcal Conjugate Vaccine in Preventing Invasive Pneumococcal Disease in Children Aged 7-59 Months. A Matched Case-Control Study. *PLoS ONE* **2017**, *12*, e0183191. [CrossRef]
31. van der Linden, M.; Falkenhorst, G.; Perniciaro, S.; Fitzner, C.; Imöhl, M. Effectiveness of Pneumococcal Conjugate Vaccines (PCV7 and PCV13) against Invasive Pneumococcal Disease among Children under Two Years of Age in Germany. *PLoS ONE* **2016**, *11*, e0161257. [CrossRef]

32. Deceuninck, G.; De Wals, P.; Boulianne, N.; De Serres, G. Effectiveness of Pneumococcal Conjugate Vaccine Using a 2+1 Infant Schedule in Quebec, Canada. *Pediatr. Infect. Dis. J.* **2010**, *29*, 546–549. [CrossRef]
33. Savulescu, C.; Hanquet, G. PCV13 Effectiveness and Overall Effect of PCV10/13 Vaccination Programmes in Children under Five Years of Age: SpIDnet Multicentre Studies. In Proceedings of the ESCAIDE 2016 Parallel Session on Vaccine-Preventable Diseases, Stockholm, Sweden, 19 November 2016.
34. Lodi, L.; Ricci, S.; Nieddu, F.; Moriondo, M.; Lippi, F.; Canessa, C.; Mangone, G.; Cortimiglia, M.; Casini, A.; Lucenteforte, E.; et al. Impact of the 13-Valent Pneumococcal Conjugate Vaccine on Severe Invasive Disease Caused by Serotype 3 Streptococcus Pneumoniae in Italian Children. *Vaccines* **2019**, *7*, 128. [CrossRef] [PubMed]
35. Peckeu, L.; van der Ende, A.; de Melker, H.E.; Sanders, E.A.M.; Knol, M.J. Impact and Effectiveness of the 10-Valent Pneumococcal Conjugate Vaccine on Invasive Pneumococcal Disease among Children under 5 Years of Age in the Netherlands. *Vaccine* **2021**, *39*, 431–437. [CrossRef]
36. Hammitt, L.L.; Etyang, A.O.; Morpeth, S.C.; Ojal, J.; Mutuku, A.; Mturi, N.; Moisi, J.C.; Adetifa, I.M.; Karani, A.; Akech, D.O.; et al. Effect of Ten-Valent Pneumococcal Conjugate Vaccine on Invasive Pneumococcal Disease and Nasopharyngeal Carriage in Kenya: A Longitudinal Surveillance Study. *Lancet* **2019**, *393*, 2146–2154. [CrossRef]
37. Rinta-Kokko, H.; Palmu, A.A.; Auranen, K.; Nuorti, J.P.; Toropainen, M.; Siira, L.; Virtanen, M.J.; Nohynek, H.; Jokinen, J. Long-Term Impact of 10-Valent Pneumococcal Conjugate Vaccination on Invasive Pneumococcal Disease among Children in Finland. *Vaccine* **2018**, *36*, 1934–1940. [CrossRef] [PubMed]
38. Serum Institute of India. A New Pneumococcal Vaccine Is Here! Why This Matters. Available online: https://www.seruminstitute.com/news_pneumococcal_vaccine.php (accessed on 10 December 2020).
39. Pfizer. U.S. FDA Accepts for Priority Review the Biologics License Application for Pfizer's Investigational 20-Valent Pneumococcal Conjugate Vaccine for Adults 18 Years of Age and Older. Available online: https://investors.pfizer.com/investor-news/press-release-details/2020/U.S.-FDA-Accepts-for-Priority-Review-the-Biologics-License-Application-for-Pfizers-Investigational-20-valent-Pneumococcal-Conjugate-Vaccine-for-Adults-18-Years-of-Age-and-Older/default.aspx (accessed on 10 December 2020).
40. Pfizer. Data from Pfizer's Adult and Pediatric Clinical Trial Programs for 20-Valent Pneumococcal Conjugate Vaccine Presented at IDWeek 2020. Available online: https://investors.pfizer.com/investor-news/press-release-details/2020/Data-From-Pfizers-Adult-and-Pediatric-Clinical-Trial-Programs-for-20-Valent-Pneumococcal-Conjugate-Vaccine-Presented-at-IDWeek-2020/default.aspx (accessed on 10 December 2020).
41. Soeters, H.M.; Kambiré, D.; Sawadogo, G.; Ouédraogo-Traoré, R.; Bicaba, B.; Medah, I.; Sangaré, L.; Ouédraogo, A.-S.; Ouangraoua, S.; Yaméogo, I.; et al. Impact of 13-Valent Pneumococcal Conjugate Vaccine on Pneumococcal Meningitis, Burkina Faso, 2016–2017. *J. Infect. Dis.* **2019**, *220*, S253–S262. [CrossRef] [PubMed]
42. Ousmane, S.; Kobayashi, M.; Seidou, I.; Issaka, B.; Sharpley, S.; Farrar, J.L.; Whitney, C.G.; Ouattara, M. Characterization of Pneumococcal Meningitis before and after Introduction of 13-Valent Pneumococcal Conjugate Vaccine in Niger, 2010–2018. *Vaccine* **2020**, *38*, 3922–3929. [CrossRef]
43. Bozio, C.H.; Abdul-Karim, A.; Abenyeri, J.; Abubakari, B.; Ofosu, W.; Zoya, J.; Ouattara, M.; Srinivasan, V.; Vuong, J.T.; Opare, D.; et al. Continued Occurrence of Serotype 1 Pneumococcal Meningitis in Two Regions Located in the Meningitis Belt in Ghana Five Years after Introduction of 13-Valent Pneumococcal Conjugate Vaccine. *PLoS ONE* **2018**, *13*, e0203205. [CrossRef] [PubMed]
44. Bennett, J.C.; Hetrich, M.K.; Garcia Quesada, M.; Sinkevitch, J.N.; Deloria Knoll, M.; Feikin, D.R.; Zeger, S.L.; Kagucia, E.W.; Cohen, A.L.; Ampofo, K.; et al. Changes in Invasive Pneumococcal Disease Caused by Streptococcus Pneumoniae Serotype 1 Following Introduction of PCV10 and PCV13: Findings from the PSERENADE Project. *Microorganisms* **2021**, *9*, 696. [CrossRef]

 *microorganisms*

Article

Changes in Invasive Pneumococcal Disease Caused by *Streptococcus pneumoniae* Serotype 1 following Introduction of PCV10 and PCV13: Findings from the PSERENADE Project

Julia C. Bennett [1,*], Marissa K. Hetrich [1], Maria Garcia Quesada [1], Jenna N. Sinkevitch [1], Maria Deloria Knoll [1,*], Daniel R. Feikin [2], Scott L. Zeger [1], Eunice W. Kagucia [3], Adam L. Cohen [4,†], Krow Ampofo [5], Maria-Cristina C. Brandileone [6], Dana Bruden [7], Romina Camilli [8], Jesús Castilla [9,10], Guanhao Chan [11], Heather Cook [12], Jennifer E. Cornick [13,14], Ron Dagan [15], Tine Dalby [16], Kostas Danis [17], Sara de Miguel [18], Philippe De Wals [19], Stefanie Desmet [20,21], Theano Georgakopoulou [22], Charlotte Gilkison [23], Marta Grgic-Vitek [24], Laura L. Hammitt [1,3], Markus Hilty [25], Pak-Leung Ho [26], Sanjay Jayasinghe [27], James D. Kellner [28], Jackie Kleynhans [29,30], Mirjam J. Knol [31], Jana Kozakova [32], Karl G. Kristinsson [33], Shamez N. Ladhani [34], Laura MacDonald [35], Grant A. Mackenzie [36,37,38], Lucia Mad'arová [39], Allison McGeer [40], Jolita Mereckiene [41], Eva Morfeldt [42], Tuya Mungun [43], Carmen Muñoz-Almagro [9,44,45], J. Pekka Nuorti [46,47], Metka Paragi [48], Tamara Pilishvili [49], Rodrigo Puentes [50], Samir K. Saha [51], Aalisha Sahu Khan [52], Larisa Savrasova [53,54], J. Anthony Scott [3], Anna Skoczyńska [55], Shigeru Suga [56], Mark van der Linden [57], Jennifer R. Verani [49,58], Anne von Gottberg [29,59], Brita A. Winje [60], Inci Yildirim [61], Khalid Zerouali [62,63], Kyla Hayford [1,‡] and the PSERENADE Team [§]

Citation: Bennett, J.C.; Hetrich, M.K.; Garcia Quesada, M.; Sinkevitch, J.N.; Deloria Knoll, M.; Feikin, D.R.; Zeger, S.L.; Kagucia, E.W.; Cohen, A.L.; Ampofo, K.; et al. Changes in Invasive Pneumococcal Disease Caused by *Streptococcus pneumoniae* Serotype 1 following Introduction of PCV10 and PCV13: Findings from the PSERENADE Project. *Microorganisms* **2021**, *9*, 696. https://doi.org/10.3390/microorganisms9040696

Academic Editor: James Stuart

Received: 3 March 2021
Accepted: 23 March 2021
Published: 27 March 2021

1 Johns Hopkins Bloomberg School of Public Health, Baltimore, MD 21205, USA; mhetric2@jhmi.edu (M.K.H.); mgarci64@jhmi.edu (M.G.Q.); jsinkev1@jhu.edu (J.N.S.); sz@jhu.edu (S.L.Z.); lhammitt@jhu.edu (L.L.H.); kylahayford@jhu.edu (K.H.)
2 Independent Consultant, 1296 Coppet, Switzerland; drf3217@gmail.com
3 KEMRI-Wellcome Trust Research Programme, Epidemiology and Demography Department, Centre for Geographic Medicine-Coast, P.O. Box 230-80108 Kilifi, Kenya; EKagucia@kemri-wellcome.org (E.W.K.); anthony.scott@lshtm.ac.uk (J.A.S.)
4 World Health Organization, 1202 Geneva, Switzerland; dvj1@cdc.gov
5 Division of Pediatric Infectious Diseases, Department of Pediatrics, University of Utah Health Sciences Center, Salt Lake City, UT 84132, USA; Krow.Ampofo@hsc.utah.edu
6 National Laboratory for Meningitis and Pneumococcal Infections, Center of Bacteriology, Institute Adolfo Lutz (IAL), São Paulo 01246-902, Brazil; maria.brandileone@ial.sp.gov.br
7 Arctic Investigations Program, Division of Preparedness and Emerging Infections, National Center for Emerging and Zoonotic Infectious Diseases, Centers for Disease Control and Prevention, Anchorage, AK 99508, USA; zkg9@cdc.gov
8 Department of Infectious Diseases, Italian National Institute of Health (Istituto Superiore di Sanità, ISS), 00161 Rome, Italy; romina.camilli@iss.it
9 CIBER Epidemiología y Salud Pública (CIBERESP), 28029 Madrid, Spain; jcastilc@navarra.es (J.C.); cma@sjdhospitalbarcelona.org (C.M.-A.)
10 Instituto de Salud Pública de Navarra—IdiSNA, 31003 Pamplona, Navarra, Spain
11 Singapore Ministry of Health, Communicable Diseases Division, Singapore 308442, Singapore; CHAN_Guanhao@moh.gov.sg
12 Centre for Disease Control, Department of Health and Community Services, Darwin, NT 8000, Australia; hcdarwin@gmail.com
13 Institute of Infection, Veterinary & Ecological Sciences, University of Liverpool, Liverpool CH64 7TE, UK; jcornick@mlw.mw
14 Malawi-Liverpool-Wellcome Trust Clinical Research Programme, Chichiri, P.O. Box 30096 Blantyre, Malawi
15 Faculty of Health Sciences, Ben-Gurion University of the Negev, 8410501 Beer-Sheva, Israel; rdagan@bgu.ac.il
16 Bacteria, Parasites and Fungi, Statens Serum Institut, DK-2300 Copenhagen, Denmark; TID@ssi.dk
17 Santé Publique France, the French National Public Health Agency, Saint Maurice CEDEX, 94415 Paris, France; Costas.DANIS@santepubliquefrance.fr
18 Epidemiology Department, Dirección General de Salud Pública, 28009 Madrid, Spain; sarade.miguel@salud.madrid.org
19 Department of Social and Preventive Medicine, Laval University, Québec, QC G1V 0A6, Canada; philippe.dewals@criucpq.ulaval.ca
20 Department of Microbiology, Immunology and Transplantation, KU Leuven, BE-3000 Leuven, Belgium; stefanie.desmet@uzleuven.be

[21] National Reference Centre for Streptococcus Pneumoniae, University Hospitals Leuven, 3000 Leuven, Belgium
[22] National Public Health Organisation, 15123 Athens, Greece; t.georgakopoulou@eody.gov.gr
[23] Epidemiology Team, Institute of Environmental Science and Research, Porirua, Wellington 5240, New Zealand; Charlotte.Gilkison@esr.cri.nz
[24] Communicable Diseases Centre, National Institute of Public Health, 1000 Ljubljana, Slovenia; Marta.Vitek@nijz.si
[25] Swiss National Reference Centre for Invasive Pneumococci, Institute for Infectious Diseases, University of Bern, 3012 Bern, Switzerland; Markus.Hilty@ifik.unibe.ch
[26] Department of Microbiology and Carol Yu Centre for Infection, The University of Hong Kong, Hong Kong, China; plho@hku.hk
[27] National Centre for Immunisation Research and Surveillance and Discipline of Child and Adolescent Health, Children's Hospital Westmead Clinical School, Faculty of Medicine and Health, University of Sydney, Westmead, NSW 2145, Australia; sanjay.jayasinghe@health.nsw.gov.au
[28] Department of Pediatrics, University of Calgary, and Alberta Health Services, Calgary, AB T3B 6A8, Canada; kellner@ucalgary.ca
[29] Centre for Respiratory Diseases and Meningitis, National Institute for Communicable Diseases of the National Health Laboratory Service, Johannesburg 2192, South Africa; JackieL@nicd.ac.za (J.K.); annev@nicd.ac.za (A.v.G.)
[30] School of Public Health, Faculty of Health Sciences, University of the Witwatersrand, Johannesburg 2000, South Africa
[31] National Institute for Public Health and the Environment, 3721 MA Bilthoven, The Netherlands; mirjam.knol@rivm.nl
[32] National Institute of Public Health (NIPH), 100 42 Praha, Czech Republic; jana.kozakova@szu.cz
[33] Department of Clinical Microbiology, Landspitali—The National University Hospital, Hringbraut, 101 Reykjavik, Iceland; karl@landspitali.is
[34] Immunisation and Countermeasures Division, Public Health England, London NW9 5EQ, UK; shamez.ladhani@phe.gov.uk
[35] Public Health Scotland, Glasgow G2 6QE, UK; laura.macdonald4@phs.scot
[36] Faculty of Infectious and Tropical Diseases, London School of Hygiene & Tropical Medicine, Keppel St, London WC1E 7HT, UK; gmackenzie@mrc.gm
[37] Medical Research Council Unit the Gambia at London School of Hygiene & Tropical Medicine, P.O. Box 273 Banjul, The Gambia
[38] New Vaccines Group, Murdoch Children's Research Institute, Parkville, Melbourne, VIC 3052, Australia
[39] National Reference Centre for Pneumococcal and Haemophilus Diseases, Regional Authority of Public Health, 975 56 Banská Bystrica, Slovakia; madarova@vzbb.sk
[40] Toronto Invasive Bacterial Diseases Network, Laboratory Medicine and Pathobiology, University of Toronto, Toronto, ON M5S 1A8, Canada; Allison.McGeer@sinaihealthsystem.ca
[41] HSE Health Protection Surveillance Centre, Mountjoy, Dublin D01 A4A3, Ireland; jolita.mereckiene@hse.ie
[42] Department of Microbiology, Public Health Agency of Sweden, 171 82 Solna, Sweden; eva.morfeldt@folkhalsomyndigheten.se
[43] National Center of Communicable Diseases (NCCD), Ministry of Health, Bayanzurkh District, Ulaanbaatar 13336, Mongolia; tuya_mungun@yahoo.com
[44] Medicine Department, Universitat Internacional de Catalunya, 08017 Barcelona, Spain
[45] Molecular Microbiology Department, Hospital Sant Joan de Déu Research Institute, 08950 Esplugues de Llobregat, Barcelona, Spain
[46] Department of Health Security, Finnish Institute for Health and Welfare, 00271 Helsinki, Finland; pekka.nuorti@tuni.fi
[47] Health Sciences Unit, Faculty of Social Sciences, University of Tampere, 33100 Tampere, Finland
[48] Centre for Medical Microbiology, National Laboratory of Health, Environment and Food, 2000 Maribor, Slovenia; metka.paragi@nlzoh.si
[49] National Center for Immunizations and Respiratory Diseases, Centers for Disease Control and Prevention, Atlanta, GA 30333, USA; tpilishvili@cdc.gov (T.P.); qzr7@cdc.gov (J.R.V.)
[50] Instituto de Salud Pública de Chile, Santiago 7780050, Santiago Metropolitan, Chile; rpuentes@ispch.cl
[51] Child Health Research Foundation, Dhaka 1207, Bangladesh; samirk.sks@gmail.com
[52] Ministry of Health and Medical Services, Suva, Fiji; aalisha@gmail.com
[53] Centre for Disease Prevention and Control of Latvia, 1005 Riga, Latvia; larisa.savrasova@spkc.gov.lv
[54] Doctoral Studies Department, Riga Stradinš University, 1007 Riga, Latvia
[55] National Reference Centre for Bacterial Meningitis, National Medicines Institute, 00-725 Warsaw, Poland; a.skoczynska@nil.gov.pl
[56] Infectious Disease Center and Department of Clinical Research, National Hospital Organization Mie Hospital, Tsu, Mie 514-0125, Japan; suga.shigeru.ke@mail.hosp.go.jp

[57] National Reference Center for Streptococci, Department of Medical Microbiology, University Hospital RWTH Aachen, 52074 Aachen, Germany; mlinden@ukaachen.de
[58] Centers for Disease Control and Prevention (CDC), Center for Global Health (CGH), Division of Global Health Protection (DGHP), P.O. Box 606-00621 Nairobi, Kenya
[59] School of Pathology, Faculty of Health Sciences, University of the Witwatersrand, Braamfontein, Johannesburg 2000, South Africa
[60] Department of Infection Control and Vaccine, Norwegian Institute of Public Health, 0456 Oslo, Norway; Brita.Askeland.Winje@fhi.no
[61] Department of Pediatrics, Yale New Haven Children's Hospital, New Haven, CT 06504, USA; inci.yildirim@yale.edu
[62] Bacteriology-Virology and Hospital Hygiene Laboratory, Ibn Rochd University Hospital Centre, Casablanca 20250, Morocco; khalid.zerouali2000@gmail.com
[63] Department of Microbiology, Faculty of Medicine and Pharmacy, Hassan II University of Casablanca, Casablanca 20000, Morocco
* Correspondence: jbenne63@jhu.edu (J.C.B.); mknoll2@jhu.edu (M.D.K.)
† Present address: Centers for Disease Control and Prevention, Atlanta, GA, USA. Affiliation at the time this work was completed was World Health Organization.
‡ Present address: Pfizer, Inc. Collegeville, PA, USA. Affiliation at the time this work was completed was Johns Hopkins Bloomberg School of Public Health.
§ Members are listed in Appendix A Table A1.

Abstract: *Streptococcus pneumoniae* serotype 1 (ST1) was an important cause of invasive pneumococcal disease (IPD) globally before the introduction of pneumococcal conjugate vaccines (PCVs) containing ST1 antigen. The Pneumococcal Serotype Replacement and Distribution Estimation (PSERENADE) project gathered ST1 IPD surveillance data from sites globally and aimed to estimate PCV10/13 impact on ST1 IPD incidence. We estimated ST1 IPD incidence rate ratios (IRRs) comparing the pre-PCV10/13 period to each post-PCV10/13 year by site using a Bayesian multi-level, mixed-effects Poisson regression and all-site IRRs using a linear mixed-effects regression (N = 45 sites). Following PCV10/13 introduction, the incidence rate (IR) of ST1 IPD declined among all ages. After six years of PCV10/13 use, the all-site IRR was 0.05 (95% credibility interval 0.04–0.06) for all ages, 0.05 (0.04–0.05) for <5 years of age, 0.08 (0.06–0.09) for 5–17 years, 0.06 (0.05–0.08) for 18–49 years, 0.06 (0.05–0.07) for 50–64 years, and 0.05 (0.04–0.06) for ≥65 years. PCV10/13 use in infant immunization programs was followed by a 95% reduction in ST1 IPD in all ages after approximately 6 years. Limited data availability from the highest ST1 disease burden countries using a 3 + 0 schedule constrains generalizability and data from these settings are needed.

Keywords: invasive pneumococcal disease; pneumococcal conjugate vaccines; serotypes; vaccine impact

1. Introduction

Streptococcus pneumoniae is a major cause of pneumonia, meningitis, and pleural effusion in children and adults [1–4]. There are at least 100 known serotypes of pneumococci [5]. Before the introduction of pneumococcal conjugate vaccines (PCVs), serotype 1 (ST1) was one the most common causes of invasive pneumococcal disease (IPD), especially in Asia and Africa, and globally was responsible for approximately 9% of IPD among children <5 years of age [6]. ST1 is distinct from other serotypes in that it has a high invasiveness potential, is not commonly carried in the nasopharynx [7,8], and in some settings occurs in a cyclical pattern, approximately every 3–9 years [9–11]. Additionally, ST1 can cause large pneumococcal outbreaks among all ages, including older children and young adults, in the African meningitis belt and other outbreak-prone settings with up to 10–30-fold increases in ST1 cases compared to pre-outbreak baselines [12–15].

The first PCV licensed for use in infants, seven-valent PCV (Prevenar/Prevnar, Pfizer), did not include ST1 antigen. Since then, the introduction of PCVs containing ST1 antigen (PCV10 [Synflorix, GlaxoSmithKline], PCV13 [Prevenar13/Prevnar13, Pfizer]) into many national infant immunization programs since 2009 has been shown to substantially reduce

ST1 IPD and end pneumococcal outbreaks caused by ST1. These effects have been demonstrated among directly immunized children and also unvaccinated older children and adults, through indirect effects, in both high and low IPD burden settings [9,10,12,16–20]. However, in some PCV10/13 using settings ST1 outbreaks continued to occur or ST1 IPD incidence rates did not substantially decline in the early years immediately following PCV10/13 introduction [21–24].

Evaluating the impact of PCV10/13 vaccination on ST1 IPD is challenging in a single surveillance site. In many settings, annual ST1 incidence rates are unstable because case counts are small, particularly after vaccine introduction. Many sites are also limited by short pre- and post-vaccine introduction surveillance periods, further limiting inferences that can be drawn from a single site. Assessing vaccine impact is also confounded by the cyclic nature of ST1 in which it is common to observe multiple years of zero ST1 cases prior to vaccine use. Quantifying the impact of PCV10/13 on ST1, which has several unique characteristics compared to other vaccine-type serotypes included in currently licensed PCVs, is important for policymakers seeking to reduce the burden of ST1 IPD through immunization. The Pneumococcal Serotype Replacement and Distribution Estimation (PSERENADE) project evaluated all available published and unpublished serotype-specific IPD data to estimate the impact of PCV10/PCV13 on ST1 IPD incidence at the global scale.

2. Materials and Methods

2.1. Data Collection and Eligibility Criteria

IPD surveillance sites with eligible data contributed annual serotype-specific IPD case data and population denominators to the project. A systematic approach to identify eligible sites and request data is described in detail elsewhere [25]. ST1 IPD was defined as the isolation of *Streptococcus pneumoniae* from a normally sterile site or detection of pneumococcus in cerebrospinal fluid (CSF) or pleural fluid using *lytA*-based polymerase chain reaction (PCR), or antigen testing confirmed as ST1. Sites with ST1 IPD case counts and population denominators that met eligibility criteria were included in the analysis (Box 1, Table 1, Table S1).

Box 1. Inclusion criteria.

1. Site reports annual ST1 IPD incidence data:
   - ST1 case counts by age group, and
   - Population-based denominators by age group.
2. At least 50% of isolates serotyped for included years by age group.
3. At least one complete year of data post-PCV10/13 introduction, excluding the year of introduction.
4. At least 50% uptake for primary PCV series at 12 months of age in at least one year post-PCV10/13 introduction.
5. PCV10 or PCV13 is universally recommended for all infants in the national infant immunization schedule.
6. No major changes or biases in surveillance that would affect estimates of ST1 incidence rates.

Two PSERENADE coordinators conducted a standard data quality review for each site to evaluate if surveillance system changes or other factors besides PCV introductions influenced incidence rates (IR) of IPD over available years of surveillance data [25]. After review and discussion with site investigators, certain site-year-age group data were excluded if determined to fall within periods of differential surveillance capture or if the impact of changes in surveillance protocols on IPD IRs could not be accounted for in the analysis. For all sites, we defined the year of PCV introduction as the year PCV10/13 was universally introduced if PCV was introduced in the first three quarters of the year, or as the following calendar year if otherwise. For data submitted in epidemiologic years rather than calendar years, the introduction year was defined accordingly. For all sites, the year of PCV10/13 introduction was defined as 'year 0' for the analyses.

Table 1. Description of infant pneumococcal conjugate vaccine program and surveillance data for included sites. Ordered by vaccine product and schedule.

Site	PCV10 Period	PCV13 Period	PCV10/13 Schedule	PCV7 Use	PCV10/13 Catch-Up	Mean PCV10/13 Uptake (%)		Included in ST1 Analysis		ST1 Cases Included in Analysis (n)	Surveillance Years Pre- and Post-PCV10/13 (n)	Proportion ST1 IPD Cases from CSF (%)
						Primary Series *	WUENIC PCV3 ***	0–17 Years	≥18 Years			
Finland	2010–	–	2 + 1	N	N	95	90	Y	Y	46	Pre: 6 Post: 8	4.3
Iceland	2011–	–	2 + 1	N	N	97	89	Y	Y	22	Pre: 16 Post: 8	0.0
Latvia	2012–	–	2 + 1	Y	N	91	83	N [b]	Y; ≥50y	5	Pre: 0 Post: 7	20.0
Slovenia	2015–2019	2019–	2 + 1	N	N	55	55	Y	Y	259	Pre: 6 Post: 4	0.0
Netherlands	2011–	–	3 + 1/2 + 1	Y	N	95	94	Y	Y	642	Pre: 7 Post: 8	2.2
Asembo, Kenya	2011–	–	3 + 0	N	Y	86	78	Y	Y; 18–49y	43	Pre: 1 Post: 8	NA
Kilifi, Kenya	2011–	–	3 + 0	N	Y	82	78	Y	Y; 18–64y	204	Pre: 11 Post: 6	19.6
Japan	–	2013–	3 + 1	Y	N	94 **	98	Y	Y; ≥65y	11	Pre: 4 Post: 5	0.0
ABCs, USA	–	2010–	3 + 1	Y	Y	88	93	Y	Y	664	Pre: 12 Post: 8	0.6
Alaska, USA	–	2010–	3 + 1	Y	Y	83	93	Y	Y	92	Pre: 19 Post: 8	0.0
Massachusetts, USA	–	2010–	3 + 1	Y	Y	94	93	Y; <5y	NA	1	Pre: 8 Post: 8	0.0
Southwest, USA (Indigenous)	–	2010–	3 + 1	Y	Y	82	93	Y	Y	180	Pre: 15 Post: 9	2.2
Alberta, Canada	–	2010–	2 + 1	Y	N	88 **	77	Y; <5y	Y	16	Pre: 10 Post: 8	0.0
Denmark	–	2010–	2 + 1	Y	N	91 **	93	Y	Y	2089	Pre: 10 Post: 9	2.2
France	–	2010–	2 + 1	Y	N	93	91	Y	Y	1346	Pre: 9 Post: 9	5.9
Ireland	–	2010–	2 + 1	Y	N	91	91	Y	Y	58	Pre: 3 Post: 8	0.0
Israel	–	2010–	2 + 1	Y	N	95	93	Y	Y	677	Pre: 8 Post: 8	3.4
Italy	–	2010–	2 + 1	Y	N	86 **	87	Y	Y	193	Pre: 0 Post: 9	6.7

Table 1. *Cont.*

Site	PCV10 Period	PCV13 Period	PCV10/13 Schedule	PCV7 Use	PCV10/13 Catch-Up	Mean PCV10/13 Uptake (%)		Included in ST1 Analysis		ST1 Cases Included in Analysis (n)	Surveillance Years Pre- and Post-PCV10/13 (n)	Proportion ST1 IPD Cases from CSF (%)
						Primary Series *	WUENIC PCV3 ***	0–17 Years	≥18 Years			
Norway	–	2011–	2 + 1	Y	N	93	93	Y	Y	637	Pre: 7 Post: 7	1.4
Singapore	–	2011–	2 + 1	Y	Y	84	74	N [d]	Y; ≥50y	8	Pre: 2 Post: 8	0.0
South Africa	–	2011–	2 + 1	Y	Y	77 **	77	Y	Y	3292	Pre: 6 Post: 8	38.2
Madrid, Spain	–	2010–	2 + 1	Y	N	98	93	Y	Y	479	Pre: 3 Post: 9	0.8
Switzerland	–	2010–	2 + 1	Y	Y	79 **	77	Y	Y	436	Pre: 8 Post: 7	0.5
England, UK	–	2010–	2 + 1	Y	N	94	92	Y	Y	4214	Pre: 10 Post: 10	1.5
Scotland, UK	–	2010–	2 + 1	Y	N	97	92	Y	Y	578	Pre: 10 Post: 9	NA
Germany	–	2009–	3 + 1/2 + 1	Y	N	85	93	Y	Y	760	Pre: 5 Post: 9	4.1
Catalonia, Spain	–	2010–2015 [a] 2016–	3 + 1/2 + 1	Y [a]	N	70	93	Y	Y	1111	Pre: 4 Post: 8	1.5
Navarra, Spain	–	2010–2015 [a] 2016–	3 + 1/2 + 1	Y [a]	N	71	93	Y	Y	93	Pre: 9 Post: 9	0.0
Australia (Non-Indigenous)	–	2011–	3 + 0	Y	Y	92	92	Y	Y	371	Pre: 9 Post: 7	0.8
Basse, The Gambia	–	2011–	3 + 0	Y	N	77	95	Y	N [b]	71	Pre: 2 Post: 7	1.4
Blantyre District, Malawi	–	2011–	3 + 0	N	Y	92	88	Y	Y	229	Pre: 5 Post: 7	55.5
Northern Territory, Australia	2009–2011	2011–	3 + 1	Y	Y	88	92	Y	Y	97	Pre: 16 Post: 8	1.0
Quebec-Nunavik, Canada	2009–2010	2011–	3 + 1	Y	N	97	75	Y; <5y	N [c]	1	Pre: 9 Post: 10	0.0
Hong Kong	2010–2011	2011-	3 + 1	Y	N	98	–	N [d]	Y; 18–49y	1	Pre: 0 Post: 5	0.0
New Zealand	2011–2014 2017–	2014–2017	3 + 1	Y	N	93	93	Y	Y	334	Pre: 9 Post: 8	0.6
Belgium	2015–2019	2011–2015 2019–	2 + 1	Y	N	95 **	94	Y	NA	872	Pre: 5 Post: 8	1.3

Table 1. *Cont.*

Site	PCV10 Period	PCV13 Period	PCV10/13 Schedule	PCV7 Use	PCV10/13 Catch-Up	Mean PCV10/13 Uptake (%)		Included in ST1 Analysis		ST1 Cases Included in Analysis (n)	Surveillance Years Pre- and Post-PCV10/13 (n)	Proportion ST1 IPD Cases from CSF (%)
						Primary Series *	WUENIC PCV3 ***	0–17 Years	≥18 Years			
Poland	2017–	2017– [e]	2 + 1	N	N	94	60	Y	N [b]	69	Pre: 9 Post: 2	4.3
Quebec (excluding Nunavik), Canada	2009–2010 2018–	2011–2018	2 + 1	Y	N	97	75	Y	Y	43	Pre: 8 Post: 10	0.0
Metropolitan Region, Chile	2011–2015	2016–	2 + 1	Y	N	97	88	Y	Y	437	Pre: 9 Post: 8	2.7
Non-Metropolitan Regions, Chile	2011–2017	2017–	2 + 1	N	N	97	89	Y	Y	69	Pre: 0 Post: 7	0.0
Grand Casablanca, Morocco	2012–	2010–2012	2 + 1	N	N	91	90	Y	Y; 18–49y	29	Pre: 4 Post: 7	37.9
Slovakia	2011–	2011–	2 + 1	Y	Y	97	97	Y	Y	20	Pre: 0 Post: 7	5.0
Sweden	2010–	2010–2019	2 + 1	Y	N	97 **	97	Y	Y	84	Pre: 1 Post: 5	NA
Ontario, Canada	2009–2010	2010–	3 + 1/2 + 1	Y	Y	72 **	79	N [d]	Y	9	Pre: 3 Post: 9	0.00
Czech Republic	2010–	2010–	3 + 1/2 + 1	N	N	74 **	–	Y	Y	227	Pre: 2 Post: 8	2.2

PCV: Pneumococcal conjugate vaccines. ST1: Serotype 1. CSF: Cerebrospinal fluid. – Not universally used. Y: Yes; N: No; NA: Not applicable. [a] Recommended for high-risk populations only but had substantial (≥50% annually) private market uptake among the general population. [b] Biases in surveillance system over time that could not be accounted for. [c] Low proportion of cases serotyped. [d] Zero ST1 cases in all years. [e] Private market uptake of approximately 30% annually. * Annual PCV uptake estimates provided by the surveillance site for the primary series of PCV by 12 months of age (if available, for some sites up to 15 months of age), excluding year of vaccine rollout. ** Annual PCV uptake estimates provided by the surveillance site for the primary series plus the booster dose by 23 months of age, excluding year of vaccine rollout. *** WHO and UNICEF Estimates of National Immunization Coverage (WUENIC) PCV3 uptake, excluding the year of vaccine rollout (PCV3 represents the third dose whether given before 12 months or at or after 12 months, but in some cases uptake estimates may reflect the percentage of surviving infants who received two doses of PCV prior to the first birthday).

2.2. *Data Analysis*

2.2.1. Adjustments for Missing Data

Adjustments for missing serotype data assume that missing serotype data are missing completely at random, that is the serotype distribution of serotyped cases is not biased or different from the serotype distribution of cases that were not serotyped or not fully serotyped. Site-year-age group strata that violated this assumption or reported serotypes for less than 50% of cases were excluded from the ST1 analysis for that stratum. For cases that were reported as not serotyped (serotyping was not attempted for any reason), the population denominators were adjusted by the proportion of cases that were serotyped (i.e., annual denominator * percent of cases that were serotyped in that year) for each site by year and age group. Because the proportion of cases serotyped varies across sites, population denominators were adjusted rather than reapportioning serotypes to unknown serotype cases in order to give appropriate weight to sites in the model based on serotype data reported. If ST1 and a second serotype was reported for a case, it was included as an ST1 case. Cases reported as a serotype pool which includes ST1 (e.g., pool A) were excluded. For cases with unknown age, the population denominators were adjusted by the proportion of cases with known age (i.e., annual denominator * percent of cases with known age in that year) for each year and age group. Minor changes were made to the cut-offs for age groups when standard age categories used for analyses were not available from the site.

2.2.2. Statistical Analysis

Annual ST1 IPD incidence rate ratios (IRRs) comparing the pre-PCV10/13 period to each post-PCV10/13 year were estimated by age group and for all ages in a three-step process. First, ST1 IR curves were estimated over years of available data for each site using a Bayesian multi-level, mixed-effects Poisson regression using the MCMCglmm package in R [26]. The model included data from all sites (using either PCV10 or PCV13) with an offset for population denominator and random effects for all of the site-specific regression coefficients, which allows for heterogeneity among sites in the shapes of their incidence curves. Sites using PCV10 and PCV13 were modeled together to increase sample size and as no difference in impact on ST1 IPD was observed by product (Figure S3). The regression identified commonalities within and across sites in the direction of change over time and smoothed out observed annual variability. Data points from the same site were treated as repeated measures over time and sites with small case counts or few years of data had less influence than sites with larger case counts and many years of data.

ST1 outbreaks tended to occur in a cyclical pattern prior to the introduction of PCV10/13. The model did not account for outbreaks occurring in a cyclical pattern. Therefore, in order to generate an expected baseline ST1 IPD IR in any given year, the regression modeled pre-PCV10/13 IRs as a single mean rate with a slope of zero to capture an 'average' pre-PCV10/13 ST1 IR. PCV7 years of use were included in the pre-PCV10/13 period as no consistent impact of PCV7 on ST1 IRs, either increases (i.e., serotype replacement) or decreases, were observed across sites, as expected given pre-PCV10/13 ST1 carriage patterns [7]. This increased the number of pre-PCV10/13 years included in the analysis and better captured the baseline ST1 IR. For each site, a non-linear break (allowing an abrupt hinge in the curve) was included in the model one year prior to PCV10/13 introduction to capture the change from the pre-PCV10/13 period to the year of PCV10/13 introduction and cubic splines knots (allowing a smooth change in the slope) were included for each site at years +1 and +3 (the second and fourth year of PCV10/13 use) to allow for flexibility in the IR of ST1 over time for each site following PCV10/13 introduction. Site-specific modeled ST1 IR curves were visually inspected for model fit and approved by site investigators with expertise in IPD surveillance at each site.

Second, the pre-PCV10/13 ST1 IR was used as a counterfactual ST1 IR (i.e., an expected ST1 IR in any given post-PCV10/13 year in the absence of PCV10/13 introduction) for sites with both pre- and post-PCV10/13 data. The site-specific modeled ST1 IR and counterfactual IR were used to estimate site-specific annual IRRs in each post-PCV10/13 year (reported as the mean of the posterior distribution of rate ratios) for each site. Site-specific IRRs were not generated for sites without pre-PCV10/13 years of data. Credibility intervals (CIs, Bayesian confidence interval analog) were estimated using the 2.5 and 97.5 percentiles of the posterior distribution of the IRs (Figure S1).

Finally, modeled site-specific IRRs were used to estimate all-site weighted average IRRs in each post-PCV10/13 year using a linear mixed-effects regression where site-specific IRRs were regressed on time since PCV10/13 introduction and weighted to give more influence to sites whose IRR standard errors were smaller. In sensitivity analyses, the all-site weighted average IRRs were estimated restricting to sites with data in all age groups and after adjusting the counterfactual IR by all-serotype IPD pre-PCV trends. All analyses were conducted in R (R Core Team, 2019).

3. Results

3.1. Description of Sites and Included Data

Of the 52 sites that met data collection eligibility criteria and contributed data to the PSERENADE project, 45 were included in the serotype 1 analysis (41 for children <5 years of age, 38 for 5–17 years of age, 37 for 18–49 years of age, 36 for 50–64 years of age, and 36 for ≥65 years of age). Two sites were excluded due to their population-based surveillance being restricted to pneumococcal meningitis, four sites were excluded due to a combination of biases in the surveillance system over time, such as changed to surveillance protocols, that could not be accounted for in the analysis and/or less than 50% of cases being serotyped, and one site was excluded due to zero ST1 cases being reported in all years of available data. Additionally, several age groups from included sites did not meet eligibility criteria and were excluded (Table S1).

Seven sites (16%) included in the analysis used PCV10, 24 (53%) used PCV13, and 14 (31%) used a combination of PCV10 and PCV13 in the infant PCV program. Only 14 (31%) sites introduced PCV10 or PCV13 into the routine immunization schedule with a catch-up campaign. The majority of sites used a PCV schedule including a booster dose (40, 89% used a 2 + 1 or 3 + 1 schedule and 5, 11% used a 3 + 0 schedule). Nearly half were from Europe (22 (49%)), 8 (18%) were from North America, 5 (11%) from Sub-Saharan Africa, 3 (7%) from Oceania, 3 (7%) from Asia, 2 (4%) from Latin America and the Caribbean and 2 (4%) from Northern Africa and Western Asia. The median PCV10/13 uptake for all years of available data after PCV10/13 introduction was 92% (range: 55–98%) (Table 1).

Of included sites with available data on specimen type, the median proportion of all ST1 IPD cases from CSF was 1.4% (range: 0–55.5%). Annual site-specific ST1 IRRs were estimated for 40 (89%) sites with both pre- and post-PCV10/13 ST1 surveillance data. The median number of surveillance years included in the analysis was 7 (range: 0–19) prior to the introduction of PCV10/13 and 8 (range: 2–10) after the introduction of PCV10/13 (including the year of PCV10/13 introduction). The median proportion of cases serotyped annually was 94% (range: 50–100%). The median number of ST1 cases included in the analysis per site was 29 (range: 1–499) for children <5 years of age, 46 (range: 2–768) for 5–17 years of age, 51 (range: 1–1776) for 18–49 years of age, 25 (range: 1–753) for 50–64 years of age, and 26 (range: 1–748) for ≥65 years of age (Table 1, Figure 1).

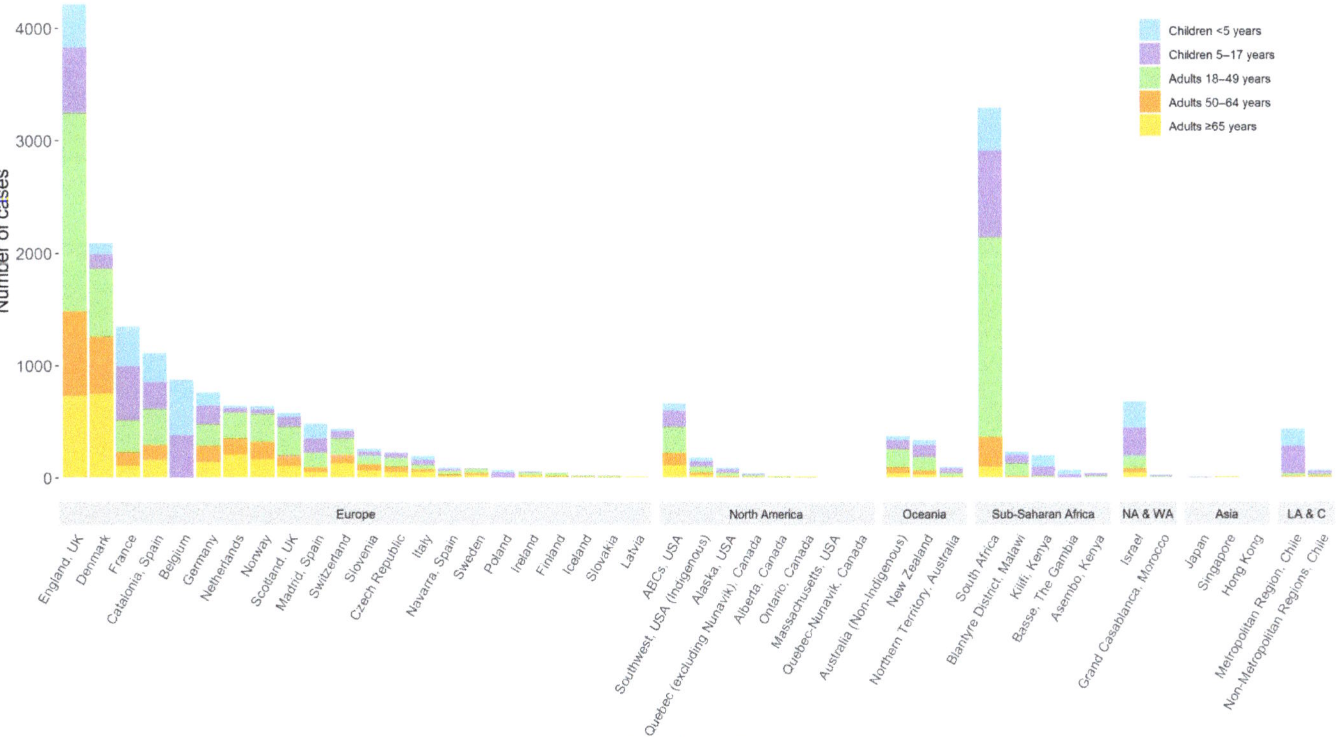

Figure 1. Number of serotype 1 cases per site included in the analysis by region and age group. NA & WA–Northern Africa and Western Asia; LA & C–Latin America and the Caribbean. Not all age groups were included for all sites (Table S1). Analyses were done with minor changes to age groups for certain sites to align with availability of population denominators and age groups provided by sites in aggregate: the <5 years age group includes 0–5 years from Morocco; the 5–17 years age group included 5–14 years from Japan and Kilifi, Kenya, 5–15 years from Germany, 6–14 years from Morocco, and 5–19 years from Australia and Malawi; and the 18–49 years age group includes 15–49 years from Japan and Kilifi, Kenya, 15–59 years from Morocco, 16–49 years from Germany, and 20–49 years from Australia and Malawi.

3.2. Impact of PCV10/13 on ST1 Incidence

All-site weighted average ST1 IPD IRRs comparing the pre-PCV10/13 period to each post-PCV10/13 year are shown in Table 2 and Figure 2. The all-site weighted average IRRs in the year of PCV10/13 introduction by age group ranged from 0.82 to 1.09 and was 1.09 (95% CI: 0.92–1.29) for children <5 years of age, 1.06 (0.88–1.28) for 5–17 years of age, 0.94 (0.73–1.22) for 18–49 years of age, 0.85 (0.70–1.04) for 50–64 years of age, and 0.82 (0.68–0.99) for ≥65 years of age. The ST1 IRR declined for every age group in each subsequent post-PCV10/13 year. By the sixth year of PCV10/13 use (year +5 post-PCV10/13 introduction), the all-site weighted average IRR compared to the pre-PCV10/13 period was 0.05 (0.04–0.06) for all ages, or a 95% relative reduction in ST1 IPD compared to the pre-PCV10/13 period. The reduction in ST1 IPD for each age group ranged from 92% to 95% in the sixth year of PCV10/13 use: IRR 0.05 (0.04–0.05) for children <5 years of age, 0.08 (0.06–0.09) for 5–17 years of age, 0.06 (0.05–0.08) for 18–49 years of age, 0.06 (0.05–0.07) for 50–64 years of age, and 0.05 (0.04–0.06) for ≥65 years of age.

In the early years of PCV10/13 use, site-specific IRRs were heterogeneous. Some sites reported outbreaks or had elevated levels of ST1 IPD around the time of PCV10/13 introduction, including two sites with very small sample sizes and large proportion increases in ST1 IRs. Other sites had little to no ST1 disease at the time of PCV10/13 introduction compared to the pre-PCV10/13 ST1 IRs. After five years of PCV10/13 use (year +4 post-PCV10/13), the impact of PCV10/13 on ST1 IPD was homogeneous across all included sites and age groups. No ST1 outbreaks were observed after five or more years of PCV10/13 use in any site (Figure 3). Results were similar when analyses were restricted to sites with data in all age groups (results not shown), when sites with very small sample size were excluded (results not shown), and after adjusting the counterfactual IR by all-serotype IPD pre-PCV trends (Figure S2). No differences in ST1 impact were observed by visual inspection among the included sites by PCV product, region, infant PCV schedule, or adult pneumococcal polysaccharide vaccine recommendation (Figures S3–S6). One site, which was excluded from the analytic model because the dataset was limited to meningitis cases, observed declines in ST1 pneumococcal meningitis IRs after PCV10 introduction that were consistent with declines seen in ST1 IPD in the other sites (Figure S7).

Table 2. Serotype 1 invasive pneumococcal disease all-site weighted average incidence rate ratios comparing the annual post-PCV10/13 incidence rate to the average pre-PCV10/13 incidence rate by age group.

	Year Post-PCV10/13 Introduction									
	0 *	1	2	3	4	5	6	7	8	9
Children <5 Years										
Nnumber of Sites [a]	37	37	36	36	35	34	33	27	10	3
IRR (95% CI)	1.09 (0.92–1.29)	0.57 (0.48–0.67)	0.29 (0.25–0.35)	0.15 (0.13–0.18)	0.08 (0.07–0.09)	0.05 (0.04–0.05)	0.03 (0.02–0.03)	0.02 (0.02–0.02)	0.01 (0.01–0.02)	0.01 (0.01–0.01)
Children 5–17 Years										
Number of Sites [a]	34	34	33	33	32	31	30	24	9	2
IRR (95% CI)	1.06 (0.88–1.28)	0.67 (0.55–0.80)	0.41 (0.34–0.49)	0.24 (0.20–0.29)	0.14 (0.11–0.16)	0.08 (0.06–0.09)	0.04 (0.04–0.05)	0.03 (0.02–0.03)	0.01 (0.01–0.02)	0.01 (0.01–0.01)
Adults 18–49 Years										
Numbers of Sites [a]	29	29	29	29	28	28	27	22	9	2
IRR (95% CI)	0.94 (0.73–1.22)	0.57 (0.44–0.74)	0.34 0.26–0.44)	0.20 (0.15–0.25)	0.11 (0.09–0.14)	0.06 (0.05–0.08)	0.03 (0.03–0.04)	0.02 (0.01–0.02)	0.01 (0.01–0.01)	0.01 (0.00–0.01)
Adults 50–64 Years										
Number of Sites [a]	29	29	29	29	27	27	27	22	9	2
IRR (95% CI)	0.85 (0.70–1.04)	0.54 (0.44–0.65)	0.33 (0.27–0.40)	0.19 (0.15–0.23)	0.10 (0.08–0.12)	0.06 (0.05–0.07)	0.03 (0.03–0.04)	0.02 (0.02–0.02)	0.01 (0.01–0.01)	0.01 (0.01–0.01)
Adults ≥65 Years										
Number of Sites [a]	28	28	28	28	27	27	27	22	9	2
IRR (95% CI)	0.82 (0.68–0.99)	0.56 (0.46–0.67)	0.36 (0.30–0.43)	0.20 (0.17–0.24)	0.10 (0.08–0.12)	0.05 (0.04–0.06)	0.03 (0.02–0.03)	0.02 (0.01–0.02)	0.01 (0.01–0.01)	0.01 (0.00–0.01)
All ages										
Number of Sites [a]	39	39	38	38	37	36	35	29	11	3
IRR (95% CI)	0.98 (0.79–1.21)	0.57 (0.47–0.71)	0.33 (0.27–0.40)	0.18 (0.15–0.22)	0.10 (0.08–0.12)	0.05 (0.04–0.06)	0.03 (0.02–0.04)	0.02 (0.01–0.02)	0.01 (0.01–0.01)	0.01 (0.01–0.01)

PCV: Pneumococcal conjugate vaccine. * Year of PCV10/13 introduction. [a] Number of sites with both pre- and post-PCV10/13 data in each post-PCV10/13 year. All-site weighted average IRRs estimated by post-PCV10/13 year and age group using linear mixed-effects regression.

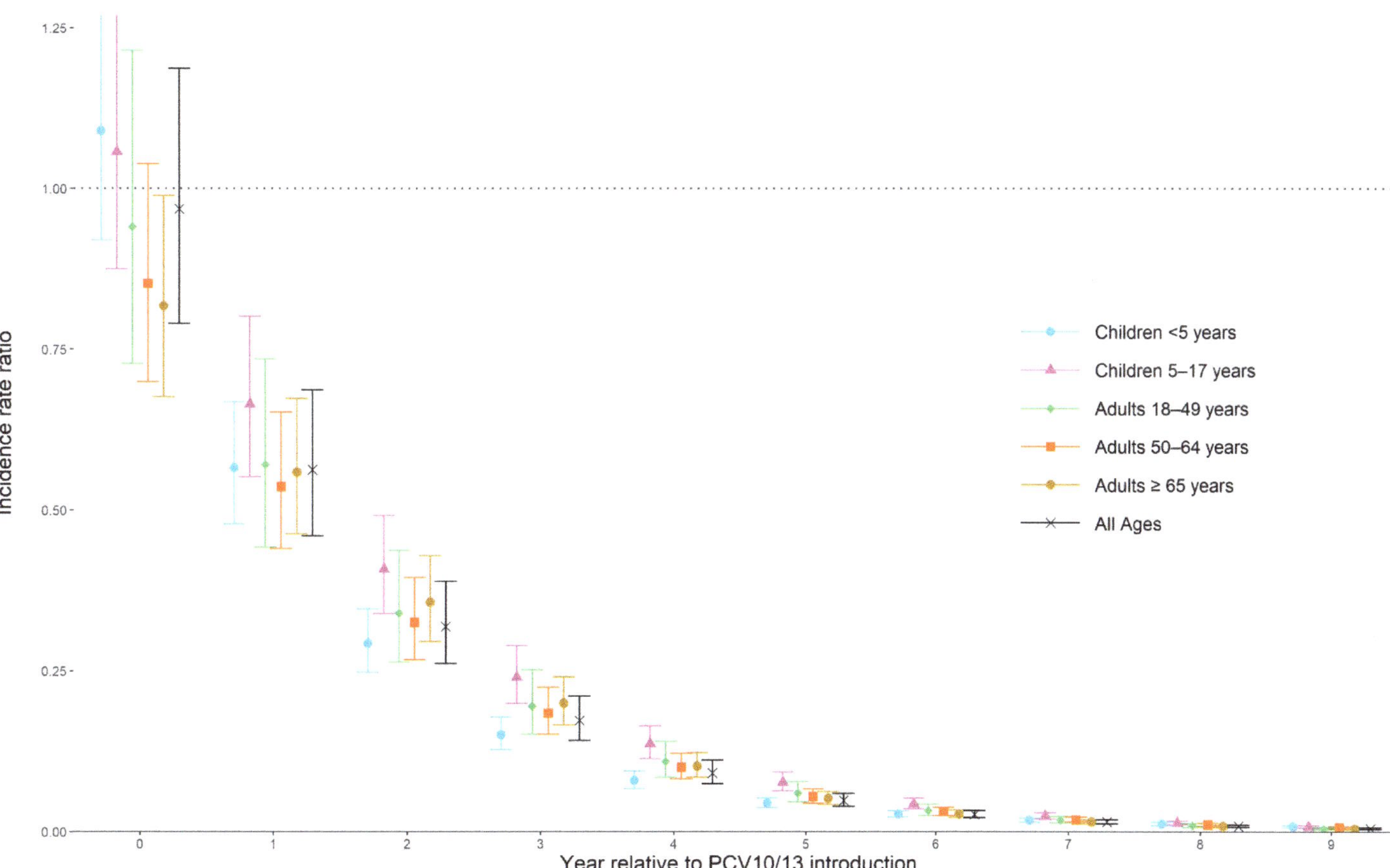

Figure 2. All-site weighted average incidence rate ratios for serotype 1 invasive pneumococcal disease for all ages and by age group. All ages' analysis (in black) is not an average of each age-specific estimate in each year but rather a re-analysis of the total cases from all ages reporting at each site.

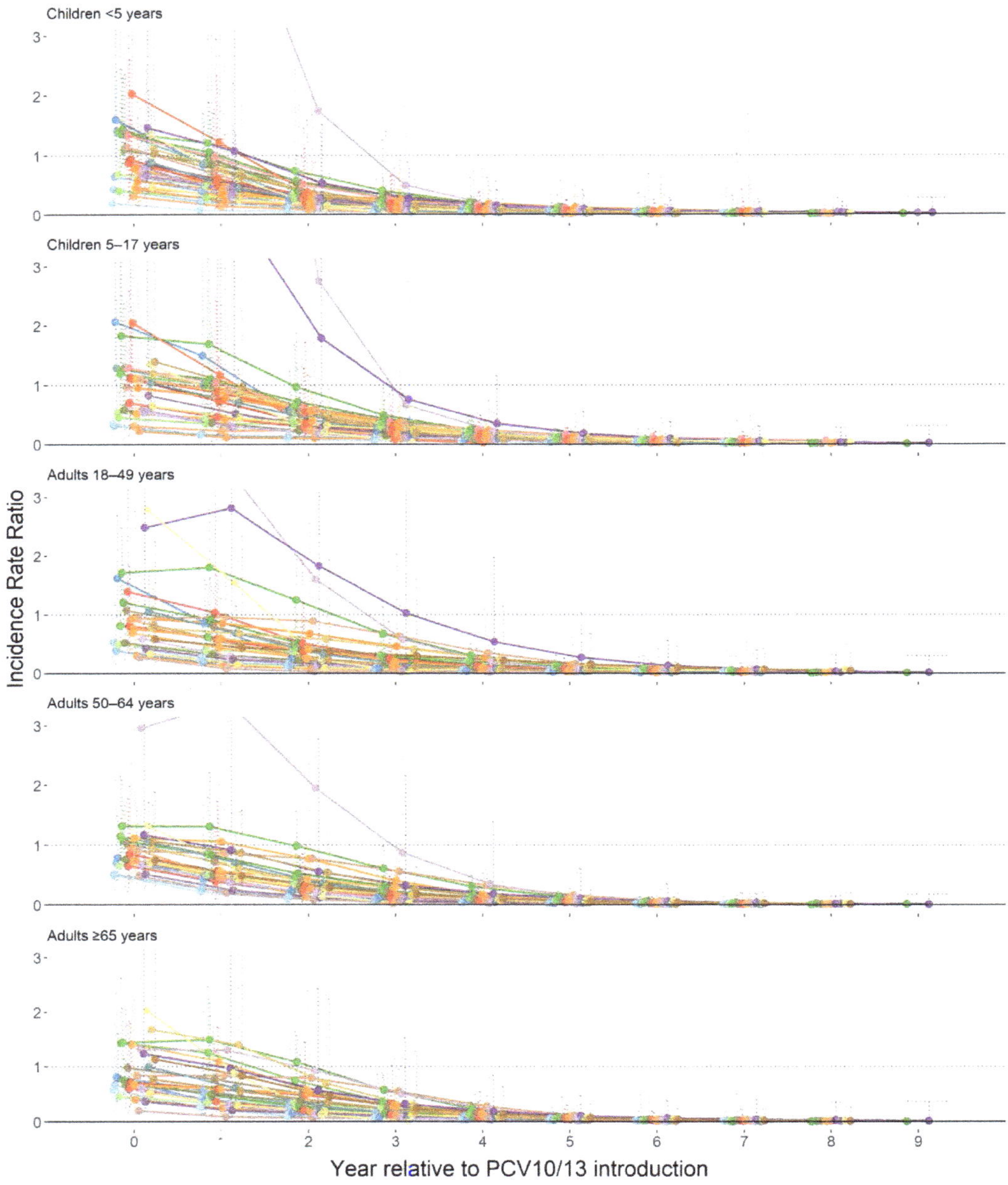

Figure 3. Site-specific modeled serotype 1 invasive pneumococcal disease incidence rate ratios comparing each post-PCV10/13 year to pre-PCV10/13 average, by age group.

4. Discussion

Our analysis demonstrates that there have been large and sustained decreases in ST1 IPD among both children targeted for immunization and among unvaccinated older children and adults through indirect effects. We used a standardized approach to analyze data from 45 surveillance sites and analytic methods that strengthened predictions from sites with few years of data and small sample sizes by borrowing strength from the overall trends observed across all sites. This allowed sites with few years of data and small sample sizes to still contribute proportionately to the analysis where data were available. As a result, this analysis is the most comprehensive assessment of changes in ST1 IPD after PCV10/13 introduction and demonstrates homogeneity in long-term impact of PCV10/13 on ST1 IPD across sites. These results were used to inform global vaccine policy recommendations around the use of pneumococcal vaccines in community outbreak settings [27].

The all-site weighted average IRRs are consistent with findings from individual surveillance sites on the long-term impact of PCV10/13 on ST1 IPD [9,10,12,16–20]. In the first several years of PCV10/13 use, the observed impact of PCV10/13 on ST1 IPD was heterogeneous, in part, due to the cyclic and outbreak nature of ST1 IPD and likely reflects heterogeneity in pre-PCV10/13 temporal trends with respect to the timing of PCV10/13 introduction. In some sites, ST1 IPD rates in the early years were greater than the pre-PCV10/13 average (because cyclical increases or outbreaks occurred at the time of or immediately following PCV10/13 introduction or because of noise in small datasets) and in other sites ST1 IPD rates were lower than the pre-PCV10/13 average immediately following PCV10/13 introduction. However, further into the PCV10/13 period, every site had sustained reductions in ST1 IPD below the pre-PCV10/13 rate. Prior to PCV10/13 introduction ST1 was known to cause severe disease to a greater degree in older children and younger adults compared to other serotypes [3,13,28] and importantly, we observed substantial reductions in ST1 IPD for all age groups. There was concern prior to the widespread introduction of PCV10/13 regarding the immunogenicity of PCV10/13 when used without a booster dose against ST1 [29]. Although only five sites using a 3 + 0 schedule were included in the analysis, the direct and indirect effects for ST1 IPD after several years of PCV10/13 use in these sites were consistent with patterns observed in sites using a booster dose schedule.

Although not observed in all sites and CIs overlap, our results showed slightly smaller declines in ST1 IPD for children <18 years compared to adults ≥18 years in the year of PCV10/13 introduction, which is contradictory to expected patterns of indirect effects among non-immunized adults following introduction of an infant vaccine [30]. This may reflect secular trends unrelated to vaccine introduction or differences in the hospital and surveillance systems between adults and pediatrics and an increased focus on pediatric surveillance around the time of pediatric vaccine introduction leading to greater detection of pediatric cases compared to adults. Ninety-two percent of sites with adult ST1 data included in the analysis have an adult pneumococcal polysaccharide vaccine recommendation. Although this may have reduced the burden of ST1 IPD among vaccinated adults prior to infant PCV10/13 programs, this does not explain observed patterns in the year of PCV10/13 introduction. The majority of adult polysaccharide vaccine programs began many years prior to the introduction of PCV10/13, recommendations vary by site for adult pneumococcal vaccine use, and data on vaccine uptake among adults was limited. We were not able to detect differences by adult pneumococcal vaccine program recommendation. Despite this, we see substantial and sustained declines in ST1 IPD for all age groups in the following years of PCV10/13 use.

To understand the impact of PCV10/13 introduction, data were restricted to sites with at least 50% uptake for the primary PCV series at 12 months of age in at least one-year post-PCV10/13 introduction and majority of included sites had high PCV uptake. This resulted in most data coming from high-income countries and limited inferences can be made to other regions or areas with lower vaccine uptake. Further, the majority of the data are from sites that used a booster dose. Among the five sites with a 3 + 0 schedule, four introduced PCV10/13 with a catch-up program. Therefore, any added effects of a booster dose and catch-up programs could not be assessed, and results may not be reflective of other settings. In particular, data were limited from areas prone to pneumococcal meningitis outbreaks, such as the African meningitis belt. Only one site from the African meningitis belt, The Gambia, was included in the analysis where a 3 + 0 schedule of PCV13 was introduced without a catch-up program. Although there were few ST1 cases (n = 71), ST1 trends for children <18 years of age were consistent with other non-meningitis belt countries in Africa and other regions. In the 4 other sites that used a 3 + 0 schedule (all of which introduced PCV10/13 with a catch-up campaign), ST1 trends were also similar to those observed in sites using a 2 + 1 or 3 + 1 schedule among both children and adults. Two meningitis belt countries with documented pneumococcal outbreaks after PCV13 introduction with a 3 + 0 schedule, Ghana and Burkina Faso, did not contribute data to the PSERENADE project. As in The Gambia, the proportion of ST1 cases occurring among children <5 years of age decreased compared to the pre-PCV13 period in Ghana and Burkina Faso [22–24]. However, pneumococcal meningitis outbreaks in persons >5 years of age were documented four years after PCV13 introduction in the Brong-Ahafo region of Ghana (outside of the traditional meningitis belt) [22] and five years after introduction in the Upper West and Northern regions of Ghana (within the traditional meningitis belt) [23]. In both of these outbreaks a large proportion of cases were due to ST1 (between 62–80%) [22,23]. PCV13 uptake in these specific communities was undocumented and national PCV13 uptake in the first two years of use was low in Ghana (41–68%) [22]. In Burkina Faso after 3 years of PCV13 use, ST1 meningitis rates declined by 59% for children <1 year of age, by 25% for children 1–4 years of age, and by 8–17% for individuals ≥5 years of age. Slightly larger declines were observed for all PCV13 serotype meningitis (76% decline for children <1 year, 58% decline for children 1–4 years, and 14–20% decline for individuals ≥5 years of age) [24]. The remaining PCV13 serotype meningitis among individuals ≥1 year of age indicates that indirect effects have not been fully achieved for all vaccine serotypes, including but not limited to ST1, and the 59% decline in ST1 disease among children <1 year of age suggests that after 3 years of use the PCV program has not yet sufficiently protected children targeted for immunization. Although the association between PCV uptake and indirect effects are not well understood, this may indicate low vaccine uptake. The persistence of ST1 IPD in unvaccinated persons in the first five years of PCV10/13 use is consistent with our results, as ST1 outbreaks were still observed in some sites during the first five years of PCV10/13 use and significant declines in ST1 IPD were not observed for some sites until after 5 years of PCV10/13 use (Figure 3). As recommended by WHO, continuation of comprehensive, high-quality serotype-specific IPD surveillance and vaccine uptake monitoring in the African meningitis belt sites still experiencing ST1 outbreaks in the post-PCV period and in countries with suboptimal PCV10/13 uptake could improve understanding of ST1 in these settings with schedules lacking a booster dose or with low PCV10/13 uptake [31].

This analysis was also limited in its ability to model the counterfactual ST1 IR in the absence of PCV10/13. An ideal ST1 counterfactual IR would have modeled the cyclical pattern of ST1 IPD in the absence of PCV10/13 introduction as a baseline comparison for each post-PCV10/13 year, as has been done for single site analyses, but is challenging without monthly data [11]. Due to the number of available years of pre-PCV data and small ST1 sample size, this was not possible for the majority of sites and instead an average pre-PCV10/13 ST1 IR was used as the counterfactual ST1 IR. Using the average pre-PCV10/13 ST1 IR would most likely lead to less valid effect estimates in the early years of PCV10/13 use and may contribute to unexplained differences in IRRs between age groups in the year of PCV10/13 introduction. However, this would have limited impact on the estimates in later post-PCV10/13 later years. Although a high proportion of the cases from included sites were fully serotyped, another limitation of this analysis, which cannot be tested, is the assumption that the prevalence of ST1 among cases that were serotyped is not biased from the prevalence of ST1 cases among cases that were not serotyped or not fully serotyped. Finally, the number of sites with post-PCV10/13 data declined over time and sites with longer follow-up periods tend to be from high-income countries that generally introduced PCV10/13 earlier than low- and middle-income countries. Eleven sites had data through the ninth year of PCV10/13 use and only three sites had data in the tenth year of PCV10/13 use.

These results can provide important context for evaluating the impact of PCV10/13 on other individual serotypes. ST1 is unique from other vaccine-serotypes in its invasiveness potential, carriage patterns, ability to cause large outbreaks among all ages, and association with meningitis [7,8,12–15]. Future analyses using the PSERENADE dataset will evaluate the impact of PCV10/13 on other individual vaccine and non-vaccine serotypes.

5. Conclusions

The introduction of PCV10/13 into infant immunization programs has been associated with the near elimination of ST1 IPD in all ages after approximately 6 years of use, including in settings without a booster dose schedule but with high PCV10/13 uptake, where data are available. Improved population-level serotype-specific IPD surveillance for all ages, including for meningitis, is needed from settings using a 3 + 0 schedule with a history of ongoing ST1 outbreaks in the post-PCV10/13 period, particularly the African meningitis belt, and in countries with suboptimal PCV10/13 uptake. This would allow for a more comprehensive evaluation of the indirect effects of PCV10/13 in older children and adults living in high burden settings using a 3 + 0 schedule or with low PCV10/13 uptake.

Supplementary Materials: The following are available online at https://www.mdpi.com/article/10.3390/microorganisms9040696/s1, Figure S1: Method for estimating annual ST1 IPD IRRs comparing the pre-PCV10/13 period to each post-PCV10/13 year for each included site with pre- and post-PCV10/13 data: example for children <5 years from one site, Figure S2: All-site weighted average IRRs for ST1 IPD for all ages and by age group adjusted for pre-PCV trends in all-serotype IPD (n = 30 sites), Figure S3: Site-specific modeled ST1 IPD IRRs by PCV product and age group, Figure S4: Site-specific modeled ST1 IPD IRRs by region and age group, Figure S5: Site-specific modeled ST1 IPD IRRs by infant PCV10/13 schedule and age group, Figure S6: Site-specific modeled ST1 IPD IRRs for adults $\geq$ 65 years by adult pneumococcal polysaccharide vaccine recommendation, Figure S7: Incidence rate of ST1 meningitis cases from cerebrospinal fluid (CSF) in Brazil by age group and year relative to PCV10 introduction (n = 51 cases), Table S1: Sites included in PSERENADE evaluated for the serotype 1 analysis by age group.

Author Contributions: Conceptualization, J.C.B., K.H., D.R.F., M.D.K.; methodology, J.C.B., S.L.Z., K.H., M.D.K., E.W.K.; software and analysis, J.C.B., S.L.Z.; writing—original draft preparation, J.C.B., K.H., M.K.H., J.N.S.; writing—review and editing, D.R.F., M.D.K., S.L.Z., M.G.Q., E.W.K., K.A., M.-C.C.B., D.B., R.C., J.C., G.C., H.C., J.E.C., R.D., T.D., K.D., S.d.M., P.D.W., S.D., T.G., C.G., M.G.-V., L.L.H., M.H., P.-L.H., S.J., J.D.K., J.K. (Jackie Kleynhans), M.J.K., J.K. (Jana Kozakova), K.G.K., S.N.L., L.M. (Laura MacDonald), G.A.M., L.M. (Lucia MadaarovA), A.M., J.M., E.M., T.M., C.M.-A., J.P.N., M.P., T.P., R.P., S.K.S., A.S.K., L.S., J.A.S., A.S., S.S., M.v.d.L., J.R.V., A.v.G., B.A.W., I.Y., K.Z.; visualization, J.C.B., M.K.H.; supervision, K.H., M.D.K., A.L.C.; project administration, J.C.B., M.G.Q.; funding acquisition, A.L.C., M.D.K., K.H.; data curation, E.W.K., K.A., M.-C.C.B., D.B., R.C., J.C., G.C., H.C., J.E.C., R.D., T.D., K.D., S.d.M., P.D.W., S.D., T.G., C.G., M.G.-V., L.L.H., M.H., P.-L.H., S.J., J.D.K., J.K. (Jackie Kleynhans), M.J.K., J.K. (Jana Kozakova), K.G.K., S.N.L., L.M. (Laura MacDonald), G.A.M., L.M. (Lucia MadaarovA), A.M., J.M., E.M., T.M., C.M.-A., J.P.N., M.P., T.P., R.P., S.K.S., A.S.K., L.S., J.A.S., A.S., S.S., M.v.d.L., J.R.V., A.v.G., B.A.W., I.Y., K.Z. All authors provided input for the analytic methodology and critically reviewed results. All authors have read and agreed to the published version of the manuscript.

Funding: The PSERENADE project is funded by the Bill and Melinda Gates Foundation as part of the World Health Organization Pneumococcal Vaccines Technical Coordination Project, grant number INV-010429/OPP1189065.

Institutional Review Board Statement: This study was determined to not quality as human subjects research as defined by DHHS regulations 45 CFR 46.102 by the Johns Hopkins Bloomberg School of

Public Health Institutional Review Board, due to the use of existing, de-identified data. Therefore, Institutional Review Board oversight was not required, and ethical review and approval were waived for this study.

Informed Consent Statement: Not applicable.

Data Availability Statement: Restrictions apply to the availability of these data. Data were obtained under data sharing agreements from contributing surveillance sites and can only be shared by contributing organizations with their permission.

Acknowledgments: This work was only possible through the contributions of many other individuals and organizations. The list of names of key individuals is provided in the Table A2. We thank all individuals involved in the surveillance activities at the sites contributing to the PSERENADE project for trusting us with their data and for the assistance along the way to ensure this work provides value. This includes the surveillance network coordinators at EpiConcept, European Centre for Disease Prevention and Control, Pan American Health Organization and WHO who facilitated access to and understanding of the data. We thank the Johns Hopkins University librarians and research assistants for providing literature review and data collection support. We thank the technical and strategic advisors from the WHO for providing guidance on the objectives and for facilitating introductions to the many surveillance sites globally. Finally, we are greatly indebted to the members of our PSERENADE Technical Advisory Group for providing us with expert technical review and strategic advice on all aspects of the methods, presentation and interpretation throughout the project.

Conflicts of Interest: KH conducted the study and analyses while working at the Johns Hopkins School of Public Health but is an employee at Pfizer, Inc. as of 26 October 2020. MDK reports grants from Merck, personal fees from Merck, and grants from Pfizer, outside the submitted work. JCB reports funding from Pfizer in the past year, unrelated to the submitted work. JAS reports grants from the Bill & Melinda Gates Foundation, the Wellcome Trust, the UK MRC, National Institute of Health Research, outside the submitted work. MCB reports lectures fee from MSD outside from submitted work. AS reports grants and personal fees from Pfizer and personal fees from MSD and Sanofi Pasteur, outside the submitted work. ML has been a member of advisory boards and has received speakers honoraria from Pfizer and Merck. German pneumococcal surveillance has been supported by Pfizer and Merck. SD reports grant from Pfizer, outside the submitted work. KA reports a grant from Merck, outside the submitted work. AvG as received researching funding from Pfizer (last year 2017, Pfizer Investigator-Initiated Research [IIR] Program IIR WI 194379); attended advisory board meetings for Pfizer and Merck. CMA reports grants and personal fees from Pfizer, Qiagen and BioMerieux and grants from Genomica SAU, outside the submitted work. AM-research support to my institution from Pfizer and Merck; honoraria for advisory board membership from GlaxoSmithKline, Merck and Pfizer. SNL performs contract research for GSK, Pfizer, Sanofi Pasteur on behalf of St. George's University of London, but receives no personal remuneration. IY stated she was a member of mRNA-1273 study group and has received funding to her institution to conduct clinical research from BioFire, MedImmune, Regeneron, PaxVax, Pfizer, GSK, Merck, Novavax, Sanofi-Pasteur, and Micron. RD has received grants/research support from Pfizer, Merck Sharp & Dohme and Medimmune; has been a scientific consultant for Pfizer, MeMed, Merck Sharp & Dohme, and Biondvax; had served on advisory boards of Pfizer, Merck Sharp & Dohme and Biondvax and has been a speaker for Pfizer. LLH reports research grants to her institution from GSK, Pfizer and Merck. JDK has received an unrestricted grant-in-aid from Pfizer Canada that supports, in part, the CASPER invasive pneumococcal disease surveillance project. MH received an educational grant from Pfizer AG for partial support of this project. However, Pfizer AG had no role in the data analysis and content of the manuscript. MC has previously received a professional fee from Pfizer (Ireland), an unrestricted research grant from Pfizer Ireland (2007–2016) and an Investigator Initiated Reward from Pfizer Ireland in 2018 (W1243730). CLB, MD has intellectual property in BioFire Diagnostics and receives royalties through the University of Utah. CLB is an advisor to IDbyDNA. AK reports personal fees from Pfizer, outside the submitted work. MT reports grants from GlaxoSmithKline and grants from Pfizer Inc. to the Finnish Institute for Health and Welfare for research projects outside the submitted work, in which she has been a co-investigator. JCS reports had received assistance from Pfizer for attending to scientific meetings outside the submitted work. SCGA received travel grant from Pfizer. BL had two research grants from Pfizer on Streptococcus pneumoniae. EV reports grants from French public health agency, during the conduct of the study; grants from Pfizer, grants from Merck, outside the submitted work. NBZ has received investigator-initiated research grants from

GlaxoSmithKline, Takeda Pharmaceuticals, Merck and the Serum Institute of India, all unrelated to this research. CGS reports grant funding from Pfizer, Merck, and AstraZeneca in the past 3 years. NMvS reports grants and fee for service from Pfizer, fee for service from MSD and GSK, outside the submitted work; In addition, NMvS has a patent WO 2013/020090 A3 with royalties paid to University of California San Diego (inventors: Nina van Sorge/Victor Nizet). All other authors did not declare any conflicts of interest. The funders had no role in the design of the study; in the collection, analyses, or interpretation of data; in the writing of the manuscript, or in the decision to publish the results.

Disclaimer: The findings and conclusions in this report are those of the authors and do not necessarily represent the official position of the Centers for Disease Control and Prevention or the World Health Organization (WHO).

Appendix A

Table A1. PSERENADE Team.

Name	Affiliation
Pedro Alarcon	Instituto de Salud Pública de Chile, 7780050 Santiago, Santiago Metropolitan, Chile
Samanta C. G. Almeida	National Laboratory for Meningitis and Pneumococcal Infections, Center of Bacteriology, Institute Adolfo Lutz (IAL), São Paulo 01246-902, Brazil
Zahin Amin-Chowdhury	Immunisation and Countermeasures Division, Public Health England, NW9 5EQ, London, United Kingdom
Michelle Ang	National Public Health Laboratory, National Centre for Infectious Diseases, 308442, Singapore
Mária Avdicová	National Reference Centre for Pneumococcal and Haemophilus Diseases, Regional Authority of Public Health, 975 56 Banská Bystrica, Slovak Republic
Naor Bar-Zeev	Malawi-Liverpool-Wellcome Trust Clinical Research Programme, P.O. Box 30096, Chichiri, Blantyre 3, Malawi; Johns Hopkins Bloomberg School of Public Health, Baltimore, MD 21205, United States
Godfrey Bigogo	Centre for Global Health Research, Kenya Medical Research Institute, P.O. Box: 1578-40100, Kisumu, Kenya
Rita Born	Federal Office of Public Health, 3097 Liebefeld, Switzerland
Michael G. Bruce	Arctic Investigations Program, Division of Preparedness and Emerging Infections, National Center for Emerging and Zoonotic Infectious Diseases, Centers for Disease Control and Prevention, Anchorage, AK 99508, United States
Carrie L. Byington	University of Utah Department of Pediatrics (emeritus), Salt Lake City, UT 84108, United States; University of California Health System, Oakland, CA 94607, United States
Kin-Hung Chow	Department of Microbiology and Carol Yu Centre for Infection, The University of Hong Kong, Hong Kong SAR, China
Urtnasan Chuluunbat	National Center of Communicable Diseases (NCCD), Ministry of Health, Bayanzurkh district, 13336 Ulaanbaatar, Mongolia
Pilar Ciruela	Surveillance and Public Health Emergency Response, Public Health Agency of Catalonia, 08005 Barcelona, Spain; CIBER Epidemiología y Salud Pública, (CIBERESP), 28029 Madrid, Spain
Cheryl Cohen	Centre for Respiratory Diseases and Meningitis, National Institute for Communicable Diseases of the National Health Laboratory Service, Sandringham, 2192 Johannesburg, South Africa; School of Public Health, Faculty of Health Sciences, University of the Witwatersrand, Johannesburg, South Africa
Mary Corcoran	Irish Meningitis and Sepsis Reference Laboratory, Children's Health Ireland at Temple Street, Rotunda, Dublin 1, D01 XD99, Ireland
Geneviève Deceuninck	Quebec University Hospital Research Centre, Québec, QC G1V 4G2, Canada
Martina Del Manso	Department of Infectious Diseases, Italian National Institute of Health (Istituto Superiore di Sanità, ISS), 00161 Rome, Italy
Idrissa Diawara	Faculty of Sciences and Health Techniques, Mohammed VI University of Health Sciences (UM6SS) of Casablanca, Casablanca, Morocco; National Reference Laboratory, Mohammed VI University of Health Sciences (UM6SS), 82403 Casablanca, Morocco

Table A1. *Cont.*

Name	Affiliation
Janepsy Díaz	Instituto de Salud Pública de Chile, 7780050 Santiago, Santiago Metropolitan, Chile
Elina Dimina	Centre for disease prevention and control of Latvia, Riga, 1005, Latvia
Mignon du Plessis	Centre for Respiratory Diseases and Meningitis, National Institute for Communicable Diseases of the National Health Laboratory Service, Sandringham, 2192 Johannesburg, South Africa
Helga Erlendsdottir	Department of Clinical Microbiology, Landspitali—The National University Hospital, Hringbraut, 101 Reykjavik, Iceland
Yvonne Galloway	Epidemiology Team, Institute of Environmental Science and Research, Porirua, 5240 Wellington, New Zealand
Ryan Gierke	National Center for Immunizations and Respiratory Diseases, Centers for Disease Control and Prevention, Atlanta, GA 30333, United States
Noga Givon-Lavi	Pediatric Infectious Disease Unit and Clinical Microbiology Laboratory, Soroka University Medical Center, Ben-Gurion University of the Negev, Beer-Sheva, 8410501, Israel
Marcela Guevara	Instituto de Salud Pública de Navarra—IdiSNA, 31003 Pamplona, Navarra, Spain; CIBER Epidemiología y Salud Pública, (CIBERESP), 28029 Madrid, Spain
Ilias Hossain	Medical Research Council, The Gambia Unit, Fajara, The Gambia
Vicki Krause	Centre for Disease Control, Department of Health and Community Services, Darwin City NT 8000, Australia
Pavla Krizova	National Institute of Public Health (NIPH), 100 42, Praha 10, Czech Republic
Alicja Kuch	National Reference Centre for Bacterial Meningitis, National Medicines Institute, 00-725 Warsaw, Poland
Brigitte Lefebvre	Laboratoire de Santé Publique du Québec, Sainte-Anne-de-Bellevue, Quebec H9X 3R5, Canada
Tiia Lepp	Department of Communicable Disease and Control and Health Protection, Public Health Agency of Sweden, 171 82 Solna, Sweden
Ioanna Magaziotou	National Public Health Organisation, 15123 Athens, Greece
Kazunori Oishi	Toyama Institute of Health, Imizu, 939-0363 Toyama, Japan
Stephen I. Pelton	Boston University Schools of Medicine and Public Health, Boston, MA 02118, United States
Kate Pennington	Communicable Disease Epidemiology and Surveillance Section, Office of Health Protection, Australian Government Department of Health, 2606 Canberra ACT, Australia
Marie-Cecile Ploy	University Hospital Centre Limoges, Regional Observatories for Pneumococci, 87000 Limoges, France
Hafizur Rahman	Child Health Research Foundation, Dhaka 1207, Bangladesh
Rita Reyburn	Murdoch Children's Research Institute, Parkville VIC 3052, Australia
Leah J. Ricketson	Department of Pediatrics, University of Calgary, and Alberta Health Services, Calgary Alberta T3B 6A8, Canada
Juan Carlos Sanz	Laboratorio Regional de Salud Pública, Dirección General de Salud Pública, Comunidad de Madrid, 28053 Madrid, Spain
Kevin Scott	Bacterial Respiratory Infection Service, Scottish Microbiology Reference Laboratory, NHS GG&C, G31 2ER Glasgow, United Kingdom
Catherine G. Sutcliffe	Johns Hopkins Bloomberg School of Public Health, Baltimore, MD 21205, United States
Koh Cheng Thoon	KK Women's and Children's Hospital, 229899, Singapore
Maija Toropainen	Department of Health Security, Finnish Institute for Health and Welfare, 00271 Helsinki, Finland
Georgina Tzanakaki	National Meningitis Reference Laboratory, National School of Public Health Athens, Athens, Greece
Palle Valentiner-Branth	Infectious Disease Epidemiology and Prevention, Statens Serum Institut, DK-2300 Copenhagen S, Denmark
Nina M. van Sorge	Medical Microbiology and Infection Prevention, Netherlands Reference Laboratory for Bacterial Meningitis, Amsterdam University Medical Centers, location AMC, University of Amsterdam, 1105 AZ Amsterdam, The Netherlands

Table A1. *Cont.*

Name	Affiliation
Emmanuelle Varon	National Reference Centre for Pneumococci, Centre Hospitalier Intercommunal de Créteil, 94000 Créteil, France
Didrik F. Vestrheim	Department of Infection Control and Vaccine, Norwegian Institute of Public Health, 0456 Oslo, Norway
Toronto Invasive Bacterial Diseases Network	

Table A2. Acknowledgement List.

PSERENADE Technical Advisory Group
Thomas Cherian
William P. Hausdorff
Marc Lipsitch
Shabir A. Madhi
Elizabeth Miller
Catherine Satzke
Cynthia G. Whitney
World Health Organization
Katherine L. O'Brien
Jenny A. Walldorf
Johns Hopkins University
Yunfeng Cao
Peggy Gross
Donna Hesson
Ananya Kumar
Kate Perepezko
Meagan E. Peterson
Francesca Schiaffino Salazar
Daniel Stephens
Yangyupei Yang
EpiConcept
Camelia Savulescu
European Centre for Disease Prevention and Control
Edoardo Colzani
Pan American Health Organization
Lúcia Helena de Oliveira
Dirección General de Salud Pública, Comunidad de Madrid, Spain
Luis García Comas
Maria Ordobás Gavín
Department of Infectious Diseases, Italian National Institute of Health (Istituto Superiore di Sanità, ISS), Rome, Italy
Flavia Riccardo

Table A2. *Cont.*

Department of Microbiology, Faculty of Medicine and Pharmacy, Hassan II University of Casablanca, Casablanca, Morocco; Bacteriology-Virology and Hospital Hygiene Laboratory, Ibn Rochd University Hospital Centre, Casablanca, Morocco
Néhémie Nzoyikorera
Surveillance and Public Health Emergency Response, Public Health Agency of Catalonia, Barcelona, Spain
Sonia Broner
Conchita Izquierdo
Australian National Notifiable Diseases Surveillance data were provided by the Office of Health Protection, Australian Government Department of Health, on behalf of the Communicable Diseases Network Australia and the Enhanced Invasive Pneumococcal Disease Surveillance Working Group.

References

1. Wahl, B.; O'Brien, K.L.; Greenbaum, A.; Majumder, A.; Liu, L.; Chu, Y.; Lukšić, I.; Nair, H.; McAllister, D.A.; Campbell, H.; et al. Burden of Streptococcus Pneumoniae and Haemophilus Influenzae Type b Disease in Children in the Era of Conjugate Vaccines: Global, Regional, and National Estimates for 2000–15. *Lancet Glob. Health* **2018**, *6*, e744–e757. [CrossRef]
2. Troeger, C.; Blacker, B.; Khalil, I.A.; Rao, P.C.; Cao, J.; Zimsen, S.R.M.; Albertson, S.B.; Deshpande, A.; Farag, T.; Abebe, Z.; et al. Estimates of the Global, Regional, and National Morbidity, Mortality, and Aetiologies of Lower Respiratory Infections in 195 Countries, 1990–2016: A Systematic Analysis for the Global Burden of Disease Study 2016. *Lancet Infect. Dis.* **2018**, *18*, 1191–1210. [CrossRef]
3. Hausdorff, W.P.; Feikin, D.R.; Klugman, K.P. Epidemiological Differences among Pneumococcal Serotypes. *Lancet Infect. Dis.* **2005**, *5*, 83–93. [CrossRef]
4. Ritchie, N.D.; Mitchell, T.J.; Evans, T.J. What Is Different about Serotype 1 Pneumococci? *Future Microbiol.* **2011**, *7*, 33–46. [CrossRef] [PubMed]
5. Ganaie, F.; Saad, J.S.; McGee, L.; van Tonder, A.J.; Bentley, S.D.; Lo, S.W.; Gladstone, R.A.; Turner, P.; Keenan, J.D.; Breiman, R.F.; et al. A New Pneumococcal Capsule Type, 10D, Is the 100th Serotype and Has a Large Cps Fragment from an Oral Streptococcus. *mBio* **2020**, *11*. [CrossRef]
6. Johnson, H.L.; Deloria-Knoll, M.; Levine, O.S.; Stoszek, S.K.; Hance, L.F.; Reithinger, R.; Muenz, L.R.; O'Brien, K.L. Systematic Evaluation of Serotypes Causing Invasive Pneumococcal Disease among Children Under Five: The Pneumococcal Global Serotype Project. *PLoS Med.* **2010**, *7*, e1000348. [CrossRef] [PubMed]
7. Brueggemann, A.B.; Peto, T.E.A.; Crook, D.W.; Butler, J.C.; Kristinsson, K.G.; Spratt, B.G. Temporal and Geographic Stability of the Serogroup-Specific Invasive Disease Potential of Streptococcus Pneumoniae in Children. *J. Infect. Dis.* **2004**, *190*, 1203–1211. [CrossRef] [PubMed]
8. Bricio-Moreno, L.; Chaguza, C.; Yahya, R.; Shears, R.K.; Cornick, J.E.; Hokamp, K.; Yang, M.; Neill, D.R.; French, N.; Hinton, J.C.D.; et al. Lower Density and Shorter Duration of Nasopharyngeal Carriage by Pneumococcal Serotype 1 (ST217) May Explain Its Increased Invasiveness over Other Serotypes. *mBio* **2020**, *11*. [CrossRef]
9. Ladhani, S.N.; Collins, S.; Djennad, A.; Sheppard, C.L.; Borrow, R.; Fry, N.K.; Andrews, N.J.; Miller, E.; Ramsay, M.E. Rapid Increase in Non-Vaccine Serotypes Causing Invasive Pneumococcal Disease in England and Wales, 2000-17: A Prospective National Observational Cohort Study. *Lancet Infect. Dis.* **2018**, *18*, 441–451. [CrossRef]
10. Hammitt, L.L.; Etyang, A.O.; Morpeth, S.C.; Ojal, J.; Mutuku, A.; Mturi, N.; Moisi, J.C.; Adetifa, I.M.; Karani, A.; Akech, D.O.; et al. Effect of Ten-Valent Pneumococcal Conjugate Vaccine on Invasive Pneumococcal Disease and Nasopharyngeal Carriage in Kenya: A Longitudinal Surveillance Study. *Lancet Lond. Engl.* **2019**, *393*, 2146–2154. [CrossRef]
11. Harboe, Z.B.; Dalby, T.; Weinberger, D.M.; Benfield, T.; Mølbak, K.; Slotved, H.C.; Suppli, C.H.; Konradsen, H.B.; Valentiner-Branth, P. Impact of 13-Valent Pneumococcal Conjugate Vaccination in Invasive Pneumococcal Disease Incidence and Mortality. *Clin. Infect. Dis.* **2014**, *59*, 1066–1073. [CrossRef]
12. Cook, H.M.; Giele, C.M.; Jayasinghe, S.H.; Wakefield, A.; Krause, V.L. An Outbreak of Serotype-1 Sequence Type 306 Invasive Pneumococcal Disease in an Australian Indigenous Population. *Commun. Dis. Intell.* **2020**, *44*. [CrossRef] [PubMed]
13. Leimkugel, J.; Adams Forgor, A.; Gagneux, S.; Pflüger, V.; Flierl, C.; Awine, E.; Naegeli, M.; Dangy, J.; Smith, T.; Hodgson, A.; et al. An Outbreak of Serotype 1 *Streptococcus Pneumoniae* Meningitis in Northern Ghana with Features That Are Characteristic of *Neisseria Meningitidis* Meningitis Epidemics. *J. Infect. Dis.* **2005**, *192*, 192–199. [CrossRef]
14. Yaro, S.; Lourd, M.; Traoré, Y.; Njanpop-Lafourcade, B.-M.; Sawadogo, A.; Sangare, L.; Hien, A.; Ouedraogo, M.S.; Sanou, O.; Parent du Châtelet, I.; et al. Epidemiological and Molecular Characteristics of a Highly Lethal Pneumococcal Meningitis Epidemic in Burkina Faso. *Clin. Infect. Dis. Off. Publ. Infect. Dis. Soc. Am.* **2006**, *43*, 693–700. [CrossRef]
15. Proulx, J.; Déry, S.; Jetté, L.; Ismaël, J.; Libman, M.; De Wals, P. Pneumonia Epidemic Caused by a Virulent Strain of Streptococcus Pneumoniae Serotype 1 in Nunavik, Quebec. *Can. Commun. Dis. Rep.* **2002**, *28*, 129–131. [PubMed]

16. Von Gottberg, A.; de Gouveia, L.; Tempia, S.; Quan, V.; Meiring, S.; von Mollendorf, C.; Madhi, S.A.; Zell, E.R.; Verani, J.R.; O'Brien, K.L.; et al. Effects of Vaccination on Invasive Pneumococcal Disease in South Africa. *N. Engl. J. Med.* **2014**, *371*, 1889–1899. [CrossRef] [PubMed]
17. Angoulvant, F.; Levy, C.; Grimprel, E.; Varon, E.; Lorrot, M.; Biscardi, S.; Minodier, P.; Dommergues, M.A.; Hees, L.; Gillet, Y.; et al. Early Impact of 13-Valent Pneumococcal Conjugate Vaccine on Community-Acquired Pneumonia in Children. *Clin. Infect. Dis.* **2014**, *58*, 918–924. [CrossRef]
18. Camilli, R.; D'Ambrosio, F.; Del Grosso, M.; Pimentel de Araujo, F.; Caporali, M.G.; Del Manso, M.; Gherardi, G.; D'Ancona, F.; Pantosti, A. Impact of Pneumococcal Conjugate Vaccine (PCV7 and PCV13) on Pneumococcal Invasive Diseases in Italian Children and Insight into Evolution of Pneumococcal Population Structure. *Vaccine* **2017**, *35*, 4587–4593. [CrossRef] [PubMed]
19. Ciruela, P.; Izquierdo, C.; Broner, S.; Muñoz-Almagro, C.; Hernández, S.; Ardanuy, C.; Pallarés, R.; Domínguez, A.; Jané, M.; Esteva, C.; et al. The Changing Epidemiology of Invasive Pneumococcal Disease after PCV13 Vaccination in a Country with Intermediate Vaccination Coverage. *Vaccine* **2018**, *36*, 7744–7752. [CrossRef]
20. Andrews, N.J.; Waight, P.A.; Burbidge, P.; Pearce, E.; Roalfe, L.; Zancolli, M.; Slack, M.; Ladhani, S.N.; Miller, E.; Goldblatt, D. Serotype-Specific Effectiveness and Correlates of Protection for the 13-Valent Pneumococcal Conjugate Vaccine: A Postlicensure Indirect Cohort Study. *Lancet Infect. Dis.* **2014**, *14*, 839–846. [CrossRef]
21. Mackenzie, G.A.; Hill, P.C.; Jeffries, D.J.; Hossain, I.; Uchendu, U.; Ameh, D.; Ndiaye, M.; Adeyemi, O.; Pathirana, J.; Olatunji, Y.; et al. Effect of the Introduction of Pneumococcal Conjugate Vaccination on Invasive Pneumococcal Disease in The Gambia: A Population-Based Surveillance Study. *Lancet Infect. Dis.* **2016**, *16*, 703–711. [CrossRef]
22. Kwambana-Adams, B.A.; Asiedu-Bekoe, F.; Sarkodie, B.; Afreh, O.K.; Kuma, G.K.; Owusu-Okyere, G.; Foster-Nyarko, E.; Ohene, S.-A.; Okot, C.; Worwui, A.K.; et al. An Outbreak of Pneumococcal Meningitis among Older Children ($\geq$5 Years) and Adults after the Implementation of an Infant Vaccination Programme with the 13-Valent Pneumococcal Conjugate Vaccine in Ghana. *BMC Infect. Dis.* **2016**, *16*. [CrossRef] [PubMed]
23. Bozio, C.H.; Abdul-Karim, A.; Abenyeri, J.; Abubakari, B.; Ofosu, W.; Zoya, J.; Ouattara, M.; Srinivasan, V.; Vuong, J.T.; Opare, D.; et al. Continued Occurrence of Serotype 1 Pneumococcal Meningitis in Two Regions Located in the Meningitis Belt in Ghana Five Years after Introduction of 13-Valent Pneumococcal Conjugate Vaccine. *PLoS ONE* **2018**, *13*, e0203205. [CrossRef]
24. Kambiré, D.; Soeters, H.M.; Ouédraogo-Traoré, R.; Medah, I.; Sangaré, L.; Yaméogo, I.; Sawadogo, G.; Ouédraogo, A.-S.; Ouangraoua, S.; McGee, L.; et al. Early Impact of 13-Valent Pneumococcal Conjugate Vaccine on Pneumococcal Meningitis—Burkina Faso, 2014–2015. *J. Infect.* **2018**, *76*, 270–279. [CrossRef]
25. Deloria-Knoll, M.; Bennett, J.C.; Garcia Quesada, M.; Kagucia, E.W.; Peterson, M.E.; Feikin, D.R.; Cohen, A.L.; Hetrich, M.K.; Yangyupei, Y.; Sinkevitch, J.N.; et al. Global Landscape Review of Serotype-Specific Invasive Pneumococcal Disease Surveillance among Countries Using PCV10/13: The Pneumococcal Serotype Replacement and Distribution Estimation (PSERENADE) Project. *Microorganisms* **2021**, in press.
26. Hadfield, J.D. MCMC Methods for Multi-Response Generalized Linear Mixed Models: The MCMCglmm R Package. *J. Stat. Softw.* **2010**, *33*, 1–22. [CrossRef]
27. World Health Organization. *Strategic Advisory Group of Experts on Immunization Yellow Book*; World Health Organization Department of Immunization, Vaccines and Biologicals (IVB): Geneva, Switzerland, 2020; p. 8.
28. Traore, Y.; Tameklo, T.A.; Njanpop-Lafourcade, B.-M.; Lourd, M.; Yaro, S.; Niamba, D.; Drabo, A.; Mueller, J.E.; Koeck, J.-L.; Gessner, B.D. Incidence, Seasonality, Age Distribution, and Mortality of Pneumococcal Meningitis in Burkina Faso and Togo. *Clin. Infect. Dis.* **2009**, *48*, S181–S189. [CrossRef]
29. Klugman, K.P.; Madhi, S.A.; Adegbola, R.A.; Cutts, F.; Greenwood, B.; Hausdorff, W.P. Timing of Serotype 1 Pneumococcal Disease Suggests the Need for Evaluation of a Booster Dose. *Vaccine* **2011**, *29*, 3372–3373. [CrossRef] [PubMed]
30. Tsaban, G.; Ben-Shimol, S. Indirect (Herd) Protection, Following Pneumococcal Conjugated Vaccines Introduction: A Systematic Review of the Literature. *Vaccine* **2017**, *35*, 2882–2891. [CrossRef]
31. World Health Organization. Pneumococcal Conjugate Vaccines in Infants and Children under 5 Years of Age: WHO Position Paper. *Wkly. Epidemiol. Rec.* **2019**, *94*, 85–104.

Article

The Epidemiology of Meningitis in Infants under 90 Days of Age in a Large Pediatric Hospital

Timothy A. Erickson [1,2], Flor M. Munoz [3], Catherine L. Troisi [2], Melissa S. Nolan [4], Rodrigo Hasbun [5], Eric L. Brown [2] and Kristy O. Murray [1,*]

1 Department of Pediatrics, Section of Pediatric Tropical Medicine, William T. Shearer Center for Human Immunobiology, Baylor College of Medicine and Texas Children's Hospital, Houston, TX 77030, USA; timothy.erickson@bcm.edu
2 School of Public Health, University of Texas Health Science Center, Houston, TX 77030, USA; Catherine.L.Troisi@uth.tmc.edu (C.L.T.); eric.l.brown@uth.tmc.edu (E.L.B.)
3 Department of Pediatrics, Section of Infectious Diseases, Baylor College of Medicine and Texas Children's Hospital, Houston, TX 77030, USA; florm@bcm.edu
4 Department of Epidemiology and Biostatistics, Arnold School of Public Health, University of South Carolina, Columbia, SC 29208, USA; msnolan@mailbox.sc.edu
5 McGovern Medical School, University of Texas, Houston, TX 77030, USA; Rodrigo.Hasbun@uth.tmc.edu
* Correspondence: Kmurray@bcm.edu

Abstract: Background: Meningitis is associated with substantial morbidity and mortality, particularly in the first three months of life. Methods: We conducted a retrospective review of patients <90 days of age with meningitis at Texas Children's Hospital from 2010–2017. Cases were confirmed using the National Healthcare Safety Network (NHSN) definition of meningitis. Results: Among 694 infants with meningitis, the most common etiology was viral (n = 351; 51%), primarily caused by enterovirus (n = 332; 95%). A quarter of cases were caused by bacterial infections (n = 190; 27%). The most common cause of bacterial meningitis was group B *Streptococcus* (GBS, n = 60; 32%), followed by Gram-negative rods other than *E. coli* (n = 40; 21%), and *E. coli* (n = 37; 19%). The majority of Gram-negative organisms (63%) were resistant to ampicillin, and nearly one-fourth of Gram-negative rods (23%) other than *E. coli* and 2 (6%) *E. coli* isolates were resistant to third-generation cephalosporins. Significant risk factors for bacterial meningitis were early preterm birth and the Black race. Conclusions: Enteroviruses most commonly caused viral meningitis in infants; GBS was the most common bacterial cause despite universal screening and intrapartum prophylaxis. The emergence of MRSA and resistance to third-generation cephalosporins in Gram-negative bacterial meningitis challenges the options for empirical antimicrobial therapy.

Keywords: neonatal infections; meningitis; epidemiology; antibiotic resistance; etiologic diagnosis; enterovirus; group B *Streptococcus*

Citation: Erickson, T.A.; Munoz, F.M.; Troisi, C.L.; Nolan, M.S.; Hasbun, R.; Brown, E.L.; Murray, K.O. The Epidemiology of Meningitis in Infants under 90 Days of Age in a Large Pediatric Hospital. *Microorganisms* **2021**, *9*, 526. https://doi.org/10.3390/microorganisms9030526

Academic Editor: James Stuart

Received: 15 January 2021
Accepted: 9 February 2021
Published: 4 March 2021

1. Introduction

Meningitis is associated with substantial morbidity and mortality, particularly in infants. Numerous causes of meningitis exist, with viral and bacterial infectious agents the most common. About one-quarter of the cases of meningitis lacks an identified cause [1]. Infants, particularly those under 90 days of age, historically have the highest attack rates for bacterial meningitis, with *Haemophilus influenzae* type b (Hib) being most common in the era prior to routine vaccination and group B *Streptococcus* (GBS) and *E. coli* most commonly reported since. Other leading causes include viral infection with enterovirus and herpes simplex virus (HSV) [2–9]. Only a few studies have investigated the epidemiology of infant meningitis since the introduction of antenatal screening and antepartum treatment for GBS in pregnant women and the advent of molecular diagnosis [7,8,10,11].

Contemporary studies examining the etiology and epidemiology of infant meningitis are needed, given advances in diagnostic methods available to clinicians in the last decade. Similarly, studies describing antimicrobial resistance patterns are needed [9]. This information is particularly important as antibiotic resistance has increased in the past two decades among the most common bacterial pathogens known to cause infant meningitis, most particularly among Gram-negatives [12,13]. To address these gaps, we conducted an investigation into the epidemiology of meningitis in infants 0 to 90 days of life at a large pediatric hospital in Houston, Texas.

2. Materials and Methods

We conducted a retrospective medical record review of all patients with an ICD-9 or -10 diagnosis code corresponding to meningitis admitted to Texas Children's Hospital from 1 January 2010 to 30 December 2017 (Table S1). Only infants <90 days of age at the time of diagnosis were retained for analysis. We employed the National Healthcare Safety Network (NHSN) definition of meningitis, which states that a case of meningitis requires having either (1) an organism identified from the cerebrospinal fluid (CSF) by culture (e.g., bacteria, fungi), PCR (e.g., viruses), or confirmatory level of IgM (e.g., arbovirus), OR (2) CSF pleocytosis and compatible signs and symptoms [14,15]. If a patient with pleocytosis and clinically compatible signs and symptoms had no organism identified in the CSF but had an organism isolated from a different sterile site (i.e., blood or urine) that is a known cause of meningitis, then they were classified as meningitis caused by that organism. Meningitis occurring after intravenous immunoglobulin (IVIG) was classified as meningitis due to IVIG, a well-known phenomenon [16]. Patients were classed into groups and subgroups by cause (viral, bacterial, fungal, and unknown).

Medical records were abstracted to acquire demographic information, pre- and post-natal metrics, including gestational age at birth, microbiology and laboratory values at the time of admission, and clinical outcomes. Patient length of stay was calculated from the date of hospital admission to date of discharge except in the case of those patients initially admitted for a non-meningeal cause; these were calculated from the date of the first identification of an abnormal CSF result. Patients were divided by age into the following categories: <7 days of age, 7–14 days, 15–21 days, 21–28 days, and >28 days of age; and by gestational age at birth as term ($\geq$37 weeks gestation) or preterm (<37 weeks gestation). Patients with preterm birth were defined as either early preterm ($\leq$34 weeks gestation) or late preterm (35 or 36 weeks gestation). Maternal age, GBS screen status (conducted, positive/negative), type of delivery (spontaneous vaginal delivery or cesarean section), and maternal antibiotic treatment were also obtained.

Categorical variables were compared using a Pearson's chi-square test with statistical significance set at the 0.05 level. All statistics were calculated using STATA version 14.2 (StataCorp, College Station, TX, USA). This study was reviewed and approved by the Baylor College of Medicine Institutional Review Board (H-35069).

3. Results

From 2010–2017, 1501 cases of meningitis and encephalitis were admitted to Texas Children's Hospital. Almost half (n = 694; 46%) of all patients presented in the first 90 days of life (Figure 1).

These infant patients accounted for 9518 days of hospital stay, with a median length of stay of 3 days. Fourteen patients (2%) died from their infections (Table 1). An etiologic cause could be determined for 547 patients in this population (79%). Overall, viral causes were most common (n = 351; 51%), followed by bacteria (n = 190; 27%), fungi (n = 5; 1%); and one case occurring after IVIG treatment (Table 2). One of every five patients had no known cause identified (n = 147; 21%). Twenty-five patients (4%) had healthcare-associated ventriculitis and meningitis (HCAVM). Among the 190 bacterial meningitis cases, the majority were culture-positive in the CSF (n = 150; 79%), followed by blood (n = 29; 15%), urine (n = 9; 5%), wound site on brain (n = 1; <1%), and one unknown.

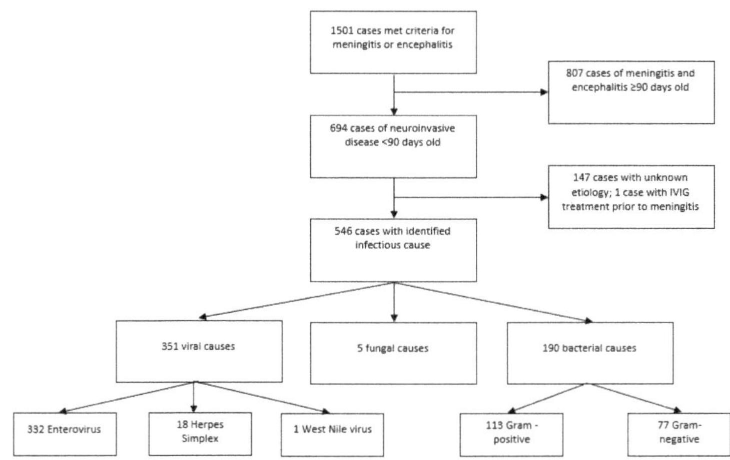

Figure 1. Selection flowchart.

Table 1. Demographic and clinical characteristics of meningitis by etiology in infants under 90 days of age, Houston, TX 2010–2017 *.

Demographics and Clinical Findings	Enterovirus and West Nile Virus n = 333	HSV n = 18	GBS n = 60	*Staphylococcus* Species n = 18	Gram-Positive Others n = 35	*E. coli* n = 37	Gram Negative Others n = 40	Fungal n = 5	Unknown n = 147
Male (%)	186 (56)	10 (56)	28 (47)	12 (66)	18 (51)	25 (68)	24 (60)	4 (80)	87 (59)
Race/ethnicity									
White (%)	135 (41)	8 (44)	14 (23)	7 (39)	5 (14)	10 (27)	16 (40)	3 (60)	49 (33)
Hispanic (%)	129 (39)	5 (28)	28 (47)	6 (33)	17 (49)	17 (46)	10 (25)	1 (20)	66 (45)
Black (%)	36 (11)	4 (22)	15 (25)	5 (28)	11 (31)	6 (16)	7 (18)	1 (20)	20 (14)
Other (%)	33 (10)	1 (6)	3 (5)	0	2 (6)	4 (11)	7 (18)	0	12 (8)
Private insurance (%)	165 (50)	5 (28)	17 (28)	8 (44)	19 (54)	13 (35)	17 (43)	3 (60)	57 (39)
Age									
<7 days	18 (5)	5 (28)	10 (17)	3 (17)	10 (29)	9 (24)	10 (25)	3 (60)	17 (12)
7–13 days (%)	33 (10)	3 (17)	9 (15)	2 (11)	1 (3)	3 (8)	3 (8)	1 (20)	14 (10)
14–20 days (%)	36 (11)	5 (28)	4 (7)	3 (17)	5 (14)	5 (14)	7 (18)	1 (20)	7 (5)
20–27 days (%)	59 (18)	2 (11)	6 (10)	2 (11)	1 (3)	4 (11)	5 (13)	0	15 (10)
>28 days (%)	187 (56)	3 (17)	31 (52)	8 (44)	18 (51)	16 (43)	15 (38)	0	94 (64)
CSF Findings									
Median CSF leukocytes (range)	202 (1–7099)	36.5 (0–247)	1185 (5–24,900)	119 (2–1290)	915 (1–13,726)	831 (16–161,500)	129 (7–7350)	549 (35–1063)	227 (8–3848)
Percent neutrophils	26 (0–95)	4 (0–83)	76.5 (3–97)	43.5 (0–85)	80 (0–98)	70.5 (2–97)	58.5 (7–95)	75.5 (71–80)	29 (0–93)
Protein mg/dL	81 (21–6000)	112.5 (59–1702)	263.5 (51–6000)	172 (58–1397)	146.5 (54–2412)	261 (46–3587)	193 (43–1498)	317 (72–1134)	90 (16–1456)
Glucose mg/dL	41 (27–112)	37 (20–102)	26.5 (20–98)	36 (20–96)	42 (20–126)	35 (20–121)	40 (20–97)	47 (20–55)	42 (20–83)
Death (%)	0	1 (6)	6 (10)	2 (11)	0	2 (5)	3 (8)	0	0

* One case of IVIG meningitis is not reported in this table.

Table 2. Subclassifications of causes of infant meningitis, Houston, TX 2010–2017.

Enterovirus and West Nile Virus *n*= 333	HSV *n* = 18	GBS *n* = 60	*Staphylococcus* Species *n* = 18	Other Gram-Positive *n* = 35	*E. coli* *n* = 37	Other Gram-Negative *n* = 40	Fungal *n* = 5	Un-known *n* = 147	IVIG *n* = 1
Enterovirus (332)	HSV-1 (4)	GBS (60)	*Staphylococcus aureus* (8)	Gram-positive bacteria, no species (13)	*E. coli* (37)	*Enterobacter cloacae* (7)	*Candida albicans* (5)		
West Nile virus (1)	HSV-2 (14)		*Staphylococcus epidermidis* (8)	*Streptococcus gallolyticus* (9)		*Klebsiella pneumoniae* (7)			
			Staphylococcus warneri (1)	*Enterococcus faecalis* (6)		*Salmonella enterica* (6)			
			Staphylococcus hominis (1)	*Streptococcus pneumoniae* (3)		*Acinetobacter baumannii* (3)			
				Streptococcus mitis (2)		*Serratia marcescens* (3)			
				Clostridium species (1)		Gram-negative rods, no species (2)			
				Streptococcus infantarius (1)		*Neisseria meningitidis* (2)			
						Proteus mirabilis (2)			
						Citrobacter braakii (1)			
						Citrobacter freundii (1)			
						Haemophilus influenzae (1)			
						Klebsiella oxytoca (1)			
						Pantoea species (1)			
						Pseudomonas aeruginosa (1)			
						Pseudomonas fluorescens (1)			
						Morganella morganii (1)			

3.1. Viral Infections

The vast majority of viral meningitis cases were caused by enteroviral infection (332/351, 95% of all viruses identified). Enterovirus infections exhibited seasonal trends, with cases occurring most commonly in the summer and with peaks in 2014 and 2015 (Figure 2). Infection with herpesviruses (*n* = 18; 5% of all viruses), most frequently HSV-2 (*n* = 14; 78%), were the next most common. A single, 41-day-old patient was also identified with NHSN confirmed meningitis, CSF pleocytosis, and positive West Nile virus IgM in the serum.

3.2. Group B Streptococcus

The leading bacterial cause of meningitis in our population was group B *Streptococcus*, with 60 cases, representing one-third (32%) of all bacterial meningitis and 9% of all-cause meningitis (Tables 2 and 3). Most cases (*n* = 45; 75%) occurred in term infants, with one of every four (*n* = 15; 25%) GBS patients born premature. The majority of these (*n* = 13; 87%) were born early preterm. Complete data on GBS screening during pregnancy was unavailable for 6 of these 60 patients; 10 patients involved premature births that were likely too early for screening. The majority of mothers of patients with maternal GBS screening data available (n = 29/44; 66%) had a confirmed negative result. None of these mothers had any course of antibiotics administered prior to delivery. The mothers of 15 infants with GBS meningitis were positive for GBS when screened prior to delivery. Of these, one did not have a record of antibiotic treatment, two refused treatment, and two were not treated with no explanation available. The remaining 10 all received appropriate prophylactic treatment prior to delivery. Most of the infants with GBS meningitis developed disease >7 days after birth (n = 50; 83%) (late-onset), while 10 (17%) developed an infection within a week of delivery (early-onset). Mothers of infants with early-onset GBS meningitis were more likely to have screened negative for GBS when compared to infants with late-onset GBS, although the comparison (89% vs. 60%) was not statistically significant (p = 0.10).

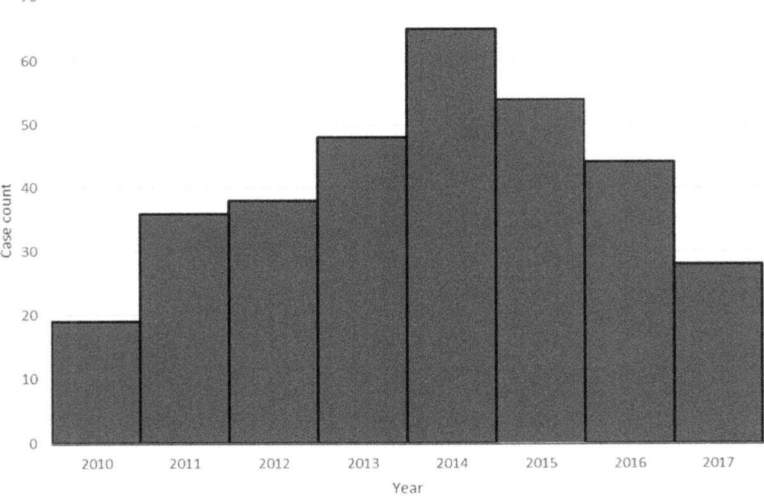

Figure 2. Enteroviral meningitis by year.

Table 3. Characteristics of cases of group B *Streptococcus* (n = 60) meningitis among infants 0 to 90 days old, by age at presentation, Houston 2010–2017.

	Early (0–7 Days) (n = 10)	Late (>7 Days) (n = 50)
Maternal screen	9	35
Positive for GBS	1 (11%)	14 (40%)
Negative for GBS	8 (89%)	21 (60%)
GBS prophylaxis	2 (22%)	10 (20%)
Concurrent bacteremia	8 (89%)	35 (69%)
Death	1 (11%)	5 (10%)

3.3. *Escherichia coli*

The next most common cause of infection was *E. coli*, which was responsible for 37 cases of meningitis in the first 90 days of life (19% of bacterial meningitis and 5% of all meningitis). Almost half (n = 16; 43%) of *E. coli* patients were premature. *E. coli* was responsible for 24% of premature and for 17% of term infants with bacterial meningitis; this difference was not statistically significant (p = 0.25). The majority of these cases (76%) presented after the first week of life. Approximately 2/3 of the *E. coli* isolates were resistant to ampicillin, while resistance to 3rd generation cephalosporins was seen in 6%.

3.4. Gram-Negative Organisms Other Than E. coli

Gram-negative organisms other than *E. coli* ($n = 40$) caused 21% of bacterial meningitis and 6% of all meningitis. Similar to *E. coli*, almost half ($n = 18$; 45%) of patients with meningitis caused by these organisms were premature; 15/18 were in early premature patients. We identified 14 species of Gram-negative organisms causing meningitis, including *Enterobacter cloacae* ($n = 7$; 18%), *Klebsiella pneumoniae* ($n = 7$; 18%), and *Salmonella enterica* ($n = 6$; 15%) (Table 2). The majority of these Gram-negative organisms were isolated from the CSF ($n = 33/40$, 83%). Approximately one-quarter (23%) of these organisms were resistant to 3rd generation cephalosporins; only one was tested for resistance to cefepime and was susceptible.

3.5. Gram-Positive Organisms Other Than GBS and Fungi

The most common single cause of bacterial meningitis outside of group B *Streptococcus* and *E. coli* was the *Streptococcus bovis* group ($n = 10/190$; 5%). *Staphylococcus* species, including *aureus* and *epidermidis* ($n = 8$; 4%), as well as *Enterococcus* ($n = 6$; 3%), accounted for the majority of remaining cases of bacterial meningitis. A number of patients with Gram-positive organisms that were not speciated were also noted ($n = 13$). The proportion of preterm ($n = 19$; 28%) and term ($n = 34$; 28%) infants were the same with regards to bacterial meningitis due to these organisms. Only five fungal meningitis cases were documented; these were all preterm patients, and all were caused by *Candida* species, specifically *Candida albicans* ($n = 4$) and *Candida lusitaneae* ($n = 1$) (Table 2).

3.6. Antimicrobial Resistance

Antimicrobial resistance was observed for all etiologies of bacterial meningitis with the exception of GBS. We observed resistance to both third-generation cephalosporins and ampicillin in *E. coli* ($n = 2/34$, 6%, 23/35, 66%, respectively) and in other Gram-negatives (8/35, 23%, 9/16, 56%, respectively) (Table 4 and Figure 3). Group B *Streptococcus* is reliably susceptible to penicillin, and, therefore, routine susceptibility testing is not performed at our hospital. On the other hand, one-third ($n = 3/10$, 30%) of *Streptococcus bovis* group isolates were resistant to penicillin. However, among those positive for *Staphylococcus* species, oxacillin resistance was observed in ($n = 11/17$, 65%), including two *S. aureus*. No isolates were found to be resistant to either vancomycin (Gram-positives) or fourth-generation cephalosporins (Gram-negatives).

Table 4. Gram-negative bacterial antimicrobial resistance.

Resistance Spectra (Total)	Ampicillin	3rd Generation Cephalosporin	Gentamicin	Piperacillin	Ciprofloxacin	4th Generation Cephalosporin (Cefepime)
E. coli (37)	23/35 (66%)	2/34 (6%)	3/33 (9%)	16/27 (59%)	5/30 (20%)	0/5
Other Gram-negative (38)	9/16 (56%)	8/35 (23%)	1/30 (3%)	12/24 (50%)	0/22	0/1

3.7. Risk Factors for Meningitis

We examined the proportion of patients with prematurity as a risk factor for meningitis. Preterm birth is associated with an increased risk for bacterial meningitis; 115 (17%) of all patients in this study with meningitis were preterm, including 68 (10%) who were early preterm (Table 5). More than one-third ($n = 68$; 36%) of infants < 90 days with bacterial meningitis were preterm, with 55 of these early preterm (<34 weeks). Bacterial meningitis represented more than three quarters (81%) of meningitis causes in all early preterm births. Meningitis patients born early preterm were statistically significantly more likely to have a bacterial etiology than those born late preterm ($p < 0.0001$; OR = 11.1, 95% CI 4.4–31.5) or full term ($p < 0.0001$; OR = 15.8, 95% CI = 8.1–32.3), while there was no significant difference when late preterm were compared to full-term infants ($p = 0.29$; OR = 1.4; 95% CI= 0.6–2.8). Fungal infections (*Candida* sp.) were only found in early preterm ($n = 4$) and late preterm ($n = 1$) infants with meningitis. Patients born early preterm were less likely to have an unknown cause of meningitis than patients born full-term or late preterm ($p < 0.0001$; OR = 0.2, 95% CI = 0.1–0.5).

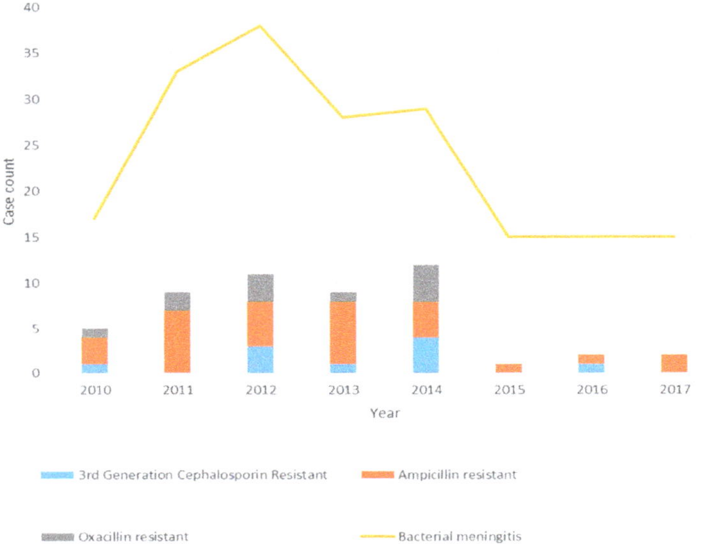

Figure 3. Bacterial meningitis and resistance patterns by year.

Table 5. Distribution of causes of meningitis by gestational age at birth.

	Early Preterm (≤34 Weeks Gestation) ($n = 68$)	Late Preterm (35–36 Weeks Gestation) ($n = 47$)	Full Term (37+ Weeks) ($n = 579$)
Viral	6 (9%)	23 (49%)	322 (56%)
Enterovirus and other viruses *	4 (6%)	20 (43%)	309 (53%)
HSV	2 (3%)	3 (6%)	13 (2%)
Bacterial	55 (81%)	13 (28%)	123 (21%)
Streptococcus agalactiae	13 (19%)	2 (4%)	45 (8%)
Staphylococcus species	8 (12%)	1 (2%)	9 (2%)
Other Gram-positive **	8 (12%)	2 (4%)	25 (4%)
E. coli	11 (16%)	5 (11%)	21 (4%)
Other Gram-negative **	15 (22%)	3 (6%)	22 (4%)
Fungal	4 (6%)	1 (2%)	0
Unknown	3 (4%)	10 (21%)	134 (23%)
Other ***	0 (0%)	0 (0%)	1 (0.2%)

* only one other virus was identified: West Nile virus in a full-term infant. ** Gram-positives include: Gram-positive bacteria, no species (13), *Streptococcus gallolyticus* (9), *Enterococcus faecalis* (6), *Streptococcus pneumoniae* (3), *Streptococcus mitis* (2), *Clostridium* species (1) *Streptococcus infantarius* (1); Gram-negatives include: *Enterobacter cloacae* (7), *Klebsiella pneumoniae* (7), *Salmonella enterica* (6), *Acinetobacter baumannii* (3), *Serratia marcescens* (3), Gram-negative rods, no species (2), *Neisseria meningitidis* (2), *Proteus mirabilis* (2), *Citrobacter braakii* (1), *Citrobacter freundii* (1), *Haemophilus influenzae* (1), *Klebsiella oxytoca* (1), *Pantoea* species (1), *Pseudomonas aeruginosa* (1), *Pseudomonas fluorescens* (1), *Morganella morganii* (1) *** other was attributed to IVIG meningitis.

A disproportionately high number of bacterial meningitis cases occurred in Black patients when compared to all other races ($p < 0.001$, 42% vs. 25%, OR = 2.2, 95% CI = 1.4–3.4) and correspondingly lower number of viral etiologies were observed in the Black population (38% in Black patients vs. 53% in non-Black patients, $p < 0.01$, OR = 0.55, 95%

CI = 0.4–0.9) (Figure 4). While Black patients did have a significantly higher proportion of early preterm births than did non-Black patients ($p < 0.01$, 17% vs. 8%, OR = 2.2, 95% CI = 1.2–4.1), Black patients remained more likely to have bacterial causes of meningitis even when controlling for premature status ($p < 0.01$, 34% vs. 19%, OR = 2.2, 95% CI = 1.3–3.8). No statistically significant differences were noted between races in the odds that bacterial meningitis was caused by GBS ($p = 0.68$, 34% vs. 31%, OR 1.16, 95% CI = 0.5–2.5). The proportion of cases of unknown etiology did not differ significantly across races/ethnicities (ranging from a low of 19% in Black patients to a high of 23% in White, Hispanic patients).

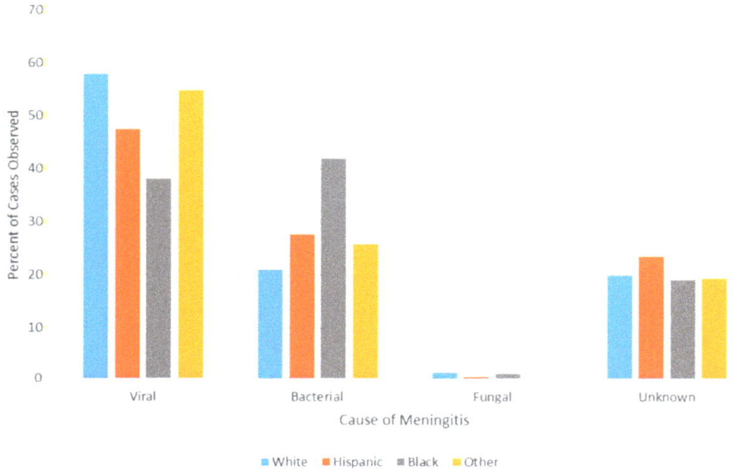

Figure 4. Meningitis cause by race.

3.8. Fatal Cases

Of the 14 deaths, nearly all occurred in patients with bacterial meningitis (13/14, 93%). Leading causes included GBS (6 deaths/60 GBS infections; 10%), *Staphylococcus aureus* (2/8; 25%), and *E. coli* (2/37, 5%). Single deaths occurred in cases of *Enterobacter*, *Serratia*, and *Pantoea* (3/40 of the non-*E. coli* Gram-negatives; 8%). The single fatality associated with viral infection was the result of HSV-2. The majority of deaths occurred in preterm infants (10/14, 71%).

4. Discussion

Overall, among infants 0 to <90 days of life treated for meningitis at Texas Children's Hospital from 2010 to 2017, viral meningitis was more common than bacterial meningitis. Enteroviruses were responsible for the majority of meningitis cases, and GBS and *E. coli* remained the most common causes of bacterial meningitis across all age and race categories. The absolute percentage of bacterial meningitis due to GBS (32%) and *E. coli* (19%) is lower than the percentages reported in other contemporary studies of meningitis or combined sepsis and meningitis in patients of equivalent age. Higher proportions of meningitis due to Gram-negative organisms other than *E. coli* were also noted [6–8]. Interestingly, *Streptococcus bovis* group pathogens were the third most common cause of bacterial meningitis, a finding much higher than other studies that have reported this pathogen [9,17]. Finally, the diversity of organisms responsible for infant meningitis in our patient population is greater than that found in other studies [7–9].

The substantial role played by GBS in the context of routine universal screening and intrapartum prophylaxis at our institution cannot be understated. The 60% prevalence of negative maternal screens in infants that later developed GBS meningitis was remarkably similar to prior studies of GBS disease [18]. Consistently, 71% of late-onset GBS meningitis cases whose mothers had a positive GBS screen and received treatment prior to delivery developed meningitis, as GBS screening and peripartum treatment is known to be ineffective for preventing late-onset GBS meningitis [19]. Other transmission routes for GBS infection have been suggested, and it is possible these contribute to late-onset disease [20,21]. A number of studies have called for a maternal vaccine for GBS to prevent infant sepsis or meningitis [11,22]. Such a vaccine, if effective, could have prevented more than 1/5th of all days of the length of stay due to meningitis, a substantial amount of ICU utilization and associated costs, as well as almost half of all fatalities in our study. Our findings support the need for a GBS vaccine for maternal immunization for the prevention of late-onset GBS disease in infants.

Racial disparities have been and continue to be a matter of substantial concern for meningitis patients. The higher frequency of bacterial meningitis compared to the typically more benign viral meningitis in the Black population leads to more severe disease outcomes. While previous studies have observed differences in the proportion of the five most commonly identified causes of bacterial meningitis in a pediatric population between Black and non-Black populations, large-scale, holistic assessment of all-cause meningitis has been lacking in the infant population and have not managed to so readily elucidate the racial disparity of these conditions [6]. Our observation of no difference in the proportion of bacterial meningitis caused by GBS between racial groups was somewhat surprising, given that Black women have been shown to have a higher rate of carriage of GBS and that GBS disease is also linked to the Black race [23–25].

The role of prematurity as a risk factor for bacterial meningitis should be noted. Bacterial meningitis was more frequently observed in early preterm patients, but we found no difference between late preterm and term infants in regard to the proportion of bacterial meningitis. In Texas, the number of preterm births has increased in recent years. In our study, 17% of patients with meningitis were born preterm, compared to the Texas state average of 10.6% for 2017 [26,27]. This may reflect the role of our institution as a referral center for newborns requiring a higher level of care, and our hospital includes a Pavilion for Women, an obstetric and maternal–fetal medicine referral center for high-risk pregnancies.

Etiologic diagnosis of meningitis patients, even in the case of the less severe enterovirus, is critical. Current Infectious Diseases Society of America (IDSA) guidelines recommend the empiric use of 3rd generation cephalosporins in combination with ampicillin in neonates or vancomycin in infants 2–3 months of age to treat bacterial meningitis. Obtaining rapid viral diagnoses such as enterovirus that accounts for ~50% of all cases can be helpful in ruling out bacterial meningitis and discontinuing unnecessary treatment. Inappropriate use of antibiotic therapy contributes to antibiotic resistance and is associated with increased toxicity such as renal dysfunction and prolonged hospitalization, resulting in increased cost.

Empiric treatment for neonatal meningitis in most parts of the world includes ampicillin and gentamicin or a 3rd generation cephalosporin. Given that we observed resistance to third-generation cephalosporins in Gram-negative organisms, the use of fourth-generation cephalosporins to provide adequate antimicrobial coverage for known and suspected cases of bacterial meningitis is warranted in some cases, such as in the empiric treatment of meningitis in early preterm and preterm infants, until organism identification and susceptibility testing results are available.

One limitation of our study is that retrospective reviews lack the ability to verify data first-hand. Another potential limitation could be related to selection bias and generalizability to other populations, as our status as a large referral hospital may have resulted in an unusual distribution of cases and causes of meningitis. This limitation could also be viewed as a strength, as we could evaluate a large sample size of patients from a diverse population. Our patient population had a high degree of racial diversity, allowing comparisons of the causes and outcomes of meningitis between races and ethnicities. Other strengths are also worth noting. This study examined all causes of meningitis in infants, as opposed to studies that focus solely on one etiology (bacterial, viral, or fungal). Additionally, given the availability of molecular diagnostic testing in addition to routine bacterial cultures, our hospital was able to identify the different causes of meningitis at a relatively high rate, which was valuable for determining the true epidemiology of meningitis in infants 0 to 90 days of age.

This study reports the findings of a large investigation into the etiology and epidemiology of meningitis in infants in the first 90 days of life. The majority of cases were caused by viral pathogens, predominantly enterovirus. However, mortality was primarily associated with bacterial causes, with changes in antimicrobial resistance patterns over time suggesting a need to consider broader spectrum coverage for Gram-negative meningitis in preterm infants given the possibility of non-*E. coli* Gram-negative infection.

Supplementary Materials: The following is available online at https://www.mdpi.com/2076-2607/9/3/526/s1, Table S1: oICD codes used to acquire case-patients.

Author Contributions: Conceptualization: T.A.E. and K.O.M.; methodology, T.A.E., K.O.M., R.H. and C.L.T.; validation, F.M.M.; writing—original draft preparation, T.A.E.; writing—review and editing, F.M.M., C.L.T., M.S.N., R.H., E.L.B. and K.O.M.; supervision, K.O.M.; funding acquisition, K.O.M. All authors have read and agreed to the published version of the manuscript.

Funding: This study was funded in part by the Chao Foundation.

Institutional Review Board Statement: The study was conducted according to the guidelines of the Declaration of Helsinki, and approved by the Institutional Review Board (or Ethics Committee) of Baylor College of Medicine (protocol code H-35069).

Informed Consent Statement: Patient consent was waived due to this being a retrospective chart review.

Data Availability Statement: These data are not made publicly available.

Acknowledgments: We wish to thank Stacia DeSantis for her kind input.

Conflicts of Interest: The authors declare no conflict of interest. The funders had no role in the design of the study; in the collection, analyses, or interpretation of data; in the writing of the manuscript, or in the decision to publish the results.

References

1. Hasbun, R.; Wootton, S.H.; Rosenthal, N.; Balada-Llasat, J.M.; Chung, J.; Duff, S.; Bozzette, S.; Zimmer, L.; Ginocchio, C.C. Epidemiology of Meningitis and Encephalitis in Infants and Children in the United States, 2011–2014. *Pediatr. Infect. Dis. J.* **2019**, *38*, 37–41. [CrossRef]
2. Cherry, J.D.; Demmler-Harrison, G.; Kaplan, S.L.; Steinbach, W.J.; Hotez, P.J. *Feigin and Cherry's Textbook of Pediatric Infectious Diseases*, 7th ed.; Saunders/Elsevier: Philadelphia, PA, USA, 2014.
3. Rotbart, H.A. Enteroviral Infections of the Central Nervous System. *Clin. Infect. Dis.* **1995**, *20*, 971–981. [CrossRef]
4. Lee, B.E.; Dele Davies, H. Aseptic meningitis. *Curr. Opin. Infect. Dis.* **2007**, *20*, 272–277. [CrossRef]
5. Murphy, T.V.; White, K.E.; Pastor, P.; Gabriel, L.; Medley, F.; Granoff, D.M.; Osterholm, M.T. Declining incidence of Haemophilus influenzae type b disease since introduction of vaccination. *JAMA* **1993**, *269*, 246–248. [CrossRef] [PubMed]
6. Thigpen, M.C.; Whitney, C.G.; Messonnier, N.E.; Zell, E.R.; Lynfield, R.; Hadler, J.L.; Harrison, L.H.; Farley, M.M.; Reingold, A.; Bennett, N.M.; et al. Bacterial Meningitis in the United States, 1998–2007. *N. Engl. J. Med.* **2011**, *364*, 2016–2025. [CrossRef]
7. Gaschignard, J.; Levy, C.; Romain, O.; Cohen, R.; Bingen, E.; Aujard, Y.; Boileau, P. Neonatal Bacterial Meningitis: 444 Cases in 7 Years. *Pediatr. Infect. Dis. J.* **2011**, *30*, 212–217. [CrossRef] [PubMed]
8. Okike, I.O.; Johnson, A.P.; Henderson, K.L.; Blackburn, R.M.; Muller-Pebody, B.; Ladhani, S.N.; Anthony, M.; Ninis, N.; Heath, P.T. Incidence, etiology, and outcome of bacterial meningitis in infants aged <90 days in the United kingdom and Republic of Ireland: Prospective, enhanced, national population-based surveillance. *Clin. Infect. Dis.* **2014**, *59*, e150–e157. [CrossRef] [PubMed]
9. Ouchenir, L.; Renaud, C.; Khan, S.; Bitnun, A.; Boisvert, A.-A.; McDonald, J.; Bowes, J.; Brophy, J.; Barton, M.; Ting, J.; et al. The Epidemiology, Management, and Outcomes of Bacterial Meningitis in Infants. *Pediatrics* **2017**, *140*, e20170476. [CrossRef]
10. MacNeil, J.R.; Cohn, A.C.; Farley, M.; Mair, R.; Baumbach, J.; Bennett, N.; Gershman, K.; Harrison, L.H.; Lynfield, R.; Petit, S.; et al. Current Epidemiology and Trends in Invasive Haemophilus influenzae Disease—United States, 1989–2008. *Clin. Infect. Dis.* **2011**, *53*, 1230–1236. [CrossRef] [PubMed]
11. Verani, J.R.; McGee, L.; Schrag, S.J. Prevention of perinatal group B streptococcal disease—Revised guidelines from CDC, 2010. *MMWR. Recomm. Rep.* **2010**, *59*, 1–36.

12. Collignon, P. Resistant Escherichia coli—We are what we eat. *Clin. Infect. Dis.* **2009**, *49*, 202–204. [CrossRef]
13. Chambers, H.F.; DeLeo, F.R. Waves of resistance: Staphylococcus aureus in the antibiotic era. *Nat. Rev. Genet.* **2009**, *7*, 629–641. [CrossRef]
14. *CDC/NHSN Surveillance Definitions for Specific Types of Infections*; National Healthcare Safety Network: Atlanta, GA, USA, 2018.
15. Arboviral Encephalitis or Meningitis 2001 Case Definition. Available online: https://wwwn.cdc.gov/nndss/conditions/arboviral-encephalitis-or-meningitis/case-definition/2001/ (accessed on 10 January 2021).
16. Bharath, V.; Eckert, K.; Kang, M.; Chin-Yee, I.H.; Hsia, C.C. Incidence and natural history of intravenous immunoglobulin-induced aseptic meningitis: A retrospective review at a single tertiary care center. *Transfusion* **2015**, *55*, 2597–2605. [CrossRef] [PubMed]
17. Beneteau, A.; Levy, C.; Foucaud, P.; Béchet, S.; Cohen, R.; Raymond, J.; Dommergues, M.-A. Childhood meningitis caused by Streptococcus bovis group: Clinical and biologic data during a 12-year period in France. *Pediatr. Infect. Dis. J.* **2015**, *34*, 136–139. [CrossRef] [PubMed]
18. Van Dyke, M.K.; Phares, C.R.; Lynfield, R.; Thomas, A.R.; Arnold, K.E.; Craig, A.S.; Mohle-Boetani, J.; Gershman, K.; Schaffner, W.; Petit, S.; et al. Evaluation of Universal Antenatal Screening for Group B Streptococcus. *N. Engl. J. Med.* **2009**, *360*, 2626–2636. [CrossRef] [PubMed]
19. Perinatal group B streptococcal disease after universal screening recommendations—United States, 2003–2005. In *Morbidity and Mortality Weekly Report*; Centers for Disease Control and Prevention: Atlanta, GA, USA, 2007; pp. 701–705.
20. Jawa, G.; Hussain, Z.; Da Silva, O. Recurrent Late-Onset Group B Streptococcus Sepsis in a Preterm Infant Acquired by Expressed Breastmilk Transmission: A Case Report. *Breastfeed. Med.* **2013**, *8*, 134–136. [CrossRef]
21. Zimmermann, P.; Gwee, A.; Curtis, N. The controversial role of breast milk in GBS late-onset disease. *J. Infect.* **2017**, *74*, S34–S40. [CrossRef]
22. Seale, A.C.; Bianchi-Jassir, F.; Russell, N.J.; Kohli-Lynch, M.; Tann, C.J.; Hall, J.; Madrid, L.; Blencowe, H.; Cousens, S.; Baker, C.J.; et al. Estimates of the Burden of Group B Streptococcal Disease Worldwide for Pregnant Women, Stillbirths, and Children. *Clin. Infect. Dis.* **2017**, *65* (Suppl. 2), S200–S219. [CrossRef]
23. Spiel, M.H.; Hacker, M.R.; Haviland, M.J.; Mulla, B.; Roberts, E.; Dodge, L.E.; Young, B.C. Racial disparities in intrapartum group B Streptococcus colonization: A higher incidence of conversion in African American women. *J. Perinatol.* **2019**, *39*, 433–438. [CrossRef] [PubMed]
24. Newton, E.R.; Butler, M.C.; Shain, R.N. Sexual behavior and vaginal colonization by group B streptococcus among minority women. *Obstet. Gynecol.* **1996**, *88 Pt 1*, 577–582. [CrossRef]
25. Schuchat, A. Epidemiology of Group B Streptococcal Disease in the United States: Shifting Paradigms. *Clin. Microbiol. Rev.* **1998**, *11*, 497–513. [CrossRef]
26. Martin, J.A.; Hamilton, B.E.; Osterman, M.J.K.; Driscoll, A.K.; Mathews, T.J. Births: Final Data for 2015. *Natl Vital Stat. Rep.* **2017**, *66*, 1. [PubMed]
27. Martin, J.A.; Hamilton, B.E.; Osterman, M.J.K.; Driscoll, A.K.; Drake, P. Births: Final Data for 2017. *Natl Vital Stat. Rep.* **2018**, *67*, 1–50. [PubMed]

 *microorganisms*

Review

A Narrative Review of the Molecular Epidemiology and Laboratory Surveillance of Vaccine Preventable Bacterial Meningitis Agents: *Streptococcus pneumoniae*, *Neisseria meningitidis*, *Haemophilus influenzae* and *Streptococcus agalactiae*

Raymond S. W. Tsang

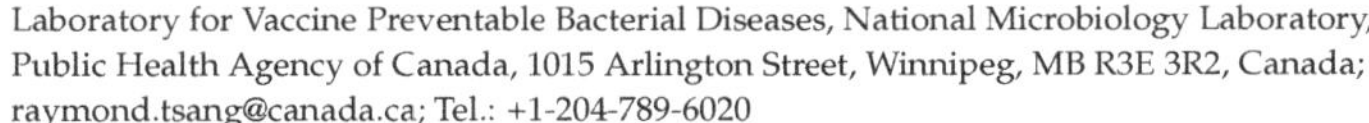

Laboratory for Vaccine Preventable Bacterial Diseases, National Microbiology Laboratory, Public Health Agency of Canada, 1015 Arlington Street, Winnipeg, MB R3E 3R2, Canada; raymond.tsang@canada.ca; Tel.: +1-204-789-6020

Abstract: This narrative review describes the public health importance of four most common bacterial meningitis agents, *Streptococcus pneumoniae*, *Neisseria meningitidis*, *Haemophilus influenzae*, and *S. agalactiae* (group B *Streptococcus*). Three of them are strict human pathogens that normally colonize the nasopharynx and may invade the blood stream to cause systemic infections and meningitis. *S. agalactiae* colonizes the genito-gastrointestinal tract and is an important meningitis agent in newborns, but also causes invasive infections in infants or adults. These four bacteria have polysaccharide capsules that protect them against the host complement defense. Currently licensed conjugate vaccines (against *S. pneumoniae*, *H. influenza*, and *N. meningitidis* only but not *S. agalactiae*) can induce protective serum antibodies in infants as young as two months old offering protection to the most vulnerable groups, and the ability to eliminate carriage of homologous serotype strains in vaccinated subjects lending further protection to those not vaccinated through herd immunity. However, the serotype-specific nature of these vaccines have driven the bacteria to adapt by mechanisms that affect the capsule antigens through either capsule switching or capsule replacement in addition to the possibility of unmasking of strains or serotypes not covered by the vaccines. The post-vaccine molecular epidemiology of vaccine-preventable bacterial meningitis is discussed based on findings obtained with newer genomic laboratory surveillance methods.

Keywords: bacterial meningitis; *S. pneumoniae*; *N. meningitidis*; *H. influenzae*; *S. agalactiae*; conjugate vaccines; post-vaccine surveillance

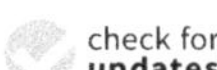

Citation: Tsang, R.S.W. A Narrative Review of the Molecular Epidemiology and Laboratory Surveillance of Vaccine Preventable Bacterial Meningitis Agents: *Streptococcus pneumoniae*, *Neisseria meningitidis*, *Haemophilus influenzae* and *Streptococcus agalactiae*. *Microorganisms* **2021**, *9*, 449. https://doi.org/10.3390/microorganisms9020449

Academic Editor: James Stuart

Received: 11 January 2021
Accepted: 16 February 2021
Published: 22 February 2021

1. Introduction

Pyogenic bacterial meningitis is a life threatening condition that can progress rapidly leading to death. When the disease happens in infants, children, and young adults, it may instill fear due to the contagious and potentially deadly nature of the disease especially in outbreak situation. The three most common causes of acute bacterial meningitis are *Streptococcus pneumoniae*, *Neisseria meningitidis*, and *Haemophilus influenzae* [1]. This group of bacterial meningitis agents can cause disease in all ages of life from newborn to the elderly. The global burden of meningitis disease in 2016 was estimated to be 2.82 million cases, and 318,400 deaths were attributed to meningitis. The three most common pathogens (*S. pneumoniae*, *N. meningitidis*, and *H. influenzae*) were responsible for 55.7% and 57.2% of the meningitis cases and deaths, respectively [2]. Besides meningitis, *S. pneumoniae*, *H. influenzae*, and *N. meningitidis* can cause other forms of invasive diseases such as bacteremic pneumonia, septicemia, septic arthritis, pericarditis, etc. The risk of developing a major (such as hearing loss, seizures, motor deficit, cognitive impairment, hydrocephalus, and visual disturbance) or a minor (learning difficulties, language impairment, developmental delay) sequela from bacterial meningitis was estimated to be 12.8% and 8.6%, respectively [3]. Meningitis caused by *S. pneumoniae* carried the highest risk with a major sequela (24.7%), followed by *H. influenzae* (9.5%) and *N. meningitidis* (7.2%) [3]. Using meningococcal disease (which carries the lowest risk of developing a major sequela) as an example, the cost to care for a case who developed a major sequela was estimated to be £160,000 (US$214,096) to £200,000 (US$267,620) for the first year alone; and the corresponding figure over the lifetime of a case may be as high as £590,000 (US$789,479) to £1,090,000 (US$1,458,529) [4]. Since the incidence of meningitis and the risk of developing sequela are much higher in low- and middle-income countries, and the resources to care for those meningitis patients who develop severe sequela are often lacking in these countries, vaccines are probably the most cost-effective strategy for the control and potentially elimination of this devastating and fearful disease.

Although a number of other bacterial agents can cause meningitis, such as *Listeria monocytogenes*, *Escherichia coli*, and other enteric bacteria, group B *Streptococcus* (*S. agalactiae*) is gaining attention as a frequent cause of either early or late onset of invasive diseases such as pneumonia, sepsis, or meningitis in the newborn [5,6] as well as various forms of invasive diseases in pregnant women and non-pregnant adults [5,7]. The World Health Organization (WHO) has also identified group B *Streptococcus* together with *S. pneumoniae*, *N. meningitidis*, and *H. influenzae* as the four major bacterial meningitis agents to be included in its work plan and global vision to defeat meningitis by 2030 [8].

Capsule-based protein-conjugate vaccines that target the major serogroups of *N. meningitidis* and serotypes of *H. influenzae* and *S. pneumoniae* causing invasive diseases are now available and implemented in vaccination programs in many countries [9–11]. As a result, the epidemiology of bacterial meningitis has changed with the number of cases caused by strains covered by the vaccine decreased dramatically but at the same time disease due to serogroups or serotypes of the pathogens not included in the vaccine has emerged [12]. Since disease surveillance has been described by the WHO as one of the five major pillars on the road map to defeat meningitis [8], the objectives of this report are to describe (i) features of *S. pneumoniae*, *N. meningitidis*, *H. influenzae*, and *S. agalactiae* that may have implications for vaccination and surveillance; (ii) currently licensed vaccines against *S. pneumoniae*, *N. meningitidis*, and *H. influenzae*; (iii) changes in the epidemiology of invasive diseases caused by these three pathogens; (iv) traditional and newer laboratory surveillance methods; and (v) how lessons learned from surveillance of the three most common bacterial meningitis agents can inform the pre- and post-vaccine licensure surveillance of invasive group B *Streptococcus* (GBS) disease when capsule polysaccharide conjugate vaccines against GBS have been developed and are in clinical trials [5,13].

2. Characteristics of *S. pneumoniae*, *N. meningitidis*, *H. influenzae*, and *S. agalactiae* Important for Vaccination and Surveillance

S. pneumoniae, *N. meningitidis*, and *H. influenzae* are respiratory pathogens that normally colonize the human respiratory tract where they serve as a reservoir of infection [14–16]. Another common characteristic of these three invasive bacterial agents is the polysaccharide capsules on their cell surface, which serve as serotyping antigens. The serotypes are traditionally identified by anti-capsular antibodies using agglutination methods (or the Quellung reaction for *S. pneumoniae*). The capsules also serve as protective antigens shielding the bacteria from the human host defense like phagocytosis and complement activation [17,18]. As the protective antigen, vaccines based on the capsule have been developed to target the most common serotypes of *H. influenzae*, *N. meningitidis*, and *S. pneumoniae* causing invasive infections [9–11]. Another feature that makes these bacteria successful pathogens is the plasticity of their genome and their recombinant nature [19–21].

Unlike *S. pneumoniae*, *N. meningitidis*, and *H. influenzae*, *S. agalactiae* colonizes the human genito-gastrointestinal tract. Not only does it cause meningitis in the newborn and various forms of invasive diseases in infants and adults, *S. agalactiae* is also known to cause disease in cattle [22,23] and may have the potential to transmit to human as a zoonotic pathogen [24]. Similar to *S. pneumoniae*, *N. meningitidis*, and *H. influenzae*, *S. agalactiae* also has a surface polysaccharide capsule that acts as virulence factor and protective antigen [5,13]. Its genome is also prone to participate in recombination events [25].

3. Currently Licensed Vaccines for Control of Bacterial Meningitis

Currently there are 6 serotypes of *H. influenzae* recognized [26], 10 serotypes of *S. agalactiae* [5,13], 12 serogroups for *N. meningitidis* [27], and 100 serotypes for *S. pneumoniae* [28,29] (Table 1). Non-encapsulated strains also exist in all three species, and are termed non-typeable (for *H. influenzae* and *S. pneumoniae*) or non-groupable (for *N. meningitidis*). Currently licensed vaccines to control some strains of *S. pneumoniae*, *N. meningitidis*, and *H. influenzae* are listed in Table 2, while vaccines for protection against *S. agalactiae* are not licensed yet but are in advanced stages of clinical trials for maternal immunization [30,31].

Table 1. Capsular antigens of *Haemophilus influenzae*, *Neisseria meningitidis*, *Streptococcus agalactiae*, and *S. pneumoniae*.

Organism	Capsular Serotype/Serogroup Antigens	Reference
H. influenzae	Serotypes a, b, c, d, e, and f,	Pittman, 1931 [26]
N. meningitidis	Serogroups A, B, C, E, H, I, K, L, W, X, Y, and Z	Harrison et al., 2013 [27]
S. agalactiae	Serotypes Ia, Ib, II, III, IV, V, VI, VII, VIII, and IX	Song et al., 2018 [5]; Lin et al., 2018 [13]
S. pneumonia *	100 serotypes have been identified and only a few are listed here; Serotypes 1, 2, 3, 4, 5, 6A, 6B, 6C, ... 11E, 20B, . . . 35D, 7D, 10D (complete list can be found in the references provided)	Geno et al., 2015 [28]; Ganaie et al., 2020 [29]

* 100 different serotypes identified, please see references for full list.

Table 2. Licensed vaccines * against *Haemophilus influenzae*, *Neisseria meningitidis* and *Streptococcus pneumoniae*.

Origanism	Vaccine Type	Serotype/Serogroup Targets	Protein Carrier †	Year First Licensed
H. influenzae	Hib conjugate	b	TT, OMP	1987
N. meningitidis	tetravalent polysaccharide	A, C, Y and W	None	1974
N. meningitidis	monovalent C conjugate	C	CRM_{197}, TT	1999
N. meningitidis	monovalent A conjugate	A	TT	2010
N. meningitidis	tetravalent conjugate	A, C, Y and W	CRM_{197}, DT, TT	2005
N. meningitidis	4 component MenB	B	protein base vaccine	2013
N. meningitidis	factor H binding protein	B	protein base vaccine	2018
S. pneumoniae	PCV7 conjugate	4, 6B, 9V, 14, 18C, 19F, 23F	CRM_{197}	2000
S. pneumoniae	PCV10 conjugate	1, 4, 5, 6B, 7F, 9V, 14, 18C, 19F, 23F	CRM_{197}, TT, DT	2009
S. pneumoniae	PCV13 conjugate	1, 3, 4, 5, 6A, 6B, 7F, 9V, 14, 18C, 19A, 19F, 23F	CRM_{197}	2011
S. pneumoniae	PPV23 plain polysaccharide	1, 2, 3, 4, 5, 6B, 7F, 8, 9N, 9V, 10A, 11A, 12F, 14, 15B, 17F, 18C, 19A, 19F, 20, 22F, 23F, 33F	None	1983

* Except for the 4 component MenB and the factor H binding protein vaccines, all the other vaccines described in this Table are polysaccharide based vaccines. † protein carriers include TT (tetanus toxoid), DT (diphtheria toxoid), CRM_{197} (mutant diphtheria toxoid), OMP (outer membrane protein of *N. meningitides*).

The first bacterial meningitis vaccine developed was solely polysaccharide-based vaccine against serotype b *H. influenzae* (Hib) but it was soon discovered that plain polysaccharide vaccines are T-cell-independent antigens and do not induce protective antibodies in infants less than two years old [32], the most vulnerable age group for developing meningitis and invasive disease [33,34]. Coupling of the capsular polysaccharide to a protein carrier converts the vaccine to a T-cell dependent antigen that induces protective antibodies in infants as young as two months of age [10]. Another characteristic of the capsular polysaccharide vaccines is they are serotype-specific and offer protection against infection by the homologous serotype and do not offer protection against heterologous serotypes. Besides preventing invasive infections, the conjugate vaccines also reduce or eliminate respiratory carriage of and hence offer herd immunity to the larger community for the serotypes of these pathogens included in the vaccines [35–37]. The fact that conjugate vaccines are serotype-specific and can eliminate nasopharyngeal carriage of the homologous serotypes means their protective coverage is limited to the serotypes included in the vaccine and they can also alter the bacterial flora in the nasopharynx of vaccinated subjects.

The choice of which serotypes or serogroups to be included in the vaccines are based on the fact that not all serotypes or serogroups are equally virulent nor have the same prevalence in causing invasive diseases. For example, animal infection with isogenic mutants of *H. influenzae* that expressed different capsule serotype antigens has shown that serotype b is the most virulent, followed by serotype a [38]. In addition, most *N. meningitidis* isolates recovered from normally sterile body sites of invasive meningococcal disease (IMD) patients belong to six of the 12 recognized serogroups (A, B, C, W, X, and Y) [39,40]. Before the introduction of pneumococcal conjugate vaccines (PCVs), 10 serotypes (1, 4, 5, 6A 6B, 14, 18C, 19A, 19F, and 23F) were responsible for at least 50% of all invasive pneumococcal disease isolates from six different parts of the world; and in one region, they were responsible for over 80% of their invasive pneumococci [41].

Even before vaccine introduction, temporal and geographical variations in the serogroups of *N. meningitidis* responsible for IMD is well documented [39,40]. Differences in the serotypes involved in invasive pneumococcal disease (IPD) have also been reported from different parts of the world [42,43]. Before Hib conjugate vaccines were introduced, most invasive *H. influenzae* diseases were caused by Hib [33,34].

4. Effects of Vaccine Pressure, and Immune and/or Antibiotic Selection

Since currently licensed conjugate vaccines against *S. pneumoniae*, *N. meningitidis*, and *H. influenzae* do not offer universal coverage for all the serotypes or serogroups, immune pressure and selection against these pathogens can be expected to happen in their natural habitat as well as in the serotypes and serogroups that will cause invasive disease in the post-vaccine period.

In the presence of vaccine pressure, these bacterial pathogens may evolve to adapt by mainly two mechanisms that affect their capsule antigens (the vaccine targets): Capsule or serotype switching, and capsule or serotype replacement. Capsule switching involves two strains of a species exchanging their capsule polysaccharide synthesis (*cps*) genes resulting in a swap of their capsule antigens. For example, as shown in Figure 1a, a strain of genetic linage 1 and with a vaccine type capsule (depicted in green) exchanges its *cps* genes with a strain of genetic lineage 2 and with a non-vaccine capsule type (depicted in red). The end result will be the genetic linage 1 strain now carries *cps* genes for non-vaccine capsule type and expresses the non-vaccine capsule (red); while conversely the genetic linage 2 strain now expresses vaccine type capsule (green). Both *S. pneumoniae* and *N. meningitidis* have been reported to have capsule switching occurring spontaneously in the absence of vaccine pressure or, i.e., such capsule switching events have been reported prior to conjugate vaccine introduction [44,45]. Capsule switched strains can also be selected for by vaccine induced immune pressure and/or by wide spread antibiotic use if the capsule switched strain carries antibiotic resistance genes [41]. After capsule switching, the recipient strain will retain its original genetic background (usually determined by multilocus sequence typing) [46] but expresses a different, e.g., non-vaccine type of capsule. Frequent capsule switch in *N. meningitidis* from serogroup C to serogroup B, if it happens in a hypervirulent clone like ST-11, may be problematic since there are no capsule-based serogroup B meningococcal vaccines and protein-based meningococcal vaccines against serogroup B may not provide universal coverage against all serogroup B strains [47].

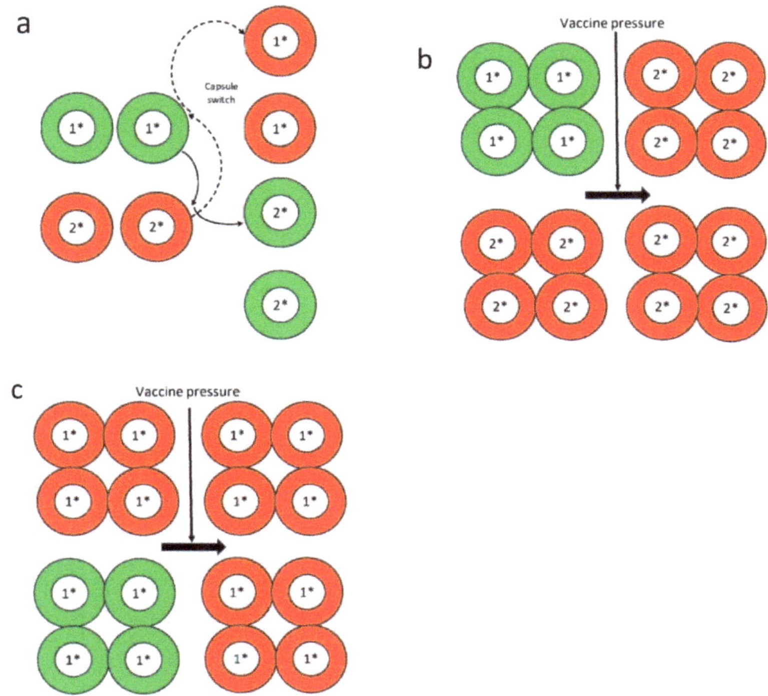

Figure 1. (**a**) Depiction of capsule switching between a genetic lineage 1* strain with a vaccine-type capsule (colored green) and a genetic lineage 2* strain with a non-vaccine type capsule (colored red). Diagram based on information from Swartley et al. [48]. (**b**) Illustration of capsule replacement when the vaccine capsule type (colored green) strain of genetic lineage 1* is removed by the vaccine, leaving the strain of genetic lineage 2* with non-vaccine capsule (colored red) to remain and proliferate. Diagram based on information from Lipstich M [49]. (**c**) Another scenario of capsule replacement when strain of genetic lineage 1* with both vaccine (colored green) and non-vaccine (colored red) capsule types are present before vaccine introduction and after vaccine use, only the non-vaccine capsule type of genetic lineage 1* strain remains. Diagram based on information from Lipstich M [49].

Capsule replacement happens when strains with vaccine capsule types that used to inhabit the nasopharynx have been eliminated by the conjugate vaccines and are now being replaced by strains expressing the non-vaccine capsule types. As a result, strains of non-vaccine capsule types may increase in prevalence and eventually cause disease. Strains with vaccine capsule type (depicted as green) and non-vaccine capsule type (depicted as red) can co-exist prior to the use of conjugate vaccines and the two capsule types may be of the different (Figure 1b) or the same (Figure 1c) genetic lineage. After selection by vaccine pressure, only strains of the non-vaccine capsule type (red) remain and expand to fill the void that used to be occupied by strains of the vaccine type (Figure 1b). If strains of the same genetic lineage that expressed both vaccine and non-vaccine capsule types exist prior to vaccine use, then strains of the same genetic lineage still persist after vaccine use but they only carry the non-vaccine capsule type (Figure 1c). The phenomenon illustrated in

Figure 1b,c is sometimes referred to as "unmasking" versus true replacement when only the vaccine capsule type exists in the nasopharynx before vaccine use, and after vaccine use, the vaccine capsule type strain is removed and the void in the nasopharynx is being replaced by a non-vaccine capsule type strain. Strains of non-vaccine capsule types can also be further selected by widespread antibiotic use if they carry the corresponding antibiotic resistance genes [41].

5. Molecular Epidemiology of Invasive Pneumococcal Disease (IPD), Invasive Meningococcal Disease (IMD), and Invasive *H. influenzae* Disease in the Post Conjugate Vaccine Era

5.1. IPD in the Post PCV Era

Although PCVs are very effective in reducing the burden of IPD caused by vaccine serotypes in many countries, IPD due to non-vaccine serotypes is still a concern. The emergence of IPD due to non-vaccine serotypes may be due to either unmasking effect when vaccine serotypes are removed from their natural habitat of the human nasopharynx, allowing non-vaccine serotypes (which already exist) to expand and occupy the nasopharynx; or by replacement due to the non-vaccine serotypes which do not exist in the nasopharynx prior to PCV introduction but emerge to fill the void in the nasopharynx left behind by the vaccine type [50]. For example, prior to PCV7 introduction, serotypes 19A (a non-PCV7 vaccine serotype) existed at a level of about 7.5% in 2001 and increased to 16% in 2007 after PCV7 was introduced in 2000, before declining to about 3% in 2014 after PCV13 was introduced in 2010 [51]. The other mechanism responsible for the emergence of non-vaccine serotypes is genetic recombination between strains leading to capsule switching. This was illustrated in the emergence of some serotype 19A strains after PCV7 introduction by a genetic recombination between a vaccine covered serotype 4, sequence type (ST)-595 recipient strain with a donor strain of non-vaccine serotype 19 ST-199, providing the recipient ST-595 strain with the non-vaccine capsule serotype 19A [52]. Another study also demonstrated that serotype 19A may again escape the PCV13 selection by a further switch to serotype 15B [53]. Although in this later study the genetic recombination occurred in strains from pre-PCV7 period; nevertheless the mechanism of genetic recombination with a non-vaccine capsule type is present in pneumococci. The pneumococcal capsule locus is a hotspot for mutation including exhibiting a higher rate of genetic recombination compared to the rest of the pneumococcal genome [54]. However, pneumococcal capsule locus recombination that leads to capsule serotype switch does not appear to be random. For example, capsule switch between strains within a serogroup occurred more often than serotype switch involving strains between different serogroups [55]. Since many factors may govern the pneumococcal population structure and the associated serotypes, some have suggested the existence of epistatic factor contributing to the dynamic of the pneumococcal capsule genetics [55,56].

Another mechanism may explain the persistence of some vaccine serotype in the post PCV period. For example, serotype 3 (included in the PCV13 vaccine) persisted in the nasopharyngeal samples as well as in specimens from IPD patients despite PCV13 usage [51,57,58]. Genome sequencing of serotype 3 isolates obtained prior to and after introduction of PCV13 showed a different clade of serotype 3 has emerged in the post PCV13 period despite the fact that both pre- and post-PCV13 isolates were typed by MLST to belong to the same ST-180 clonal complex (CC). However, the new clade has been shown to have sub-capsular protein antigen changes, which could explain strains of the new clade have adapted to exist despite the presence of immunity induced by PCV13 [57].

Regardless of the vaccine escape mechanism, various non-PCV serotypes have emerged in places where PCV immunization programs have been implemented reflecting geographical differences in serotype prevalence and distribution [59]. Non-PCV serotypes like serotype 2, 8, 10A, 11A, 12F, 15A, 15B/C. 16F, 22F, 24F, 33F, and 35B/D, have been described as causes of IPD [41,59–63]. To deal with this increase in non-PCV serotypes, 15-valent and 20-valent PCVs have been developed and are now in early clinical trials [64,65]. However, in the post PCV era, predominance by a single or a few serotypes as causes of IPD was not observed. Instead, increase serotype diversity of invasive pneumococci recovered from IPD cases has been observed [66], which may challenge the usefulness of increasing the valency of PCVs. Two editorials in 2007, "Invasive pneumococcal disease, the target is moving" [67] and "Serotype replacement in invasive pneumococcal disease: where do we go from here?" [68] appear to be just as relevant today after two decades of PCV use. Indeed, expert comments in 2021 still wrestle with the changing epidemiology of IPD due to shifting serotypes, and identify continuous surveillance as an important function in the control of IPD [69,70]. Ideally, a pan-pneumococcal universal vaccine would solve the problem of chasing after the emergence of non-vaccine serotypes as causes of IPD.

5.2. IMD in the Post Conjugate Vaccine Era

In the US, quadrivalent (A, C, W, Y) meningococcal conjugate vaccine was licensed in 2005 and recommended for the 11 to 18 years age group [71]. In the post-quadrivalent conjugate vaccine period of 2006-2010, no capsule or serogroup replacement was detected [72]. In the Canadian province of Quebec, outbreak due to a serogroup B strain of ST-269 appeared in 2013 [73] after two rounds of province wide vaccination against serogroup C meningococci (MenC) (first with plain polysaccharide vaccine in 1992-1993 and then with the MenC-conjugate vaccine in 2011) for control of outbreaks due to the hyper-virulent strain of ET-15 (ST-11) [74,75].

In Europe, serogroup B *N. meningitidis* was responsible for most IMD (73.6% in 2011) while an increase in serogroup Y IMD has been reported in a number of European countries [40]. Beginning in 2013, an increase in IMD due to serogroup W meningococci (MenW) has been reported from across Europe with both incidence rates of disease and the proportion of IMD isolates due to serogroup W showing yearly increase [76]. This increase in MenW disease in Europe was due to the introduction of a new ST-11 strain (different from the Hajj strain, which emerged during the 2000 Hajj pilgrimage in Saudi Arabia) from South America into the UK with further diversification to the 2013 UK strain, which spread through Europe [76,77]. This new MenW strain has also been reported to cause an increase in IMD in both Australia and Canada [78,79]. Expansion of a penicillin-resistant MenW ST-11 clone has also been described [80].

Before introduction of the monovalent meningococcal serogroup A conjugate vaccine, MenAfriVac, in 2010 [81], most meningococcal epidemics in Africa were mostly caused by serogroup A *N. meningitidis* (MenA). However, a serogroup X meningococcus (MenX) epidemic was reported from southwest Niger in 2004-2006 [82], and subsequently MenX outbreaks had occurred in Burkina Faso, Niger, Togo, and Uganda [83]. The MenX outbreak strain has been characterized as ST-181 CC with high experimental animal pathogenicity [84]. After introduction of the MenAfriVac, epidemics due to serogroups C, W, and X meningococci have been reported in the African meningitis belt countries. [85,86]. The MenC strain appeared to be a new strain typed as ST-10217, which has been shown to have arisen from a nongroupable strain recovered from a healthy carrier in Burkina Faso in 2012 prior to the emergence of the MenC ST-10217 and the MenC outbreak in 2013 [87]. The MenW strain causing outbreaks in Africa has been studied, and it appeared to be related to, and to have diversified from the 2000 Hajj strain [88].

Longitudinal carriage studies have been carried out in Africa to understand the epidemic nature of meningococcal meningitis in recent years both before and after the introduction of MenAfriVac in 2010. In the study carried out in one district of northern Ghana over the period of 1998 to 2005 before MenAfriVac was introduced, it was found that the colonized meningococcal population changed with time and matched temporally with the strain causing epidemics in the region [89]. Three successive waves of colonized meningococci were observed with ST-5 MenA, followed by ST-751 MenX, and ST-7 MenA. In the study to assess the effect of immunization with MenAfriVac, the carriage study has shown that the vaccine was both effective in control of MenA disease and in elimination of MenA from the respiratory tract of healthy carriers up to six or seven years after vaccine introduction. Like the other longitudinal carriage study in Ghana, a small percentage of oropharyngeal samples contained MenW of ST-11 CC (0.48%) and MenC of ST-10217 CC (0.10%) [90]. These studies certainly pointed to the importance of meningococci in healthy carriers as contributors of infection and potential sources of epidemics; and that conjugate vaccination may further change the population of meningococci in the normal habitat of the human upper respiratory tract.

5.3. *Invasive H. influenzae Disease in the Post Hib Conjugate Vaccine Era*

Following introduction of the Hib conjugate vaccine in the early 1990s, the epidemiology of invasive *H. influenzae* disease in those countries with Hib vaccination programs have changed substantially in the past three decades. Non-typeable or non-encapsulated *H. influenzae* (NT-Hi) is now the most frequent cause of invasive *H. influenzae* disease worldwide [91]. In Europe, during the period of 2007 to 2014, NT-Hi was the most common type identified but 74.1% of their invasive encapsulated *H. influenzae* were typed as serotype f, followed by serotype e (21.4%) [92]. In contrast, in the U.S., although NT-Hi is also the most common cause of invasive *H. influenzae* disease, the incidence of serotype a invasive *H. influenzae* disease has increased by 13% annually during the period of 2002 to 2015 [93].The incidence of invasive disease caused by NT-Hi has increased by 3% annually while incidence of invasive *H. influenzae* disease due to other serotypes was either stable or decreasing. The global presence of serotype a *H. influenzae* (Hia) has been documented [94,95], and the severity of invasive Hia disease has been described [96–98] which called for a Hia vaccine development [99].

Genetic analysis of Hia has revealed a population biology very similar to Hib, i.e., (a) with two phylogenetic populations similar to the clonal divisions I and II descried for Hib; and (b) with most invasive Hia isolates clustered together in a phylogenetic population (named clonal division I as for the majority of invasive Hib strains), represented by isolates typed by MLST as ST-23 and many STs related to ST-23 as single, double, or triple locus variants [100,101]. Another clone within this larger genetic population of clonal division I and identified by MLST as ST-4 has been reported in Brazil to be associated with more severe disease and higher case fatality rate [102]. In contrast to clonal division I Hia, clonal division II Hia is rarely isolated from invasive disease cases in Canada [101] and has not been found associated with invasive disease in Alaska [100]. However, clonal division II Hia identified by MLST as ST-62 has been found in 75% (21/28) of the Hia invasive disease case isolates from children < 18 years old in Utah, United States [103]. This may suggest unique geographical distribution of Hia genotypes.

To understand the emergence of NT-Hi as a cause of invasive disease in the post Hib conjugate vaccine era, comparative genome studies have revealed that NT-Hi showed much higher genetic diversity when compared to Hib or other serotypes that have been regarded as more genetically conserved or clonal [104,105]. Non-encapsulated *S. pneumoniae* have also been reported to have higher genetic diversity probably as a result of higher rates of genetic recombination as the capsule may serve as a barrier for foreign DNA uptake [21,106]. The higher genetic diversity of non-encapsulated *H. influenzae* may offer better adaptation to the host, e.g., by evading host immunity.

6. Laboratory Surveillance of *S. pneumoniae*, *N. meningitidis*, and *H. influenzae*

It is with this background of a changing microbial ecology in the nasopharynx of the human host as the bacterial pathogens are adapting to the vaccine pressure that laboratory surveillance of invasive bacterial meningitis pathogens are becoming increasingly important as well as challenging. The molecular typing methods for outbreak detection and surveillance of IMD, IPD, and invasive *H. influenzae* disease have been reviewed a decade ago with a focus on DNA sequencing methods [107]. They can be briefly summarized below as:

(1) Serogrouping and serotyping by the conventional method of using antisera to detect the capsular antigens, through either bacterial agglutination or Quelling reactions, has the value of detecting expression of the capsule antigens albeit the method may

sometimes be inaccurate [108]. Molecular method like PCR has been introduced to improve the detection and identification of serotypes, including that for *S. agalactiae* [109–112].

(2) Clonal analysis by multilocus sequence typing (MLST) is available for *S. pneumoniae*, *N. meningitidis*, and *H. influenzae* [46,113,114]. Strains are typed as STs and related STs are grouped together to form a CC.

(3) Target gene sequencing for fine typing:

The following genes have been proposed for typing of meningococci: *fetA*, which encodes an iron-regulated outer membrane protein; *porA*, which encodes the class 1 outer membrane protein PorA; *porB*, which encodes the class 2/3 outer membrane protein PorB [107]; and the newer protein-based MenB vaccine target genes, *fHbp*, *nhba*, and *nadA*, which encode for factor H binding protein, Neisseria heparin binding antigen, and the Neisseria adhesion A, respectively [115,116].

For *S. pneumoniae*, the *ply* and *lytA* genes, which encode for the pneumolysin and the autolysin, respectively; as well as the *pspA* gene, which encodes for the pneumococcal surface protein A, have been proposed as targets for potential typing purposes [107]. PspA has been reported to be associated with virulence and invasiveness of pneumococci [117]. Other gene markers associated with virulence have been suggested, including *pspC*, which encodes for the pneumococcal choline-binding protein C (PspC), for association with invasiveness of strains [117]. The *slaA* gene, which encodes for phospholipase A2, and four contiguous genes, one of which predicted as *pblB* that encodes a prophage tail protein, were either associated with the clinical disease of meningitis, or 30-day mortality rate, respectively [118].

The following genes have been proposed for typing of *H. influenzae*: *ompP2*, *ompP5*, *hmw1*, and *hmw2* [107]. The *omp2* and *omp5* genes encode for two different outer membrane proteins, a porin and a OmpA family protein, respectively. The *hmw1* and *hmw2* encode for HMW1 and HMW2, which are surface adhesion proteins (HMW stands for high molecular weight). A number of potential vaccine candidates have also been identified and they may have potential as further typing targets [119].

(4) Antibiotic susceptibility profile: Antibiogram can serve as a typing tool but more usefully in direct patient care as well as for surveillance purpose. Testing can be done by the disk diffusion method or quantitatively by the dilution assays (broth or agar dilution methods). Guidelines for the testing methods including the culture media, classes of antibiotics to be tested, as well as the interpretation of results have been published by both The European Committee on Antimicrobial Susceptibility Testing (EUCAST) [120] and the Clinical Laboratory Standards Institute's (CLSI's) Subcommittee on Antimicrobial Susceptibility Testing (AST) [121].

Besides the phenotypic methods, genetic prediction of antibiotic susceptibility has also been described by Harrison et al. [107]. Of the genes associated with decreased susceptibility or resistance to different types of antibiotics, the penicillin binding protein genes of *S. pneumoniae* that determines susceptibility towards penicillin are of special interest. The *cps* locus of *S. pneumoniae* is flanked by two of the penicillin binding protein genes, *pbp1a* and *pbp2x*; their juxtaposition sometimes allow capsule switching and transfer of the penicillin resistant genes to occur in a single recombination event.

7. Whole Genome Sequencing (WGS) for Molecular Epidemiology and Genomic Surveillance of Vaccine-Preventable Bacterial Meningitis Agents

For quite some years, MLST has been proven useful to classify isolates into clonal types and it has been applied to identify hypervirulent clones [122,123] and capsule switching events between serogroups or serotypes [44,45]. However, intra-clonal variations have been described, which may have implications in our understanding of the changing epidemiology of these vaccine-preventable diseases [57,117,124].

With the first bacterial genome sequenced and published in 1995 [125], there has been a very rapid development over the last two decades in sequencing technologies that include cost reduction as well as web-based bioinformatics platforms and pipelines to assemble and analyze genome sequences. As such, genome sequencing has now become a standard laboratory tool to study microbes. Many of our current understanding of the molecular epidemiology of a number of infectious diseases, including the common bacterial meningitis agents of *S. pneumoniae, N. meningitidis, H. influenzae,* and *S, agalactiae* are based on data obtained by whole-genome sequencing projects.

However, to make WGS a routine laboratory tool for the global surveillance of vaccine-preventable bacterial meningitis, additional work may still be required on standardization and harmonization of methodology, data analysis and nomenclature on top of the issues of data ownership, and confidentiality. A global partnership to study the genomes of *S. pneumoniae* has published an international definition of pneumococcal lineage [126]. For *N. meninigitidis*, nomenclature is currently based on historical convention [127] and an international committee is responsible for naming clonal complexes (https://pubmlst.org/organisms/neisseria-spp/further-info (accessed on 15 December 2020)). Similar development for *H. influenzae* appears to be lacking for now. Traditional analysis of the population biology of encapsulated *H. influenzae* divided them into two clonal divisions [128], and WGS analysis of the recently emerged serotype a *H. influenzae* also revealed two populations [101] like the two clonal divisions described using multilocus enzyme electrophoresis of Hib [128]. However, the definition of lineages of non-typeable *H. influenzae* may need further study and discussion because their genetic background appear to be much more diverse than the encapsulated or serotypeable strains of *H. influenzae* [129].

WGS data can be used to predict results obtained by the traditional surveillance methods. Use of WGS to predict serotype of *S. pneumoniae* [130] and *H. influenzae* [131] as well as serogroup of *N. meningitidis* [132] have been described. A platform that uses WGS data for determination of MLST ST and clonal analysis has also been developed [133]. Use of WGS data to identify genetic typing markers and virulence factors has also been published [105,118]. Pipelines to apply WGS to predict antibiotic susceptibility of bacterial pathogens have been developed [134,135]. Improved sequencing technology has allowed direct non-culture genome sequencing from clinical specimens to identify the cause of culture negative fulminant fever [136]. This metagenomics approach has been applied to investigate a meningococcal outbreak in Liberia and the genome data identified the outbreak strain as identical to the unique serogroup C meningococcal strain causing outbreaks in West Africa [137]. The experience gained from WGS studies of *S. pneumoniae, N. meningitidis,* and to a lesser extent *H. influenzae,* would help to inform and prepare for the pre- and post-vaccine introduction surveillance of *S. agalactiae.* The platforms built for *S. pneumoniae, N. meningitidis,* and *H. influenzae* will likely shorten the deployment of these technologies to study *S. agalactiae.*

To enable a whole genome nucleotide sequence-based surveillance tool to complement conjugate vaccines in the global effort to defeat meningitis, a WHO-led partnership called Global Meningitis Genome Partnership (GMGP) was formed to coordinate, assist, and develop guidelines for using WGS data to identify and track the global epidemiology of common bacterial meningitis agents of *S. pneumoniae, N. meningitidis, H. influenzae,* and *S. agalactiae* [138]. This collaborative approach has the potential of building synergy between the international partners to achieve the goal of defeating vaccine-preventable bacterial meningitis by 2030.

8. Chemoprophylaxis, Corticosteroids, and Experimental Immune Modulating Approaches for Prevention and as Adjuvant Therapeutic Agents of Bacterial Meningitis

Although vaccines remain the primary tool to offer active protection against infections, chemoprophylaxis can prevent secondary cases by offering protection to close contacts and household members of index cases. Chemoprophylaxis can also offer protection to those immunized subjects before adequate level of adaptive immunity can be developed. Guidelines that define household members and close contacts of index cases of Hib and IMD as well as the choice and dosage of prophylactic antibiotics have been published [139,140]. For those requiring chemoprophylaxis to prevent IMD, a single dose of ciprofloxacin is recommended, or rifampicin given twice daily for two days as an alternative. Other prophylactic antibiotics may include ceftriaxone, cefixime, and azithromycin. IMD patients treated with benzylpenicillin (which may not eliminate pharyngeal meningococci) are recommended to receive chemoprophylaxis that can eliminate nasopharyngeal carriage of meningococci before hospital discharge to prevent potential transmission to household members. Rifampicin once a day for four days or ciprofloxacin twice a day for five days are recommended prophylactic antibiotics for contacts of index cases of Hib. Other effective antibiotics may include ceftriaxone and azithromycin. Chemoprophylaxis is generally not recommended for close contacts of IPD patients. However, children with increased risk of IPD such as those with asplenia or sickle cell disease should receive daily prophylaxis with oral penicillin [141]. Public Health England also has guidelines of infection control, vaccination, and chemoprophylaxis (with rifampicin, penicillin, or azithromycin) for high risk individuals living in closed settings when outbreak or cluster of severe pneumococcal disease occur [142]. To prevent early onset of GBS in neonates, pregnant women should be offered screening for GBS and intrapartum antibiotic prophylaxis in indicated situations [143]. Besides chemoprophylaxis, immunization with the recommended vaccines for IMD, Hib, and IPD should be the primary tool for prevention of these vaccine preventable diseases.

Early treatment with dexamethasone reduced mortality and improved the outcome of adult patients with acute meningitis [144]. However, in a Cochrane review to study corticosteroids as an adjuvant therapy of bacterial meningitis, the authors found that corticosteroids did not reduce the overall mortality in meningitis patients but can reduce hearing loss and neurological sequelae [145]. The effect of corticosteroids on meningitis mortality and sequelae varied according to the bacterial agent causing meningitis [145]. Benefits of corticosteroids in treatment of meningitis patients have led to hypothesis and experimental approaches to modulate the immune response in order to decrease the harmful effects of inflammation and to improve the outcome of bacterial meningitis [146]. In one study, the benefit of prophylactic palmitoylethanolamide (a natural fatty acid amide) was demonstrated in a mouse model of *E. coli* meningitis to prolong survival and reduce symptoms by reducing inflammation and slowing the progression of infection [147]. Despite success as immunomodulation therapy for a number of auto-immune diseases such as arthritis and psoriasis, this approach, other than the use of dexamethasone, as adjuvant therapy of bacterial meningitis remain elusive and at the pre-clinical stages of development.

9. Looking Ahead and What to Expect in the Post-Genomic Era of Meningitis Control

The conjugate vaccines currently in use to control *S. pneumoniae, N. meningitidis*, and *H. influenzae* invasive infections have undoubtedly saved tens of thousands of lives [148] but there is no room for complacency because these vaccines do not offer universal coverage against all serotypes or serogroups of these pathogens. Therefore, laboratory surveillance couples with good epidemiological work remain important to monitor the trends of vaccine-preventable bacterial meningitis. We need to stay vigilant for diseases due to strains arising from the phenomenon of capsule switching and capsule replacement. For *H. influenzae*, the most common invasive strains now are non-encapsulated [91]. An increase in the detection of non-encapsulated *S. pneumoniae* has recently been reported [149]. Although most non-encapsulated *S. pneumoniae* do not cause IPD, their increase in prevalence may be concerning since they may serve as reservoirs of gene pools, including antibiotic resistance genes, for transfer into encapsulated *S. pneumoniae*. Another concern is the finding of hybrid capsules [150,151] including new capsule types due to recombination with a different *Streptococcus* species, for example, *S. mitis* [29]. Transfer of a *S. pneumoniae* capsule into a normally non-pathogenic or non-invasive *S. mitis* strain has also been reported [152]. With a large repertoire of capsule genes in *S pneumoniae*, and related *Streptococcus* species, there may be endless combinations for the organisms to take advantage of to evade vaccine immunity.

The genomic era seems to have opened up new opportunities like "reverse vaccinology" to quickly identify potential vaccine candidates [153]. Machine-learning and artificial intelligence have also been proposed to mine genomes for useful data and genes for potential applications [154].

10. Conclusions

Nowadays, we have powerful conjugate vaccines that target the most common bacterial meningitis agents (at least the most common invasive serotypes or serogroups) to not only prevent infections in the vulnerable age group, but also by eliminating nasopharyngeal carriage, to provide herd immunity to the non-vaccinated individuals. Conjugate vaccines have prevented millions of deaths from bacterial meningitis over the last two decades [2]. We now also have genomic tools that can read the complete coding sequences of bacteria for a never-before-seen gene-by-gene comparison at the nucleotide sequence level to identify and track the movement of strains (including new strains) and infections globally [76,84,97,133] in order to either quickly deploy vaccines or to develop newer vaccines for control. Nevertheless, we cannot be complacent as we have witnessed changes in the three bacterial meningitis agents after vaccine introduction. The significant increase of invasive *H. influenzae* disease due to non-encapsulated or non-typeable strains or the increase in Hia in some population in recent years are of concern [94–98]. The epidemiology of IMD in Africa has changed with much success in the deployment of the monovalent MenAfriVac leading to dramatic decreases in incidences of serogroup A diseases [81]. However, other vaccine-preventable serorgroups like W and C still continue to cause significant amount of disease when vaccines against these serogroups have not been deployed yet. The most problematic may be related to IPD due to non-vaccine serotypes emerging to cause disease after the sequential introduction of PCV7, PCV10 and PCV13 [50,52,53,57,58]. Whether this is related to the large number of serotypes of *S. pneumoniae* in contrast to the much smaller number of serotypes of *H. influenzae* or serogroups of *N. meningitidis* is unknown, but mathematical modelling suggested the number of serotypes might have an effect on strain replacement in nasopharyngeal carriage after vaccination [49]. Even though only 10 serotypes of *S. agalactiae* have been identified, its different ecology (genito-gastrointestinal colonizer versus pharyngeal colonizer) may make the effect of conjugate vaccines on the subsequent epidemiology difficult to predict.

In summary, we are in a much better position to control bacterial meningitis than ever before and surveillance continues to have a key role to play [69,70].

Funding: The author's laboratory surveillance activities on vaccine preventable bacterial diseases are supported by the Public Health Agency of Canada.

Acknowledgments: The author thanks Mei Ling Lam for her assistance in preparation of the tables, figures and the reference section.

Conflicts of Interest: The author has no conflict of interest to declare. Opinions expressed in this manuscript are those of the author and they do not represent the view of the National Microbiology Laboratory nor the Public Health Agency of Canada.

References

1. Tunkel, A.R.; van der Beek, D.; Scheld, W.M. Acute meningitis. In *Mandell, Douglas, and Bennett's Principles and Practice of Infectious Diseases*, 7th ed.; Mandell, G.L., Bennett, J.E., Dolin, R., Eds.; Churchill Livingstone Elsevier: Philadelphia, PA, USA, 2010.
2. Kassebaum, N.J.; Zunt, J.R.; for the GBD Meningitis Collaborators. Global, regional and national burden of meningitis 1990-2016: A systematic analysis for the Global Burden of Disease Study 2016. *Lancet Neurol.* **2018**, *17*, 1061–1082. [CrossRef]
3. Edmond, K.; Clark, A.; Korczak, V.S.; Sanderson, C.; Griffiths, U.K.; Rudan, I. Global and regional risk of disabling sequelae from bacterial meningitis: A systematic review and meta-analysis. *Lancet Infect. Dis.* **2010**, *10*, 317–328. [CrossRef]
4. Wright, C.; Wordsworth, R.; Glennie, L. Counting the cost of meningococcal disease, scenarios of severe meningitis and septicemia. *Pediatr. Drugs* **2013**, *15*, 49–58. [CrossRef]
5. Song, J.Y.; Lim, J.H.; Lim, S.; Yong, Z.; Seo, S.H. Progress towards a group B streptococcal vaccine. *Hum. Vaccines Immunother.* **2018**, *14*, 2669–2681. [CrossRef]

6. Nanduri, S.A.; Petit, S.; Smelser, C.; Apostol, M.; Alden, N.B.; Harrison, L.H.; Lynfield, R.; Vagnone, P.S.; Burzlaff, K.; Spina, N.L.; et al. Epidemiology of invasive early-onset and late-onset group B streptococcal disease in the United States, 2006 to 2015: Multistate laboratory and population-based surveillance. *JAMA Pediatr.* **2019**, *173*, 224–233. [CrossRef] [PubMed]
7. Watkins, L.K.F.; McGee, L.; Schrag, S.J.; Beall, B.; Jain, J.H.; Ponda, T.; Farley, M.M.; Harrison, L.H.; Zansky, S.M.; Baumbach, J.; et al. Epidemiology of invasive group B streptococcal infections among non-pregnant adults in the United States, 2008–2016, *JAMA Intern. Med.* **2019**, *179*, 479–488.
8. World Health Organization. Defeating meningitis by 2030: A Global Road Map. 2018. Available online: https://www.who.int/immunization/research/development/DefeatingMeningitisRoadmap.pdf (accessed on 25 December 2020).
9. Black, S.; Shinefield, H.; Fireman, B.; Lewis, E.; Ray, P.; Hansen, J.R.; Elvin, L.; Ensor, K.M.; Hackell, J.; Siber, G.; et al. Efficacy, safety and immunogenicity of heptavalent pneumococcal conjugate vaccine in children. Northern California Kaiser Permanente Vaccine Study Center Group. *Pediatr. Infect. Dis. J.* **2000**, *19*, 187–195. [CrossRef] [PubMed]
10. Kelly, D.F.; Moxon, E.R.; Pollard, A.J. *Haemophilus influenzae* type b conjugate vaccines. *Immunology* **2004**, *113*, 163–174. [CrossRef] [PubMed]
11. Harrison, L.H. A multivalent conjugate vaccine for prevention of meningococcal disease in infants. *JAMA* **2009**, *299*, 217–219. [CrossRef] [PubMed]
12. McIntyre, P.B.; O'Brien, K.L.; Greenwood, B.; van de Beek, D. Bacterial meningitis: Effect of vaccines on bacterial meningitis worldwide. *Lancet* **2012**, *380*, 1703–1711. [CrossRef]
13. Lin, S.M.; Zhi, Y.; Ahn, K.B.; Lim, S.; Seo, H.S. Status of group B streptococcal vaccine development. *Clin. Exp. Vaccine Res.* **2018**, *7*, 76–81. [CrossRef] [PubMed]
14. Jacups, S.P. The continuing role of *Haemophilus influenzae* type b carriage surveillance as a mechanism for early detection of invasive disease activity. *Human Vaccines* **2011**, *7*, 1254–1260. [CrossRef] [PubMed]
15. Bogaert, D.; de Groot, R.; Hermans, P.W.M. *Streptococcus pneumoniae* colonisation: The key to pneumococcal disease. *Lancet Infect. Dis.* **2004**, *4*, 144–154. [CrossRef]
16. Caugant, D.A.; Maiden, M.C.J. Meningococcal carriage and disease—Population biology and evolution. *Vaccine* **2009**, *27*, B64–B70. [CrossRef]
17. Carlin, A.F.; Chang, A.C.; Areschoug, T.; Lindahl, G.; Hurtado-Ziola, N.; King, C.C.; Varki, A.; Nizet, V. Group B *Streptococcus* suppression of phagocyte functions by protein-mediated engagement of human Siglec-5. *J. Exp. Med.* **2009**, *206*, 1691–1699. [CrossRef]
18. Hallstrom, T.; Riesbeck, K. *Haemophilus influenzae* and the complement system. *Trends Microbiol.* **2010**, *18*, 258–265. [CrossRef]
19. Hogg, J.S.; Hu, F.Z.; Janto, B.; Boissy, R.; Hayes, J.; Keefe, R.; Post, J.C.; Ehrlich, G.D. Characterization and modeling of the *Haemophilus influenzae* core and supragenomes based on the complete genomic sequences of Rd and 12 clinical nontypeable strains. *Genome Biol.* **2007**, *8*, R103. [CrossRef]
20. Schoen, C.; Tettelin, H.; Parkhill, J.; Frosch, M. Genome flexibility in *Neisseria meningitidis*. *Vaccine* **2009**, *27S*, B103–B111. [CrossRef]
21. Chaguza, C.; Andam, C.P.; Harris, S.R.; Cornick, J.E.; Yang, M.; Bricio-Moreno, L.; Kamng'ona, A.W.; Parkhill, J.; French, N.; Heyderman, R.S.; et al. Recombination in *Streptococcus pneumoniae* lineages increase with carriage duration and size of the polysaccharide capsule. *mBio* **2016**, *7*, e01053. [CrossRef]
22. Keefe, G.P. Streptococcus agalactiae mastitis: A review. *Can. Vet. J.* **1997**, *38*, 429–437. [PubMed]
23. Carvalho-Castro, G.A.; Silvaa, J.R.; Paivab, L.V.; Custódio, D.A.C.; Moreirab, R.O.; Miana, G.F.; Ingrid, A.; Prado, I.A.; Chalfun-Junior, A.C.; Costa, G.M. Molecular epidemiology of *Streptococcus agalactiae* isolated from mastitis in Brazilian dairy herds. *Braz. J. Microbiol.* **2017**, *48*, 551–559. [CrossRef] [PubMed]
24. Manning, S.D.; Springman, C.; Million, A.D.; Milton, N.R.; McNamara, S.E.; Somsel, P.A.; Bartlett, P.; Davies, H.D. Association of group B *Streptococcus* colonization and bovine exposure: A prospective cross-sectional cohort study. *PLoS ONE* **2010**, *5*, e8795. [CrossRef] [PubMed]
25. Neemuchwala, A.; Teatero, S.; Athey, T.B.T.; McGeer, A.; Fittipaldi, N. Capsular switching and other large-scale recombination events in invasive sequence type 1 group B *Streptococcus*. *Emerg. Infect. Dis.* **2016**, *22*, 1941–1944. [CrossRef] [PubMed]
26. Pittman, M. Variation and type specificity in the bacterial species *Haemophilus influenzae*. *J. Exp. Med.* **1931**, *53*, 471–492. [CrossRef] [PubMed]
27. Harrison, O.B.; Claus, H.; Jiang, Y.; Bennett, J.S.; Bratcher, H.B.; Jolley, K.A.; Corton, C.; Care, R.; Poolman, J.T.; Zollinger, W.D. Description and nomenclature of *Neisseria meningitidis* capsule locus. *Emerg. Infect. Dis.* **2013**, *19*, 566–573. [CrossRef] [PubMed]
28. Geno, K.A.; Gilbert, G.L.; Song, J.Y.; Skovsted, I.C.; Klugman, K.P.; Jones, C.; Konradsen, H.B.; Nahm, M.H. Pneumococcal capsules and their types: Past, present, and future. *Clin. Microbiol. Rev.* **2015**, *28*, 871–899. [CrossRef]
29. Ganaie, F.; Saad, J.S.; McGee, L.; van Tonder, A.J.; Bentley, S.D.; Lo, S.W.; Gladstone, R.A.; Turner, P.; Keenan, J.D.; Breiman, R.F.; et al. A new pneumococcal capsule type, 10D, is the 100th serotype and has a large *cps* fragment from an oral Streptococcus. *mBio* **2020**, *11*, e00937. [CrossRef] [PubMed]
30. Madhi, S.A.; Dangor, Z.; Heath, P.T.; Schrag, S.; Izu, A.; Meulen, A.S.T.; Dull, P.M. Considerations for a phase-III trial to evaluate a group B *Streptococcus* polysaccharide-protein conjugate vaccine in pregnant women for the prevention of early- and late-onset invasive disease in young-infants. *Vaccine* **2013**, *31* (Suppl. 40), D52–D57. [CrossRef]

31. Buurman, E.T.; Timofeyeva, Y.; Gu, J.; Kim, J.H.; Kodali, S.; Liu, Y.; Mininni, T.; Moghazeh, S.; Pavliakova, D.; Singer, C.; et al. A novel hexavalent capsular polysaccharide conjugate vaccine (GBS6) for the prevention of neonatal group B streptococcal infections by maternal immunization. *J. Infect. Dis* **2019**, *220*, 105–115. [CrossRef] [PubMed]
32. Vos, Q.; Lees, A.; Wu, Z.Q.; Snapper, C.M.; Mond, J.J. B-cell activation by T-cell-independent type 2 antigens as an integral part of the humoral immune response to pathogenic microorganisms. *Immunol. Rev.* **2000**, *176*, 154–170.
33. Mason, E.O.; Kaplan, S.L.; Lamberthm, L.B.; Hinds, D.B.; Kvernland, S.J.; Loiselle, E.M.; Feigin, R.D. Serotype and ampicillin susceptibility of *Haemophilus influenzae* causing systemic infections in children: 3 years of experience. *J. Clin. Microbiol.* **1982**, *15*, 543–546. [CrossRef]
34. Falla, T.J.; Dobson, S.R.; Crook, D.W.; Kraak, W.A.; Nichols, W.W.; Anderson, E.C.; Jordens, L.J.; Slack, M.P.; Mayon-White, D.; Moxon, E.R. Population-based study of non-typable *Haemophilus influenzae* invasive disease in children and neonates. *Lancet* **1993**, *341*, 851–864. [CrossRef]
35. Barbour, M.L.; Mayon-White, R.T.; Coles, C.; Crook, D.W.; Moxon, E.R. The impact of conjugate vaccine on carriage of *Haemophilus influenzae* type b. *J. Infect. Dis* **1995**, *171*, 93–98. [CrossRef]
36. Huang, S.S.; Platt, R.; Rifas-Shiman, S.L.; Pelton, S.I.; Goldmann, D.; Finkelstein, J.A. Post-PCV7 changes in colonizing pneumococcal serotypes in 16 Massachusetts communities, 2001 and 2004. *Pediatrics* **2005**, *116*, e408–e413. [CrossRef] [PubMed]
37. Pizza, M.; Bekkat-Berkani, R.; Rappuoli, R. Vaccines against meningococcal diseases. *Microorganism* **2020**, *8*, 1521. [CrossRef]
38. Zwahlen, A.; Kroll, J.S.; Rubin, L.G.; Moxon, E.R. The molecular basis of pathogenicity in *Haemophilus influenzae*: Comparative virulence of genetically-related capsular transformants and correlation with changes at the capsulation locus cap. *Microb. Pathog.* **1989**, *7*, 225–235. [CrossRef]
39. Jafri, R.Z.; Ali, A.; Messonnier, N.E.; Tevi-Benissan, C.; Durrheim, D.; Eskola, J.; Fermon, F.; Klugman, K.P.; Ramsay, M.; Sow, S.; et al. Global epidemiology of invasive meningococcal disease. *Popul. Health Metr.* **2013**, *11*, 17. [CrossRef]
40. Pelton, S.I. The global evolution of meningococcal epidemiology following the introduction of meningococcal vaccines. *J. Adolesc. Health* **2016**, *59*, S3–S11. [CrossRef]
41. Lo, S.W.; Gladstone, R.A.; van Tonder, A.J.; Lees, J.A.; du Plessis, M.; Benisty, R.; Givon-Lavi, N.; Hawkins, P.A.; Cornick, J.E.; Kwambana-Adams, B.; et al. The Global Pneumococcal Sequencing Consortium. Pneumococcal lineages associated with serotype replacement and antibiotic resistance in childhood invasive pneumococcal disease in the post-PCV13 era: An international whole-genome sequencing study. *Lancet Infect. Dis.* **2019**, *19*, 759–769. [CrossRef]
42. Farrell, D.J.; Jenkins, S.G.; Reinert, R.R. Global distribution of *Streptococcus pneumoniae* serotypes isolated from paediatric patients during 1999–2000 and the in vitro efficacy of telithromycin and comparators. *J. Med. Microbiol.* **2004**, *53*, 1109–1117. [CrossRef] [PubMed]
43. Yahiaoui, R.Y.; Bootsma, H.J.; den Heijer, C.D.J.; Pluister, G.N.; Paget, W.J.; Spreeuwenberg, P.; Trzcinski, K.; Stobberingh, E.E. Distribution of serotypes and patterns of antimicrobial resistance among commensal *Streptococcus pneumoniae* in nine European countries. *BMC Infect. Dis.* **2018**, *18*, 440. [CrossRef] [PubMed]
44. Harrison, L.H.; Shutt, K.A.; Schmink, S.E.; Marsh, J.W.; Harcourt, B.H.; Wang, X.; Whitney, A.M.; Stephens, D.S.; Cohn, A.A.; Messonnier, N.E.; et al. Population structure and capsular switching of invasive *Neisseria meningitidis* isolates in the pre-meningococcal conjugate vaccine era, United States, 2000–2005. *J. Infect. Dis* **2010**, *201*, 1208–1224. [CrossRef]
45. Hu, F.Z.; Eutsey, R.; Ahmed, A.; Frazao, N.; Powell, E.; Hiller, N.L.; Hillman, T.; Buchinsky, F.J.; Boissy, R.; Benjamin Janto, B.; et al. In vivo capsular switch in *Streptococcus pneumoniae* – analysis by whole genome sequencing. *PLoS ONE* **2012**, *7*, e47983. [CrossRef] [PubMed]
46. Maiden, M.C.; Bygraves, J.A.; Feil, E.; Morelli, G.; Russell, J.E.; Urwin, R.; Zhang, Q.; Zurth, K.; Caugant, D.A.; Feavers, I.M.; et al. Multilocus sequence typing: A portable approach to the identification of clones within population of pathogenic microorganisms. *Proc. Natl. Acad. Sci. USA* **1998**, *95*, 3140–3145. [CrossRef]
47. Lucidarme, J.; Lekshmi, A.; Parikh, S.R.; Bray, J.E.; Hill, D.M.; Bratcher, H.B.; Gray, S.J.; Carr, A.D.; Jolley, K.A.; Findlow, J.; et al. Frequent capsule switching in "ultra-virulent" meningococci—Are we ready for a serogroup B ST-11 complex outbreak? *J. Infect.* **2017**, *75*, 95–103. [CrossRef]
48. Swartley, J.S.; Marfin, A.A.; Edupuganti, S.; Liu, L.J.; Cieslar, P.; Perkins, B.; Wenger, J.D.; Stephens, D.S. Capsule switching of *Neisseria meningitidis*. *Proc. Natl. Acad. Sci. USA* **1997**, *94*, 271–276. [CrossRef] [PubMed]
49. Lipstich, M. Bacterial vaccines and serotype replacement: Lessons from *Haemophilus influenzae* and prospects for *Streptococcus pneumoniae*. *Emerg. Infect. Dis.* **1999**, *5*, 336–345. [CrossRef]
50. Weinbergera, D.M.; Malley, R.; Lipsitcha, M. Serotype replacement in disease following pneumococcal vaccination: A discussion of the evidence. *Lancet* **2011**, *378*, 1962–1973. [CrossRef]
51. Mitchell, P.R.; Azarian, T.; Croucher, N.J.; Callendrello, A.; Thompson, C.M.; Pelton, S.I.; Lipsitch, M.; Hanage1, W.P. Population genomics of pneumococcal carriage in Massachusetts children following introduction of PCV-13. *Microb. Genom.* **2019**, *5* e000252. [CrossRef]
52. Brueggemann, A.B.; Pai, R.; Crook, D.W.; Beall, B. Vaccine escape recombinants emerge after pneumococcal vaccination in the United States. *PLoS Pathog.* **2007**, *3*, e168. [CrossRef] [PubMed]
53. Makarewicz, O.; Lucas, M.; Brandt, C.; Herrmann, L.; Albersmeier, A.; Ruckert, C.; Blom, J.; Goesmann, A.; van der Linden, M.; Kalinowski, J.; et al. Whole genome sequencing of 39 invasive *Streptococcus pneumoniae* sequence type 199 isolates revealed switches from serotype 19A to 15B. *PLoS ONE* **2017**, *12*, e0169370. [CrossRef] [PubMed]

54. Mostowy, R.; Croucher, N.J.; De Maio, N.; Chewapreecha, C.; Salter, S.J.; Turner, P.; Aanensen, D.M.; Bentley, S.D.; Didelot, X.; Fraser, C. Pneumococcal capsule synthesis locus *cps* as evolutionary hotspot with potential to generate novel serotypes by recombination. *Mol. Biol. Evol.* **2017**, *34*, 2537–2554. [CrossRef]
55. Croucher, N.J.; Kagedan, L.; Thompson, C.M.; Parkhill, J.; Bentley, S.D.; Finkelstein, J.A.; Lipsitch, M.; Hanage, W.P. Selective and genetic constraints on pneumococcal serotype switching. *PLoS Genet.* **2015**, *11*, e1005095. [CrossRef] [PubMed]
56. Wyres, K.L.; Lambertsen, L.M.; Croucher, M.J.; McGee, L.; von Gottberg, A.; Liñares, J.; Jacobs, M.R.; Kristinsson, K.G.; Beall, B.W.; Klugman, K.P.; et al. Pneumococcal capsular switching: A historical perspective. *J. Infect. Dis.* **2013**, *207*, 439–449. [CrossRef]
57. Groves, N.; Sheppard, C.L.; Litt, D.; Rose, S.; Silva, A.; Njoku, N.; Rodrigues, S.; Amin-Chowdhury, Z.; Andrews, N.; Ladhani, S.; et al. Evolution of *Streptococcus pneumoniae* serotype 3 in England and Wales: A major vaccine evader. *Genes* **2019**, *10*, 845. [CrossRef] [PubMed]
58. Kandasamy, R.; Voysey, M.; Collins, S.; Berbers, G.; Robinson, H.; Noel, I.; Hughes, H.; Ndimah, S.; Gould, K.; Fry, N.; et al. Persistent circulation of vaccine serotypes and serotype replacement after 5 years of infant immunization with 13-valent pneumococcal conjugate vaccine in the United Kingdom. *J. Infect. Dis.* **2020**, *221*, 1361–1370. [CrossRef] [PubMed]
59. Balsells, E.; Guillot, L.; Nair, H.; Kyaw, M.H. Serotype distribution of *Streptococcus pneumoniae* causing invasive disease in children in the post-PCV era: A systematic review and meta-analysis. *PLoS ONE* **2017**, *12*, e0177113. [CrossRef]
60. Dagan, R.; Ben-Shimol, S.; Benisty, R.; Regev-Yochay, G.; Ron, M.; Givon-Lavi, N.; Rokney, A. 1888. A nationwide outbreak of invasive pneumococcal disease (IPD) caused by a novel *Streptococcus pneumoniae* serotype 2 (SP2) clone in the PCV13 era, in Israel. *Open Forum Infect. Dis.* **2019**, *6*, S54. [CrossRef]
61. Levy, C.; Ouldali, N.; Caeymaex, L.; Angoulvant, F.; Varon, E.; Cohen, R. Diversity of serotype replacement after pneumococcal conjugate vaccine implementation in Europe. *J. Pediatr.* **2019**, *213*, 252–256. [CrossRef] [PubMed]
62. Neves, F.P.G.; Cardoso, N.T.; Cardoso, C.A.C.; Teixeira, L.M.; Riley, L.W. Direct effect of the 13-valent pneumococcal conjugate vaccine use on pneumococcal colonization among children in Brazil. *Vaccine* **2019**, *37*, 5265–5269. [CrossRef] [PubMed]
63. Pick, H.; Daniel, P.l.; Rodrigo, C.; Bewick, T.; Ashton, D.; Lawrence, H.; Baskaran, V.; Edwards-Pritchard, R.C.; Sheppard, C.; Seyi, D.; et al. Pneumococcal serotype trends, surveillance and risk factors in UK adult pneumonia, 2013–2018. *Thorax* **2020**, *75*, 38–49. [CrossRef] [PubMed]
64. Rupp, R.; Hurley, D.; Grayson, S.; Li, J.; Nolan, K.; McFetridge, R.D. A dose ranging study of 2 different formulations of 15-valent pneumococcal conjugate vaccine (PCV15) in healthy infants. *Hum. Vaccines Immunother.* **2019**, *15*, 549–559. [CrossRef] [PubMed]
65. Thompson, A.; Lamberth, E.; Severs, J.; Scully, I.; Tarabar, S.; Ginis, J. Phase 1 trial of a 20 valent pneumococcal conjugate vaccine in healthy adults. *Vaccine* **2019**, *37*, 6201–6207. [CrossRef]
66. Løchen, A.; Croucher, N.J.; Anderson, R.M. Divergent serotype replacement trends and increasing diversity in pneumococcal disease in high income settings reduce the benefit of expanding vaccine valency. *Sci. Rep.* **2020**, *10*, 18977. [CrossRef] [PubMed]
67. Peters, T.R.; Poehling, K.A. Invasive pneumococcal disease, the target is moving. *JAMA* **2017**, *297*, 1825–1926. [CrossRef] [PubMed]
68. Hanage, W.P. Serotype replacement in invasive pneumococcal disease: Where do we go from here? (editorial commentary). *J. Infect. Dis.* **2007**, *196*, 1282–1284. [CrossRef]
69. Knol, M.J.; van der Ende, A. Continuous surveillance of invasive pneumococcal disease is key. *Lancet Infect. Dis.* **2021**, *21*, 13–14. [CrossRef]
70. Klugman, K.P.; Rodgers, G.L. Time for a third-generation pneumococcal vaccine. *Lancet Infect. Dis.* **2021**, *21*, 14–16. [CrossRef]
71. Advisory Committee on Immunization Practices, Centers for Disease Control and Prevention (CDC). Revised recommendations of the Advisory Committee on Immunization Practices to vaccinate all persons aged 11–18 years with meningococcal conjugate vaccine. *Morb. Mortal. Wkly. Rep.* **2007**, *56*, 794–795.
72. Wang, X.; Shutt, K.A.; Vuong, J.T.; Cohn, A.; MacNeil, J.; Schmink, S.; Plikaytis, B.; Messonnoer, N.E.; Harrison, L.H.; Clark, T.A.; et al. Changes in the population structure of invasive *Neisseria meningitidis* in the United States after quadrivalent meningococcal conjugate vaccine licensure. *J. Infect. Dis.* **2015**, *211*, 1887–1894. [CrossRef]
73. Law, D.K.S.; Lorange, M.; Ringuette, L.; Dion, R.; Giguère, M.; Henderson, A.M.; Stoltz, J.; Zollinger, W.D.; De Wals, P.; Tsang, R.S.W. Invasive meningococcal disease in Quebec, Canada, due to an emerging clone of ST-269 serogroup B meningococci with serotype antigen 17 and serosubtype antigen P1.19 (B:17:P1.19). *J. Clin. Microbiol.* **2006**, *44*, 2743–2749. [CrossRef]
74. Wals, P.; Dionne, M.; Douville-Fradet, M.; Boulianne, N.; Drapeau, J.; De Serres, D. Impact of a mass immunization campaign against serogroup C meningococcus in the Province of Quebec, Canada. *Bull. WHO* **1996**, *74*, 407–411. [PubMed]
75. Wals, P.; Deceuninck, G.; Boulianne, N.; De Serres, G. Effectiveness of a mass immunization campaign using serogroup C meningococcal conjugate vaccine. *JAMA* **2004**, *292*, 2491–2494. [CrossRef]
76. Krone, M.; Gray, S.; Abad, R.; Skoczyńska, A.; Stefanelli, P.; van der Ende, A.; Tzanakaki, G.; Mölling, P.; Simões, M.J.; Křížová, P.; et al. Increase of invasive meningococcal serogroup W disease in Europe, 2013 to 2017. *Eurosurveill* **2019**, *24*, 1800245. [CrossRef]
77. Taha, M.K.; Deghmane, A.E.; Knol, M.; van der Ende, A. Whole genome sequencing reveals Trans-European spread of an epidemic *Neisseria meningitidis* serogroup W clone. *Clin. Microbiol. Infect.* **2019**, *25*, 765–767. [CrossRef] [PubMed]
78. Martin, N.V.; Ong, K.S.; Howden, B.P.; Lahra, M.M.; Lambert, S.B.; Beard, F.H.; Dowse, G.K.; Saul, N. Communicable Diseases Network Australia MenW Working Group. Rise in invasive serogroup W meningococcal disease in Australia 2013–2015. *Commun. Dis. Intell. Q. Rep.* **2016**, *40*, E454–E459. [PubMed]

79. Tsang, R.S.W.; Hoang, L.; Tyrrell, G.J.; Horsman, G.; Caeseele, P.; Jamieson, F.B.; Lefebvre, B.; Haldane, D.; Gad, R.R.; German, G.J.; et al. Increase in *Neisseria meningitidis* serogroup W invasive disease in Canada: 2009-2016. *Can. Commun. Dis. Rep.* **2017**, *43*, 144–149. [CrossRef] [PubMed]
80. Mowlaboccus, S.; Jolley, K.A.; Bray, J.E.; Pang, S.; Lee, Y.T.; Bew, J.D.; Speers, D.S.; Keil, A.D.; Coombs, G.W.; Kahler, C.M. Clonal expansion of new penicillin-resistant clade of *Neisseria meningitidis* serogroup W clonal complex 11, Australia. *Emerg. Infect. Dis.* **2017**, *23*, 1364–1367. [CrossRef]
81. Trotter, C.L.; Lingani, C.; Fernandez, K.; Cooper, L.V.; Bita, A.; Tevi-Benissan, C.; Ronveaux, O.; Préziosi, M.P.; Stuart, J.M. Impact of MenAfriVac in nine countries of the African meningitis belt, 2010–2015: An analysis of surveillance data. *Lancet Infect. Dis.* **2017**, *17*, 867–872. [CrossRef]
82. Boisier, P.; Nicolas, P.; Djibo, S.; Taha, M.K.; Jeanne, J.; Maïnassara, H.B.; Tenebray, B.; Kairo, K.K.; Giorgini, D.; Chanteau, S. Meningococcal meningitis: Unprecedented incidence of serogroup X-related cases in 2006 in Niger. *Clin. Infect. Dis.* **2007**, *44*, 657–663. [CrossRef]
83. Xie, O.; Pollard, A.J.; Mueller, J.E.; Norheim, G. Emergence of serogroup X meningococcal disease in Africa: Need for a vaccine. *Vaccine* **2013**, *31*, 2852–2861. [CrossRef]
84. Agnememel, A.; Hong, E.; Giorgini, D.; Nuñez-Samudio, V.; Deghmane, A.E.; Taha, M.K. *Neisseria meningitidis* serogroup X in Sub-Saharan Africa. *Emerg. Infect. Dis.* **2016**, *22*, 698–702. [CrossRef]
85. Soeters, H.M.; Diallo, A.O.; Bicaba, B.W.; Kadadé, G.; Dembélé, A.Y.; Acyl, M.A.; Nikiema, C.; Sadji, A.Y.; Poy, A.N.; Lingani, C.; et al. Bacterial meningitis epidemiology in 5 countries in the meningitis belt of sub-Saharan Africa, 2015–2017. *J. Infect. Dis.* **2019**, *220* (Suppl. 4), S165–S174. [CrossRef] [PubMed]
86. Fernandez, K.; Lingani, C.; Aderinola, O.M.; Goumbi, K.; Bicaba, B.; Edea, Z.A.; Glèlè, C.; Sarkodie, B.; Tamekloe, A.; Ngomba, A.; et al. Meningococcal meningitis outbreaks in the African meningitis belt after meningococcal serogroup A conjugate vaccine introduction, 2011–2017. *J. Infect. Dis.* **2019**, *220* (Suppl. 4), S225–S232. [CrossRef]
87. Brynildsruda, O.B.; Eldholma, V.; Bohlina, J.; Uadialeb, K.; Obaro, S.; Cauganta, D.A. Acquisition of virulence genes by a carrier strain gave rise to the ongoing epidemics of meningococcal disease in West Africa. *Proc. Natl. Acad. Sci. USA* **2018**, *115*, 5510–5515. [CrossRef]
88. Retchless, A.C.; Hu, F.; Ouédraogo, A.; Diarra, S.; Knipe, K.; Sheth, M.; Rowe, L.A.; Sangaré, L.; Ba, A.K.; Ouangraoua, S.; et al. The establishment and diversification of epidemic-associated serogroup W meningococcus in the African meningitis belt, 1994 to 2012. *mSphere* **2016**, *1*, e00201–e00216. [CrossRef] [PubMed]
89. Leimkugel, J.; Hodgson, A.; Forgor, A.A.; Pfluger, V.; Dangy, J.P.; Smith, T.; Achtman, M.; Sebastien Gagneux, S.; Gerd Pluschke, G. Clonal waves of *Neisseria* colonisation and disease in the African meningitis belt: Eight year longitudinal study in Northern Ghana. *PLoS Med.* **2007**, *4*, e101. [CrossRef]
90. Mbaeyi, S.; Sampo, E.; Dinanibè, K.; Yaméogo, I.; Congo-Ouédraogo, M.; Tamboura, M.; Sawadogo, G.; Ouattara, K.; Sanou, M.; Kiemtoré, K.; et al. Meningococcal carriage 7 years after introduction of a serogroup A meningococcal conjugate vaccine in Burkina Faso: Results from four cross-sectional carriage surveys. *Lancet Infect. Dis.* **2020**, *20*, 1418–1425. [CrossRef]
91. Langereis, J.D.; de Jonge, M.I. Invasive disease caused by non-typeable *Haemophilus influenzae*. *Emerg. Infect. Dis.* **2015**, *21*, 1711–1718. [CrossRef] [PubMed]
92. Whittaker, R.; Economopoulou, A.; Dias, J.G.; Bancroft, E.; Ramliden, M.; Celentano, L.P. European Centre for Disease Prevention and Control Country Experts for Invasive *Haemophilus influenzae* Disease. Epidemiology of invasive *Haemophilus influenzae* disease, Europe, 2007–2014. *Emerg. Infect. Dis.* **2017**, *23*, 396–404. [CrossRef] [PubMed]
93. Soeters, H.M.; Blain, A.; Pondo, T.; Doman, B.; Farley, M.M.; Harrison, L.H.; Lynfield, R.; Miller, L.; Petit, S.; Reinglod, A.; et al. Current epidemiology and trends in invasive *Haemophilus influenzae* disease—United States, 2009-2015. *Clin. Infect. Dis.* **2018**, *67*, 881–889. [CrossRef] [PubMed]
94. Ulanova, M. Global epidemiology of invasive *Haemophilus influenzae* type a disease: Do we need a new vaccine? *J. Vaccine* **2013**, *14*, 941461. [CrossRef]
95. Tsang, R.S.W.; Ulanova, M. The changing epidemiology of invasive *Haemophilus influenzae* disease: Emergence and global presence of serotype a strains that may require a new vaccine for control. *Vaccine* **2017**, *35*, 4270–4275. [CrossRef]
96. Boisvert, A.A.; Moore, D. Invasive disease due to *Haemophilus influenzae* type a in Canada's north: A priority for prevention. *Can. J. Infect. Dis. Med. Microbiol.* **2015**, *26*, 291–292. [CrossRef] [PubMed]
97. Soeters, H.M.; Oliver, S.E.; Plumb, I.D.; Blain, A.E.; Zulz, T.; Simons, B.C.; Barnes, M.; Farley, M.M.; Harrison, L.H.; Lynfield, R.; et al. Epidemiology of invasive *Haemophilus influenzae* serotype a disease—United States, 2008–2017. *Clin. Infect. Dis.* **2020**, ciaa875. [CrossRef] [PubMed]
98. Bozio, C.H.; Blain, A.; Edge, K.; Farley, M.M.; Harrison, L.H.; Poissant, T.; Schaffner, W.; Scheuer, T.; Torres, S.; Triden, L.; et al. Clinical characteristics and adverse clinical outcomes of invasive *Haemophilus influenzae* serotype a cases—United States, 2011–2015. *Clin. Infect. Dis.* **2020**, ciaa990. [CrossRef]
99. Cox, A.D.; Williams, D.; Cairns, C.; St. Michael, F.; Fleming, P.; Vinogradov, E.; Arbour, M.; Masson, L.; Zou, W. Investigating the candidacy of a capsular polysaccharide-based glycoconjugate as a vaccine to combat *Haemophilus influenzae* type a disease: A solution for an unmet public health need. *Vaccine* **2017**, *35*, 6129–6136. [CrossRef] [PubMed]

100. Bruce, M.G.; Zulz, T.; DeByle, C.; Singleton, R.; Hurlburt, D.; Bruden, D.; Rudolph, K.; Hennessy, T.; Klejka, J.; Wenger, J.D. *Haemophilus influenzae* serotype a invasive disease, Alaska, USA, 1983–2011. *Emerg. Infect. Dis.* **2013**, *19*, 932–937. [CrossRef] [PubMed]
101. Tsang, R.S.W.; Shuel, M.; Ahmad, T.; Hayden, K.; Knox, N.; Van Domselaar, G.; Hoang, L.; Tyrrell, G.; Minion, J.; Van Caeseele, P.; et al. Whole genome sequencing to study the phylogenetic structure of serotype a *Haemophilus influenzae* recovered from patients in Canada. *Can. J. Microbiol.* **2019**, *66*, 99–110. [CrossRef]
102. Lima, J.B.; Ribeiro, G.S.; Cordeiro, S.M.; Gouveia, E.L.; Salgado, K.; Spratt, B.G.; Godoy, D.; Reis, M.G.; Ko, A.I.; Reis, J.N. Poor clinical outcome for meningitis caused by *Haemophilus influenzae* serotype a strains containing the IS*1016-bexA* deletion. *J. Infect. Dis.* **2010**, *202*, 1577–1584. [CrossRef] [PubMed]
103. Crandall, H.; Christiansen, J.; Varghese, A.A.; Russon, A.; Korgenski, E.K.; Bengtson, E.K.; Dickey, M.; Killpack, J.; Knackstedt, E.D.; Daly, J.A.; et al. Clinical and molecular epidemiology of invasive *Haemophilus influenzae* serotype a infections in Utah children. *J. Pediatr. Infect. Dis. Soc.* **2019**, piz088. [CrossRef] [PubMed]
104. LaCross, N.C.; Marrs, C.F.; Patel, M.; Sandstedt, S.A.; Gilsdorf, J.R. High genetic diversity of nontypeable *Haemophilus influenzae* isolates from two children attending a day care center. *J. Clin. Microbiol.* **2008**, *46*, 3817–3821. [CrossRef]
105. Pinto, M.; González-Díaz, A.; Machado, M.P.; Duarte, S.; Vieira, L.; Carriço, J.A.; Marti, S.; Bajanca-Lavado, M.P.; Gomes, J.P. Insights into the population structure and pan-genome of *Haemophilus influenzae*. *Infect. Gene Evol.* **2019**, *67*, 126–135. [CrossRef]
106. Chewapreecha, C.; Harris, S.R.; Croucher, N.J.; Turner, C.; Marttinen, P.; Cheng, L.; Pessia, A.; Aanensen, D.M.; Mather, A.E.; Page, A.J.; et al. Dense genomic sampling identifies highways of pneumococcal recombination. *Nat. Genet.* **2014**, *46*, 305–309. [CrossRef]
107. Harrison, O.B.; Brueggemann, A.B.; Caugant, D.A.; van der Ende, A.; Frosch, M.; Gray, S.; Heuberger, S.; Krizova, P.; Olcen, P.; Slack, M.; et al. Molecular typing methods for outbreak detection and surveillance of invasive disease caused by *Neisseria meningitidis*, *Haemophilus influenzae* and *Streptococcus pneumoniae*, a review. *Microbiology* **2011**, *157*, 2181–2195. [CrossRef] [PubMed]
108. LaClaire, L.L.; Tondella, M.L.C.; Beall, D.S.; Noble, C.A.; Raghunathan, P.I.; Rosenstein, N.E.; Popovic, T. The Active Bacterial Core Surveillance Team Members. Identification of *Haemophilus influenzae* serotypes by standard slide agglutination serotyping and PCR-based capsule typing. *J. Clin. Microbiol.* **2003**, *41*, 393–396. [CrossRef]
109. Falla, T.J.; Crook, D.W.M.; Brophy, L.N.; Maskell, D.; Kroll, J.S.; Moxon, E.R. PCR for capsular typing of *Haemophilus influenzae*. *J. Clin. Microbiol.* **1994**, *32*, 2382–2386. [CrossRef] [PubMed]
110. Kong, F.; Gowan, S.; Martin, D.; James, G.; Gilbert, G.L. Serotype identification of group B streptococci by PCR and sequencing. *J. Clin. Microbiol.* **2002**, *40*, 216–226. [CrossRef]
111. Brito, D.A.; Ramirez, M.; de Lencastre, H. Serotyping *Streptococcus pneumoniae* by multiplex PCR. *J. Clin. Microbiol.* **2003**, *41*, 2378–2384. [CrossRef]
112. Zhu, H.; Wang, Q.; Wen, L.; Xu, J.; Shao, Z.; Chen, M.; Chen, M.; Reeves, P.R.; Cao, B.; Wanga, L. Development of a multiplex PCR assay for detection and genogrouping of *Neisseria meningitidis*. *J. Clin. Microbiol.* **2012**, *50*, 46–51. [CrossRef]
113. Enright, M.C.; Spratt, B.G. A multilocus sequence typing scheme for *Streptococcus pneumoniae*: Identification of clones associated with serious invasive disease. *Microbiology* **1998**, *144*, 3049–3060. [CrossRef]
114. Meats, E.; Feil, E.J.; Stringer, S.; Cody, A.J.; Goldstein, R.; Kroll, J.S.; Popovic, T.; Spratt, B.G. Characterization of encapsulated and non-encapsulated *Haemophilus influenzae* and determination of phylogenetic relationships by multilocus sequence typing. *J. Clin. Microbiol.* **2003**, *41*, 1623–1626. [CrossRef]
115. Serruto, D.; Bottomley, M.J.; Ram, S.; Giuliliani, M.M.; Rappuoli, R. The new multicomponent vaccine against meningococcal serogroup B, 4CMenB: Immunological, functional and structural characterization of the antigens. *Vaccine* **2012**, *30* (Suppl. 2), B87–B97. [CrossRef] [PubMed]
116. Gandhi, A.; Balmer, P.; York, L.J. Characteristics of a new meningococcal serogroup B vaccine, bivalent rLP2086 (MenB-FHbp; Trumenba®). *Postgrad. Med.* **2016**, *128*, 548–556. [CrossRef] [PubMed]
117. Browall, S.; Norman, M.; Tångrot, J.; Galanis, I.; Sjöström, K.; Dagerhamn, J.; Hellberg, C.; Pathak, A.; Spadafina, T.; Sandgren, A.; et al. Intraclonal variations among *Streptococcus pneumoniae* isolates influence the likelihood of invasive disease in children. *J. Infect. Dis* **2014**, *209*, 377–388. [CrossRef] [PubMed]
118. Cremers, A.J.H.; Mobegi, F.M.; van der Gaast–de Jongh, C.; van Weert, M.; van Opzeeland, F.J.; Vehkala, M.; Knol, M.J.; Bootsma, H.J.; Välimäki, N.; Croucher, N.J.; et al. The contribution of genetic variation of *Streptococcus pneumoniae* to the clinical manifestation of invasive pneumococcal disease. *Clin. Infect. Dis.* **2019**, *68*, 61–69. [CrossRef]
119. Pettigrewa, M.M.; Ahearnb, C.P.; Gentd, J.F.; Konge, Y.; Gallob, M.C.; Munroh, J.B.; D'Melloh, A.; Sethic, S.; Tettelinh, H.; Murphy, T.F. *Haemophilus influenzae* genome evolution during persistence in the human airways in chronic obstructive pulmonary disease. *Proc. Natl. Acad. Sci. USA* **2018**, *115*, E3256–E3265. [CrossRef] [PubMed]
120. The European Committee on Antimicrobial Susceptibility Testing—EUCAST. Breakpoint Tables for Interpretation of MICs and Zone Diameters, Version 10.0. 2020. Available online: http://www.eucast.org/clinical_breakpoints/ (accessed on 15 December 2020).
121. Clinical and Laboratory Standards Institute (CLSI). *Performance Standards for Antimicrobial Susceptibility Testing*, 30th ed.; Supplement M100-S30; CLSI: Wayne, PA, USA, 2020.

122. Yazdankhah, S.P.; Kriz, P.; Tzanakaki, G.; Kremastinou, J.; Kalmusova, J.; Musilek, M.; Alvestad, T.; Jolley, K.A.; Wilson, D.J.; McCarthy, N.D.; et al. Distribution of serogroups and genotypes among disease-associated and carried isolates of *Neisseria meningitidis* from the Czech Republic, Greece, and Norway. *J. Clin. Microbiol.* **2004**, *42*, 5146–5153. [CrossRef]
123. Sandgren, A.; Sjostrom, K.; Olsson-Liljequist, O.; Christensson, B.; Samuelsson, A.; Kronvall, G.; Normark, B.H. Effect of clonal and serotype-specific properties on the invasive capacity of *Streptococcus pneumoniae*. *J. Infect. Dis.* **2004**, *189*, 784–796. [CrossRef]
124. Zhou, J.; Jamieson, F.; Dolman, S.; Hoang, L.M.; Rawte, P.; Tsang, R.S.W. Genetic and antigenic analysis of invasive serogroup C *Neisseria meningitidis* in Canada: A decrease in the electrophoretic type (ET)-15 clonal type and an increase in the proportion of isolates belonging to the ET-37 (but not ET-15) clonal type during the period from 2002 to 2009. *Can. J. Infect. Dis Med. Microbiol.* **2012**, *23*, e55–e59. [PubMed]
125. Fleischmann, R.D.; Adams, M.D.; White, O.; Clayton, R.A.; Kirkness, E.F.; Kerlavage, A.R.; Bult, C.J.; Tomb, J.F.; Dougherty, B.A.; Merrick, J.M.; et al. Whole-genome random sequencing and assembly of *Haemophilus influenzae* Rd. *Science* **1995**, *269*, 496–512. [CrossRef]
126. Gladstone, R.A.; Lo, S.W.; Lees, J.A.; Croucher, N.J.; van Tonder, A.J.; Corander, J.; Page, A.J.; Marttinen, P.; Bentley, L.J.; Ochoa, T.J.; et al. International genomic definition of pneumococcal lineages, to contextualize disease, antibiotic resistance and vaccine impact. *EBioMedicine* **2019**, *43*, 338–346. [CrossRef]
127. Caugant, D.A. Population genetics and molecular epidemiology of *Neisseria meningitidis*. *APMIS* **1998**, *106*, 505–525. [CrossRef] [PubMed]
128. Musser, J.M.; Kroll, J.S.; Moxon, E.R.; Selander, R.K. Clonal population structure of encapsulated *Haemophilus influenzae*. *Infect. Immun.* **1988**, *56*, 1837–1845. [CrossRef] [PubMed]
129. De Chiaraa, M.; Hood, D.; Muzzia, A.; Pickardd, D.J.; Perkinse, T.; Pizzaa, M.; Dougand, G.; Rappuolia, R.; Moxxon, E.R.; Soriani, M.; et al. Genome sequencing of disease and carriage isolates of nontypeable *Haemophilus influenzae* identifies discrete population structure. *Proc. Natl. Acad. Sci. USA* **2014**, *111*, 5439–5444. [CrossRef]
130. Epping, L.; Tonder, A.J.; Gladstone, R.A.; The Global Pneumococcal Sequencing Consortium; Bentley, S.D.; Page, A.J.; Keane, J.A. SeroBA: Rapid high-throughput serotyping of *Streptococcus pneumoniae* from whole genome sequence data. *Microb. Genom.* **2018**, *4*, e000204.
131. Watts, S.C.; Holt, K.E. Hicap: In Silico serotyping of the *Haemophilus influenzae* capsule locus. *J. Clin. Microbiol.* **2019**, *57*, e00190-19. [CrossRef] [PubMed]
132. Marjuki, H.; Topaz, N.; Rodriguez-Rivera, L.D.; Ramos, E.; Potts, C.A.; Chen, A.; Retchless, A.C.; Doho, G.H.; Wang, X. Whole-genome sequencing for characterization of capsule locus and prediction of serogroup of invasive meningococcal isolates. *J. Clin. Microbiol.* **2019**, *57*, e01609–e01618. [CrossRef]
133. Jolley, K.A.; Maiden, M.C.J. BIGSdb: Scalable analysis of bacterial genome variation at the population level. *BMC Bioinform.* **2010**, *11*, 595. [CrossRef] [PubMed]
134. Davis, J.J.; Boisvert, S.; Brettin, T.; Kenyon, R.W.; Mao, C.; Olson, R.; Overbeek, R.; Santerre, J.; Shukla, M.; Wattam, A.R.; et al. Antimicrobial resistance prediction in PATRIC and RAST. *Sci. Rep.* **2015**, *6*, 27930. [CrossRef]
135. Metcalf, B.J.; Chochua, S.; Gertz, R.E., Jr.; Li, Z.; Walker, H.; Tran, T.; Hawkins, P.A.; Glennen, A.; Lynfield, R.; Li, Y.; et al. Active Bacterial Core surveillance team. Using whole genome sequencing to identify resistance determinants and predict antimicrobial resistance phenotypes for year 2015 invasive pneumococcal disease isolates recovered in the United States. *Clin. Microbiol. Infect.* **2016**, *22*, 1002.e1–1002.e8. [CrossRef]
136. Conteville, L.C.; de Filippis, A.M.B.; Nogueira, R.M.R.; de Mendonça, M.C.L.; Vicente, A.C.P. Metagenomics in undiagnosed cases of fulminant fever. *J. Infect.* **2018**, *77*, 160–161. [CrossRef] [PubMed]
137. Bozio, C.H.; Vuong, J.; Dokubo, E.K.; Fallah, M.P.; McNamara, L.A.; Potts, C.C.; Doedeh, J.; Gbanya, M.; Retchless, A.C.; Jaymin, C.P.; et al. Outbreak of *Neisseria meningitidis* serogroup C outside the meningitis belt—Liberia, 2017: An epidemiological and laboratory investigation. *Lancet Infect. Dis.* **2018**, *18*, 1360–1367. [CrossRef]
138. Rodgers, E.; Bentley, S.D.; Borrow, R.; Bratcher, H.B.; Brisse, S.; Brueggemann, A.B.; Caugant, D.A.; Findlow, J.; Fox, L.; Glennie, L.; et al. The global meningitis genome partnership. *J. Infect.* **2020**, *81*, 510–520. [CrossRef] [PubMed]
139. Public Health England. Guidance for Public Health Management of Meningococcal Disease in the UK. August 2019. Available online: https://assets.publishing.service.gov.uk/government/uploads/system/uploads/attachment_data/file/829326/PHE_meningo_disease_guideline.pdf (accessed on 5 February 2021).
140. Public Health England. Revised Recommendations for the Prevention of Secondary Haemophilus influenzae type b (Hib) Disease. July 2013. Available online: https://assets.publishing.service.gov.uk/government/uploads/system/uploads/attachment_data/file/231009/Revised_recommendations_for_the_preventions_of_secondary_Haemophilus_influenzae_type_b_disease.pdf (accessed on 5 February 2021).
141. American Academy of Pediatrics. *Red Book: 2018 Report of the Committee on Infectious Diseases*, 31st ed.; Kimberlin, D.W., Brady, M.T., Jackson, M.A., Long, S.S., Eds.; American Academy of Pediatrics: Elk Grove Village, IL, USA, 2018.
142. Public Health England. Guidelines for the Public Health Management of Clusters of Severe Pneumococcal Disease in Closed Settings. Updated January 2020. Available online: https://assets.publishing.service.gov.uk/government/uploads/system/uploads/attachment_data/file/867175/Pneumococccal_cluster_guidelines.pdf (accessed on 5 February 2021).

143. Hughes, R.G.; Brocklehurst, P.; Steer, P.J.; Heath, P.; Stenson, B.M.; On behalf of the Royal College of Obstetricians and Gynaecologists. Prevention of early-onset neonatal group B streptococcal disease. Green-top guideline No. 36. *BJOG* **2017**, *124*, e280–e305. [CrossRef]
144. de Gans, J.; van de Beek, D. For the European Dexamethasone in Adulthood Bacterial Meningitis Study Investigators. Dexamethasone in adults with bacterial meningitis. *N. Engl. J. Med.* **2002**, *347*, 1549–1556. [CrossRef]
145. Brouwer, M.C.; McIntyre, P.; Prasad, K.; van de Beek, D. Corticosteroids for acute bacterial meningitis (review). *Cochrane Database Syst. Rev.* **2015**, CD004405. [CrossRef]
146. van der Flier, M.; Geelen, S.P.M.; Kimpen, J.L.L.; Hoepelman, I.M.; Tuomanen, E.I. Reprogramming the host response in bacterial meningitis: How best to improve outcome? *Clin. Microbiol. Rev.* **2003**, *16*, 415–429. [CrossRef]
147. Heide, E.C.; Bindila, L.; Post, J.M.; Malzahn, D.; Lutz, B.; Seele, J.; Nau, R.; Ribes, S. Prophylactic palmitoylethanolamide prolongs survival and decreases detrimental inflammation in aged mice with bacterial meningitidis. *Front. Immunol.* **2018**, *9*, 2671. [CrossRef]
148. Makwana, N.; Riordan, A.I. Bacterial meningitis: The impact of vaccination. *CNS Drugs* **2007**, *21*, 355–366. [CrossRef]
149. Bradshaw, J.L.; McDaniel, L.S. Selective pressure: Rise of the non-encapsulated pneumococcus. *PLoS Pathog.* **2019**, *15*, e1007911. [CrossRef]
150. Tsang, R.S.W.; Tsai, C.M.; Henderson, A.M.; Tyler, S.; Law, D.K.S.; Zollinger, W.; Jamieson, F. Immunochemical studies and genetic background of two *Neisseria meningitidis* isolates expressing unusual capsule polysaccharide antigens with specificities of both serogroup Y and W135. *Can. J. Microbiol.* **2008**, *54*, 229–234. [CrossRef]
151. Oliver, M.B.; van der Linden, M.P.G.; Küntzel, S.A.; Saad, J.S.; Nahm, M.H. Discovery of *Streptococcus pneumoniae* serotype 6 variants with glycosyltransferases synthesizing two differing repeating units. *J. Biol. Chem.* **2013**, *288*, 25976–25985. [CrossRef]
152. Lessa, F.C.; Milucky, J.; Rouphael, N.G.; Bennett, N.M.; Talbot, H.K.; Harrison, L.H.; Farley, M.M.; Walston, J.; Pimenta, F.; Gertz, R.E.; et al. *Streptococcus mitis* expressing pneumococcal serotype 1 capsule. *Sci. Rep.* **2018**, *8*, 17959. [CrossRef] [PubMed]
153. Sette, A.; Rappuoli, R. Reverse vaccinology: Developing vaccines in the era of genomics. *Immunity* **2010**, *33*, 530–541. [CrossRef]
154. Obolski, U.; Gori, A.; Lourenço, J.; Thompson, C.; Thompson, R.; French, N.; Heyderman, R.S.; Gupta, S. Identifying genes associated with invasive disease in *S. pneumoniae* by applying a machine learning approach to whole genome sequence typing data. *Sci. Rep.* **2019**, *9*, 4049. [CrossRef]

 MDPI

Review

Bacterial Meningitis in Children: Neurological Complications, Associated Risk Factors, and Prevention

Abdulwahed Zainel [1], Hana Mitchell [1,2] and Manish Sadarangani [1,2,*]

[1] Department of Pediatrics, Division of Infectious Diseases, University of British Columbia, Vancouver, BC V6t 1Z4, Canada; abdulwahed.zainel@cw.bc.ca (A.Z.); hana.mitchell@cw.bc.ca (H.M.)
[2] Vaccine Evaluation Center, BC Children's Hospital, Vancouver, BC V6H 3N1, Canada
* Correspondence: msadarangani@bcchr.ubc.ca; Tel.: +1-604-875-2422

Abstract: Bacterial meningitis is a devastating infection, with a case fatality rate of up to 30% and 50% of survivors developing neurological complications. These include short-term complications such as focal neurological deficit and subdural effusion, and long-term complications such as hearing loss, seizures, cognitive impairment and hydrocephalus. Complications develop due to bacterial toxin release and the host immune response, which lead to neuronal damage. Factors associated with increased risk of developing neurological complications include young age, delayed presentation and *Streptococcus pneumoniae* as an etiologic agent. Vaccination is the primary method of preventing bacterial meningitis and therefore its complications. There are three vaccine preventable causes: *Haemophilus influenzae* type b (Hib), *S. pneumoniae,* and *Neisseria meningitidis.* Starting antibiotics without delay is also critical to reduce the risk of neurological complications. Additionally, early adjuvant corticosteroid use in Hib meningitis reduces the risk of hearing loss and severe neurological complications.

Keywords: bacterial meningitis; neurological sequelae; hearing loss; seizure; epilepsy; hydrocephalus; focal neurological deficit; vaccine; corticosteroid; dexamethasone

Citation: Zainel, A.; Mitchell, H.; Sadarangani, M. Bacterial Meningitis in Children: Neurological Complications, Associated Risk Factors, and Prevention. *Microorganisms* **2021**, *9*, 535. https://doi.org/10.3390/microorganisms9030535

Academic Editor: James Stuart

Received: 27 January 2021
Accepted: 3 March 2021
Published: 5 March 2021

Publisher's Note: MDPI stays neutral with regard to jurisdictional claims in published maps and institutional affiliations.

1. Introduction

Acute bacterial meningitis is the most common bacterial central nervous system (CNS) infection. It is a devastating illness, especially in neonates (age < one month) and infants (age < one year). Bacterial meningitis has a high case-fatality rate of up to 30% [1–5], and as many as 50% of survivors develop neurological complications [1,3,6–10]—with outcomes highly dependent on patient's age and the infecting organism. This article provides an overview of the short and long-term neurological complications of bacterial meningitis, the associated risk factors, and available vaccines and therapies which may reduce the risk of complications.

2. Epidemiology and Etiology

The incidence of bacterial meningitis in children differs by age group and is highest in infants aged younger than two months [11,12]. In the United States, the incidence rate during 2006–2007 in children under two months was 81 cases per 100,000, compared with 0.4 cases per 100,000 in children aged 11–17 years. Bacterial meningitis is more common in low and middle income countries (LMICs) compared to high income countries (HICs) [13,14]. For example, the incidence rate of meningitis in 2016 in all ages in South Sudan was 270 per 100,000 whereas in Australia it was 0.5 per 100,000 [14].

The most common organisms causing bacterial meningitis vary by age group (Table 1). Introduction of vaccines against *Haemophilus influenzae type b* (Hib), *Neisseria meningitidis* and *Streptococcus pneumoniae* over the last three decades has led to a drastic decrease in the incidence rate of bacterial meningitis beyond the neonatal period in countries with these vaccines included as part of their routine infant and children immunization programs [15]. However, the case fatality rate has not changed significantly [16–20].

Table 1. Most common organism for different Age Groups.

Age Group	Most Common Organisms	References
Pre-term neonate	*Escherichia coli*, GBS *	[12,21,22]
Term neonate and infants < three months	GBS *, *E. coli*, *Streptococcus pneumoniae*, *Listeria monocytogenes*	[12]
Children ≥ three months to ten years	*S. pneumoniae*, *Neisseria meningitidis*, *Haemophilus influenzae* type b	[3,13,23,24]
Adolescent until 19 years old	*N. meningitidis*, *S. pneumoniae*	[4,25,26]

* GBS: Group B Streptococcus.

Hib was the leading cause of bacterial meningitis in the 1990s, but became uncommon in countries that introduced Hib immunization. However, it is still a frequent cause of bacterial meningitis in the countries were Hib vaccine is not universal or in unvaccinated children due to vaccine hesitancy [27]. In countries where Hib is now rare, *S. pneumoniae* has become the leading cause of bacterial meningitis outside the neonatal period, except in some European and Sub-Saharan African countries, where it is the second most common cause in this age group after *N. meningitidis* [26]. Group B Streptococcus (GBS) and *Escherichia coli* remain the leading causes of bacterial meningitis in neonates with little change in incidence rates over time [4,12]. Intrapartum antibiotic prophylaxis has reduced the risk of early onset, but not late-onset GBS meningitis [28,29]. Although, *Listeria monocytogenes* is an uncommon cause, it should be considered in neonates.

3. Pathophysiology of Bacterial Meningitis

Meningitis develops after the pathogen invades the CNS either though hematogenous route (bacteremia) or by direct extension secondary to sinusitis or mastoiditis and multiplies in the subarachnoid space. The presence of bacteria in the subarachnoid space leads to activation of the immune response, resulting in bacterial lysis. The presence of bacterial particles triggers a further inflammatory response with on-going migration of neutrophils across the blood–brain barrier and continuous cytokine and chemokine release (including IL-1B or CXCL1,2,5) (Figure 1) [7,30–35]. A persistent inflammatory state subsequently leads to decreased cerebral perfusion, cerebral edema, raised intracranial pressure, metabolic disturbances, and vasculitis, all contributing to neuronal injury and ischemia [32].

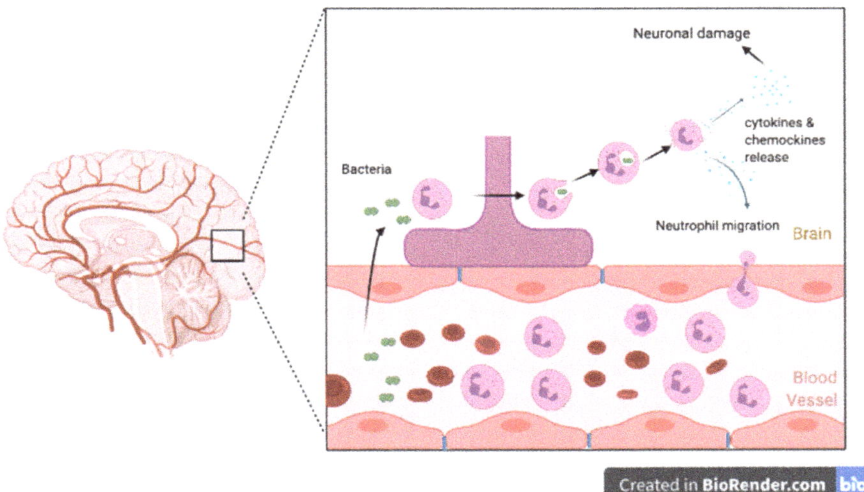

Figure 1. Pathophysiology of Neuronal Damage due to Bacterial Meningitis.

4. Bacterial Meningitis Complications

There are many complications that are associated with bacterial meningitis. These include short-term complications such as seizures, focal neurological deficits and subdural effusions, and long-term complications such as hearing loss, cognitive impairment, hydrocephalus, learning disability, and epilepsy [8,36–38]. Neurological sequelae are more likely to happen in LMICs compared with HICs [37] due to delayed presentation to medical services, lack of access to healthcare and limited resources. Additionally, complications are likely under reported in LMICs (Table 2) [7].

Table 2. Long and Short-term Neurological Complications following pneumococcal and meningococcal meningitis in Low and Middle Income Countries (LMICs) and High Income Countries (HICs).

	Pneumococcal Meningitis		Meningococcal Meningitis		References
	LMICs	**HICs**	**LMICs**	**HICs**	
Focal deficits	12%	3–14%	2–4%	3%	[13,39–42]
Hearing loss	25%	14–32%	19–23%	4%	[10,13,43–45]
Seizures	45–63%	15–48%	17–33%	2%	[40,45–47]
Cognitive impairment	4–41%	N/A *	4%	12–19%	[10,13,43,48,49]

*: Not avilable.

4.1. Short-Term Complications

4.1.1. Subdural Effusion

Subdural effusions occur in 20–39% of children with bacterial meningitis [50–52]. Subdural effusion is most common in infants (age < one year) compared to older children [8,51]. There is no significant difference in subdural effusions complication due to Hib meningitis compared to *S. pneumoniae* and *N. meningitidis* [51]. Most subdural effusions in the context of bacterial meningitis are asymptomatic, resolve spontaneously and rarely require intervention. Indications for drainage include an infected effusion or empyema, focal neurological signs or symptoms, or increased intracranial pressure [51].

4.1.2. Focal Neurological Deficit

Focal neurological deficit refers to a set of signs and symptoms resulting from a lesion localized to a specific anatomical site in central nervous system [53]. Examples include isolated limb weakness or hemiparesis, visual deficit or a speech impediment. They are estimated to occur in 3–14% of bacterial meningitis cases [41,42].

Acute focal neurological deficits after bacterial meningitis are usually due to ischemic stroke, but can also occur due to subdural empyema, cerebral abscess or intracranial bleeding [7]. Focal neurological deficits generally improve over months to years after the initial insult. Abscess collections can be surgically drained which usually leads to complete symptom resolution while focal deficits resulting from an ischemic event take longer to resolve. In a recent study, children with stroke following bacterial meningitis compared to children without stroke were less likely to have a normal neurological exam at discharge (21% vs. 76%) and within a seven years follow-up period (31% vs. 74%) [54]. Focal neurological deficits that persist after acute infection are not a very common complication after bacterial meningitis in childhood (Table 2); however, the incidence is highest for infections occurring in the first year of life [55].

4.2. Long-Term Complications

4.2.1. Hearing Loss

Sensorial hearing loss is the most widely reported neurological sequelae of bacterial meningitis [7,27,44,51,56–59]. Hearing loss may develop from both the direct spread of bacterial products and as a result of the host inflammatory response in the meninges and CSF. When bacteria reach the cochlea, a severe labyrinthitis results, which leads to blood-labyrinth barrier breakage, and ultimately meningitis-associated hearing loss [7]. Around 10% of children with bacterial meningitis develop unilateral or bilateral sensorineural hearing loss [37,60]; 5% of children develop bilateral severe or profound hearing loss [37]. Hearing loss is a more common complication in infections caused by *S. pneumoniae* (14–32%), compared with *N. meningitidis* (4–23%) and *H. influenzae* (20%) [44,60]. Children with hearing loss are at risk of further developing balance disturbances [61] and speech and language delay [62], and are therefore at higher risk of having long-term behavioral problems [63].

Reversible deafness (i.e., transient hearing loss) has been documented in long term follow up of children with pneumococcal meningitis [7,61]. For example, a study done among children with pneumococcal meningitis in Bangladesh reported that 33% of children in short-term follow-up (30–40 days) had hearing loss, but only 18% had persistent hearing loss in long-term follow-up (6–24 months) following hospital. The difference was attributed to recovery of transient hearing impairment [43].

4.2.2. Cognitive Impairment

Due to the irreversible neuronal damage that occurs during bacterial meningitis, the risk of developing long-term cognitive deficits and learning difficulties are significant [3,48]. The rates of cognitive impairment world-wide are difficult to estimate because there is no standardized method of measuring it and long-term data on meningitis survivors are rarely available.

In a Dutch study, 680 children between the age of 4 and 13 years who survived bacterial meningitis were followed up for 6 years after their meningitis episode. The study followed their educational, behavioral and general health issues. The survivors were compared to a control group of healthy school-age siblings and peers, with similar socioeconomic background. It was found that 30% of children with meningitis had problems with school achievement or concentration. Additionally, these children repeated a school year twice as often as the control group (16% vs. 8%). Moreover, the post-meningitis group were referred to special-needs school four times more frequently compared to control group [48].

In a Danish nationwide population-based cohort study, adults who had a bacterial meningitis in childhood were compared to control group that included a general population of the same age and sex, their siblings and the siblings of meningitis patients. Adults who had childhood bacterial meningitis had lower educational achievements and economic self-sufficiency compared to control group. By the age of 35 years, 11%, 10.2% and 5.5% fewer had completed high school in meningococcal pneumococcal, and Hib meningitis, respectively. Additionally, 7.9%, 8.9% and 6.5% fewer had obtained a higher education in meningococcal pneumococcal, and Hib meningitis, respectively. Additionally, 3.8%, 10.6% and 4.3% had lower economic self-sufficiency in meningococcal pneumococcal, and Hib meningitis, respectively [49]. In a study done in Bangladesh, short (30–40 days) and long-term (6–24 months) follow-up revealed that 41% in both groups had deficits in mental development and 49% and 35%, respectively, had psychomotor delay [43]. Another example, in a study that was done in Brazil, 5.88% children developed learning disabilities, and 7.35% children had developmental delay [40].

Finally, psychiatric disease including anxiety and depression is likely under-recognized and underreported in meningitis survivors and contributes to cognitive difficulties and overall quality of life [64].

4.2.3. Seizures and Epilepsy

One of the clinical presentations of bacterial meningitis is seizures [65,66]. In cases of bacterial meningitis with seizures, if seizures develop early during the illness and are easily controlled, permanent neurological complications are rarely of concern. However, if seizures are prolonged, difficult to control or develop 72 h after admission, neurological sequelae are more likely to occur and are usually suggestive of a cerebrovascular event [44,67]. In HICs, 1–5% of epilepsy cases are presumed to be due to CNS infection; including bacterial meningitis [68]. In Sub-Saharan Africa 26% of patients have epilepsy attributed to CNS infection [69].

In a neonatal bacterial meningitis study seizures have been more commonly associated with GBS then *E. coli* (41% vs. 25%) [21]. 71% of children who late seizure after bacterial meningitis had permanent focal neurological deficit [70].

4.2.4. Hydrocephalus

Hydrocephalus incidence is around 7% of bacterial meningitis in children [71] and it is more common in neonates and infants; 25% [72,73]. It is more common in neonatal Gram negative meningitis [57]. Hydrocephalus may develop at the beginning of the illness or weeks later after diagnosis with bacterial meningitis. The most common type of hydrocephalus after bacterial meningitis is communicating hydrocephalus; seen in up to 52% of cases with hydrocephalus [74]. In communicating hydrocephalus CSF flows freely between the ventricles but is not adequately reabsorbed back into the blood stream. Depending on the size of hydrocephalus and resulting neurologic impairment temporary or permanent ventricular shunt placement may be required [75].

5. Risk Factors

There are many risk factors associated with neurological complications in bacterial meningitis (Table 3).

Table 3. Risk factors for developing neurological complication in Bacterial Meningitis.

Risk Factor	% with Neurological Complications	References
Young Age (infants < 12 months)	71%	[9]
Etiology: *S. pneumoniae*	75%	[9]
Altered Level of Consciousness on Presentation	82%	[9]
Delayed Presentation	N/A *	[6,9]
Delayed initiation of antibiotics	N/A *	[76]

* N/A: Not available.

In general, infants are at higher risk of developing neurological complications compared to older children [3,9,23]. 71% of infants (aged < one year) with bacterial meningitis develop neurological complications compared to 38% in children aged one to five years and 10% in those aged six to 16 years [9]. Children younger than 12 months at time of diagnosis with bacterial meningitis are at increased risk of developing hydrocephalus, subdural effusion, seizure disorder and hearing loss [76]. Altered level of consciousness is associated with poor prognosis [6,9]. 82% of children with bacterial meningitis who developed neurological complications had altered level of consciousness on presentation; where, 39% of children with bacterial meningitis who did not develop neurological complications had altered level of consciousness [9]. The longer the duration that the child was unconscious, the worst the outcome is.

In bacterial meningitis, delayed presentation to hospital increase the risk of subdural effusion, hydrocephalus, hearing impairment and seizure disorder [76]. Although delayed presentation is one of the known risk factors for developing neurological complications, there is no universal definition for the duration of the delay. In one study, children admitted with duration of illness <48 h had a lower incidence of neurological complication (40%) compared to children who were admitted after 48 h of illness [9]. Children with *S. pneumoniae* meningitis have a higher risk of developing neurological complication (75% of *S. pneumoniae* meningitis cases) compared to *N. meningitidis* (25%) and Hib (20%) [9]. *S. pneumoniae* compared to *N. meningitidis* and Hib is associated with higher risk of symptomatic seizures, hydrocephalus, hearing loss and mental retardation [76]. Delay in starting antibiotics beyond 24–72 h has a poor prognosis and leads to increased risk of severe neurological complication such as hydrocephalus, subdural effusion, hearing loss, and seizure disorder [76]. In summary, young age, delayed presentation and *S. pneumoniae* as an etiologic agent were associated with increased risk of neurological complications in both HIC and LMIC settings [11,15,54,71,76–78].

6. Prevention of Neurological Complication

6.1. Primary Prevention

The most effective prevention of neurological complications from bacterial meningitis is preventing the infection though infant and childhood vaccination programs. Despite the development of multiple vaccines against the organisms causing bacterial meningitis, there continue to be many meningitis outbreaks caused by vaccine-preventable organisms [79,80]. There are currently vaccines against 3 of the organisms that cause bacterial meningitis: Hib, *N. meningitidis* (capsular groups A, B, C, W and Y) and 23 of the >90 serotypes of *S. pneumoniae* [4,15]. Hib conjugate vaccine targets only type b *H. influenzae*, and is given as three or four doses before 18 months of age [81]. There are two types of vaccine against *N. meningitidis*: Conjugate vaccines against capsular groups A, C, W, and Y and protein vaccines against group B. There are two types of vaccines against *S. pneumoniae*: Pneumococcal conjugate vaccines (PCV 10 against 10 serotypes, PCV 13 against 13 serotypes) and polysaccharide vaccine against 23 serotypes which is not routinely used in healthy children [82].

Routine vaccination can lead to development of community protection by indirect effect prevention of transmission within a population [83]. Since the introduction of pneumococcal conjugate vaccines (PCVs), the overall incidence of invasive pneumococcal disease (IPD) has dropped significantly, including in unimmunized children, highlighting these indirect effects [84,85]. For example, in South Africa, Morocco, Gambia, Mozambique, Kenya and Burkina Faso, 32–81% reduction in IPD has been reported after PCV introduction, with highest reduction in children aged under 24 months (55–89%) [86]. However, infections caused by non-vaccine serotypes infections have increased in some countries. In the United States, the proportion of IPD caused by non-vaccine serotype increased from 6% to 38% after the introduction of PCV7 vaccine; sometimes referred to as serotype replacement [85,87,88]. Overall, IPD incidence dropped.

The highest rate of meningococcal disease worldwide is in the Sub-Saharan Africa, specifically in the "meningitis belt" region where major epidemics occur every 5–12 years [89]. After the introduction of MenA vaccine to the Sub-Saharan Africa in 2010, there has been a 99% reduction in group A meningitis in this region [90] and capsular group W is the currently the commonest [91]. Rate of meningococcal meningitis are much lower in other parts of Sub-Saharan Africa, although longitudinal surveillance outside of the meningitis belt is limited [92]. Following wide-spread introduction of Hib vaccine, *S. pneumoniae* accounted for 65% of acute bacterial meningitis cases in Malawi, while the rates of *N. meningitidis* have remained constant at <5% [90].

6.2. Secondary Prevention of Complications

6.2.1. Antibiotic Therapy

It is important to have a high clinical suspicion of bacterial meningitis and start appropriate treatment without delay [93,94]. The empiric antibiotic choice should be based on the most likely causative agent for patient's age [73,95,96]. In children, third-generation cephalosporins, such as cefotaxime or ceftriaxone, are the usual empirical choice to cover the most common organisms—*S. pneumoniae* and *N. meningitidis*. Ampicillin should be added to cover *L. monocytogenes* in very young children; some guidelines recommended for younger than 3 months, others recommended this for those younger than 1 month [95,96].

6.2.2. Corticosteroids

Neuronal damage due to acute bacterial meningitis is not only due to bacterial invasion to the subarachnoid space, but also due to the host's inflammatory response to this invasion [36]. The only widely researched agent that can limit subarachnoid inflammation is dexamethasone. The recommendation for the use of dexamethasone in bacterial meningitis unfortunately cannot be generalized and depends on the causative organism and the ability to administer dexamethasone within 1–12 h of administration of antibiotics [27,40,44,97].

In infant and children with Hib meningitis, administration of dexamethasone with antibiotics has shown a significant reduction in neurological sequelae rate (17%) compared to antibiotic only (23.4%) [27,59]. Additionally, rates of hearing loss were lower with dexamethasone use (12.9%) compared to antibiotics only (17.4%) [59,96,98,99]. In a large study done in Malawi, children with *H. influenzae* meningitis who received dexamethasone were less likely to have neurological sequelae compared to a placebo group (27% vs. 40%) [13]. Therefore, it is generally recommended that dexamethasone be administered before or with the first dose of antibiotics [1,13,96] when *H. influenzae* is confirmed or strongly suspected.

On the other hand, the use of dexamethasone in infants and children with pneumococcal meningitis is controversial as it has not been clearly proven to change the outcome [1,13,36,98,99]. Additionally, the use of dexamethasone for meningococcal meningitis was not proven to be effective in reduce neurological sequelae [96].

Better understanding of specific microbial and host factors contributing to CNS infection and inflammatory response may help in identification of new therapeutic targets and specific immunomodulatory regimens.

7. Long-Term Follow Up

All children who are diagnosed with meningitis should have a hearing assessment done before discharge or 1 month within discharge; even if hearing loss is not clinically suspected [13,61]. It is critical to perform audiology assessment month after diagnosis or earlier if possible, as up to 90% of children's cochlea with hearing loss due to meningitis can ossify, preventing appropriate treatment with cochlear implants [100,101].

Children with seizure disorders require antiepileptic medication and should ideally have long-term follow up by a neurologist [7,8,67,70]. Children with hearing loss and/ or intellectual disability will need neurodevelopmental follow up and support for speech, language and social development. Mental health concerns and psychiatric problems are likely underreported in children who were diagnosed with meningitis and periodic mental health assessments by an appropriate specialist should be included into their long-term care [64].

8. Conclusions

Children with bacterial meningitis are at risk of developing neurological complications that include focal neurological deficits, subdural effusion, hearing loss, cognitive impairment, seizure disorder, and hydrocephalus. There is a need to optimize utilization of available vaccines and to develop vaccines for pathogens implicated in neonatal meningitis (GBS and *E. coli*). For children diagnosed with bacterial meningitis, starting antibiotic therapy without delay is critical for a good prognosis and to reduce the risk of developing neurological complications. So far, steroids are the only drug that can control inflammatory response, but effectiveness is limited to specific situations.

Author Contributions: A.Z., H.M., and M.S. conceived and designed the structure of the manuscript. A.Z. drafted the original version of the manuscript. All authors edited and approved the final version. All authors have read and agreed to the published version of the manuscript.

Funding: None.

Institutional Review Board Statement: Not applicable.

Informed Consent Statement: Not applicable.

Data Availability Statement: Not applicable.

Acknowledgments: MS is supported via salary awards from the BC Children's Hospital Foundation, the Canadian Child Health Clinician Scientist Program and the Michael Smith Foundation for Health Research.

Conflicts of Interest: MS has been an investigator on projects funded by GlaxoSmithKline, Merck, Pfizer, Sanofi-Pasteur, Seqirus, Symvivo and VBI Vaccines. All funds have been paid to his institute, and he has not received any personal payments.

References

1. Grandgirard, D.; Leib, S.L. Strategies to prevent neuronal damage in paediatric bacterial meningitis. *Curr. Opin. Pediatr.* **2006**, *18*, 112–118. [CrossRef] [PubMed]
2. Sonko, A.M.; Dube, F.S.; Okoi, C.B.; Diop, A.; Thiongane, A.; Senghore, M.; Ndow, P.; Worwui, A.; Faye, P.M.; Dieye, B.; et al. Changes in the Molecular Epidemiology of Pediatric Bacterial Meningitis in Senegal after Pneumococcal Conjugate Vaccine Introduction. *Clin. Infect. Dis.* **2019**, *69*, S156–S163. [CrossRef] [PubMed]
3. Shieh, H.H.; Ragazzi, S.L.B.; Gilio, A.E. Risk factors for neurological complications and sequelae in childhood acute bacterial meningitis. *J. Pediatr.* **2012**, *88*, 184. [CrossRef]
4. Thigpen, M.C.; Whitney, C.G.; Messonnier, N.E.; Zell, E.R.; Lynfield, R.; Hadler, J.L.; Harrison, L.H.; Farley, M.M.; Reingold, A.; Bennett, N.M.; et al. Bacterial Meningitis in the United States, 1998–2007. *N. Engl. J. Med.* **2011**, *364*, 2016–2025. [CrossRef]
5. Yogev, R.; A Guzmancottrill, J. Bacterial Meningitis in Children. *Drugs* **2005**, *65*, 1097–1112. [CrossRef]
6. Pelkonen, T.; Roine, I.; Monteiro, L.; Correia, M.; Pitkäranta, A.; Bernardino, L.; Peltola, H. Risk Factors for Death and Severe Neurological Sequelae in Childhood Bacterial Meningitis in Sub-Saharan Africa. *Clin. Infect. Dis.* **2009**, *48*, 1107–1110. [CrossRef]
7. Lucas, M.J.; Brouwer, M.C.; van de Beek, D. Neurological sequelae of bacterial meningitis. *J. Infect.* **2016**, *73*, 18–27. [CrossRef] [PubMed]
8. Namani, S.A.; Koci, B.M.; Milenković, Z.; Koci, R.; Qehaja-Buçaj, E.; Ajazaj, L.; Mehmeti, M.; Ismaili-Jaha, V. Early neurologic complications and long-term sequelae of childhood bacterial meningitis in a limited-resource country (Kosovo). *Child's Nerv. Syst.* **2012**, *29*, 275–280. [CrossRef]
9. Namani, S.; Milenković, Z.; Koci, B. A prospective study of risk factors for neurological complications in childhood bacterial meningitis. *J. Pediatr.* **2013**, *89*, 256–262. [CrossRef] [PubMed]
10. Ispahani, P.; Slack, R.C.B.; Donald, E.F.; Weston, V.C.; Rutter, N. Twenty year surveillance of invasive pneumococcal disease in Nottingham: Serogroups responsible and implications for immunisation. *Arch. Dis. Child.* **2004**, *89*, 757–762. [CrossRef]
11. Softić, I.; Tahirović, H.; Hasanhodžić, M. Neonatal bacterial meningitis: Results from a cross-sectional hospital based study. *Acta Med. Acad.* **2015**, *44*, 117–123.
12. Baud, O.; Aujard, Y. *Neonatal Bacterial Meningitis*; Elsevier: Amsterdam, The Netherlands, 2013; Volume 112, pp. 1109–1113.
13. Molyneux, E.M.; Walsh, A.L.; Forsyth, H.; Tembo, M.; Mwenechanya, J.; Kayira, K.; Bwanaisa, L.; Njobvu, A.; Rogerson, S.; Malenga, G. Dexamethasone treatment in childhood bacterial meningitis in Malawi: A randomised controlled trial. *Lancet* **2002**, *360*, 211–218. [CrossRef]

14. Zunt, J.R.; Kassebaum, N.J.; Blake, N.; Glennie, L.; Wright, C.; Nichols, E.; Abd-Allah, F.; Abdela, J.; Abdelalim, A.; A Adamu, A.; et al. Global, regional, and national burden of meningitis, 1990–2016: A systematic analysis for the Global Burden of Disease Study 2016. *Lancet Neurol.* **2018**, *17*, 1061–1082. [CrossRef]
15. McIntyre, P.B.; O'Brien, K.L.; Greenwood, B.; van de Beek, D. Effect of vaccines on bacterial meningitis worldwide. *Lancet* **2012**, *380*, 1703–1711. [CrossRef]
16. McAlpine, A.; Sadarangani, M. Meningitis vaccines in children. *Curr. Opin. Infect. Dis.* **2019**, *32*, 510–516. [CrossRef] [PubMed]
17. Mwenda, J.M.; Soda, E.; Weldegebriel, G.; Katsande, R.; Biey, J.N.-M.; Traore, T.; De Gouveia, L.; Du Plessis, M.; Von Gottberg, A.; Antonio, M.; et al. Pediatric Bacterial Meningitis Surveillance in the World Health Organization African Region Using the Invasive Bacterial Vaccine-Preventable Disease Surveillance Network, 2011–2016. *Clin. Infect. Dis.* **2019**, *69*, 49–57. [CrossRef]
18. Bettinger, J.A.; Scheifele, D.W.; Kellner, J.D.; Halperin, S.A.; Vaudry, W.; Law, B.; Tyrrell, G. The effect of routine vaccination on invasive pneumococcal infections in Canadian children, Immunization Monitoring Program, Active 2000–2007. *Vaccine* **2010**, *28*, 2130–2136. [CrossRef] [PubMed]
19. Bettinger, J.A.; Scheifele, D.W.; Halperin, S.A.; Kellner, J.D.; Vanderkooi, O.G.; Schryvers, A.; De Serres, G.; Alcantara, J. Evaluation of meningococcal serogroup C conjugate vaccine programs in Canadian children: Interim analysis. *Vaccine* **2012**, *30*, 4023–4027. [CrossRef] [PubMed]
20. Leal, J.; Vanderkooi, O.G.; Church, D.L.; MacDonald, J.; Tyrrell, G.J.; Kellner, J.D. Eradication of Invasive Pneumococcal Disease due to the Seven-valent Pneumococcal Conjugate Vaccine Serotypes in Calgary, Alberta. *Pediatr. Infect. Dis. J.* **2012**, *31*, e169–e175. [CrossRef] [PubMed]
21. Gaschignard, J.; Levy, C.; Romain, O.; Cohen, R.; Bingen, E.; Aujard, Y.; Boileau, P. Neonatal Bacterial Meningitis. *Pediatr. Infect. Dis. J.* **2011**, *30*, 212–217. [CrossRef]
22. Smith, P.B.; Garges, H.P.; Cotton, C.M.; Walsh, T.J.; Clark, R.H.; Benjamin, D.K.; Cotten, C.M. Meningitis in Preterm Neonates: Importance of Cerebrospinal Fluid Parameters. *Am. J. Perinatol.* **2008**, *25*, 421–426. [CrossRef]
23. McCormick, D.W.; Wilson, M.L.; Mankhambo, L.; Phiri, A.; Chimalizeni, Y.; Kawaza, K.; Denis, B.; Carrol, E.D.; Molyneux, E.M. Risk Factors for Death and Severe Sequelae in Malawian Children with Bacterial Meningitis, 1997–2010. *Pediatr. Infect. Dis. J.* **2013**, *32*, e54–e61. [CrossRef]
24. Schuchat, A.; Robinson, K.; Wenger, J.D.; Harrison, L.H.; Farley, M.; Reingold, A.L.; Lefkowitz, L.; Perkins, B.A. Bacterial Meningitis in the United States in 1995. *N. Engl. J. Med.* **1997**, 970–976. [CrossRef] [PubMed]
25. Cockcroft, D.W.; Nair, P. Methacholine test and the diagnosis of asthma. *J. Allergy Clin. Immunol.* **2012**, *130*, 556–557. [CrossRef]
26. Ceyhan, M.; Gürler, N.; Ozsurekci, Y.; Keser, M.; Aycan, A.E.; Gurbuz, V.; Salman, N.; Camcioglu, Y.; Dinleyici, E.C.; Ozkan, S.; et al. Meningitis caused by Neisseria Meningitidis, Hemophilus Influenzae Type B and Streptococcus Pneumoniae during 2005–2012 in Turkey. *Hum. Vaccines Immunother.* **2014**, *10*, 2706–2712. [CrossRef]
27. Duke, T.; Curtis, N.; Fuller, D.G. The management of bacterial meningitis in children. *Expert Opin. Pharmacother.* **2003**, *4*, 1227–1240. [CrossRef] [PubMed]
28. Phares, C.R.; Lynfield, R.; Farley, M.M.; Mohle-Boetani, J.; Harrison, L.H.; Petit, S.; Craig, A.S.; Schaffner, W.; Zansky, S.M.; Gershman, K.; et al. Epidemiology of Invasive Group B Streptococcal Disease in the United States, 1999–2005. *JAMA* **2008**, *299*, 2056–2065. [CrossRef]
29. Edmond, K.M.; Kortsalioudaki, C.; Scott, S.; Schrag, S.J.; Zaidi, A.K.; Cousens, S.; Heath, P.T. Group B streptococcal disease in infants aged younger than 3 months: Systematic review and meta-analysis. *Lancet* **2012**, *379*, 547–556. [CrossRef]
30. Kim, K.S. Pathogenesis of bacterial meningitis: From bacteraemia to neuronal injury. *Nat. Rev. Neurosci.* **2003**, *4*, 376–385. [CrossRef]
31. Koedel, U.; Klein, M.; Pfister, H.-W. New understandings on the pathophysiology of bacterial meningitis. *Curr. Opin. Infect. Dis.* **2010**, *23*, 217–223. [CrossRef]
32. Scheld, W.M.; Koedel, U.; Nathan, B.; Pfister, H. Pathophysiology of Bacterial Meningitis: Mechanism (s) of Neuronal Injury. *J. Infect. Dis.* **2002**, *186*, 225–233. [CrossRef]
33. Barichello, T.; Generoso, J.S.; Simões, L.R.; Goularte, J.A.; Petronilho, F.; Saigal, P.; Badawy, M.; Quevedo, J. Role of Microglial Activation in the Pathophysiology of Bacterial Meningitis. *Mol. Neurobiol.* **2016**, *53*, 1770–1781. [CrossRef]
34. Hoffman, O.; Weber, J.R. Review: Pathophysiology and treatment of bacterial meningitis. *Ther. Adv. Neurol. Disord.* **2009**, *2*, 401–412. [CrossRef]
35. Mook-Kanamori, B.B.; Geldhoff, M.; Van Der Poll, T.; Van De Beek, D. Pathogenesis and Pathophysiology of Pneumococcal Meningitis. *Clin. Microbiol. Rev.* **2011**, *24*, 557–591. [CrossRef]
36. Chaudhuri, A. Adjunctive dexamethasone treatment in acute bacterial meningitis. *Lancet Neurol.* **2004**, *3*, 54–62. [CrossRef]
37. Baraff, L.J.; Lee, S.I.; Schriger, D.L. Outcome of Bacterial Meningitis in Children a Meta-Analysis. *Pediatr. Infect. Dis. J.* **1993**, *12*, 389–394. [CrossRef]
38. Mahmoudi, S.; Zandi, H.; Pourakbari, B.; Ashtiani, M.T.H.; Mamishi, S. Acute Bacterial Meningitis among Children Admitted into an Iranian Referral Children's Hospital. *Jpn. J. Infect. Dis.* **2013**, *66*, 503–506. [CrossRef] [PubMed]
39. Smith, A.; Bradley, A.; Wall, R.; McPherson, B.; Secka, A.; Dunn, D.; Greenwood, B. Sequelae of epidemic meningococcal meningitis in Africa. *Trans. R. Soc. Trop. Med. Hyg.* **1988**, *82*, 312–320. [CrossRef]
40. Casella, E.B.; Cypel, S.; Osmo, A.A.; Okay, Y.; Lefèvre, B.H.; Lichtig, I.; Marques-Dias, M.J. Sequelae from meningococcal meningitis in children: A critical analysis of dexamethasone therapy. *Arq. Neuro-Psiquiatr.* **2004**, *62*, 421–428. [CrossRef] [PubMed]

41. Østergaard, C.; Konradsen, H.B.; Samuelsson, S. Clinical presentation and prognostic factors of Streptococcus pneumoniae meningitis according to the focus of infection. *BMC Infect. Dis.* **2005**, *5*. [CrossRef] [PubMed]
42. Klobassa, D.S.; Zoehrer, B.; Paulke-Korinek, M.; Gruber-Sedlmayr, U.; Pfurtscheller, K.; Strenger, V.; Sonnleitner, A.; Kerbl, R.; Ausserer, B.; Arocker, W.; et al. The burden of pneumococcal meningitis in Austrian children between 2001 and 2008. *Eur. J. Nucl. Med. Mol. Imaging* **2014**, *173*, 871–878. [CrossRef]
43. Saha, S.K.; Khan, N.Z.; Ahmed, A.S.M.N.U.; Amin, M.R.; Hanif, M.; Mahbub, M.; Anwar, K.S.; Qazi, S.A.; Kilgore, P.; Baqui, A.H.; et al. Neurodevelopmental Sequelae in Pneumococcal Meningitis Cases in Bangladesh: A Comprehensive Follow-up Study. *Clin. Infect. Dis.* **2009**, *48*, 90–96. [CrossRef]
44. Arditi, M.; Mason, E.O.; Bradley, J.S.; Tan, T.Q.; Barson, W.J.; Schutze, G.E.; Wald, E.R.; Givner, L.B.; Kim, K.S.; Yogev, R.; et al. Three-Year Multicenter Surveillance of Pneumococcal Meningitis in Children: Clinical Characteristics, and Outcome Related to Penicillin Susceptibility and Dexamethasone Use. *Pediatrics* **1998**, *102*, 1087–1097. [CrossRef] [PubMed]
45. Karppinen, M.; Pelkonen, T.; Roine, I.; Cruzeiro, M.L.; Peltola, H.; Pitkäranta, A. Hearing impairment after childhood bacterial meningitis dependent on etiology in Luanda, Angola. *Int. J. Pediatr. Otorhinolaryngol.* **2015**, *79*, 1820–1826. [CrossRef] [PubMed]
46. King, B.; Richmond, P. Pneumococcal meningitis in Western Australian children: Epidemiology, microbiology and outcome. *J. Paediatr. Child Health* **2004**, *40*, 611–615. [CrossRef] [PubMed]
47. Osuorah, D.; Shah, B.; Manjang, A.; Secka, E.; Ekwochi, U.; Ebenebe, J. Outbreak of serotype W135 Neisseria meningitidis in central river region of the Gambia between February and June 2012: A hospital-based review of paediatric cases. *Niger. J. Clin. Pr.* **2014**, *18*, 41–47.
48. Koomen, I.; Grobbee, D.; Jennekens-Schinkel, A.; Roord, J.; Furth, A. Parental perception of educational, behavioural and general health problems in school-age survivors of bacterial meningitis. *Acta Paediatr.* **2007**, *92*, 177–185. [CrossRef]
49. Roed, C.; Omland, L.H.; Skinhoj, P.; Rothman, K.J.; Sorensen, H.T.; Obel, N. Educational Achievement and Economic Self-sufficiency in Adults after Childhood Bacterial Meningitis. *JAMA* **2013**, *309*, 1714–1721. [CrossRef]
50. Taylor, H.G.; Schatschneider, C.; Minich, N.M. Longitudinal outcomes of Haemophilus influenzae meningitis in school-age children. *Neuropsychology* **2000**, *14*, 509–518. [CrossRef] [PubMed]
51. Snedeker, J.D.; Kaplan, S.L.; Dodge, P.R.; Holmes, S.J.; Feigin, R.D. Subdural effusion and its relationship with neurologic sequelae of bacterial meningitis in infancy: A prospective study. *Pediatrics* **1990**, *86*, 163–170.
52. Olarte, L.; Barson, W.J.; Barson, R.M.; Lin, P.L.; Romero, J.R.; Tan, T.Q.; Givner, L.B.; Bradley, J.S.; Hoffman, J.A.; Hultén, K.G.; et al. Impact of the 13-Valent Pneumococcal Conjugate Vaccine on Pneumococcal Meningitis in US Children. *Clin. Infect. Dis.* **2015**, *61*, 767–775. [CrossRef]
53. Wippold, F.J. Focal neurologic deficit. *Am. J. Neuroradiol.* **2008**, *29*, 1998–2000. [PubMed]
54. Dunbar, M.; Shah, H.; Shinde, S.; Vayalumkal, J.; Vanderkooi, O.G.; Wei, X.-C.; Kirton, A. Stroke in Pediatric Bacterial Meningitis: Population-Based Epidemiology. *Pediatr. Neurol.* **2018**, *89*, 11–18. [CrossRef]
55. Tibussek, D.; Sinclair, A.; Yau, I.; Teatero, S.; Fittipaldi, N.; Richardson, S.E.; Mayatepek, E.; Jahn, P.; Askalan, R. Late-Onset Group B Streptococcal Meningitis Has Cerebrovascular Complications. *J. Pediatr.* **2015**, *166*, 1187–1192. [CrossRef] [PubMed]
56. Masri, A.; Alassaf, A.; Khuri-Bulos, N.; Zaq, I.; Hadidy, A.; Bakri, F.G. Recurrent meningitis in children: Etiologies, outcome, and lessons to learn. *Child's Nerv. Syst.* **2018**, *34*, 1541–1547. [CrossRef]
57. Unhanand, M.; Mustafa, M.M.; McCracken, G.H.; Nelson, J.D. Gram-negative enteric bacillary meningitis: A twenty-one-year experience. *J. Pediatr.* **1993**, *122*, 15–21. [CrossRef]
58. Brouwer, M.C.; Wijdicks, E.F.M.; Van De Beek, D. What's new in bacterial meningitis. *Intensiv. Care Med.* **2016**, *42*, 415–417. [CrossRef]
59. Wang, Y.; Liu, X.; Wang, Y.; Liu, Q.; Kong, C.; Xu, G. Meta-analysis of adjunctive dexamethasone to improve clinical outcome of bacterial meningitis in children. *Child's Nerv. Syst.* **2017**, *34*, 217–223. [CrossRef]
60. Dodge, P.R.; Davis, H.; Feigin, R.D.; Holmes, S.J.; Kaplan, S.L.; Jubelirer, D.P.; Stechenberg, B.W.; Hirsh, S.K. Prospective Evaluation of Hearing Impairment as a Sequela of Acute Bacterial Meningitis. *N. Engl. J. Med.* **1984**, *311*, 869–874. [CrossRef]
61. Sáez-Llorens, X.; McCracken, G.H. *Acute Bacterial Meningitis beyond the Neonatal Period*; Elsevier: Amsterdam, The Netherlands, 2008; pp. 284–291.
62. Yoshinaga-Itano, C.; Sedey, A.L.; Coulter, D.K.; Mehl, A.L. Language of Early- and Later-identified Children with Hearing Loss. *Pediatrics* **1998**, *102*, 1161–1171. [CrossRef]
63. Hall, W.C.; Li, N.; Dye, T.D.V. Influence of Hearing Loss on Child Behavioral and Home Experiences. *Am. J. Public Health* **2018**, *108*, 1079–1081. [CrossRef] [PubMed]
64. Kostenniemi, U.J.; Bazan, A.; Karlsson, L.; Silfverdal, S.-A. Psychiatric Disabilities and Other Long-term Consequences of Childhood Bacterial Meningitis. *Pediatr. Infect. Dis. J.* **2020**, *40*, 26–31. [CrossRef]
65. Kostenniemi, U.J.; Norman, D.; Borgström, M.; Silfverdal, S.A. The clinical presentation of acute bacterial meningitis varies with age, sex and duration of illness. *Acta Paediatr.* **2015**, *104*, 1117–1124. [CrossRef] [PubMed]
66. Curtis, S.; Stobart, K.; VanderMeer, B.; Simel, D.L.; Klassen, T. Clinical Features Suggestive of Meningitis in Children: A Systematic Review of Prospective Data. *Pediatrics* **2010**, *126*, 952–960. [CrossRef]
67. Namani, S.A.; Kuchar, E.; Koci, R.; Mehmeti, M.; Dedushi, K. Early symptomatic and late seizures in Kosovar children with bacterial meningitis. *Child's Nerv. Syst.* **2011**, *27*, 1967–1971. [CrossRef] [PubMed]
68. Murthy, J.M.K.; Prabhakar, S. Bacterial meningitis and epilepsy. *Epilepsia* **2008**, *49*, 8–12. [CrossRef]

69. Preux, P.-M.; Druet-Cabanac, M. Epidemiology and aetiology of epilepsy in sub-Saharan Africa. *Lancet Neurol.* **2005**, *4*, 21–31. [CrossRef]
70. Pomeroy, S.L.; Holmes, S.J.; Dodge, P.R.; Feigin, R.D. Seizures and Other Neurologic Sequelae of Bacterial Meningitis in Children. *N. Engl. J. Med.* **1990**, *323*, 1651–1657. [CrossRef]
71. De Jonge, R.C.; Van Furth, A.M.; Wassenaar, M.; Gemke, R.J.; Terwee, C.B. Predicting sequelae and death after bacterial meningitis in childhood: A systematic review of prognostic studies. *BMC Infect. Dis.* **2010**, *10*, 232. [CrossRef]
72. Pong, A.; Bradley, J.S. Bacterial Meningitis and the Newborn Infant. *Infect. Dis. Clin. N. Am.* **1999**, *13*, 711–733. [CrossRef]
73. Ouchenir, L.; Renaud, C.; Khan, S.; Bitnun, A.; Boisvert, A.-A.; McDonald, J.; Bowes, J.; Brophy, J.; Barton, M.; Ting, J.; et al. The Epidemiology, Management, and Outcomes of Bacterial Meningitis in Infants. *Pediatrics* **2017**, *140*, e20170476. [CrossRef]
74. Huo, L.; Fan, Y.; Jiang, C.; Gao, J.; Yin, M.; Wang, H.; Yang, F.; Cao, Q. Clinical Features of and Risk Factors for Hydrocephalus in Childhood Bacterial Meningitis. *J. Child Neurol.* **2018**, *34*, 11–16. [CrossRef]
75. Tamber, M.S.; Klimo, P.; Mazzola, C.A.; Flannery, A.M. Pediatric hydrocephalus: Systematic literature review and evidence-based guidelines. Part 8: Management of cerebrospinal fluid shunt infection. *J. Neurosurg. Pediatr.* **2014**, *14*, 60–71. [CrossRef]
76. Teixeira, D.C.; Diniz, L.M.O.; Guimarães, N.S.; Moreira, H.M.D.A.S.; Teixeira, C.C.; Romanelli, R.M.D.C. Risk factors associated with the outcomes of pediatric bacterial meningitis: A systematic review. *J. Pediatr.* **2020**, *96*, 159–167. [CrossRef]
77. Feavers, I.; Pollard, A.J.; Sadarangani, M. *Handbook of Meningococcal Disease Management*; Springer International Publishing: Berlin/Heidelberg, Germany, 2016. [CrossRef]
78. Pace, D.; Pollard, A.J. Meningococcal disease: Clinical presentation and sequelae. *Vaccine* **2012**, *30*, B3–B9. [CrossRef]
79. Aku, F.Y.; Lessa, F.C.; Asiedu-Bekoe, F.; Balagumyetime, P.; Ofosu, W.; Farrar, J.; Ouattara, M.; Vuong, J.T.; Issah, K.; Opare, J.; et al. Meningitis Outbreak Caused by Vaccine-Preventable Bacterial Pathogens—Northern Ghana, 2016. *Morb. Mortal. Wkly. Rep.* **2017**, *66*, 806–810. [CrossRef]
80. Greenwood, B. Editorial: 100 years of epidemic meningitis in West Africa—Has anything changed? *Trop. Med. Int. Health* **2006**, *11*, 773–780. [CrossRef] [PubMed]
81. US Department of Health and Human Services Center for Disease Control and Prevention. Haemophilus Influenzae Type b (Hib) Vaccine: What You Need to Know Hib Vaccine. 2019. Available online: https://www.cdc.gov/vaccines/hcp/vis/vis-statements/hib.html (accessed on 2 January 2021).
82. US Department of Health and Human Services Center for Disease Control and Prevention. Pneumococcal Polysaccharide Vaccine (PPSV23): What You Need to Know. 2019. Available online: https://www.cdc.gov/vaccines/hcp/vis/vis-statements/ppv.html (accessed on 2 January 2021).
83. Bijlsma, M.W.; Brouwer, M.C.; Spanjaard, L.; Van De Beek, D.; Van Der Ende, A. A Decade of Herd Protection after Introduction of Meningococcal Serogroup C Conjugate Vaccination. *Clin. Infect. Dis.* **2014**, *59*, 1216–1221. [CrossRef] [PubMed]
84. Whitney, C.G.; Farley, M.M.; Hadler, J.; Harrison, L.H.; Bennett, N.M.; Lynfield, R.; Reingold, A.; Cieslak, P.R.; Pilishvili, T.; Jackson, D.; et al. Decline in Invasive Pneumococcal Disease after the Introduction of Protein–Polysaccharide Conjugate Vaccine. *N. Engl. J. Med.* **2003**, *348*, 1737–1746. [CrossRef] [PubMed]
85. Nigrovic, L.E.; Kuppermann, N.; Malley, R. For the Bacterial Meningitis Study Group of the Pediatric Emergency Medicine Collaborative Research Committee of the American Academy of Pediatrics Children with Bacterial Meningitis Presenting to the Emergency Department during the Pneumococcal Conjugate Vaccine Era. *Acad. Emerg. Med.* **2008**, *15*, 522–528. [CrossRef]
86. Ngocho, J.S.; Magoma, B.; Olomi, G.A.; Mahande, M.J.; Msuya, S.E.; De Jonge, M.I.; Mmbaga, B.T. Effectiveness of pneumococcal conjugate vaccines against invasive pneumococcal disease among children under five years of age in Africa: A systematic review. *PLoS ONE* **2019**, *14*, e0212295. [CrossRef]
87. Kaplan, S.L.; O Mason, E.; Wald, E.R.; Schutze, G.E.; Bradley, J.S.; Tan, T.Q.; Hoffman, J.A.; Givner, L.B.; Yogev, R.; Barson, W.J. Decrease of invasive pneumococcal infections in children among 8 children's hospitals in the United States after the introduction of the 7-valent pneumococcal conjugate vaccine. *Pediatrics* **2004**, *113*, 443–449. [CrossRef] [PubMed]
88. Singleton, R.J.; Hennessy, T.W.; Bulkow, L.R.; Hammitt, L.L.; Zulz, T.; Hurlburt, D.A.; Butler, J.C.; Rudolph, K.; Parkinson, A. Invasive Pneumococcal Disease Caused by Nonvaccine Serotypes among Alaska Native Children with High Levels of 7-Valent Pneumococcal Conjugate Vaccine Coverage. *JAMA* **2007**, *297*, 1784–1792. [CrossRef] [PubMed]
89. Bozio, C.H.; Vuong, J.; Dokubo, E.K.; Fallah, M.P.; A McNamara, L.; Potts, C.C.; Doedeh, J.; Gbanya, M.; Retchless, A.C.; Patel, J.C.; et al. Outbreak of Neisseria meningitidis serogroup C outside the meningitis belt—Liberia, 2017: An epidemiological and laboratory investigation. *Lancet Infect. Dis.* **2018**, *18*, 1360–1367. [CrossRef]
90. Wall, E.C.; Everett, D.B.; Mukaka, M.; Bar-Zeev, N.; Feasey, N.; Jahn, A.; Moore, M.; Van Oosterhout, J.J.; Pensalo, P.; Baguimira, K.; et al. Bacterial Meningitis in Malawian Adults, Adolescents, and Children during the Era of Antiretroviral Scale-up and Haemophilus influenzae Type b Vaccination, 2000–2012. *Clin. Infect. Dis.* **2014**, *58*, e137–e145. [CrossRef] [PubMed]
91. Peterson, E.M.; Li, Y.; Bita, A.; Moureau, A.; Nair, H.; Kyaw, M.H.; Abad, R.; Bailey, F.; Garcia, I.D.L.F.; Decheva, A.; et al. Meningococcal serogroups and surveillance: A systematic review and survey. *J. Glob. Health* **2018**, *9*, 010409. [CrossRef]
92. Mustapha, M.M.; Harrison, L.H. Vaccine prevention of meningococcal disease in Africa: Major advances, remaining challenges. *Hum. Vaccines Immunother.* **2018**, *14*, 1107–1115. [CrossRef]
93. Le Saux, N. Guidelines for the management of suspected and confirmed bacterial meningitis in Canadian children older than one month of age. *Paediatr. Child Health* **2014**, *19*, 141–146. [CrossRef]

94. NICE UK. Meningitis (Bacterial) and Meningococcal Septicaemia in under 16s: Recognition, Diagnosis and Management. 2010. Available online: https://www.nice.org.uk/guidance/CG102/chapter/1-Guidance#pre-hospital-management-of-suspected-bacterial-meningitis-and-meningococcal-septicaemia (accessed on 14 January 2021).
95. Agrawal, S.; Nadel, S. Acute bacterial meningitis in infants and children: Epidemiology and management. *Pediatr. Drugs.* **2011**, *13*, 385–400. Available online: http://ovidsp.ovid.com/ovidweb.cgi?T=JS&PAGE=reference&D=emed10&NEWS=N&AN=2011577682 (accessed on 16 December 2020). [CrossRef]
96. Nudelman, Y.; Tunkel, A.R. Bacterial Meningitis. *Drugs* **2009**, *69*, 2577–2596. [CrossRef]
97. Chaudhuri, A.; Martinez-Martin, P.; Kennedy, P.G.E.; Seaton, R.A.; Portegies, P.; Bojar, M.; Steiner, I.; EFNS Task Force. EFNS guideline on the management of community-acquired bacterial meningitis: Report of an EFNS Task Force on acute bacterial meningitis in older children and adults. *Eur. J. Neurol.* **2008**, *15*, 649–659. [CrossRef] [PubMed]
98. Brouwer, M.C.; McIntyre, P.; Prasad, K.; Van De Beek, D. Corticosteroids for acute bacterial meningitis. *Cochrane Database Syst. Rev.* **2015**, *2015*, e004405. [CrossRef] [PubMed]
99. Schaad, U.B.; Kaplan, S.L.; McCracken, J.G.H. Steroid Therapy for Bacterial Meningitis. *Clin. Infect. Dis.* **1995**, *20*, 685–690. [CrossRef] [PubMed]
100. Green, K.; Nichani, J.; Hans, P.; Bruce, I.; Henderson, L.; Ramsden, R. C096 Cochlear implantation in profound hearing loss following bacterial meningitis in children. *Int. J. Pediatr. Otorhinolaryngol.* **2011**, *75*, 50. [CrossRef]
101. Bille, J.; Ovesen, T. Cochlear implant after bacterial meningitis. *Pediatr. Int.* **2014**, *56*, 400–405. [CrossRef]

Article

Field Evaluation of the Performance of Two Rapid Diagnostic Tests for Meningitis in Niger and Burkina Faso

Marc Rondy [1], Mamadou Tamboura [2], Fati Sidikou [3], Issaka Yameogo [4], Kambire Dinanibe [5], Guetwende Sawadogo [6], Chantal Kambire [7], Halima Mainassara [3], Ali Elhaj Mahamane [3], Baruani Bienvenu [8], Haladou Moussa [8], Rasmata Ouedraogo [2], Katya Fernandez [9,*], Muhamed-Kheir Taha [10] and Olivier Ronveaux [11]

1 Family, Gender and Life Course Department, Comprehensive Family Immunization Unit, Pan American Health Organization, Washington, DC 20037, USA; rondym@who.int
2 Centre Hospitalier Universitaire Pédiatrique-Charles De Gaulle, Ouagadougou 01 BP 1198, Burkina Faso; matouboura@yahoo.fr (M.T.); ramaouedtra@yahoo.fr (R.O.)
3 Centre de Recherche Médicale et Sanitaire (CERMES), Niamey 10887, Niger; fatsidik@yahoo.fr (F.S.); halima@cermes.org (H.M.); elmahamane@cermes.org (A.E.M.)
4 Direction de la Protection de la Santé de la Population, Ouagadougou 03 BP 7035, Burkina Faso; yameogoissaka@yahoo.fr
5 Unité VIH et Santé de la Reproduction (UVSR), Ouagadougou 03 BP 7192, Burkina Faso; dinanibekambire@yahoo.fr
6 Davycas International, Ouagadougou 08 BP 11593, Burkina Faso; guetasawadogo@gmail.com
7 World Health Organization, Burkina Faso Office, Ouagadougou 03 BP 7019, Burkina Faso; kambirec@who.int
8 World Health Organization, Niger Office, Niamey B.P. 10739, Niger; baruaningoyb@who.int (B.B.); haladoum@who.int (H.M.)
9 World Health Organization, CH-1211 Geneva, Switzerland
10 Institut Pasteur, 75015 Paris, France; muhamed-kheir.taha@pasteur.fr
11 Pan American Health Organization, Dominican Republic Office, Santo Domingo 1464, Dominican Republic; ronveauoli@paho.org
* Correspondence: fernandezk@who.int

Citation: Rondy, M.; Tamboura, M.; Sidikou, F.; Yameogo, I.; Dinanibe, K.; Sawadogo, G.; Kambire, C.; Mainassara, H.; Mahamane, A.E.; Bienvenu, B.; et al. Field Evaluation of the Performance of Two Rapid Diagnostic Tests for Meningitis in Niger and Burkina Faso. *Microorganisms* **2021**, *9*, 832. https://doi.org/10.3390/microorganisms9040832

Academic Editor: James Stuart

Received: 15 March 2021
Accepted: 11 April 2021
Published: 14 April 2021

Abstract: New lateral flow tests for the diagnosis of *Neisseria meningitidis* (Nm) (serogroups A, C, W, X, and Y), MeningoSpeed, and *Streptococcus pneumoniae* (Sp), PneumoSpeed, developed to support rapid outbreak detection in Africa, have shown good performance under laboratory conditions. We conducted an independent evaluation of both tests under field conditions in Burkina Faso and Niger, in 2018–2019. The tests were performed in the cerebrospinal fluid of suspected meningitis cases from health centers in alert districts and compared to reverse transcription polymerase chain reaction tests performed at national reference laboratories (NRLs). Health staff were interviewed about feasibility. A total of 327 cases were tested at the NRLs, with 26% confirmed Nm (NmC 63% and NmX 37%) and 8% Sp. Sensitivity and specificity were, respectively, 95% (95% CI: 89–99) and 90% (95% CI: 86–94) for Nm and 92% (95% CI: 75–99) and 99% (95% CI: 97–100) for Sp. Positive and negative predictive values were, respectively, 77% (95% CI: 68–85) and 98% (95% CI: 95–100) for Nm and 86% (95% CI: 67–96) and 99% (95% CI: 98–100) for Sp. Concordance showed 82% agreement for Nm and 97% for Sp. Interviewed staff evaluated the tests as easy to use and to interpret and were confident in their readings. Results suggest overall good performance of both tests and potential usefulness in meningitis outbreak detection.

Keywords: meningitis; *Neisseria meningitidis*; *Streptococcus pneumoniae*; rapid diagnostic test; national reference laboratory; cerebrospinal fluid; Niger; Burkina Faso

1. Introduction

Because of its high case fatality (around 10% [1,2]) and epidemic potential, meningococcal meningitis is a major public health threat [3], especially in the "meningitis belt". This region, which stretches from Senegal to Ethiopia, is characterized by a high seasonal incidence of meningococcal meningitis between January and June [4].

Meningococcal meningitis caused by *Neisseria meningitidis* (Nm) bacteria may cause large outbreaks and six of the 12 Nm serogroups (A, B, C, W, X, and Y) are responsible for invasive forms of meningitis [5]. The epidemiology of Nm meningitis is evolving rapidly, and new variants are continually emerging [6].

A wide choice of meningococcal vaccines covering a variable number of serogroups is currently available [7]. In a context of a diversification of serogroups with epidemic potential in the meningitis belt, it is important to strengthen microbiological surveillance to prompt the early implementation of vaccination campaigns appropriate to circulating serogroups.

Pneumococcal meningitis causes clusters of cases and, less frequently, outbreaks, in the meningitis belt, even following introduction of pneumococcal conjugate vaccines. In 2019, following massive the reduction of NmA, with MenA vaccine rollout, *Streptococcus pneumoniae* (Sp) was found to be responsible for 40% of meningitis cases in countries in the region [8]. Pneumococcal meningitis has very high case fatality rates (36–66% in the meningitis belt [9]) and, given the difficulty of treatment, requires longer treatment protocols than those for meningococcal meningitis [10].

The microbiological diagnosis of meningitis is currently based on culture or, more frequently, on polymerase chain reaction (PCR) tests carried out by national reference laboratories (NRLs) on a sample of cerebrospinal fluid (CSF) [11]. This last procedure, which requires maintaining the sample in a cold chain in primary health care centers (PHCs) and during transport to the NRL, makes the early identification of circulating strains difficult. To reduce the time needed to detect circulating strains, diagnostic tests have been developed that can be used at bedside and provide a diagnosis in a few minutes. However, currently available rapid diagnostic tests (RDTs) to detect meningococcal meningitis have a short shelf life, are sensitive to heat, and cannot detect all circulating serogroups, making them of little utility in the meningitis belt context [12].

The rapid diagnostic tests (RDT) were first developed at the Institut Pasteur, Paris, as immunochromatographic tests [13,14]. They were thereafter transferred the BioSpeedia company who then developed MeningoSpeed and PneumoSpeed that can be used between 2 and 30 °C to diagnose, respectively, the five main meningococcal serogroups (A, C, W, X and Y) [15] found in the African meningitis belt and Sp meningitis. The tests are based on the use of antibodies directed against capsular polysaccharide of Nm (serogroups A, C, W, Y, and X) and against pneumococcal cell wall polysaccharide C that is common to all Sp isolates. Laboratory based tests suggest good sensitivities and specificities of these two RDTs [12].

These RDTs require the transfer of two drops of CSF in each cassette (three for Nm groups in A/W serogroups, Y/C serogroups and X serogroup; one for Sp meningitis) and a waiting time of 15 min. The reading process is similar to any RDTs, based on control and case lines appearance and they can be performed by all staff after a short demonstration.

The objective of the study was to measure the performance of the MeningoSpeed and PneumoSpeed RDTs in diagnosing Nm and Sp meningitis in CSF at the bedside among patients in PHCs in Niger and Burkina Faso in 2018 and 2019.

2. Materials and Methods

2.1. Overview

To measure the performance of the RDTs, we compared the results of RDTs conducted at bedside in PHCs with the results of PCRs obtained at NRLs. PHCs in districts where incidence exceeded 3 suspected cases per week per 100,000 inhabitants were invited to participate, and staff were trained, supervised and supplied with RDTs.

To measure the performance of the RDTs after sample transportation, RDTs were also repeated at the NRLs and results compared with PCR tests.

The study population consisted of any consenting patient aged at least two months admitted to a participating PHC with suspected bacterial meningitis (according to the WHO case definition [16]) (sudden onset of fever and stiff neck or other meningeal signs (including bulging fontanelle for patients under 12 months)) between April 2018 and June 2019. Suspected cases with contraindications for lumbar puncture or who refused to participate were excluded from the study.

According to national guidelines for suspected bacterial meningitis case management, medical staff collected CSF through lumbar puncture on every suspected case. For each patient included in the study, the two RDTs (MeningoSpeed and PneumoSpeed) were performed on CSF at the PHC. A 1 mL CSF tube was then refrigerated and sent to the NRL where an additional RDT and a real-time PCR test was performed and used as a gold standard.

PHC staff collected information on patients' symptoms; antibiotic treatment before admission; dates of onset of symptoms, lumbar puncture, RDT reading, and dispatch of sample to the NRL; lot number; and RDT results. PHC information was merged with NRL data, including the results of the PCR tests and the RDTs performed at the NRLs, and data were analyzed using the STATA software package version 14 [17]). The producer of the tests (BioSpeedia Company, Saint Etienne, France) did not fund the trail, did not contribute to the design and did not participate in the analysis of the results. None of the authors is employed by that company.

2.2. Performance of the RDTs

MeningoSpeed test results were classified as positive or negative per serogroup A, C, W, Y, or X (positive if positive for that serogroup; negative if negative for that serogroup, regardless of other serogroup's results) and for any Nm (positive if positive for any serogroup A, C, W, Y, or X; negative if negative for all these 5 serogroups). PneumoSpeed test results were classified as positive or negative for Sp meningitis.

We compared the results of the RDTs performed in the PHCs with the results of PCR tests performed at the NRLs to calculate the sensitivity, specificity, positive predictive value (PPV) and negative predictive value (NPV) of the tests. The analyses were stratified by year of lumbar puncture (2018 vs. 2019) and epidemic period (defined as between January and June) versus non-epidemic period.

2.3. RDT Feasibility and Acceptability

PHC staff photographed the RDTs as they were reading them. The photographs were then reviewed and the RDTs blindly re-interpreted by independent readers. We compared the results of the RDTs as interpreted by the PHC staff and independent experts and quantified the level of concordance between them using the kappa coefficient [18]. We considered the agreement between the tests moderate if $\kappa < 0.4$, average or good if $\kappa = 0.4$ to 0.75 and excellent if $\kappa > 0.75$. National study coordinators conducted semi-structured interviews with RDT users to assess the acceptability and feasibility of the RDTs.

2.4. Sample Size

To estimate an expected 95% sensitivity of MeningoSpeed Nm meningitis identification, with an absolute precision of 5% and a 30% PCR positivity [19], we estimated that at least 243 suspected cases should be included in the study.

2.5. Ethical Considerations

The study followed the principles governing biomedical research involving human participation and was carried out in line with principles of the Declaration of Helsinki to ensure that the rights, integrity, and confidentiality of participants were protected. The agreements of the ethics committees of WHO (WHO ERC.0002926), Niger (deliberation N° 35/2017/CNRES) and Burkina Faso (deliberation N° 2017-10-156) were obtained before the start of the study as well as for its extension into 2019.

Written consent was obtained from suspected meningitis cases or from their parents or legal guardians. Standardized information was read to the potential participant. The confidentiality of the data collected, and the anonymity of the participants were ensured during and after the survey (no patient name appears in the databases).

3. Results

3.1. RDT Performance

Between 8 April 2018 and 30 June 2019, 421 people with suspected meningitis were admitted to the participating PHCs and tested with the RDTs. Of those people, 327 were eligible for the study, with completed questionnaires and PCR results entered into the database (246 in Niger, and 81 in Burkina Faso) and 198 (61%) were recruited during the epidemic period. The distribution of symptoms among suspected cases was typical of that found in the meningitis belt. A total of 106 (32%) and 28 (9%) patients tested positive for Nm and Sp, respectively, with the RDTs in the PHCs (Table 1).

Table 1. Characteristics and distribution of symptoms among suspected cases (N = 327), Niger and Burkina Faso, 2018–2019.

	Characteristics	**N**	**%**
	Total patients	327	100
Recruitment country	Burkina Faso	81	25
	Niger	246	75
Recruitment period	Epidemic	198	61
	Non-epidemic	129	39
Age—average (range)		9 years (3 months—86 years)	
Symptoms	Sudden onset of fever	316	97
	Abdominal pain	181	55
	Confusion and disorientation	180	55
	Joint pain	136	42
	Stiff neck	116	36
	Bulging fontanelle	10	3
	Petechial rash	1	0
	other meningeal signs	18	6
Nm RDT results	All serogroups	106	32
	NmA	9	3
	NmC	56	17
	NmW	2	1
	NmX	40	12
	NmY	1	0
	Negative	221	68
Sp RDT results	Positive	28	9
	Negative	295	91

Of the 327 cases with PCR results obtained at an NRL, 86 (26%) were confirmed with Nm (NmC 63% and NmX 37%) and 26 (8%) with Sp (Table 2). The median time between lumbar puncture and PCR result was 18 days outside the epidemic period, and 22 days during the epidemic period.

Table 2. Concordance of results obtained by RDT in PHCs and NRLs with results obtained by PCR in NRLs, and serogroup-specific performances, Burkina Faso and Niger, 2018–2019.

		PCR Negative/RDT Negative	PCR Positive/RDT Positive	PCR Negative/RDT Positive	PCR Positive/RDT Negative	Total	Sensitivity (%)	95% CI	Specificity (%)	95% CI	PPV (%)	95% CI	NPV (%)	95% CI
RDT at health center	All Nm	217	82	24	4	327	95	(89;99)	90	(86;94)	77	(68;85)	98	(95;100)
	NmA	318	0	9	0	327	NA		97	(95;99)	NA		100	(99;100)
	NmC	266	50	6	4	326	93	(82;98)	98	(95;99)	89	(78;96)	99	(96;100)
	NmW	324	0	2	0	326	NA		99	(98;100)	NA		100	(99;100)
	NmX	284	29	11	3	327	91	(75;98)	96	(93;98)	73	(56;85)	99	(97;100)
	NmY	324	0	1	0	325	NA		100	(98;100)	NA		100	(99;100)
	Sp	293	24	4	2	323	92	(75;99)	99	(97;100)	86	(67;96)	99	(98;100)
RDT at NRL	All Nm	108	28	6	2	144	93	(78;99)	95	(89;98)	82	(66;93)	98	(94;100)
	NmA	141	0	6	0	147	NA		96	(91;99)	NA		100	(97;100)
	NmC	127	16	2	1	146	94	(71;100)	98	(95;100)	89	(65;99)	99	(96;100)
	NmW	142	0	3	0	145	NA		98	(94;100)	NA		100	(97;100)
	NmX	132	12	0	1	145	92	(64;100)	100	(97;100)	100	(74;100)	99	(96;100)
	NmY	144	0	2	0	146	NA		99	(95;100)	NA		100	(98;100)
	Sp	138	8	0	1	147	89	(52;100)	100	(97;100)	100	(63;100)	99	(96;100)

RDT: rapid diagnostic test; PHC: primary health care center; NRL: national reference laboratory; PCR: polymerase chain reaction, 95% CI: 95% confidence intervel; PPV: positive predictive value; NPV: negative predictive value; Nm: Neisseria meningitidis; Sp: Streptococcus pneumoniae.

Sensitivity and specificity were, respectively, 95% (95% CI: 89–99) and 90% (95% CI: 86–94) for any Nm, 93% (95% CI: 82–98), and 98% (95% CI: 95–99) for NmC, and 91% (95% CI: 75–98) and 96% (95% CI: 93–98) for NmX.

PPV and NPV were, respectively, 77% (95% CI: 68–85) and 98% (95% CI: 95–100) for any Nm. PPV was better in 2019 than in 2018: 90% (95% CI: 81–96) vs. 49% (95% CI: 31–67), p-value < 0.01 (Table 3); and was better during epidemic months: 89% (95% CI: 79–95) versus 53% (95% CI: 35–70), p-value < 0.01 (Tables 3 and 4).

Table 3. Concordance of results obtained by RDT in PHCs with results obtained by PCR in NRLs, and serogroup specific performances, by year of lumbar puncture, Burkina Faso and Niger, 2018–2019.

		PCR Negative/RDT Negative	PCR Positive/RDT Positive	PCR Negative/RDT Positive	PCR Positive/RDT Negative	Total	Sensitivity (%)	95% CI	Specificity (%)	95% CI	PPV (%)	95% CI	NPV (%)	95% CI
2018	All Nm	80	16	17	1	114	94	(71;100)	83	(73;89)	49	(31;67)	99	(93;100)
	NmA	106	NA	8	NA	114	NA	NA	93	(87;97)	NA	NA	100	(97;100)
	NmC	110	4	NA	NA	114	100	(40;100)	100	(97;100)	100	(40;100)	100	(97;100)
	NmW	112	NA	1	NA	113	NA	NA	99	(95;100)	NA	NA	100	(97;100)
	NmX	93	12	8	1	114	92	(64;100)	92	(85;97)	60	(36;81)	99	(94;100)
	NmY	114	NA	NA	NA	114	NA	NA	100	(97;100)	NA	NA	100	(97;100)
	Sp	108	4	1		113	100	(40;100)	99	(95;100)	80	(28;100)	100	(97;100)
2019	All Nm	135	64	7	3	209	96	(88;99)	95	(90;98)	90	(81;96)	98	(94;100)
	NmA	208	NA	1	NA	209	NA	NA	100	(97;100)	NA	NA	100	(98;100)
	NmC	154	45	5	4	208	92	(80;98)	97	(93;99)	90	(78;97)	98	(94;99)
	NmW	208	NA	1	NA	209	NA	NA	100	(97;100)	NA	NA	100	(98;100)
	NmX	188	17	3	1	209	94	(73;100)	98	(96;100)	85	(62;97)	100	(97;100)
	NmY	207	NA	1	NA	208	NA	NA	100	(97;100)	NA	NA	100	(98;100)
	Sp	182	20	3	2	207	91	(71;99)	98	(95;100)	87	(66;97)	99	(96;100)

RDT: rapid diagnostic test; PHC: primary health care center; NRL: national reference laboratory; PCR: polymerase chain reaction; 95% CI: 95% confidence intervel; PPV: positive predictive value; NPV: negative predictive value; Nm: Neisseria meningitidis; Sp: Streptococcus pneumoniae.

For Sp, sensitivity and specificity were, respectively, 92% (95% CI: 75–99) and 99% (95% CI: 97–100), and PPV and NPV values were 86% (95% CI: 67–96) and 99% (95% CI: 98–100) (Table 2).

Owing to a lack of tests and a strike by staff, it was only possible to repeat the RDTs at the NRLs in 147 of the 327 patients (45%). Performances of the RDTs at the NRLs and the health centers were comparable (Table 2).

Table 4. Concordance of results obtained by RDT in PHCs with results obtained by PCR in NRLs, and serogroup specific performances, by epidemic vs. non epidemic period, Burkina Faso and Niger, 2018–2019.

		PCR Negative/RDT Negative	PCR Positive/RDT Positive	PCR Negative/RDT Positive	PCR Positive/RDT Negative	Total	Sensitivity (%)	95% CI	Specificity (%)	95% CI	PPV (%)	95% CI	NPV (%)	95% CI
Epidemic period	All Nm	123	64	8	3	198	96	(88;99)	94	(88;97)	89	(79;95)	98	(93;100)
	NmA	197	NA	1	NA	198	NA	NA	100	(97;100)	NA	NA	100	(98;100)
	NmC	143	45	5	4	197	92	(80;98)	97	(92;99)	90	(78;97)	97	(93;99)
	NmW	197	NA	1	NA	198	NA	NA	100	(97;100)	NA	NA	100	(98;100)
	NmX	176	17	4	1	198	94	(73;100)	98	(94;99)	81	(58;95)	99	(97;100)
	NmY	196	NA	1	NA	197	NA	NA	100	(97;100)	NA	NA	100	(98;100)
	Sp	174	17	3	2	196	90	(67;99)	98	(95;100)	85	(62;97)	99	(96;100)
Non epidemic period	All Nm	94	18	16	1	129	95	(74;100)	86	(78;92)	53	(35;70)	99	(94;100)
	NmA	121	NA	8	NA	129	NA	NA	94	(88;97)	NA	NA	100	(97;100)
	NmC	123	5	1	NA	129	100	(48;100)	99	(96;100)	83	(36;100)	100	(97;100)
	NmW	127	NA	1	NA	128	NA	NA	99	(96;100)	NA	NA	100	(97;100)
	NmX	108	12	7	2	129	86	(57;98)	94	(88;98)	63	(38;84)	98	(94;100)
	NmY	128	NA	NA	NA	128	NA	NA	100	(97;100)	NA	NA	100	(97;100)
	Sp	119	7	1	NA	127	100	(59;100)	99	(95;100)	88	(47;100)	100	(97;100)

RDT: rapid diagnostic test; PHC: primary health care center; NRL: national reference laboratory; PCR: polymerase chain reaction; 95% CI: 95% confidence intervel; PPV: positive predictive value; NPV: negative predictive value; Nm: Neisseria meningitidis; Sp: Streptococcus pneumoniae.

3.2. *RDT Feasibility and Acceptability*

There was an 82% agreement for Nm and 97% for Sp between RDT results registered at the PHCs and photographs reviewed by an independent expert. The kappa coefficient suggested an excellent concordance for Sp and NmX, a good concordance for all Nm and moderate concordance for NmA, NmC, and NmW (Table 5).

Table 5. Concordance of the results of RDTs as interpreted live in the PHCs and by independent reading of photographs, Burkina Faso and Niger, 2018–2019.

	Test	Photo Reading Negative/RDT Negative	Photo Reading Positive/RDT Positive	Photo Reading Negative/RDT Positive	Photo Reading Positive/RDT Negative	Total	Concordance (%)	Kappa Coefficient
RDTs at health centers	All Nm	36	17	11	1	65	82	61
	NmA	64	2	6	1	73	90	32
	NmC	71	1	3	0	75	96	39
	NmW	68	1	0	4	73	95	32
	NmX	53	13	4	0	70	94	83
	NmY	71	0	0	0	71	100	NA
	Sp	66	5	1	1	73	97	82

RDT: rapid diagnostic test; PHC: primary health care center; Nm: Neisseria meningitidis; Sp: Streptococcus pneumoniae.

Eleven NmX, nine NmA, six NmC, two NmW, and one NmY were detected positive with RDTs but were tested negative by real-time PCR.

All nine positive NmA results identified in health centers were from Burkina Faso, eight in 2018 and one in 2019. Of these, two were identified as positive for NmA on photographs by the external expert. The authors confirmed that a faint line, suggestive of a positive NmA result was indeed visible in these two photographs (Figure 1). Of the eleven false positive NmX results, nine were from Burkina Faso and two from Niger. Of those, four photographs were reviewed and classified as negative by the independent experts. One photograph of a false positive NmC result was available and suggested a misreading by the PHC medical staff.

A total of 31 staff were interviewed about their experiences in using the RDTs (20 in Niger and 11 in Burkina Faso). Overall, respondents found the RDTs easy to use (average difficulty score: 1.9/10). They found their interpretation very easy (average difficulty score: 1.2/10) and had great confidence in their result (average lack of confidence score: 1.3/10).

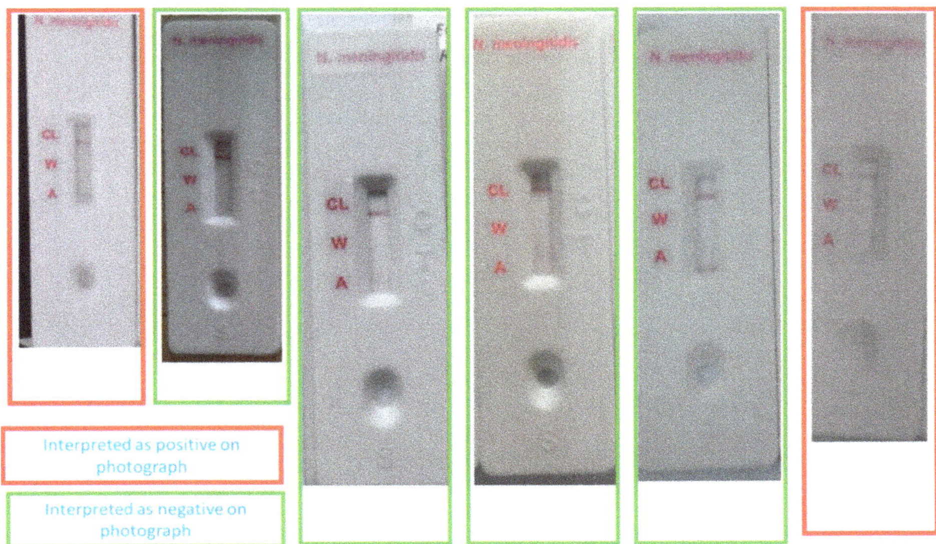

Figure 1. Photographs of six false positive NmA RDTs by independent expert classification, Burkina Faso and Niger, 2018–2019.

4. Discussion

Performance of the RDTs

We described in this article the performance results from ready-to-use bedside RDTs, usable by any medical staff, storable at 2–30 °C, allowing to detect Nm and Sp within 15 min. Our results suggest that the RDTs for diagnosing Nm and Sp performed well under field conditions at PHC level, with sensitivity and specificity above 90%. They performed similarly whether used by medical staff at a PHC or technical staff at an NRL. These RDTs were described as acceptable and easy to use in a bedside context. Serogroup-specific Nm RDT performance could be measured for NmC and NmX and were in the same range, above 90%. The two serogroups were the most prevalent in West Africa in 2018–2019 [20,21]. Further studies would be needed to validate their use in detecting NmA, NmW, and NmY due to low prevalence during our study period.

By extending the study over two years we were able to obtain a sample size large enough to measure the performances of the RDTs with acceptable precision. The study protocol was well followed, despite challenges in systematizing the capture of photographs and the use of the RDTs at the NRLs. In addition, recruitment was lower in Burkina Faso due to security issues along with a strike by health workers in 2019. Owing to these limitations, the precision of some secondary outcomes was low. Logistical issues between the PHCs and NRLs, potentially leading to alteration of samples and affecting the measured performance of the RDTs, cannot be excluded. In order to account for such a risk, the protocol included repetition of RDTs at the NRLs. Similar results between RDTs carried out at the PHCs and NRLs suggest that sample alteration was low in this study.

Our results suggest a higher sensitivity of the RDT in detecting NmC in this field context than previously reported in laboratory conditions (95% versus 65%) [15]. This could be due to the fact that the circulating clone in Niger and Burkina Faso (NmC, clonal complex ST-10217) [19,22] is particularly well detected by the antibodies used in the MeningoSpeed test.

PPV for any Nm and NmX were at 77% and 73%, respectively, corresponding to a quarter of false positive RDT readings for these outcomes. Moreover, recurrent false positives for NmA were registered. Further analyses suggested that most of these findings were attributed to specific PHCs and that constant supervision and training of RDT users led to a significant decrease in false positives between the first and the second study years. NmA RDT issues, documented with photographs, were fed back to the manufacturer.

The original study protocol included the use of PneumoSpeed tests on urine samples. However, only 31 tests were reported, precluding any interpretation related to this objective. Considering its potential impact on patients' comfort and practice safety, specific studies should be implemented to measure the performance of RDT tests on urine.

Performance measurements are the results of both the quality of the test and the interpretive capacity of the user. Measurement of the concordance between the interpretations of the nursing staff and the independent expert made it possible to discuss this additional information bias linked to the evaluator. Although the proportion of photographs that could be read was low, concordance was generally good, suggesting a good understanding of the use of the RDTs by the PHC staff.

These field performance results met the WHO target product profile acceptable values for sensitivity and specificity, of >90% [23] and were superior or comparable to field performance values for existing meningococcal rapid tests (69–80% sensitivity and 81–94% specificity for latex agglutination tests and 89–92% sensitivity and 85–99 specificity for lateral flow test) [12]. Our field evaluation confirms the good performance of the tests in laboratory conditions and suggests that the tests are suitable for use in field conditions and that they are acceptable to health personnel, but that they should be accompanied by clear instructions and effective training. Considering this performance, their longer shelf-life and improved thermostability (but still below the desired target product profile value of 40°), easier test procedures, and inclusion of all main Nm serogroups (including NmX), MeningoSpeed and PneumoSpeed are good candidate tests for the early detection of meningitis epidemics in Africa. However, this field study is only one step towards ensuring access to safe appropriate diagnostic tests of good quality. At any rate, ensuring confirmation of test results with a more specific test such as PCR will continue to be key, given the decrease in incidence of Nm A and anticipated decrease of other Nm serogroups with future vaccination efforts.

One of the priority goals identified by the Defeating meningitis by 2030 Roadmap is the improvement of diagnosis at all levels of care, through the development and access to diagnostic assays [24]. An expert group gathered by WHO in 2018 [25] identified three essential objectives for the development of in vitro diagnostic tests (IVD) for meningitis diagnosis, including the rapid detection of epidemics in the African meningitis belt. The MeningoSpeed and PneumoSpeed could potentially meet this specific need. Issues that remain to be addressed, before procuring the test more widely, are costs, further thermostability improvements, and scale-up of production capacity with a reliable quality management system. Until elimination of meningitis epidemics in the region is achieved, one of three visionary goals of the roadmap, the use of rapid diagnostic tests will remain an important tool for meningitis control in Africa.

Author Contributions: Conceptualization, M.-K.T. and O.R.; Formal analysis, M.R.; Investigation, M.T., F.S., I.Y., K.D. G.S., H.M. (Halima Mainassara), H.M. (Haladou Moussa) and R.O.; Methodology, M.R. and R.O.; Validation, O.R.; Writing—original draft, M.R.; Writing—review & editing, M.T., F.S., I.Y., K.D., G.S., C.K., H.M. (Halima Mainassara), A.E.M., B.B., H.M. (Haladou Moussa), R.O., K.F., M.-K.T. and O.R. All authors have read and agreed to the published version of the manuscript.

Funding: This research received no external funding.

Institutional Review Board Statement: The study followed the principles governing biomedical research involving human participation and was carried out in line the principles of the Declaration of Helsinki to ensure that the rights, integrity, and confidentiality of participants were protected. The agreements of the ethics committees of WHO (WHO ERC.0002926), Niger (deliberation N° 35/2017/CNRES) and Burkina Faso (deliberation N° 2017-10-156) were obtained before the start of the study as well as for its extension into 2019.

Informed Consent Statement: Written consent was obtained from suspected meningitis cases or from their parents or legal guardians. Standardized information was read to the potential participant. The confidentiality of the data collected and the anonymity of the participants were ensured during and after the survey (no patient name appears in the databases).

Data Availability Statement: Not applicable.

Conflicts of Interest: The authors declare no conflict of interest.

References

1. World Health Organization. Meningococcal vaccines: WHO position paper, November 2011. *Wkly. Epidemiol. Rec.* **2011**, *18*, 521–539.
2. Boisier, P.; Maïnassara, H.B.; Sidikou, F.; Djibo, S.; Kairo, K.K.; Chanteau, S. Case-fatality ratio of bacterial meningitis in the African meningitis belt: We can do better. *Vaccine* **2007**, *25*, A24–A29. [CrossRef] [PubMed]
3. Van de Beek, D. Progress and challenges in bacterial meningitis. *Lancet* **2012**, *380*, 1623–1624. [CrossRef]
4. Molesworth, A.M.; Thomson, M.C.; Connor, S.J.; Cresswell, M.P.; Morse, A.P.; Shears, P.; Hart, C.A.; Cuevas, L.E. Where is the meningitis belt? Defining an area at risk of epidemic meningitis in Africa. *Trans. R. Soc. Trop. Med. Hyg.* **2002**, *96*, 242–249. [CrossRef]
5. Harrison, L.H.; Pelton, S.I.; Wilder-Smith, A.; Holst, J.; Safadi, M.A.; Vazquez, J.A.; Taha, M.K.; LaForce, F.M.; Von Gottberg, A.; Borrow, R.; et al. The Global Meningococcal Initiative: Recommendations for reducing the global burden of meningococcal disease. *Vaccine* **2011**, *29*, 3363–3371. [CrossRef] [PubMed]
6. Halperin, S.A.; Bettinger, J.A.; Greenwood, B.; Harrison, L.H.; Jelfs, J.; Ladhani, S.N.; McIntyre, P.; Ramsay, M.E.; Sáfadi, M.A. The changing and dynamic epidemiology of meningococcal disease. *Vaccine* **2012**, *30*, B26–B36. [CrossRef] [PubMed]
7. World Health Organization. *Defeating Meningitis by 2030: Baseline Situation Analysis*; World Health Organization: Geneva, Switzerland, 2019; Available online: https://www.who.int/immunization/research/BSA_20feb2019.pdf?ua=1 (accessed on 13 April 2021).
8. Soeters, H.M.; Diallo, A.O.; Bicaba, B.W.; Kadadé, G.; Dembélé, A.Y.; Acyl, M.A.; Nikiema, C.; Sadji, A.Y.; Poy, A.N.; Lingani, C.; et al. Bacterial Meningitis Epidemiology in Five Countries in the Meningitis Belt of Sub-Saharan Africa, 2015–2017. *J. Infect. Dis.* **2019**, *220* (Suppl. 4), S165–S174. [CrossRef] [PubMed]
9. Gessner, B.D.; Mueller, J.E.; Yaro, S. African meningitis belt pneumococcal disease epidemiology indicates a need for an effective serotype 1 containing vaccine, including for older children and adults. *BMC Infect. Dis.* **2010**, *10*, 22. [CrossRef] [PubMed]
10. World Health Organization. Pneumococcal meningitis outbreaks in sub-Saharan Africa. *Wkly. Epidemiol. Rec.* **2016**, *91*, 298–302.
11. Lingani, C.; Bergeron-Caron, C.; Stuart, J.M.; Fernandez, K.; Djingarey, M.H.; Ronveaux, O.; Schnitzler, J.C.; Perea, W.A. Meningococcal Meningitis Surveillance in the African Meningitis Belt, 2004–2013. *Clin. Infect. Dis.* **2015**, *61* (Suppl. 5), S410–S415. [CrossRef] [PubMed]
12. Feagins, A.R.; Ronveaux, O.; Taha, M.K.; Caugant, D.A.; Smith, V.; Fernandez, K.; Glennie, L.; Fox, L.M.; Wang, X. Next generation rapid diagnostic tests for meningitis diagnosis. *J. Infect.* **2020**, *81*, 712–718. [CrossRef] [PubMed]
13. Chanteau, S.; Dartevelle, S.; Mahamane, A.E.; Djibo, S.; Boisier, P.; Nato, F. New rapid diagnostic tests for Neisseria meningitidis serogroups A, W135, C, and Y. *PLoS Med.* **2006**, *3*, e337. [CrossRef] [PubMed]
14. Agnememel, A.; Traincard, F.; Dartevelle, S.; Mulard, L.; Mahamane, A.E.; Oukem-Boyer, O.O.; Denizon, M.; Kacou, N.A.; Dosso, M.; Gake, B.; et al. Development and evaluation of a dipstick diagnostic test for Neisseria meningitidis serogroup X. *J. Clin. Microbiol.* **2015**, *53*, 449–454. [CrossRef] [PubMed]
15. Haddar, C.H.; Terrade, A.; Verhoeven, P.; Njanpop-Lafourcade, B.M.; Dosso, M.; Sidikou, F.; Mahamane, A.E.; Lombart, J.P.; Razki, A.; Hong, E.; et al. Validation of a New Rapid Detection Test for Detection of Neisseria meningitidis A/C/W/X/Y Antigens in Cerebrospinal Fluid. *J. Clin. Microbiol.* **2020**, *58*, e01699-19. [CrossRef] [PubMed]
16. World Health Organization. *Contrôle des Epidémies de Méningite en Afrique. Guide de Référence Rapide à L'intention des Autorités Sanitaires et des Soignants. Organisation Mondiale de la Santé [Control of Meningitis Epidemics in Africa. Quick Reference Guide for Health Authorities and Caregivers]*; World Health Organization: Geneva, Switzerland, 2015.
17. StataCorp. *Stata Statistical Software: Release 14*; StataCorp LP: College Station, TX, USA, 2015.
18. Nam, J. Comparison of validity of assessment methods using indices of adjusted agreement: Comparison of validity of assessment methods. *Stat. Med.* **2007**, *26*, 620–632. [CrossRef] [PubMed]
19. Sidikou, F.; Zaneidou, M.; Alkassoum, I.; Schwartz, S.; Issaka, B.; Obama, R.; Lingani, C.; Tate, A.; Ake, F.; Sakande, S.; et al. Emergence of epidemic Neisseria meningitidis serogroup C in Niger, 2015: An analysis of national surveillance data. *Lancet Infect. Dis.* **2016**, *16*, 1288–1294. [CrossRef]
20. World Health Organization. Epidemic meningitis control in countries of the African meningitis belt, 2019. *Wkly. Epidemiol. Rec.* **2020**, *95*, 133–144.
21. World Health Organization. Epidemic meningitis control in countries of the African meningitis belt, 2018. *Wkly. Epidemiol. Rec.* **2019**, *94*, 179–188.

22. Funk, A.; Uadiale, K.; Kamau, C.; Caugant, D.A.; Ango, U.; Greig, J. Sequential Outbreaks Due to a New Strain of Neisseria Meningitidis Serogroup C in Northern Nigeria, 2013–2014. *PLoS Curr.* **2014**. Available online: https://currents.plos.org/outbreaks/?p=40715 (accessed on 13 April 2021). [CrossRef] [PubMed]
23. World Health Organization. *Specifications for a Rapid Diagnostic Test for Meningitis, African Meningitis Belt*; World Health Organization: Geneva, Switzerland, 2016; Available online: https://www.who.int/publications/m/item/specifications-for-a-rapid-diagnostic-test-for-meningitis-african-meningitis-belt (accessed on 13 April 2021).
24. World Health Organization. *Defeating Meningitis by 2030: A Global Roadmap*; World Health Organization: Geneva, Switzerland, 2020; Available online: https://www.who.int/publications/m/item/defeating-meningitis-by-2030-a-global-road-map (accessed on 13 April 2021).
25. World Health Organization. *Meeting Report: Developing a New Generation RDTs for Meningitis*; World Health Organization: Geneva, Switzerland, 2018.

MDPI

Review

Defeating Paediatric Tuberculous Meningitis: Applying the WHO "Defeating Meningitis by 2030: Global Roadmap"

Robindra Basu Roy [1,2,*], Sabrina Bakeera-Kitaka [3], Chishala Chabala [4], Diana M Gibb [2], Julie Huynh [5,6], Hilda Mujuru [7], Naveen Sankhyan [8], James A Seddon [9,10], Suvasini Sharma [11], Varinder Singh [11], Eric Wobudeya [3] and Suzanne T Anderson [2]

1 Clinical Research Department, Faculty of Infectious & Tropical Diseases, London School of Hygiene & Tropical Medicine, Keppel Street, London WC1E 7HT, UK
2 MRC Clinical Trials Unit at UCL, 90 High Holborn, Holborn, London WC1V 6LJ, UK; diana.gibb@ucl.ac.uk (D.M.G.); suzanne.anderson@ucl.ac.uk (S.T.A.)
3 MUJHU Research Collaboration, Kampala, Uganda; sabrinakitaka@yahoo.co.uk (S.B.-K.); ewobudeya@mujhu.org (E.W.)
4 School of Medicine & University Teaching Hospital (UTH), University of Zambia, Lusaka, Zambia; cchabala@gmail.com
5 Oxford University Clinical Research Unit, Centre for Tropical Medicine, Hospital for Tropical Diseases, 764 Vo Van Kiet, District 5, Ho Chi Minh City, Vietnam; jhuynh@oucru.org
6 Centre for Tropical Medicine and Global Health, Nuffield Department of Medicine, Oxford University, Oxford OX3 7LG, UK
7 University of Zimbabwe Clinical Research Centre, Harare, Zimbabwe; hmujuru@mweb.co.zw
8 Post Graduate Institute of Education and Medical Research (PGI), Chandigarh 160017, India; drnsankhyan@yahoo.co.in
9 Department of Infectious Diseases, Imperial College London, Norfolk Place, London W2 1PG, UK; james.seddon@imperial.ac.uk
10 Desmond Tutu TB Centre, Department of Paediatrics and Child Health, Stellenbosch University, Cape Town 8000, South Africa
11 Department of Pediatrics, Lady Hardinge Medical College and Assoc Kalawati Saran Children's Hospital (Hospital-LHH), New Delhi 110001, India; sharma.suvasini@gmail.com (S.S.); 4vsingh@gmail.com (V.S.)
* Correspondence: robin.basu-roy@lshtm.ac.uk

Citation: Basu Roy, R.; Bakeera-Kitaka, S.; Chabala, C.; Gibb, D.M; Huynh, J.; Mujuru, H.; Sankhyan, N.; Seddon, J.A; Sharma, S.; Singh, V.; et al. Defeating Paediatric Tuberculous Meningitis: Applying the WHO "Defeating Meningitis by 2030: Global Roadmap". *Microorganisms* **2021**, *9*, 857. https://doi.org/10.3390/microorganisms9040857

Academic Editor: James Stuart

Received: 29 March 2021
Accepted: 13 April 2021
Published: 16 April 2021

Publisher's Note: MDPI stays neutral with regard to jurisdictional claims in published maps and institutional affiliations.

Abstract: Children affected by tuberculous meningitis (TBM), as well as their families, have needs that lie at the intersections between the tuberculosis and meningitis clinical, research, and policy spheres. There is therefore a substantial risk that these needs are not fully met by either programme. In this narrative review article, we use the World Health Organization (WHO) "Defeating Meningitis by 2030: global roadmap" as a starting point to consider key goals and activities to specifically defeat TBM in children. We apply the five pillars outlined in the roadmap to describe how this approach can be adapted to serve children affected by TBM. The pillars are (i) prevention; (ii) diagnosis and treatment; (iii) surveillance; (iv) support and care for people affected by meningitis; and (v) advocacy and engagement. We conclude by calling for greater integration between meningitis and TB programmes at WHO and at national levels.

Keywords: tuberculosis; tuberculous meningitis; TBM; children

1. Introduction

Tuberculous meningitis (TBM) is a devastating childhood disease, with one in five affected children dying and only one in three surviving without disability [1]. Depending upon tuberculosis prevalence, the age range being studied, and whether it is a population or hospital-based setting, TBM accounts from 1 to 10% of childhood TB cases, although there is considerable uncertainty around these estimates, given the challenges of making a microbiological diagnosis, the difficulties in discriminating TBM from other meningitides on clinical grounds, and underreporting [2,3]. The forthcoming publication of the first global road map on defeating meningitis from the World Health Organization (WHO) provides an opportune moment to highlight the specific challenges and goals needed to defeat TBM [4]. Issues affecting children with TBM and their families lie at the intersection of the tuberculosis (TB) and meningitis clinical, research, and policy spheres and, hence, there is an accompanying risk that the specific needs of this vulnerable group are not fully addressed (Table 1).

Table 1. Specific reference to tuberculous meningitis in the main text of recent key World Health Organization policy documents for either tuberculosis or meningitis *.

Document	References to TBM
Defeating meningitis by 2030: a global road map (26 October, 2020 draft) [4]	"[The road map] will complement other global control strategies, such as those addressing sepsis, pneumonia, tuberculosis and HIV." (p. 3) "Estimated cases and deaths due to tuberculous and cryptococcal meningitis are categorised under tuberculosis, HIV or other infectious diseases and are not included in these figures." (p. 4) "Although the focus of this road map is not on other important causes of meningitis, such as tuberculosis, Cryptococcus, enteric bacteria and viruses such as enterovirus, several goals aimed at reducing the burden of disease are applicable to all causes of meningitis." (p. 6)
Global Tuberculosis Report 2020 [5]	Bacille Calmette-Guerin "[BCG] provides moderate protection against severe forms of TB (TB meningitis and miliary TB) in infants and young children." (p. 125) Mention of "High-dose rifampicin for TB meningitis (ReDEFINe) study" on p. 183 and p. 186.
Roadmap towards ending TB in children and adolescents 2018 [6]	Within "specific age- and disease-related challenges" box (p. 8): "Young children are at increased risk of developing severe forms of TB disease (e.g., disseminated TB, TB meningitis) with increased risk of death (especially children < 2 years)." and "TB is frequently missed as underlying cause or co-morbidity of children presenting with pneumonia, malnutrition or meningitis." Within "Improving recording and reporting of detected TB cases, TB-related deaths and prevention" (p. 10): "Fatal cases of TB that present as severe pneumonia, HIV, malnutrition or meningitis are attributed to these conditions." Within "Implement integrated family-and community-centred strategies" (p. 15): "Ensure children and adolescents with other common co-morbidities (e.g., meningitis, malnutrition, pneumonia, chronic lung disease and HIV infection) are routinely evaluated for TB."

* Documents primarily related to meningitis were searched for "TB" or "tuberculosis" whilst documents primarily related to tuberculosis were searched for "meningitis." p denotes page.

In this article we therefore apply the framework outlined in the October 2020 draft "defeating meningitis" road map to TBM. As a team of paediatricians working on the SURE trial, the largest ever randomised controlled treatment trial in paediatric TBM, we focus specifically upon TBM in children [7]. The "defeating meningitis" roadmap explicitly places its emphasis on the four main causes of acute bacterial meningitis (meningococcus, pneumococcus, *Haemophilus influenzae*, and group B streptococcus), although it also highlights that many of the goals are applicable to reducing the burden of disease from all causes of meningitis (Table 1) [4]. The "defeating meningitis" roadmap sits within a network of interconnected issues related to TB including universal health coverage, antimicrobial resistance and inclusion of people with disabilities [4]. Its vision is "Towards a world free of meningitis," although this is qualified, as meningitis is heterogenous and not amenable to elimination or eradication. Three overarching visionary goals are proposed: (i) elimination of bacterial meningitis epidemics; (ii) reduction of cases and deaths from vaccine-preventable bacterial meningitis; (iii) reduction of disability and improved quality of life after meningitis of any cause [4]. From the outset, the limitations of applying the roadmap to paediatric TBM are clear, as although it is, to some extent, vaccine-preventable and a significant cause of disability, it is not a disease that occurs in epidemics [1,8].

We consider the five pillars of (i) prevention; (ii) diagnosis and treatment; (iii) disease surveillance; (iv) support and care for people affected by meningitis; and (v) advocacy and engagement, to outline how this approach can be adapted to serve children affected by TBM. We conclude by calling for greater integration of the approaches to eliminate TB and defeat meningitis, especially in light of the challenges and opportunities presented to healthcare systems around the world tackling the COVID-19 pandemic.

2. Pillar 1: Prevention

Prevention of paediatric TBM is closely related to the prevention of TB infection and thereby TB disease, and severe forms of disease such as TBM. Strategies to control TBM should focus on TB prevention with a special focus on vulnerable groups. In regions with high TB prevalence, children under 5 years are most commonly affected with TBM [2,9]. Others at risk of TBM include those living with HIV and other immunocompromised children. The WHO lists the three primary strategies for TB prevention, including (i) TB preventive therapy (TPT); (ii) prevention of transmission through infection control; and (iii) BCG vaccination [5]. It is recommended to use TPT to prevent TB infection from progressing to TB disease. This is particularly important in immunocompromised hosts and young children in whom the risk of severe forms of TB, including TBM, is high. The WHO recommends TPT for three high-risk groups: household contacts (especially children under five years of age), people living with HIV, and other clinical risk groups [10]. This latter group, which includes those initiating anti-TNF treatment or transplantation, prisoners, health workers, homeless people, and people who use drugs, represents a relatively small proportion of children at risk in high TB prevalence areas.

In March 2020, WHO updated recommendations on TPT [10]. Alongside established regimes, such as 6 or 9 months of daily isoniazid, 3 months of weekly rifapentine plus isoniazid, and 3 months of daily isoniazid and rifampicin, new recommendations included:

- A 1-month regimen of daily rifapentine and isoniazid (1HP);
- A 4-month regimen of rifampicin (4R);
- Isoniazid preventive treatment in pregnancy; and
- Advice on the use of rifapentine and dolutegravir for people living with HIV.

The inclusion of rifampicin and rifapentine in regimens for TB prevention allows shorter treatment and higher completion rates. There is ongoing research to improve uptake and scale-up TPT, such as the Unitaid-funded: IMPAACT 4TB and CaP TB studies [5]. Despite clear guidance, the numbers of household contacts, including children, provided with TPT globally, remain dismally low at 20–30% [5,10,11]. It is therefore not surprising that, in a case series of children with TBM, TPT was not provided to any of the children who had known exposure to an adult TB case [12].

TB infection, prevention and control is part of the End TB strategy. Particular areas of concern for preventing TB transmission through infection control include:

- Newborn care settings. There are many documented outbreaks of TB among neonates with the source case usually being a mother or a member of staff with potential to infect many babies in the neonatal care unit. Neonates are particularly vulnerable for acute onset or development of disseminated severe disease [13,14].

- Health facilities that provide care for adults, older children, and adolescents with TB, who are often infectious.
- Antenatal care settings.
- HIV clinics, including Preventing Mother to Child Transmission (PMTCT) settings.
- Facilities that care for children with severe malnutrition.
- Other congregate settings, including childcare facilities, orphanages, prisons, and schools. For older children, this includes boarding schools. School-aged children with sputum smear-positive TB should be kept from attending school until it is considered that there is a very low risk of transmission. This is usually for 2 weeks in the case of drug-susceptible TB.
- Children in displaced and mobile populations, including migrant labour camps, informal and crowded refugee camps, and temporary shelters.

In 2019, WHO released new guidance, and recommended administrative, environmental, and personal protection measures to limit the spread of TB [15]. The COVID-19 pandemic has threatened to derail the global progress in TB control. Alarmingly, it has been estimated that the COVID-19 pandemic could cause an additional 6.3 million TB cases globally between 2020 and 2025 [16,17]. However, there are some positives too. The expertise and experience in rapid testing and contact tracing for COVID-19 could be leveraged for TB case tracing and testing. Additionally, digital communication strategies developed during COVID-19 could be expanded for remote care, follow-up, and support of TB patients [18,19]. Lastly, the enhanced awareness generated during the COVID-19 pandemic could help emphasise basic infection prevention strategies in healthcare settings, such as the use of personal protective equipment (PPE), cough etiquette, and respiratory isolation [5].

The third key strategy is neonatal BCG vaccination. This has been shown to protect against TBM and disseminated TB in children with protection lasting up to the age of 10 years [8,20,21]. Data on protection beyond 15 years are limited; however, a small number of trials and observational studies suggest that BCG vaccination may protect for longer. Importantly, there are limitations to the potential for BCG to further decrease incidence of TBM in childhood, as the majority of children with TBM have been vaccinated with BCG. An investigational TB vaccine candidate (M72/AS01E) was evaluated in adults infected with *Mycobacterium tuberculosis* (*M. tuberculosis*). The vaccine efficacy at month 36 was 49.7% (90% confidence interval [CI], 12.1 to 71.2; 95% CI, 2.1 to 74.2) in prevention of TB disease [22]. It remains to be seen if this vaccine will prove to have the effectiveness needed for more widespread use (Table 2).

Table 2. Selected strategic goals reproduced from Pillar 1 of the "defeating meningitis" roadmap and suggested paediatric tuberculous meningitis (TBM)-related activities [4].

Adapted Strategic Goals from "Defeating Meningitis" Road Map	Suggested Key Activities Adapted to Paediatric TBM
Strategic Goal 1: achieve and maintain high coverage of licensed WHO vaccines with equal access in all countries and introduce these vaccines in countries that have not yet introduced them in line with WHO recommendations.	Implement locally appropriate tailored immunization strategies to achieve and maintain high BCG vaccination coverage, reinforcing and complementing existing immunization strategies, including those targeting special risk groups. Ensure effective linkages and synergies between WHO, UNICEF, Gavi, the Vaccine Alliance, and other global or regional initiatives aiming to reduce price and increase sustainable access to both BCG and future novel licensed vaccines for LMICs.

Table 2. *Cont.*

Adapted Strategic Goals from "Defeating Meningitis" Road Map	Suggested Key Activities Adapted to Paediatric TBM
Strategic Goal 2: introduce effective and affordable new WHO prequalified vaccines.	Support development, licensure, WHO prequalification and introduction of effective, affordable, and safe new TB vaccines. Improve support to vaccine manufacturers in their efforts to ensure diversification of sufficient quality-assured vaccine production capacity in more countries, including LMICs.
Strategic Goal 3: develop evidence-based policy on vaccination strategies that result in optimal individual protection and, where possible.	Enable and promote the sharing of knowledge between countries (for example, on accurate cost-effectiveness models) to support national policy decisions, particularly in low-incidence settings. Assess the overall vaccine impact, duration of protection, and indirect effects induced with BCG (and evaluation of strain-specific differences) and novel TB vaccines to inform vaccination strategies to maximise long-lasting immunity in populations and to prevent/control vaccine-preventable TB among at-risk individuals. Establish immune correlates of protection against TB Quantify the potential benefits of BCG and novel vaccines on reducing multidrug-resistant TB.

LMICs = low and middle-income countries.

3. Pillar 2: Diagnosis and Treatment

Early and accurate diagnosis of TBM remains challenging. Diagnostic and treatment delay are the most important predictors of mortality and disability [23]. The challenges are numerous: (1) non-specific symptoms leading to misdiagnosis; (2) inadequately sensitive diagnostic tools to confirm TBM; (3) suboptimal sample collection for laboratory processing; and (4) lack of standardised region-specific training of healthcare staff to identify TBM and initiate appropriate anti-TB therapy.

Mycobacterial cultures are considered the "gold standard" diagnostic test for TBM. However they take 2–6 weeks to yield a result and require specialised laboratory facilities and experience not available at all healthcare levels. The rapid molecular test, Xpert MTB/RIF, has emerged as an important diagnostic test for all forms of TB, such that WHO have recommended it as the initial TB diagnostic test, replacing the acid-fast bacilli smear [24]. Next-generation Xpert MTB/RIFUltra (Xpert Ultra) has a sensitivity of 44–77% and specificity approaching 100% in the cerebrospinal fluid (CSF) of adults with TBM [25–27]. Xpert MTB/RIF offers value for clinical decision-making as it allows early initiation of anti-TB therapy and information on the presence/absence of rifampicin resistance and, hence, whether treatment for multidrug-resistant (MDR) TBM is indicated. However, the test requires an expensive platform, access to electricity, and costly consumables. Xpert cartridges costs USD $10 in countries that are allowed concessional pricing; however, for high TB burden countries, this equates to a substantial expenditure [28]. Although the next generation Xpert Ultra has slightly higher sensitivity than Xpert MTB/RIF it does not have a sufficiently high negative predictive value to exclude TBM when the result is negative. Microbiological confirmation of MDR-TB in children is uncommon. In the absence of a microbiological diagnosis, MDR-TBM requires careful history-taking including recent exposure to an infectious MDR-TB source case or someone who failed TB treatment or died from suspected MDR-TB.

Xpert MTB/RIF and Xpert Ultra should not be used as the sole diagnostic test in TBM and every attempt should be made to obtain samples from elsewhere such as sputum, gastric aspirates, lymph node aspirates, biopsies, to support the diagnosis. Recent rapid diagnostic tests, such as TB-LAMP (loop-mediated isothermal amplification), which amplifies MTB DNA and is implementable at peripheral health centre level, and urinary TB FujiLAM, which detects *M. tuberculosis* lipoarabinomannan antigen, remove the need for advanced laboratory expertise [29–31]. However, information on their role in TBM diagnosis, diagnostic performance in the paediatric population, and impact on mortality have yet to be clearly elucidated. In the absence of an optimal and accessible diagnostic test, diagnosis of TBM still relies on clinical symptoms and signs, CSF parameters, and radiology (chest radiographs at the minimum and neuroimaging where available), and, where available, bacterial, viral and fungal microbiological tests to distinguish TBM from other infective causes of meningitis. A clinical decision tool (i.e., CHILD TB LP- altered Consciousness, caregiver HIV-infected, Illness length, Lethargy, focal neurological Deficit, failure to Thrive, Blood/serum sodium, CSF Lymphocytes, CSF Protein) developed to facilitate early diagnosis of childhood TBM has been shown to accurately classify microbiologically confirmed TBM (sensitivity 100%, specificity 90%) [32]. However, this algorithm requires prospective evaluation as a rapid diagnostic tool in children with CNS infections in numerous geographical settings with varying TB burden and risk factors.

Bacteriologic confirmation of meningitis caused by *M. tuberculosis* relies heavily on obtaining adequate volumes of CSF (at least 6 mL for mycobacterial testing and additional volume for tests to exclude other causes of meningitis, e.g., pyogenic bacterial meningitis, *Cryptococcus* sp., and viral meningoencephalitis) [33]. As TBM is paucibacillary in nature, maximising the number of mycobacteria in a sample will increase the probability of a positive result. This is achieved by obtaining large volumes of CSF with subsequent centrifugation (3000× g for 15 min) to concentrate *M. tuberculosis* into pellet form [34]. Increased focus on the importance of CSF collection and processing should be incorporated into TB guidelines, and in training of healthcare and laboratory staff (Table 3). Even in the absence of positive TB microbiology, CSF taken for microscopy and biochemistry is important for TBM diagnosis. This highlights the pivotal role of frontline clinicians in obtaining informed consent for the lumbar puncture as part of the routine diagnostic approach. Parental refusal to perform lumbar punctures in children with suspected central nervous system infection is common in low-middle income (LMIC) settings; a cross-sectional study of 215 families of Pakistani children who had indications for lumbar puncture showed that 33% refused [35]. Common reasons for refusal were lack of knowledge about the risks of the procedure (30%) and fear of paralysis of lower limbs (49%). High levels of illiteracy, stigma associated with the procedure and potential differences of opinion amongst the extended family impact on LP consent in many LMIC settings. Training of healthcare staff on how to counsel families in these settings, and guidance on the process of informed consent with real-world scenarios may reduce the frequency of parental refusal for LP, given its importance to securing a diagnosis of TBM. Current research is focussed on detecting novel blood, CSF, and urine biomarkers (transcriptome, proteome, and metabolome) to diagnose TBM in children [36–39]. Such approaches and new biomarkers of cerebral injury in CSF offer much promise for future diagnosis of TBM in children [40].

Table 3. Selected relevant strategic goals reproduced from Pillar 2 the "defeating meningitis" roadmap and suggested paediatric TBM-related activities [4].

Adapted Strategic Goals from "Defeating Meningitis" Road Map	Suggested Key Activities Adapted to Paediatric TBM
Goal 6: improve diagnosis of TBM at all levels of care.	Improving time to diagnosis. CSF rapid diagnostics tests performed in TBM. Increased availability and usage of Xpert MTB/RIF Ultra as the initial test. Further evaluation of performance of TB-LAMP in TBM which can detect DNA in less than 1 h. Further evaluation of mycobacterial antigen tests such as LAM in CSF in TBM. Active research into blood/CSF biomarkers which distinguish TB meningitis from other CNS infections. Non-CSF rapid diagnostic tests. Further evaluation of blood-based biomarkers (RNA, protein, metabolite). Further evaluation of urinary LAM in TBM. Improving sampling and laboratory processes to optimise yield. Greater emphasis on optimal CSF volumes for diagnosis and safety of obtaining these volumes. Ongoing training of laboratory staff. Maintenance and upgrading of laboratory equipment. Improving healthcare worker understanding of timely diagnosis, referral, and treatment. Development and adoption of standardised international guidelines on diagnosis, referral, and treatment of TBM particularly in LMICs. Greater emphasis on TBM within national TB programs. Accessible education of front-line healthcare professionals regarding thresholds and indications for lumbar puncture. Qualitative research to understand, from communities and health professionals, the factors contributing to low parental acceptance of lumbar puncture in different settings and evidence-based measures to increase acceptance.
Goal 7: develop and facilitate access to diagnostic assays at all levels of care to increase confirmation of TBM.	Funding mechanisms to facilitate development and uptake of novel rapid diagnostic assays. Partnering with diagnostic companies to evaluate new and rapid diagnostic tests. Partnerships with research grant funding agencies to evaluate diagnostic tests. Mechanisms for validation, production and adoption of diagnostic assays. Development of diagnostic assays to support immediate medical decision making at point-of-care.
Goal 8: develop and implement a context specific policy to identify mothers who have TB disease in pregnancy and post-partum, and for diagnosis of neonatal TB, particularly for low-resource settings.	Develop and implement a context-specific strategy for diagnosis, particularly for low-resource settings. Updating national TB guidelines to include diagnosis of maternal and neonatal TB, identification of infants. at risk of disseminated TB or TBM, and provision of TPT.

Table 3. *Cont.*

Adapted Strategic Goals from "Defeating Meningitis" Road Map	Suggested Key Activities Adapted to Paediatric TBM
Goal 9: provide and implement appropriate, context specific, quality-assured guidelines and tools for treatment and supportive care to reduce the risk of mortality, sequelae, and antimicrobial resistance.	Review evidence on potential benefit of adjunctive therapies for bacterial meningitis (steroids/aspirin/thalidomide) in LMICs. Develop and implement updated evidence-based regionally adapted guidelines and recommended tools on patient treatment and care for all age groups from early diagnosis to early identification, treatment, and care of sequelae, and addressing antimicrobial resistance and integration into existing guidelines. Ensure that recommended and quality-assured antimicrobials and medical supplies needed for supportive care are affordable and accessible at country level including paediatric formulations.

CSF = cerebrospinal fluid, WHO = World Health Organisation, MTB = mycobacterium tuberculosis, Rif = rifampicin resistance, TB = tuberculosis, TBM = TB meningitis, HIV = human immunodeficiency virus, LAM = lipoarabinomannan, LAMP = loop-mediated isothermal amplification, TPP = target product profile, LMIC = low and middle income countries.

The treatment of TBM is urgent to limit disability and death, especially in a condition with delayed presentation and diagnosis. The WHO regimen for drug susceptible TBM is two months of HRZE ((INH: 10 mg/kg), rifampicin (RMP: 15 mg/kg), pyrazinamide (PZA: 35 mg/kg), and ethambutol (EMB: 20 mg/kg) (HRZE)) intensive phase followed by a continuation phase of ten months of HR treatment, which is 6 months longer than the treatment for pulmonary TB. The 2 HRZE used in the crucial initial 2 months of TB treatment is recognised to be suboptimal in penetrating the blood–brain barrier and new regimens are being put into clinical trials [7,41,42]. The pathophysiology of TBM includes severe inflammation in the brain with potential for raised intracranial pressure, hydrocephalus, ischaemia, and infarction [43]. Adjuvant treatment with steroids in the initial 6–8 weeks has been shown in a Cochrane systematic review to improve mortality, but to have little effect on disabling sequelae [44]. However, the benefits of immunomodulatory treatment may depend on genetically regulated levels of inflammation [45]. More recently, data from an adult treatment trial of TBM suggests that addition of adjunctive aspirin may improve clinical outcomes in certain patient subgroups at high dose and this is under evaluation in an ongoing paediatric clinical trial [7,46].

The optimal dosing, choice, route and duration of antitubercular agents in TBM and the context of an inflamed blood brain barrier remains the subject of debate. In Cape Town, South Africa, children with TBM are treated with a shorter, higher dose 6 month regimen: INH: 20 mg/kg, RMP: 20 mg/kg, PZA: 40 mg/kg and ethionamide: 20 mg/kg for 6 months [47]. Although excellent outcomes have been reported with this "Cape Town" regimen it has never been subjected to a randomised controlled trial and it is not clear whether the apparent improvements are due to the regimen or high quality supportive care [47]. Given the potential to inform global TB policy with improved CSF penetration, halving the duration of treatment and hence improving concordance, the recently commenced SURE trial will compare the 12-month WHO regimen with a modified 6-month intensified regimen (RMP 30 mg/kg; INH 20 mg/kg; ethionamide replaced with levofloxacin, which penetrates CSF well) in children under 15 years [7]. The impact of adjunctive high dose aspirin vs placebo on severe disability will also be evaluated.

Successful treatment of MDR-TBM in children presents further challenges. MDR-TB meningitis is associated with high mortality owing to delayed diagnosis of drug resistance, the absence of a standardised approach to the management and poor CSF penetration of many MDR-TB drugs [48,49]. Whilst it is recommended that treatment regimens for MDR-TB should include at least 4–5 effective drugs, and careful consideration of drugs that penetrate the CSF well such as fluoroquinolones and linezolid, there is a lack of evidence to inform best antibiotic combination, duration and doses [50]. In light of the increasing threat of MDR-TB, pre-clinical studies evaluating CSF-brain penetrating properties of new TB drugs and clinical trials assessing optimal drug regimens to improve outcome are urgently needed (Table 3).

4. Pillar 3: Disease Surveillance

Systems for TB surveillance are some of the oldest in the world, having begun more than two centuries ago with the recording of TB mortality in England and Wales [51]. As defined by the WHO, the primary aim of global TB surveillance is to assess the progress of TB control activities in the context of the End TB strategy [5]. WHO collates all the annual surveillance data provided by countries and then generates global estimates. These are adjusted by correction factors to account for underreporting, over- and under-diagnosis [5]. Globally, an estimated 10 million (range, 8.9–11.0 million) people fell ill with TB in 2019 [5]. Extrapulmonary TB represented 16% of the 7.1 million incident cases that were notified in 2019 [5].

The backbone of TB surveillance is case notification, which is statutory in many countries. The most accepted international case definitions for notification are the 2013 WHO definitions, last updated in January 2020, which incorporate: bacteriological status, classification of disease as pulmonary and extra-pulmonary, history of previous TB treatment, HIV status, and drug resistance [52]. Surveillance is incorporated within the national TB control programs in most LMIC as a monitoring and evaluation tool. The focus of surveillance at present are the infectious pulmonary cases for which more data are provided, while all extra-pulmonary forms are reported collectively. For example, the Global TB report provides no further details about extra-pulmonary TB including TBM [5].

Use of electronic registers for TB cases allows the capture of more detailed information such as a breakdown of the extra-pulmonary sites of disease but these are not part of formal reporting at national and global level. The National TB Elimination Program of India kindly shared data on CNS TB from their national e-Register for this publication. They reported that CNS TB among under 18 years olds contributed to 1.06%, 1.27%, and 1.66% of the total extra-pulmonary TB cases in the years 2018, 2019 and 2020, respectively (personal communication to Varinder Singh).

Amongst the other challenges that compromise detailed surveillance for TBM is the lack of a clear case definition. As paediatric TBM is a paucibacillary disease, many infectious and non-infectious neurological diseases may be misdiagnosed as TBM, such as partially treated bacterial meningitis, viral encephalitis, autoimmune encephalitis, and subacute onset neurodegenerative diseases. This is compounded by a paucity of neuroimaging facilities or facilities for molecular TB diagnostics in low resource settings. TBM, despite its own specific diagnostic and therapeutic challenges, is neglected in most TB programmatic guidelines, which typically only address the duration of treatment and adjunctive corticosteroids [53,54]. Both under- and over-reporting of TBM, as a result of misdiagnosis, is also likely in low-resource settings.

As TBM diagnosis requires some key laboratory facilities and clinical expertise, it is often centralised to large hospitals. Thus, TBM surveillance requires data from hospitals, both in the public and private sector, and also from laboratories where microbiological testing is performed. Even in public hospitals, reporting has many challenges. In a study from China, 25% of cases that were documented in the hospital records were not reported to the public health authorities [55,56]. Factors cited that also apply in other settings were unqualified and overworked health personnel, poor supervision and accountability at local and national levels, and a complicated incohesive health information management system [55,56]. The situation is complex in the case of private practitioners as the system requires them to provide patient details from outside the state health system. Limited practitioner time is an important barrier to TB notification and therefore user-friendly interventions, such as mobile based notification, have potential to improve notification, although are constrained by technology and internet access [57]. In a pilot study of the use of a mobile interface voice-based TB notification system, only 6% of private practitioners were found to use it [57].

Although TB surveillance is one of the most well-established and prevalent surveillance systems across the globe, there is a need for a well-defined strategy for TBM surveillance, which is poorly quantified at present. The data thus collected will be helpful to understand the disease, its trends and inform appropriate action (Table 4).

Table 4. Selected relevant strategic goals reproduced from Pillar 3 the "defeating meningitis" roadmap and suggested paediatric TBM-related activities [4].

Adapted Strategic Goals from "Defeating Meningitis" Road Map	Suggested Key Activities Adapted to Paediatric TBM
Strategic Goal 10: ensure that effective systems for surveillance of meningitis and detection of the main meningitis pathogens are in place.	Improved reporting of different sites of extrapulmonary TB, including TBM. Triangulation of laboratory, public sector healthcare, and private healthcare notifications of TBM diagnoses. Specific reporting of MDR-TB TBM given the accompanying therapeutic challenges. Improved access to global genome partnerships for TB, including TBM.
Strategic Goal 12: develop and conduct surveys and studies to establish the burden of sequelae.	Develop and implement a global strategy and tools for studies and surveys to establish and monitor the burden of TBM sequelae.

5. Pillar 4: Support and Care for People Affected by Meningitis

For many childhood survivors of TBM, the majority of whom live in LMICs, severe illness results in significant neurodevelopmental sequelae. A meta-analysis on treatment outcomes in childhood TBM demonstrated neurological sequelae in 54% of survivors [1]. However, data on the physical, cognitive, and behavioural sequelae of TBM, which have lasting socioeconomic implications for patients and their families, are limited and rarely include long-term follow-up. Children living with disabilities in LMICs are likely to experience poorer health and quality of life, reduced school participation and high rates of poverty when compared with their non-disabled peers [58,59], yet this has not been evaluated in TBM survivors.

Common impairments documented post TBM are in cognition, learning, emotion, and behaviour, all potentially affecting educational attainment and future employment. Poor neurodevelopmental outcome is associated with younger age, delayed presentation and treatment initiation, clinical severity and hydrocephalus, highlighting the need for increasing awareness of TBM and better clinical and diagnostic tools for timely initiation of treatment and management of sequelae [9,60,61].

The United Nations Sustainable Development Goal (SDG) 4, Pillar 4 of the Roadmap to Defeating Meningitis by 2030, and United Nations Convention on the Rights of the Child together highlight the need for timely identification and management of sequelae, together with reliable, valid measures to evaluate preventive and interventional efforts as well as improved access to appropriate support and care services [4,62,63]. Achieving these goals, considered standard of care in most high-income countries, will be challenging for several reasons.

First, there is a paucity of robust and standardised neurodevelopmental assessment tools (NDATs) developed for, and with normal reference populations, across different geographical and cultural settings. To fully understand the burden of impairment caused by TBM will, therefore, require appropriately adapted, as well as new, locally developed NDATs to detect both early developmental and later, emergent speech, behavioural and cognitive difficulties together with adaptive function [64]. A number of NDATs that have either been adapted for use in or designed for different geographical and cultural settings are now being used [65]. However, these often require a high level of skill and training for healthcare professionals and are time-consuming. Screening tools to detect childhood disability and suitable for use by a broader range of healthcare workers are evolving. For example, the WHO Disability Assessment Schedule (DAS) 2.0, a generic instrument developed for adults to assess health and disability, has been adapted for use in children in a number of settings [66–68]. The culturally neutral, WHO Indicators of Infant and Young Child Development caregiver report tool, which monitors pre-school children across multiple LMIC settings, has been developed with feasibility testing and piloting across a number of LMIC planned [69].

Second, for children in whom development disability is identified, access to health and rehabilitation interventions to improve functioning and quality of life are limited. Even when available, the uptake is low with numerous barriers to access including cost, transport, physical inaccessibility, lack of appropriately trained healthcare workers, as well as cultural beliefs and stigma around disability [70,71]. Moreover, caregivers are likely to experience high levels of depression and anxiety with limited family and community support [72].

Third, children with developmental disabilities are often excluded from programmes and clinical trials of early child development interventions as it may be difficult to quantify improvement when developmental progress is the primary outcome, further compounding delay and reducing learning opportunities [73].

Fourth, there is a dearth of good quality information from large studies on the spectrum of evolving disabilities in children of all ages post TBM and limited information from LMICs on what interventions are effective for children with disabilities in general, their cost effectiveness and scalability. Collaborative efforts between stakeholders, funders and researchers, will help to improve outcome for children and families dealing with the sequelae of TBM. As TB clinicians and researchers, we must strongly advocate for these initiatives (Table 5).

Table 5. Selected relevant strategic goals reproduced from Pillar 4 the "defeating meningitis" roadmap and suggested paediatric TBM-related activities [4].

Adapted Strategic Goals from "Defeating Meningitis" Road Map	Suggested Key Activities Adapted to Paediatric TBM
Strategic Goal 13: strengthen early recognition and management of sequelae from meningitis in healthcare and community settings.	Conduct research on: (i) socioeconomic impact of sequelae on children, adults and their families/carers; (ii) effectiveness of aftercare/support interventions in reducing impact. Develop and implement best practice guidelines for LMICs on detection, monitoring and management of TBM sequelae after discharge from hospital, at all levels of healthcare and in community settings, for example, schools (including disability sensitization and communication skills). Promote community-based programmes to: (i) Identify sequelae and disabilities, based on standardised instruments (especially for child development and hearing), and refer for assessment and appropriate care. (ii) Provide care, support and aftercare to individuals, families and communities affected by TBM, for example, psychosocial support.

Table 5. *Cont.*

Adapted Strategic Goals from "Defeating Meningitis" Road Map	Suggested Key Activities Adapted to Paediatric TBM
Strategic Goal 14: increase the availability and access to appropriate care and support (i) for people affected by meningitis; (ii) for their families and carers.	Map out existing services and support systems by country for: (i) children and people with disabilities, including those with TBM sequelae, and (ii) for families/carers of people affected by TBM; identify barriers to access, availability, and use, with the involvement of organisations for persons with disabilities and other networks where possible, and undertake a gap analysis to improve service provision. Strengthen partnerships between government and civil society organisations, including organisations for persons with disabilities and other networks, so that people with sequelae or disabilities, their families/carers and those bereaved due to TBM have access to quality and effective services that are in line with international human rights standards and frameworks. Provide relevant, up-to-date information to people and carers affected by TBM about access to services for managing sequelae as well as about the rights of people with disabilities guaranteed under national policies and laws and through global human rights instruments.

6. Pillar 5: Advocacy and Engagement

Advocacy is key to bringing about changes to policies and practices at institutional, community and individual level [74]. Advocacy for TBM fits into the scope of the meningitis roadmap by working with partners to raise public and political awareness of TBM and its devastating effects in order to improve diagnosis, treatment, prevention and support for affected families [4].

There are existing frameworks for TB control that advocates for TBM can leverage. The End TB strategy envisions the world free of TB, with zero deaths, disease, and its catastrophic consequences [75]. It is well known that TBM has devastating sequelae and contributes to significant morbidity and mortality. The child and adolescent TB roadmap aims to draw attention to the childhood TB epidemic and has placed advocacy and fostering partnerships as one of its key actions points [6]. Within these frameworks, global partnerships and national TB programmes can be utilised to raise the profile of TBM and highlight its significant contribution of TB-related mortality in adults and children. This will ensure that strategic interventions are formulated and planned with the aim of increasing its detection, treatment, and prevention. These interventions should include promotion of operational research to better understand the burden of TBM disease, its outcomes and consequences outside what is routinely monitored in country TB programmes. Partnerships with the academic community, civil society organisations, community, and patient advocacy groups can play a key role in raising the profile of TBM and developing champions that serve as key leaders in advocacy for TBM.

Existing advocacy materials for TB can be included with enhanced messaging, raising awareness of communities about the signs and symptoms of TBM in order to improve early recognition and promote early healthcare-seeking. In addition, healthcare providers, who are fundamental to diagnostic and care pathways, should receive continued education to improve their ability to recognise and diagnose the disease, together with managing the sequelae and other neurological disabilities that often arise from TBM. Furthermore, TBM should be incorporated into advocacy for syndromic neuro-disabilities such as epilepsy and cerebral palsy. Research funders should be targeted to harness resources for research in novel, non-invasive, affordable diagnostics and more effective or shorter treatment regimens.

Community messaging should highlight the importance of BCG vaccination and counter vaccine hesitancy that is increasing worldwide. In addition, advocacy for effective household contact tracing and provision and monitoring of TPT for high-risk groups, including individuals living with HIV, should be promoted and strengthened within national programs with communities recognising their rights to this. This combination of efforts has the potential to prevent severe forms of TB in vulnerable populations and the catastrophic costs associated with TBM (Table 6).

Table 6. Selected relevant strategic goals reproduced from Pillar 5 the "defeating meningitis" roadmap and suggested paediatric TBM-related activities [4].

Adapted Strategic Goals From "Defeating Meningitis" Road Map	Suggested Key Activities Adapted to Paediatric TBM
Strategic Goal 15: ensure that funders and policymakers at the national, regional, and global levels recognise that the road map to defeat meningitis is prioritised and integrated into country plans at all levels.	Raise awareness of TBM as a health priority. Among funders and policymakers through national and international champions, civil society organizations, advocacy groups and healthcare providers, including the disability sector. Identify and create synergies between key activities on strategy, implementation, and communication with other initiatives at the global, regional, and national levels, especially for the immunisation and disability sectors. Build a business case for investment in vaccines, surveillance, diagnosis, and treatment of meningitis, and for the prevention and management of sequelae, as set out in the road map, that is targeted for use by policymakers, decision-makers, and funders at the global, regional, and national levels including the disability sector. Countries undertake needs assessment on TBM and its impact and create national action plans that address gaps and are aligned to the global road map. Develop communications and engagement strategy and improve global recognition of World Meningitis Day, World TB day and other global health dates (for example—cerebral palsy, disability), adapt messaging to policymakers as well as to the public, and raise funding to promote activities that support the road map.
Strategic Goal 16: ensure awareness, among all populations, of the symptoms, signs, and consequences of meningitis so that they seek appropriate healthcare.	Undertake integrated communication programmes and activities that increase population awareness of the risk, symptoms, signs, and consequences of TBM, and of the recommended health-seeking response, and create community awareness of TB disease and prevention. Study the community understanding of the risk of TBM, and the factors that facilitate or act as barriers to health-seeking behaviours for TBM, and integrate actions into country plans to address the issues identified.

Table 6. *Cont.*

Adapted Strategic Goals From "Defeating Meningitis" Road Map	Suggested Key Activities Adapted to Paediatric TBM
Strategic Goal 17: ensure and raise awareness of communities about the impact of meningitis and available support after meningitis.	Support global and national campaigns on the International Day of Persons with Disabilities to increase and raise awareness of communities about disability, and to address significant attitudinal barriers that lead to stigma and undignified treatment of people with disabilities. Raise awareness of new systems for data collection on sequelae/disabilities and of available support and specialist services. Identify, encourage and support civil society organisations that do or could promote the interests of those affected by TBM, including those with sequelae, and invite involvement in delivering the goals of the road map through their communities, engagement with national and regional authorities and international networks of civil society organizations. Study community understanding of BCG and new TB vaccines, TPT and other preventive strategies.
SG19: maintain high vaccine confidence	Develop risk and communication strategies to address issues of access, acceptance, and generation of demand for BCG and novel TB vaccines. Develop risk and crisis communication plans for BCG and new TB vaccines to address potential inaccurate communication of adverse events.

7. Conclusions

Both the "Defeating Meningitis by 2030: global roadmap" and "End TB Strategy" are ambitious plans, whose timelines for success are likely to be hampered by the COVID-19 pandemic. Collateral impacts from the pandemic include impact on the resilience of health systems, reporting and surveillance structures, health-seeking behaviours, together with cuts in funding from major donors [4,76]. Nevertheless, the pandemic also brings opportunities and provides an opportune moment to consider children with TBM who have amongst the highest burdens of morbidity and mortality of any of the conditions encompassed by the two plans. Although there are evident constraints and limits to how much the "Defeating meningitis roadmap" can be applied to children with TBM, creative and collaborative working between clinicians, policymakers, public health, and community and advocacy organisations can help bring us closer to defeating paediatric TBM.

Author Contributions: Conceptualisation: R.B.R.; writing—original draft preparation: R.B.R., S.B.-K., C.C., J.H., H.M., N.S., S.S., V.S., E.W., S.T.A.; writing—review and editing: R.B.R., S.B.-K., C.C., D.M.G., J.H., H.M., N.S., J.A.S., S.S., V.S., E.W., S.T.A. All authors have read and agreed to the published version of the manuscript.

Funding: The SURE trial is funded by the Joint Global Health Trials Scheme of the Department for International Development, UK (DFID), the Wellcome Trust, the Medical Research Council (MRC UK) and the National Institute for Health Research (NIHR). Grant reference number: MR/R006113/1. RB is funded by an NIHR Academic Clinical Lectureship (CL-2018-20-001). JAS is supported by a Clinician Scientist Fellowship jointly funded by the UK Medical Research Council (MRC) and the UK Department for International Development (DFID) under the MRC/DFID Concordat agreement (MR/R007942/1).

Acknowledgments: We thank and acknowledge the Deputy Director General (TB), Central TB Division, Ministry of Health and Family Welfare, Government of India for providing the data of notified cases of CNS TB.

Conflicts of Interest: The authors declare no conflict of interest. The funders had no role in the writing of the manuscript or in the decision to publish.

References

1. Chiang, S.S.; Khan, F.A.; Milstein, M.B.; Tolman, A.W.; Benedetti, A.; Starke, J.R.; Becerra, M.C. Treatment outcomes of childhood tuberculous meningitis: A systematic review and meta-analysis. *Lancet Infect. Dis.* **2014**, *14*, 947–957. [CrossRef]
2. Seddon, J.A.; Tugume, L.; Solomons, R.S.; Prasad, K.; Bahr, N.C. The current global situation for tuberculous meningitis: Epidemiology, diagnostics, treatment and outcomes. *Wellcome Open Res.* **2019**, *4*, 167. [CrossRef] [PubMed]
3. Seddon, J.; Hesseling, A.C.; Marais, B.J.; Jordaan, A.; Victor, T.; Schaaf, H.S. The evolving epidemic of drug-resistant tuberculosis among children in Cape Town, South Africa. *Int. J. Tuberc. Lung Dis.* **2012**, *16*, 928–933. [CrossRef] [PubMed]
4. World Health Organization. *Defeating Meningitis by 2030: A Global Road Map*; World Health Organization: Geneva, Switzerland, 2020.
5. World Health Organization. *Global Tuberculosis Report 2020*; World Health Organization: Geneva, Switzerland, 2020.
6. World Health Organization. Roadmap towards Ending TB in Children and Adolescents. Geneva. 2018. Available online: http://www.who.int/tb/publications/2018/tb-childhoodroadmap/en/ (accessed on 6 October 2018).
7. SURE. Short Intensive Treatment for Children with Tuberculous Meningitis. Available online: https://www.isrctn.com/ISRCTN40829906 (accessed on 14 April 2021).
8. Trunz, B.B.; Fine, P.; Dye, C. Effect of BCG vaccination on childhood tuberculous meningitis and miliary tuberculosis worldwide: A meta-analysis and assessment of cost-effectiveness. *Lancet* **2006**, *367*, 1173–1180. [CrossRef]
9. Van Well, G.T.J.; Paes, B.F.; Terwee, C.B.; Springer, P.; Roord, J.J.; Donald, P.R.; Schoeman, J.F. Twenty years of pediatric tuberculous meningitis: A retrospective cohort study in the western cape of South Africa. *Pediatrics* **2009**, *123*, e1–e8. [CrossRef] [PubMed]
10. World Health Organization. *WHO Consolidated Guidelines on Tuberculosis*; Module 1: Prevention; Tuberculosis Preventive Treatment; World Health Organization: Geneva, Switzerland, 2020.
11. Hamada, Y.; Glaziou, P.; Sismanidis, C.; Getahun, H. Prevention of tuberculosis in household members: Estimates of children eligible for treatment. *Bull. World Health Organ.* **2019**, *97*, 534D–547D. [CrossRef] [PubMed]
12. Dhawan, S.R.; Gupta, A.; Singhi, P.; Sankhyan, N.; Malhi, P.; Khandelwal, N. Predictors of Neurological Outcome of Tuberculous Meningitis in Childhood. *J. Child Neurol.* **2016**, *31*, 1622–1627. [CrossRef] [PubMed]
13. Millership, S.E.; Anderson, C.; Cummins, A.J.; Bracebridge, S.; Abubakar, I. The risk to infants from nosocomial exposure to tuberculosis. *Pediatr. Infect. Dis. J.* **2009**, *28*, 915–916. [CrossRef]
14. Ahn, J.G.; Kim, D.S.; Kim, K.H. Nosocomial exposure to active pulmonary tuberculosis in a neonatal intensive care unit. *Am. J. Infect. Control* **2015**, *43*, 1292–1295. [CrossRef]
15. World Health Organization. *WHO Guidelines on Tuberculosis Infection Prevention and Control*; World Health Organization: Geneva, Switzerland, 2019.
16. Cilloni, L.; Fu, H.; Vesga, J.F.; Dowdy, D.; Pretorius, C.; Ahmedov, S.; Arinaminpathy, N. The potential impact of the COVID-19 pandemic on the tuberculosis epidemic a modelling analysis. *EClinicalMedicine* **2020**, *28*, 100603. [CrossRef]
17. Stop TB Partnership. The Potential Impact of the Covid-19 Response on Tuberculosis in High-Burden Countries: A Modelling Analysis. 2020. Available online: http://www.stoptb.org/assets/documents/news/Modeling%20Report_1%20May%202020_FINAL.pdf (accessed on 14 April 2021).
18. Visca, D.; Tiberi, S.; Pontali, E.; Spanevello, A.; Migliori, G.B. Tuberculosis in the time of COVID-19: Quality of life and digital innovation. *Eur. Respir. J.* **2020**, *56*, 2001998. [CrossRef] [PubMed]
19. Koura, K.G.; Harries, A.D.; Fujiwara, P.I.; Dlodlo, R.A.; Sansan, E.K.; Kampoer, B.; Affolabi, D.; Combary, A.; Mbassa, V.; Gando, H.; et al. COVID-19 in Africa: Community and digital technologies for tuberculosis management. *Int. J. Tuberc. Lung Dis.* **2020**, *24*, 863–865. [CrossRef] [PubMed]
20. Mangtani, P.; Abubakar, I.; Ariti, C.; Beynon, R.; Pimpin, L.; Fine, P.E.M.; Rodrigues, L.C.; Smith, P.G.; Lipman, M.; Whiting, P.F.; et al. Protection by BCG Vaccine Against Tuberculosis: A Systematic Review of Randomized Controlled Trials. *Clin. Infect. Dis.* **2014**, *58*, 470–480. [CrossRef] [PubMed]
21. Abubakar, I.; Pimpin, L.; Ariti, C.; Beynon, R.; Mangtani, P.; Sterne, J.A.C.; Fine, P.E.M.; Smith, P.G.; Lipman, M.; Elliman, D.; et al. Systematic review and meta-analysis of the current evidence on the duration of protection by bacillus Calmette–Guérin vaccination against tuberculosis. *Heal. Technol. Assess.* **2013**, *17*, 1–372. [CrossRef]
22. Tait, D.R.; Hatherill, M.; Van Der Meeren, O.; Ginsberg, A.M.; Van Brakel, E.; Salaun, B.; Scriba, T.J.; Akite, E.J.; Ayles, H.M.; Bollaerts, A.; et al. Final Analysis of a Trial of M72/AS01E Vaccine to Prevent Tuberculosis. *N. Engl. J. Med.* **2019**, *381*, 2429–2439. [CrossRef] [PubMed]
23. He, Y.; Han, C.; Chang, K.-F.; Wang, M.-S.; Huang, T.-R. Total delay in treatment among tuberculous meningitis patients in China: A retrospective cohort study. *BMC Infect. Dis.* **2017**, *17*, 1–5. [CrossRef] [PubMed]
24. World Health Organization. WHO Consolidated Guidelines on Tuberculosis; Module 3: Diagnosis Rapid Diagnostics for Tuberculosis Detection. 2020. Available online: https://www.who.int/publications/i/item/who-consolidated-guidelines-on-tuberculosis-module-3-diagnosis---rapid-diagnostics-for-tuberculosis-detection (accessed on 14 April 2021).
25. Bahr, N.C.; Nuwagira, E.; Evans, E.E.; Cresswell, F.V.; Bystrom, P.V.; Byamukama, A.; Bridge, S.C.; Bangdiwala, A.S.; Meya, D.B.; Denkinger, C.M.; et al. Diagnostic accuracy of Xpert MTB/RIF Ultra for tuberculous meningitis in HIV-infected adults: A prospective cohort study. *Lancet Infect. Dis.* **2018**, *18*, 68–75. [CrossRef]

26. Cresswell, F.V.; Tugume, L.; Bahr, N.C.; Kwizera, R.; Bangdiwala, A.S.; Musubire, A.K.; Team, A.S.T.R.O.; Rutakingirwa, M.; Kagimu, E.; Nuwagira, E.; et al. Xpert MTB/RIF Ultra for the diagnosis of HIV-associated tuberculous meningitis: A prospective validation study. *Lancet Infect. Dis.* **2020**, *20*, 308–317. [CrossRef]
27. Donovan, J.; Thu, D.D.A.; Phu, N.H.; Dung, V.T.M.; Quang, T.P.; Nghia, H.D.T.; Oanh, P.K.N.; Nhu, T.B.; Chau, N.V.V.; Ha, V.T.N.; et al. Xpert MTB/RIF Ultra versus Xpert MTB/RIF for the diagnosis of tuberculous meningitis: A prospective, randomised, diagnostic accuracy study. *Lancet Infect. Dis.* **2020**, *20*, 299–307. [CrossRef]
28. Donovan, J.; Cresswell, F.V.; Thuong, N.T.T.; Boulware, D.R.; Thwaites, G.E.; Bahr, N.C.; Aarnoutse, R.E.; Anderson, S.T.B.; Bang, N.D.; Boyles, T.; et al. Xpert MTB/RIF Ultra for the Diagnosis of Tuberculous Meningitis: A Small Step Forward. *Clin. Infect. Dis.* **2020**, *71*, 2002–2005. [CrossRef] [PubMed]
29. Yu, G.; Shen, Y.; Zhong, F.; Ye, B.; Yang, J.; Chen, G. Diagnostic accuracy of the loop-mediated isothermal amplification assay for extrapulmonary tuberculosis: A meta-analysis. *PLoS ONE* **2018**, *13*, e0199290. [CrossRef]
30. Cresswell, F.V.; Ellis, J.; Kagimu, E.; Bangdiwala, A.S.; Okirwoth, M.; Mugumya, G.; Rutakingirwa, M.; Kasibante, J.; Quinn, C.M.; Ssebambulidde, K.; et al. Standardized Urine-Based Tuberculosis (TB) Screening With TB-Lipoarabinomannan and Xpert MTB/RIF Ultra in Ugandan Adults With Advanced Human Immunodeficiency Virus Disease and Suspected Meningitis. *Open Forum Infect. Dis.* **2020**, *7*, ofaa100. [CrossRef] [PubMed]
31. Quinn, C.M.; Kagimu, E.; Okirworth, M.; Bangdiwala, A.S.; Mugumya, G.; Ramachandran, P.S.; Wilson, M.R.; Meya, D.B.; Cresswell, F.V.; Bahr, N.C.; et al. Fujifilm SILVAMP TB LAM Assay on Cerebrospinal Fluid for the Detection of Tuberculous Meningitis in Adults With Human Immunodeficiency Virus. *Clin. Infect. Dis.* **2021**. [CrossRef] [PubMed]
32. Goenka, A.; Jeena, P.M.; Mlisana, K.; Solomon, T.; Spicer, K.; Stephenson, R.; Verma, A.; Dhada, B.; Griffiths, M.J. Rapid Accurate Identification of Tuberculous Meningitis Among South African Children Using a Novel Clinical Decision Tool. *Pediatr. Infect. Dis. J.* **2018**, *37*, 229–234. [CrossRef] [PubMed]
33. Thwaites, G.; Fisher, M.; Hemingway, C.; Scott, G.; Solomon, T.; Innes, J. British Infection Society guidelines for the diagnosis and treatment of tuberculosis of the central nervous system in adults and children. *J. Infect.* **2009**, *59*, 167–187. [CrossRef]
34. Bahr, N.C.; Tugume, L.; Rajasingham, R.; Kiggundu, R.; Williams, D.A.; Morawski, B.; Alland, D.; Meya, D.B.; Rhein, J.; Boulware, D.R. Improved diagnostic sensitivity for tuberculous meningitis with Xpert®MTB/RIF of centrifuged CSF. *Int. J. Tuberc. Lung Dis.* **2015**, *19*, 1209–1215. [CrossRef] [PubMed]
35. Ahmed, M.; Ejaz, M.S.; Jahangeer, A.; Khan, S.; Hashmi, S.S.R.; Jawaid, T.; Nasir, S. Frequency and Associated Factors of Parental Refusal to Perform Lumbar Puncture in Children with Suspected Central Nervous System Infection: A Cross-sectional Study. *Cureus* **2019**, *11*, e5653. [CrossRef]
36. Manyelo, C.M.; Solomons, R.S.; Snyders, C.I.; Manngo, P.M.; Mutavhatsindi, H.; Kriel, B.; Stanley, K.; Walzl, G.; Chegou, N.N. Application of Cerebrospinal Fluid Host Protein Biosignatures in the Diagnosis of Tuberculous Meningitis in Children from a High Burden Setting. *Mediat. Inflamm.* **2019**, *2019*, 1–11. [CrossRef]
37. Manyelo, C.M.; Solomons, R.S.; Snyders, C.I.; Mutavhatsindi, H.; Manngo, P.M.; Stanley, K.; Walzl, G.; Chegou, N.N. Potential of Host Serum Protein Biomarkers in the Diagnosis of Tuberculous Meningitis in Children. *Front. Pediatr.* **2019**, *7*, 7. [CrossRef]
38. van Zyl, C.D.W.; Loots, D.T.; Solomons, R.; van Reenen, M.; Mason, S. Metabolic characterization of tuberculous meningitis in a South African paediatric population using 1H NMR metabolomics. *J. Infect.* **2020**, *81*, 743–752. [CrossRef] [PubMed]
39. Gliddon, H.D.; Kaforou, M.; Alikian, M.; Habgood-Coote, D.; Zhou, C.; Oni, T.; Anderson, S.T.; Brent, A.J.; Crampin, A.C.; Eley, B.; et al. Identification of Reduced Host Transcriptomic Signatures for Tuberculosis Disease and Digital PCR-Based Validation and Quantification. *Front. Immunol.* **2021**, *12*, 12. [CrossRef]
40. Rohlwink, U.K.; Figaji, A.; Wilkinson, K.A.; Horswell, S.; Sesay, A.K.; Deffur, A.; Enslin, N.; Solomons, R.; Van Toorn, R.; Eley, B.; et al. Tuberculous meningitis in children is characterized by compartmentalized immune responses and neural excitotoxicity. *Nat. Commun.* **2019**, *10*, 1–8. [CrossRef] [PubMed]
41. Donald, P. Cerebrospinal fluid concentrations of antituberculosis agents in adults and children. *Tuberculosis* **2010**, *90*, 279–292. [CrossRef] [PubMed]
42. Optimizing Treatment to Improve TBM Outcomes in Children (TBM-KIDS). Available online: https://clinicaltrials.gov/ct2/show/NCT02958709 (accessed on 11 April 2021).
43. Cresswell, F.V.; Davis, A.G.; Sharma, K.; Basu Roy, R.; Ganiem, A.R.; Kagimu, E. Recent Developments in Tuberculous Meningitis Pathogenesis and Diagnostics. *Wellcome Open Res.* **2021**, *4*, 164. [CrossRef]
44. Prasad, K.; Singh, M.B.; Ryan, H. Corticosteroids for managing tuberculous meningitis. *Cochrane Database Syst. Rev.* **2016**, *4*, CD002244. [CrossRef]
45. Thwaites, G.E.; van Toorn, R.; Schoeman, J. Tuberculous meningitis: More questions, still too few answers. *Lancet Neurol.* **2013**, *12*, 999–1010. [CrossRef]
46. Mai, N.T.; Dobbs, N.; Phu, N.H.; Colas, R.A.; Thao, L.T.; Thuong, N.T.; Nghia, H.D.; Hanh, N.H.; Hang, N.T.; Heemskerk, A.D.; et al. A randomised double blind placebo controlled phase 2 trial of adjunctive aspirin for tuberculous meningitis in HIV-uninfected adults. *eLife* **2018**, *7*, e33478. [CrossRef]
47. van Toorn, R.; Schaaf, H.S.; Laubscher, J.A.; van Elsland, S.L.; Donald, P.R.; Schoeman, J.F. Short Intensified Treatment in Children with Drug-susceptible Tuberculous Meningitis. *Pediatr. Infect. Dis. J.* **2014**, *33*, 248–252. [CrossRef]

48. Seddon, J.A.; Visser, D.H.; Bartens, M.; Jordaan, A.M.; Victor, T.C.; van Furth, A.M.; Schoeman, J.F.; Schaaf, H.S. Impact of Drug Resistance on Clinical Outcome in Children with Tuberculous Meningitis. *Pediatr. Infect. Dis. J.* **2012**, *31*, 711–716. [CrossRef] [PubMed]
49. Tho, D.Q.; Török, M.E.; Yen, N.T.B.; Bang, N.D.; Lan, N.T.N.; Kiet, V.S.; Chau, N.V.V.; Dung, N.H.; Day, J.; Farrar, J.; et al. Influence of Antituberculosis Drug Resistance and Mycobacterium tuberculosis Lineage on Outcome in HIV-Associated Tuberculous Meningitis. *Antimicrob. Agents Chemother.* **2012**, *56*, 3074–3079. [CrossRef] [PubMed]
50. World Health Organization. WHO Consolidated Guidelines on Tuberculosis; Module 4: Treatment Drug-resistant Tuberculosis Treatment. 2020. Available online: https://www.who.int/publications/i/item/9789240007048 (accessed on 14 April 2021).
51. Schwoebel, V. Surveillance of tuberculosis. *Indian J. Tuberc.* **2020**, *67*, S33–S42. [CrossRef] [PubMed]
52. World Health Organization. Definitions and Reporting Framework for Tuberculosis: 2013 Revision; Updated December 2014 and January 2020. 2020. Available online: https://www.who.int/tb/publications/definitions/en/ (accessed on 14 April 2021).
53. Central TB Division Directorate General of Health Services Ministry of Health with Family Welfare Nirman Bhavan New Delhi India. National Strategic Plan for Tuberculosis Elimination 2017–2025. 2017. Available online: https://tbcindia.gov.in/WriteReadData/NSPDraft20.02.20171.pdf (accessed on 14 April 2021).
54. World Health Organization. *Guidance for National Tuberculosis Programmes on the Management of Tuberculosis in Children*, 2nd ed.; World Health Organization: Geneva, Switzerland, 2014.
55. Zhou, D.; Pender, M.; Jiang, W.; Mao, W.; Tang, S. Under-reporting of TB cases and associated factors: A case study in China. *BMC Public Health* **2019**, *19*, 1–9. [CrossRef] [PubMed]
56. Preez, K.D.; Schaaf, H.S.; Dunbar, R.; Swartz, A.; Bissell, K.; Enarson, D.A.; Hesseling, A.C. Incomplete registration and reporting of culture-confirmed childhood tuberculosis diagnosed in hospital. *Public Health Action* **2011**, *1*, 19–24. [CrossRef]
57. Velayutham, B.; Thomas, B.; Nair, D.; Thiruvengadam, K.; Prashant, S.; Kittusami, S.; Vijayakumar, H.; Chidambaram, M.; Shivakumar, S.V.B.Y.; Jayabal, L.; et al. The Usefulness and Feasibility of Mobile Interface in Tuberculosis Notification (MITUN) Voice Based System for Notification of Tuberculosis by Private Medical Practitioners A Pilot Project. *PLoS ONE* **2015**, *10*, e0138274. [CrossRef] [PubMed]
58. World Health Organization; World Bank. World Report on Disability. 2011. Available online: https://www.who.int/disabilities/world_report/2011/report.pdf (accessed on 14 April 2021).
59. Moffitt, T.E.; Arseneault, L.; Belsky, D.W.; Dickson, N.; Hancox, R.J.; Harrington, H.; Houts, R.; Poulton, R.; Roberts, B.W.; Ross, S.A.; et al. A gradient of childhood self-control predicts health, wealth, and public safety. *Proc. Natl. Acad. Sci. USA* **2011**, *108*, 2693–2698. [CrossRef] [PubMed]
60. Schoeman, J.; Wait, J.; Hons, M.B.; Zyl, F.; Fertig, G.; Rensburg, A.J.; Springer, P.; Donald, P. Long-term follow up of childhood tuberculous meningitis. *Dev. Med. Child Neurol.* **2007**, *44*, 522–526. [CrossRef]
61. Humphries, M.; Teoh, R.; Lau, J.; Gabriel, M. Factors of prognostic significance in Chinese children with tuberculous meningitis. *Tuberculosis* **1990**, *71*, 161–168. [CrossRef]
62. United Nations. Transforming Our World: The 2030 Agenda for Sustainable Development A/RES/70/1; 2015. Available online: https://www.un.org/en/development/desa/population/migration/generalassembly/docs/globalcompact/A_RES_70_1_E.pdf (accessed on 14 April 2021).
63. Basu Roy, R.; Brandt, N.; Moodie, N.; Motlagh, M.; Rasanathan, K.; Seddon, J.A.; Kampmann, B. Why the Convention on the Rights of the Child must become a guiding framework for the realization of the rights of children affected by tuberculosis. *BMC Int. Health Hum. Rights* **2016**, *16*, 32. [CrossRef] [PubMed]
64. Davis, A.G.; Nightingale, S.; Springer, P.E.; Solomons, R.; Arenivas, A.; Wilkinson, R.J.; Anderson, S.T.; Chow, F.C.; Tuberculous Meningitis International Research Consortium; Davis, A.G. Neurocognitive and functional impairment in adult and paediatric tuberculous meningitis. *Wellcome Open Res.* **2019**, *4*, 178. [CrossRef] [PubMed]
65. Bodeau-Livinec, F.; Davidson, L.L.; Zoumenou, R.; Massougbodji, A.; Cot, M.; Boivin, M.J. Neurocognitive testing in West African children 3–6 years of age: Challenges and implications for data analyses. *Brain Res. Bull.* **2019**, *145*, 129–135. [CrossRef] [PubMed]
66. Üstün, T.B.; Chatterji, S.; Kostanjsek, N.; Rehm, J.; Kennedy, C.; Epping-Jordan, J.; Saxena, S.; Von Korff, M.; Pull, C. Developing the World Health Organization Disability Assessment Schedule 2.0. *Bull. World Health Organ.* **2010**, *88*, 815–823. [CrossRef] [PubMed]
67. Scorza, P.; Stevenson, A.; Canino, G.; Mushashi, C.; Kanyanganzi, F.; Munyanah, M.; Betancourt, T. Validation of the "World Health Organization Disability Assessment Schedule for Children, WHODAS-Child" in Rwanda. *PLoS ONE* **2013**, *8*, e57725. [CrossRef] [PubMed]
68. Hamdani, S.U.; Huma, Z.-E.; Wissow, L.; Rahman, A.; Gladstone, M. Measuring functional disability in children with developmental disorders in low-resource settings: Validation of Developmental Disorders-Children Disability Assessment Schedule (DD-CDAS) in rural Pakistan. *Glob. Ment. Health* **2020**, *7*, e17. [CrossRef] [PubMed]
69. Lancaster, G.A.; McCray, G.; Kariger, P.; Dua, T.; Titman, A.; Chandna, J.; McCoy, D.; Abubakar, A.; Hamadani, J.D.; Fink, G.; et al. Creation of the WHO Indicators of Infant and Young Child Development (IYCD): Metadata synthesis across 10 countries. *BMJ Glob. Health* **2018**, *3*, e000747. [CrossRef] [PubMed]
70. Zuurmond, M.; Seeley, J.; Nyant, G.G.; Baltussen, M.; Abanga, J.; Polack, S.; Bernays, S.; Shakespeare, T. Exploring caregiver experiences of stigma in Ghana: They insult me because of my child. *Disabil. Soc.* **2020**, 1–21. [CrossRef]

71. Adugna, M.B.; Nabbouh, F.; Shehata, S.; Ghahari, S. Barriers and facilitators to healthcare access for children with disabilities in low and middle income sub-Saharan African countries: A scoping review. *BMC Health Serv. Res.* **2020**, *20*, 1–11. [CrossRef] [PubMed]
72. Scherer, N.; Verhey, I.; Kuper, H. Depression and anxiety in parents of children with intellectual and developmental disabilities: A systematic review and meta-analysis. *PLoS ONE* **2019**, *14*, e0219888. [CrossRef] [PubMed]
73. Smythe, T.; Zuurmond, M.; Tann, C.J.; Gladstone, M.; Kuper, H. Early intervention for children with developmental disabilities in low and middle-income countries the case for action. *Int. Health* **2020**. [CrossRef] [PubMed]
74. The Advocacy Partnership. TB/MDR-TB Advocacy Tool Kit. 2011. Available online: http://stoptb.org/assets/documents/global/awards/cfcs/TB_MDR%20Advocacy%20Tool%20Kit.pdf (accessed on 14 April 2021).
75. Uplekar, M.; Weil, D.; Lonnroth, K.; Jaramillo, E.; Lienhardt, C.; Dias, H.M.; Raviglione, M.; Floyd, K.; Gargioni, G.; Getahun, G.; et al. WHO's new end TB strategy. *Lancet* **2015**, *385*, 1799–1801. [CrossRef]
76. World Health Organization. *The End TB Strategy*; World Health Organization: Geneva, Switzerland, 2015.

MDPI
St. Alban-Anlage 66
4052 Basel
Switzerland
Tel. +41 61 683 77 34
Fax +41 61 302 89 18
www.mdpi.com

Microorganisms Editorial Office
E-mail: microorganisms@mdpi.com
www.mdpi.com/journal/microorganisms

www.ingramcontent.com/pod-product-compliance
Lightning Source LLC
LaVergne TN
LVHW070148170726
843515LV00007B/1874
9783036518077